Synthetic Pyrethroids

ACS SYMPOSIUM SERIES **991**

Synthetic Pyrethroids

Occurrence and Behavior in Aquatic Environments

Jay Gan, Editor
University of California

Frank Spurlock, Editor
California Department of Pesticide Regulation

Paul Hendley, Editor
Syngenta Crop Protection, Inc.

Donald P. Weston, Editor
University of California

Sponsored by the
ACS Division of Agrochemicals

American Chemical Society, Washington, DC

ISBN: 978-0-8412-7433-4

The paper used in this publication meets the minimum requirements of American National Standard for Information Sciences—Permanence of Paper for Printed Library Materials, ANSI Z39.48-1984.

Distributed by Oxford University Press

PRINTED IN THE UNITED STATES OF AMERICA

Foreword

The ACS Symposium Series was first published in 1974 to provide a mechanism for publishing symposia quickly in book form. The purpose of the series is to publish timely, comprehensive books developed from ACS sponsored symposia based on current scientific research. Occasionally, books are developed from symposia sponsored by other organizations when the topic is of keen interest to the chemistry audience.

Before agreeing to publish a book, the proposed table of contents is reviewed for appropriate and comprehensive coverage and for interest to the audience. Some papers may be excluded to better focus the book; others may be added to provide comprehensiveness. When appropriate, overview or introductory chapters are added. Drafts of chapters are peer-reviewed prior to final acceptance or rejection, and manuscripts are prepared in camera-ready format.

As a rule, only original research papers and original review papers are included in the volumes. Verbatim reproductions of previously published papers are not accepted.

ACS Books Department

Contents

Overview and Occurrence

Analytical Methodology

Fate and Transport Processes

Aquatic Toxicology

Risk Mitigation and Stewardship

Indexes

Overview and Occurrence

Chapter 1

Synthetic Pyrethroid Use Patterns, Properties, and Environmental Effects

Frank Spurlock and Marshall Lee

Environmental Monitoring Branch, California Department of Pesticide Regulation, 1001 I Street, Sacramento, CA 95812–4015

In this paper we present a broad overview of the class of insecticides known as synthetic pyrethroids. The discussion includes a summary of agricultural and urban pyrethroid use patterns and trends, pyrethroid chemical structure and properties, the significance of photostability to pyrethroid environmental fate, and hydrophobicity, persistence and relative aquatic toxicity as compared to other pesticides. Finally we provide a brief summary of California's regulatory response to recent detections of pyrethroids in aquatic sediments and a discussion of scientific and regulatory issues associated with ongoing pyrethroid aquatic exposure assessments and mitigation efforts.

Introduction

California leads the nation in agricultural production, so it's no surprise that the state accounts for approximately 20% of all agricultural insecticides applied to U.S. crop lands (*1*). Insecticides are also used extensively in California's urban areas. For example, synthetic pyrethroids are one of the most widely used families of insecticides, and we estimate that approximately 70% of California's total pyrethroid use occurs in urban areas. The importance of synthetic pyrethroids as a pest management tool in California is evidenced by the number of registered pyrethroid products. In 2006 there were 1255 California registered synthetic pyrethroid products from 128 registrants. These products accounted for more than 40% of all registered insecticide products in the state. Total California synthetic pyrethroid sales in 2004 were approximately 1.4 million lbs active ingredient (AI).

Pyrethroids are used in nearly all agricultural crops, nurseries, various urban structural and landscaping sites, construction sites (pre-construction termiticides), the home/garden environment, and many other sites. Several desireable characteristics contribute to the commercial success of pyrethroids, including their efficacy against a broad range of insect pests and mites, low mammalian and avian toxicities, low potential to contaminate ground water, and relatively low application rates. However, there have been numerous recent reports of pyrethroid detections in California aquatic sediments, and toxicity to the sediment dweller *Hyallela azteca* has been observed in concomitant bioassays (*2-5*). Coupled with steadily increasing use of pyrethroids, these observations have led to renewed interest in the environmental fate and transport of these insecticides. This chapter provides an overview of synthetic pyrethroid environmental fate characteristics, summarizes pyrethroid use patterns and trends in California, and summarizes some of the unique issues associated with synthetic pyrethroid aquatic risk assessment.

Background

To understand different synthetic pyrethroids and their uses, it is instructive to review the major milestones in synthetic pyrethroid synthesis and development. Only a brief overview is given here. Readers seeking more information should consult the excellent review by Davies (*6*).

Pyrethrins

Pyrethrum is a preparation of dried *Chrysanthemum cinerariaefolium* and/or *Chrysanthemum cineum* flower heads that contains the six insecticidally active chemicals known as pyrethrins. Each of the six naturally-occurring pyrethrins is comprised of a cyclopropane-carboxylic acid group and a cyclopentenolone (alcohol) group joined by an ester linkage (Table I). The various synthetic pyrethroid analogues are generally similar in structure to the pyrethrins, although there are some deviations from the basic chryanthemic acid ester structure.

Table I. Structures of the six naturally-occurring pyrethrin esters

acid moieties / *alcohol moieties*	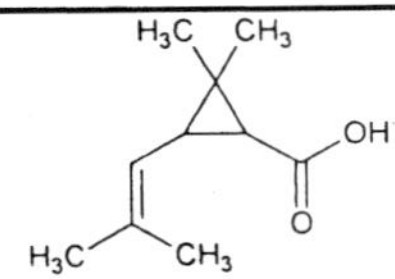 *chrysanthemic acid esters*	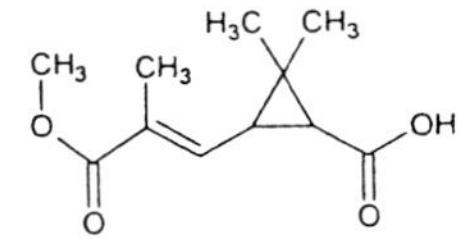 *pyrethric acid esters*
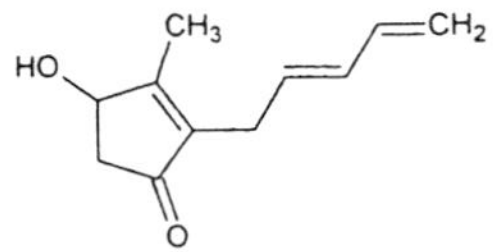 *pyrethrolone esters*	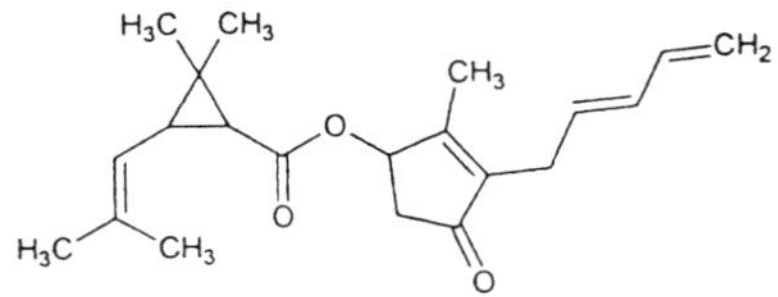 pyrethrin I	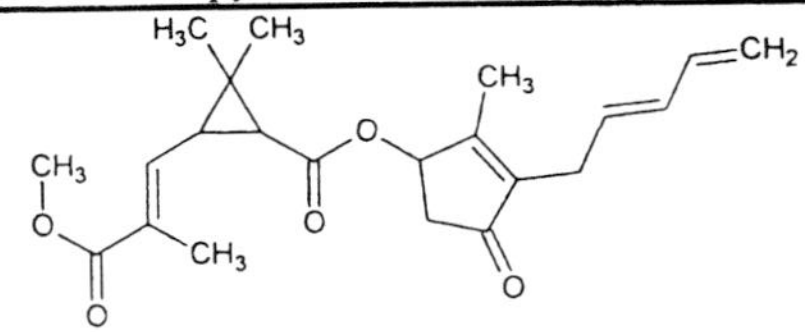 pyrethrin II
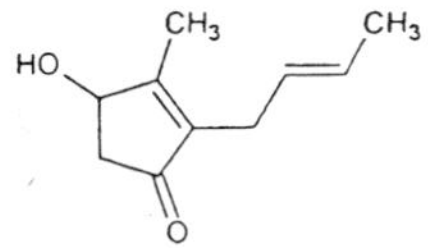 *cinerolone esters*	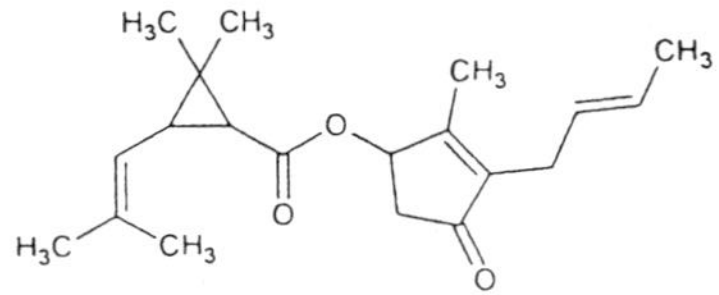 cinerin I	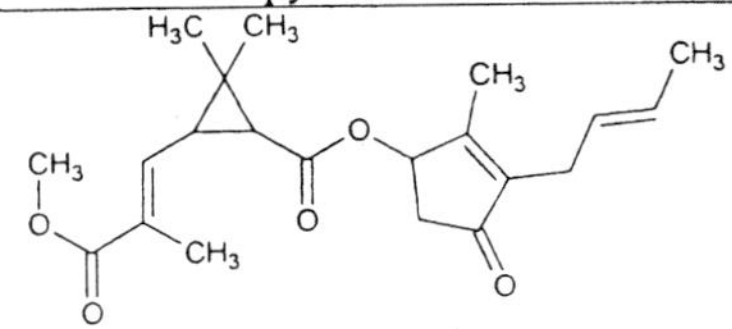 cinerin II
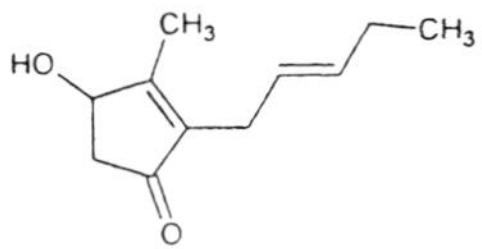 *jasmolone esters*	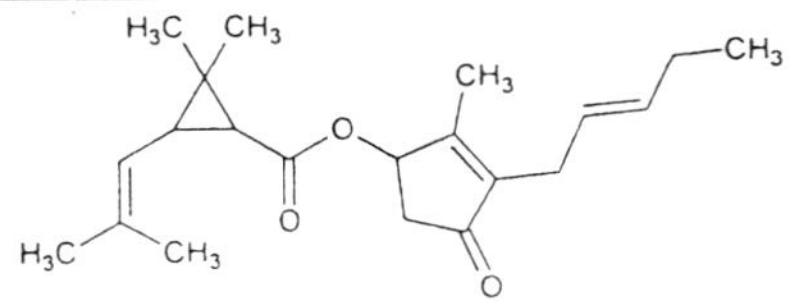 jasmolin I	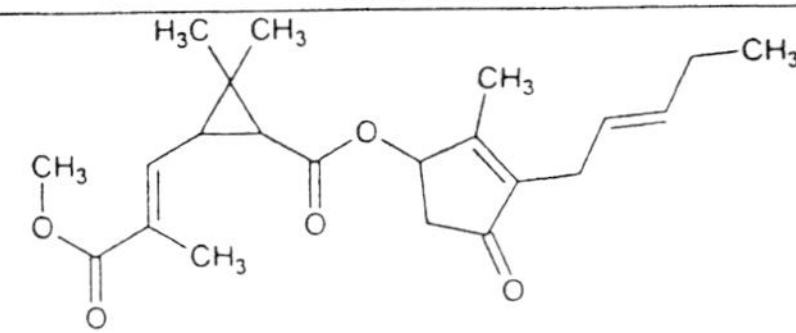 jasmolin II

First-generation Photolabile Synthetic Pyrethroids

Many early attempts at pyrethroid synthesis focused on substitutions to the alcohol portion of the molecule (*6, 7*). Allethrin was one of the earliest synthetic analogues to eventually achieve commercial success, and allethrin-containing products are still marketed throughout the world today. Numerous other "alcohol-substituted" chrysanthemic acid esters were synthesized between the 1950s to the early 1970s, and many are still registered for use in the United States today, including resmethrin, tetramethrin, and phenothrin (Table II). These chrysanthemic acid derivatives are often called "first-generation" synthetic pyrethroids. The first-generation synthetic pyrethroids are similar to naturally-occurring pyrethrins in that they photolyze relatively easily (*6*). While their photolysis half-lives vary depending on measurement method and experimental conditions, half-lives on surfaces exposed to sunlight or simulated sunlight are generally on the order of hours (*8-10*).

Photostable Type I and Type II synthetic pyrethroids

Modifications to the chrysanthemic acid portion of the pyrethroid molecule improved photostabilities. In particular, esters of chrysanthemic acid dihalovinyl analogues were found to display much improved photostabilities compared to the esters of chrysanthemic acid (*11,12*). The first commercial photostable synthetic pyrethroid based on this approach was permethrin, synthesized in the early 1970s. Permethrin is still the most widely used synthetic pyrethroid in California today. While various photostable synthetic pyrethroids have since been developed based on different structural modifications to the basic chrysanthemate ester moiety, the halogenated vinylcyclopropylcarboxylates are among the most important in agriculture today, and include the various cypermethrins, cyfluthrins, and cyhalothrins (Table III). Reported aqueous and soil photolysis half-lives are generally on the order of tens to occasionally hundreds of days for the various photostable pyrethroids (*13*).

Type I vs. Type II

An additional structural feature common to several commercially successful synthetic pyrethroids is the "α-cyano" group. These pyrethroids are α-cyano-3-phenoxybenzyl pyrethroid esters and are commonly referred to as "type II pyrethroids". Type II pyrethroids display markedly increased biological activity relative to their type I 3-phenoxybenzyl analogues (cf. type II cypermethrin vs. type I permethrin, Table III) and also demonstrate certain differences in their mode of toxic action (*14*). Other type II pyrethroids include cyfluthrin, cyhalothrin and esfenvalerate (Table III).

Isomeric Enrichment

Most synthetic pyrethroids are comprised of several stereoisomers due to the presence of multiple asymmetric carbons, often in the cyclopropane ring as in the case of cypermethrin and cyhalothrin (Table III). In addition, several of the pyrethroids also possess an alkene moiety, giving rise to cis/trans isomerism (e.g. permethrin, Table III). In general the biological activity of different stereoisomers varies substantially (*15-18*), so that enrichment of the most active isomer(s) yields a product with greatly enhanced insecticidal activity. In recent years several isomerically enriched pyrethroid active ingredients have been introduced into commercial use. One of the first such pyrethroids registered in California was esfenvalerate in 1988. Esfenvalerate is now widely used and there are no longer any registered products containing the original racemate fenvalerate. Numerous other isomerically enriched synthetic pyrethroids have since been introduced, including lambda cyhalothrin, gamma cyhalothrin, beta cyfluthrin and (S)-cypermethrin (zeta-cypermethrin). One consequence is that application rates expressed on an AI basis are lower for the more active enriched products due to their enhanced activity. However, several recent articles have reported differences in persistence also, likely due to differences in biodegradability among different isomers (*16-18*). Data on stereoselective biodegradation of pyrethroids are relatively sparse, so the practical signficance of stereoselective biodegradation is not well understood.

General Use Patterns and Trends

Due to their instability in sunlight, chrysanthemate ester pyrethroids such as allethrin are not used in agriculture. These pyrethroids are formulated primarily as indoor or residential products such as aerosol ant and roach sprays, foggers, pet products, carpet and upholstery sprays, and commercial/institutional uses such as in food preparation or storage facilities. In California, the chrysenthamate esters account for 10 of 24 synthetic pyrethroid active ingredients in registered products, where isomerically-enriched mixtures (e.g. allethrin, d-allethrin, bioallethrin) are considered different active ingredients. These chrysanthemate esters accounted for approximately 8% of total synthetic pyrethroid sales in California in 2004 (*19*).

Table II. Chyrysanthemate ester "first-generation" photolabile pyrethroids registered in California as of 2006 (stereochemistry not shown).

Chemical	*Uses/Product Types*
allethrin, esbiothrin, d-trans-allethrin, d-allethrin 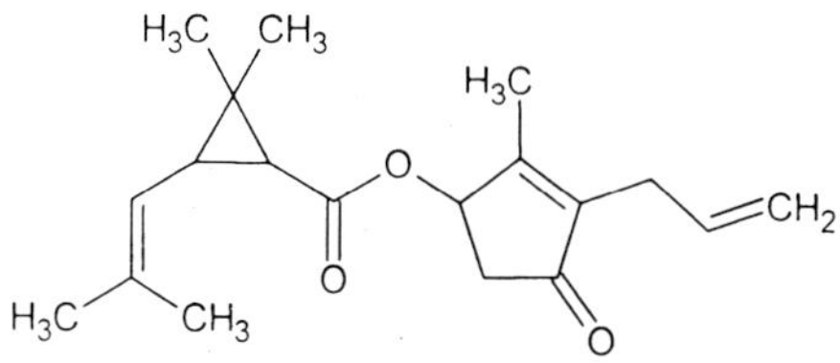	Products of allethrin and it's various isomers are used indoor/ outdoor in household, industrial, commercial and institutional settings. Many are pressurized aerosols or foggers; a few pet shampoos. Most often co-formulated with other pyrethroids and/or synergists.
cyphenothrin 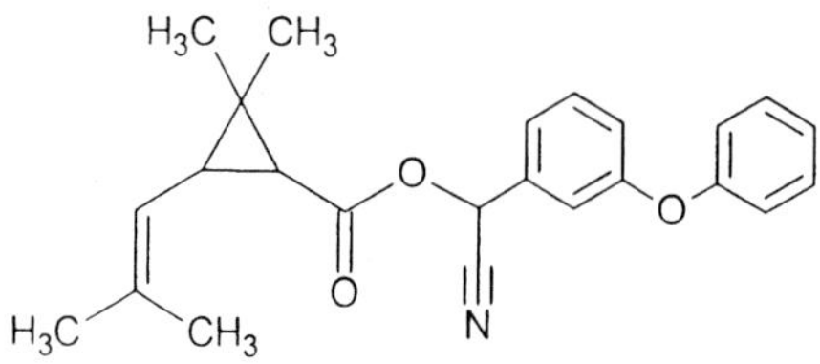	Aerosol or fogger insecticide; animal husbandry premises or indoor/outdoor household use.
imiprothrin 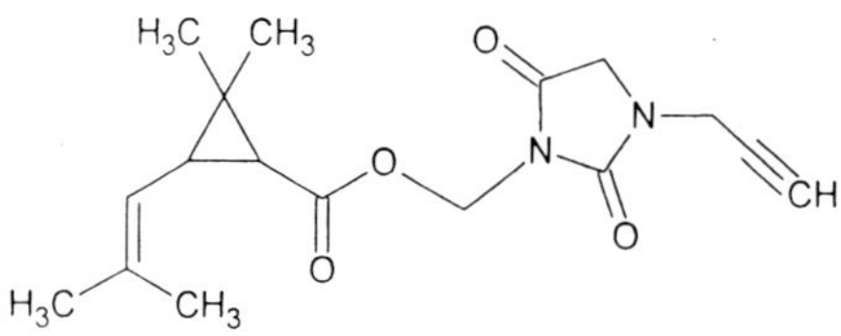	Mostly pressurized aerosols, used in household, industrial, commercial and institutional settings. A few crack/crevice products. Typically co-formulated with other pyrethroids and/or synergists.

Table II. *Continued*

phenothrin 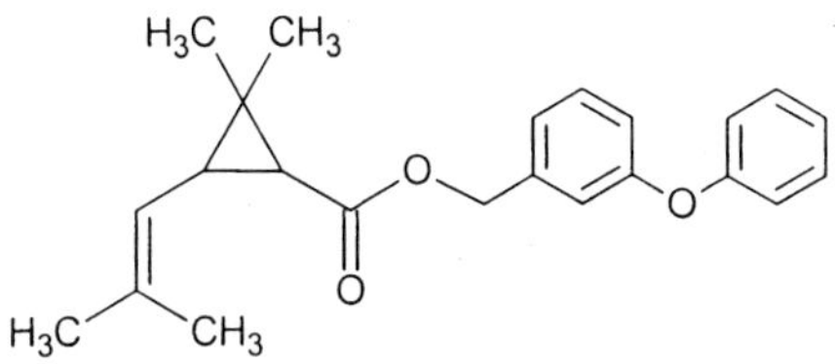	Two main uses: household indoor/outdoor flying insect control and flea control products (collars, direct application drops). Typically co-formulated with other pyrethroids and/or synergists.
prallethrin 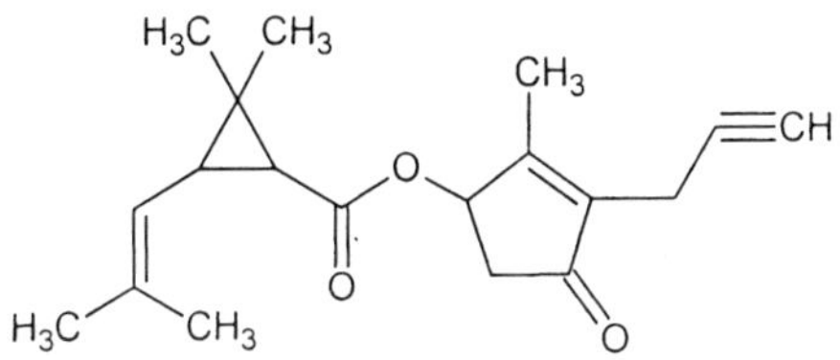	Primarily pressurized aerosols used in household, industrial, commercial and institutional settings. Often co-formulated with other pyrethroids and/or synergists.
resmethrin 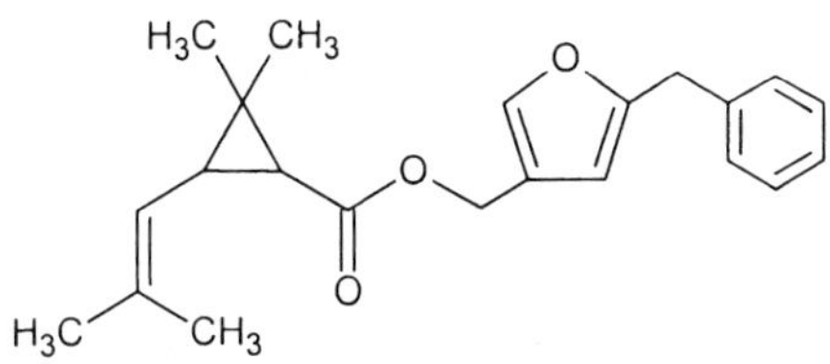	Liquids or pressurized aerosols used in household, industrial, commercial and institutional settings. Some outdoor garden and ornamental uses. Used in animal husbandry premises. Often co-formulated.
tetramethrin 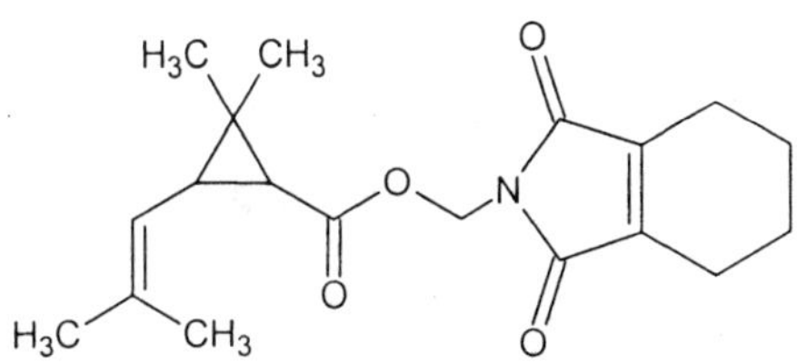	Primarily pressurized aerosols used in household, industrial, commercial and institutional settings. Often co-formulated with other pyrethroids and/or synergists.

Table III. Photostable Type I and Type II pyrethroids registered in California as of 2006 and their general use patterns (stereochemistry not shown).

Structure	*Use Pattern (2004)*
bifenthrin 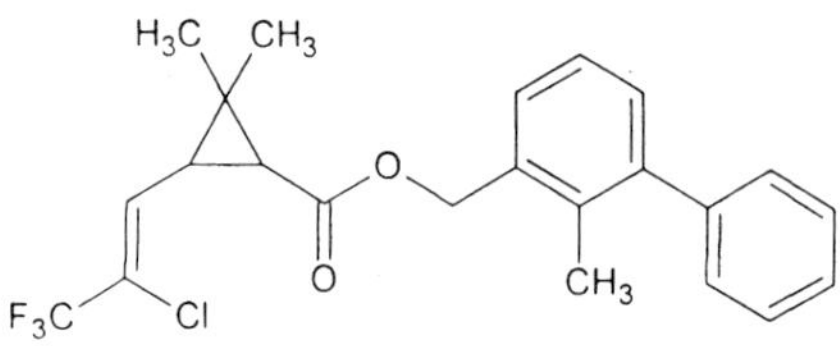 	bifenthrin use: 110,000 lbs 20% agricultural use (cotton, corn); 40% commercial structural and landscape; 40% home and garden.
cyfluthrin, beta-cyfluthrin 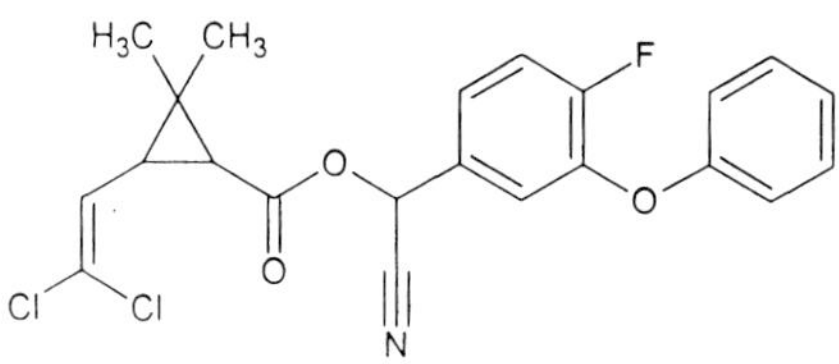 	cyfluthrin use: 50,000 lbs 30% agricultural (alfalfa, citrus, cotton), 70% commercial structural, <5% other urban. beta-cyfluthrin use: 16,000 lbs <5% agricultural use; 70% commercial structural and landscape; 30% home and garden
cypermethrin, (S)-cypermethrin 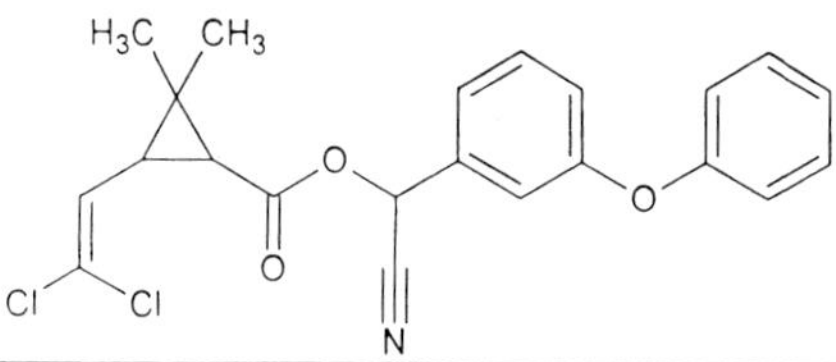 	cypermethrin use: 200,000 lbs 5% agricultural, 95% commercial structural and landscape, 5% home and garden. (S)-cypermethrin use: 25,000 lbs >95% agricultural use (lettuce, alfala, onions)
lambda cyhalothrin, gamma cyhalothrin 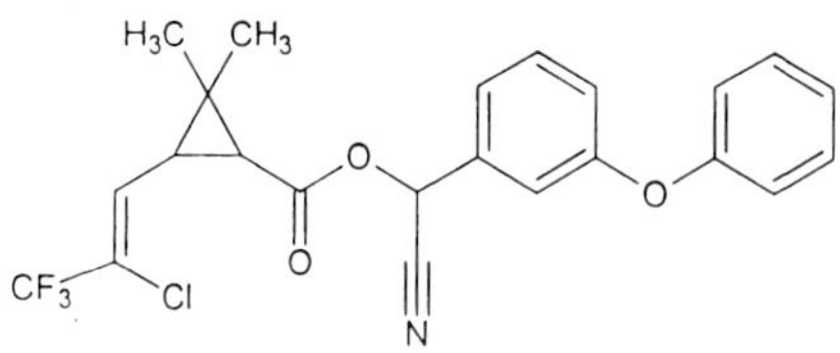 	lambda-cyhalothrin use: 40,000 lbs 50% agricultural (alfalfa, lettuce, tomatoe, rice), 40% commercial structural and landscape, 10% home and garden. gamma-cyhalothrin: new AI registered 2005, use data not available. All currently registered products are agricultural.

Table III. *Continued*

deltamethrin 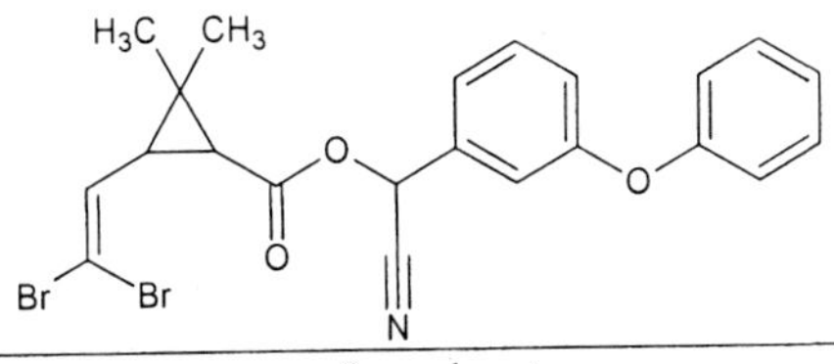	deltamethrin use: 10,000 lbs >95% commercial structural and landscape maintenance
esfenvalerate 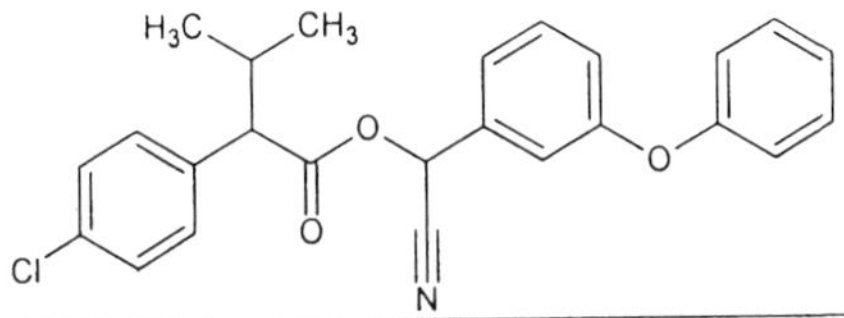	esfenvalerate use: 60,000 lbs 60% agricultural (almonds, peaches, tomato, artichoke); <5% commercial structural and landscape; 40% home and garden
etofenprox 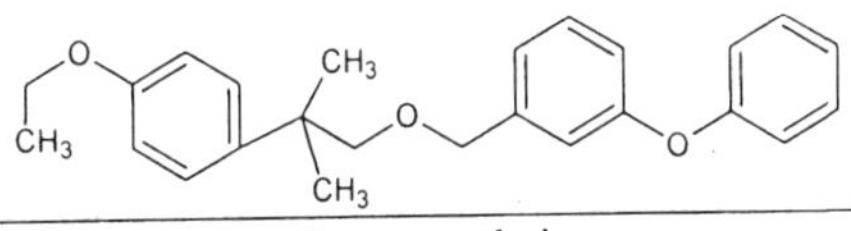	etofenprox use: <100 lbs; >95% home uses (pet and foggers).
fenpropathrin 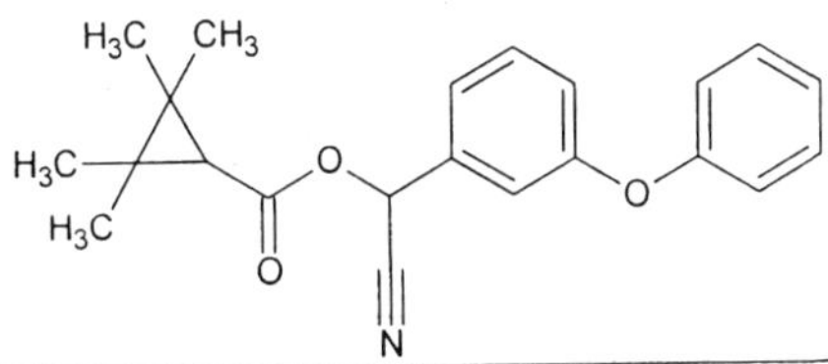	fenpropathrin use: 40,000 lbs >95% agricultural use (grapes, citrus, strawberry, cotton).
tau-fluvalinate 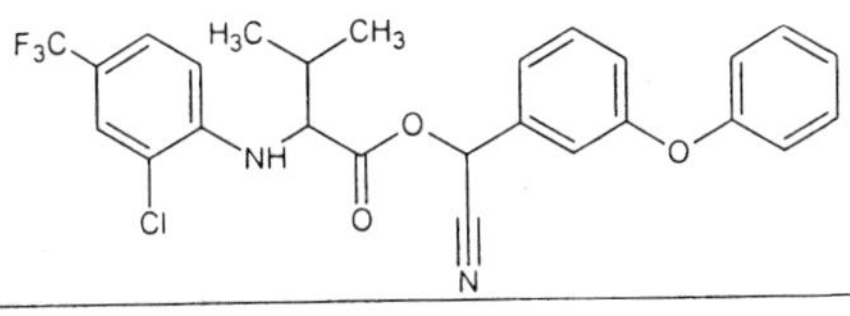	tau-fluvalinate use: 2,000 lbs 90% percent agricultural use (nursery-outdoor, nursery-greenhouses), 10% commercial structural and landscape .

Continnued on next page.

Table III. *Continued*

permethrin 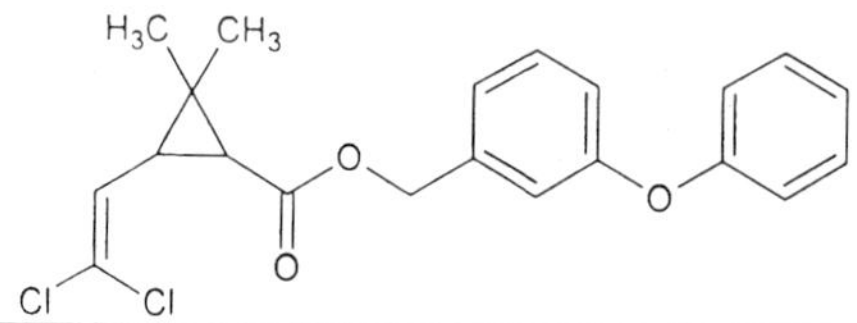	permethrin use: 500,000 lbs 30% agricultural (pistachio, lettuce, almond, celery), 60% commercial structural and landscape, 10% home and garden.
tralomethrin 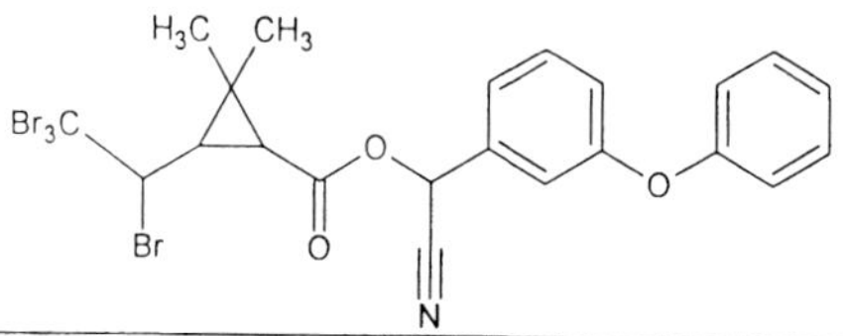	tralomethrin use: 4,000 lbs <5% agricultural use, 5% commercial structural and landscape, 90% home and garden.

NOTE: Approximate use and category percentages estimated from 2004 California use and sales data.

In contrast, photostable pyrethroids are registered for use in a wide variety of agricultural and nonagricultural sites in California. These include essentially all crops grown in California, greenhouse and field-grown nursery plantings, pre-construction soil treatments, structural applications, turf and sod, institutional and commercial application sites, pet products and shampoos, animal husbandry premises, landscape maintenance in parks, golf courses and around buildings, and various home and garden uses, including lawns, ornamental plantings and indoor uses. The photostable pyrethroids account for more than 90% of synthetic pyrethroids sold in California. Because of their greater persistence, higher toxicity, higher total use and their outdoor use patterns, the photostable synthetic pyrethroids pose the greatest water and sediment quality concerns. Consequently the remainder of this chapter will focus largely on the photostable synthetic pyrethroids (Table III).

Historically, the organophosphate (OP) insecticides diazinon and chlorpyrifos were extensively used on a wide variety of agricultural and nonagricultural application sites, but overall use of the two OPs has decreased markedly since the mid-1990s (Figure 1) for two main reasons. The first reason for sharp decreases in total diazinon and chlorpyrifos use has been the phase-out of essentially all residential diazinon and chlorpyrifos uses, including most chlorpyrifos termiticide uses (*21*). The USEPA agreements with registrants to phase-out these uses took effect in the early 2000s, and residential home and

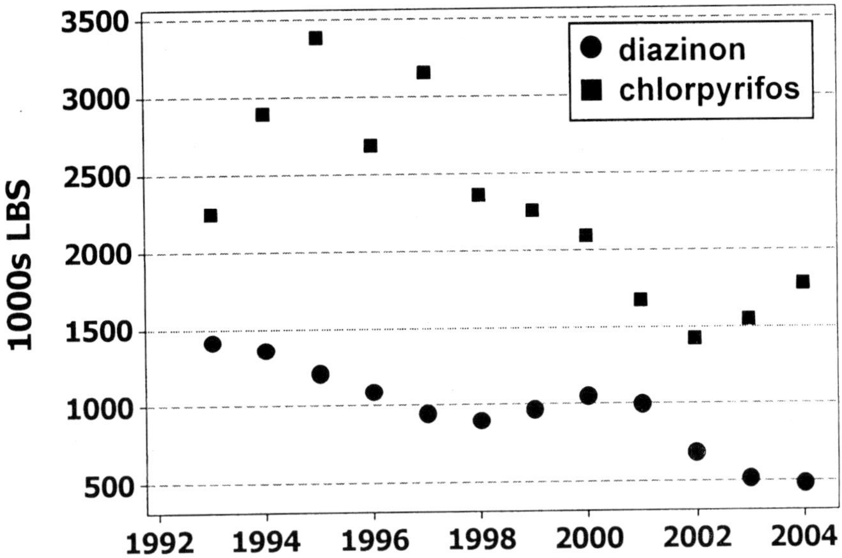

Figure 1. Overall trend in California reported use (16) of diazinon and chlorpyrifos. Includes agricultural, commercial structural, and landscape maintenance applications.

garden use and structural uses of the two pesticides have now been essentially eliminated.

The second reason for decreased California use of diazinon and chlorpyrifos in the late 1990s and early 2000s was widespread recognition of OP surface water quality problems throughout California. Numerous California waterways have been listed as impaired due to frequent widespread OP detections and associated toxicity under Federal Clean Water Act section 303(d) provisions in recent years (*22*).

California has a statewide sales database that tracks total sales of all registered pesticide products in the state. We are also fortunate to have an extensive pesticide use reporting system (PUR) that requires reporting of all agricultural applications and some non-agricultural applications such as commercial structural and landscape maintenance applications (*20*). Although residential homeowner applications are not included in the PUR, it is possible to infer general trends in residential homeowner use by comparing the statewide sales and PUR databases as shown below.

One consequence of decreased OP use has been a corresponding increase in reported uses of synthetic pyrethroids because they are effective OP substitutes in most types of applications. The large increase in agricultural and commercial structural pyrethroid use is evident in the total PUR-reported annual applications, where total pounds pyrethroid applied doubled between 1994 and 2005 (Figure 2). The general trend in residential home and garden use has similarly shown an increase since the beginning of the OP phase-out in the early 2000s (Figure 3). That trend is based on comparison of PUR data to DPR's statewide sales database (which includes all registered products) and the assumption that all non-PUR use is homeowner use. While the latter assumption is only approximate, the general trend is clear: The discrepancy between total pyrethroid sales and PUR-reported use has increased markedly since the OP phase-out began. The increased sales relative to reported use is largely due to residential home and garden uses. DPR has funded shelf surveys in an effort to improve our understanding of urban pesticide use (*23-26*). These surveys also demonstrate the increase in homeowner pyrethroid products concomittant with the OP phase-out. In summary, the available data show that both PUR-reported and home and garden synthetic pyrethroid use in California have increased markedly in recent years.

Pyrethroid Environmental Fate Characteristics

Figures 4-6 illustrate general differences in fate and toxicity characteristics among organochlorine (OC), photostable pyrethroid (PY), organophosphate (OP), carbamate (CB), and "other" pesticides. The latter group is essentially a random sample of various nonionic herbicides, miticides, insecticides, and other miscellaneous pesticides for which reliable data were available. The synthetic pyrethroid data are limited to the photostable Type I and Type II pyrethroids because they are of the greatest interest due to their use patterns and greater persistence. The octanol/water partition coefficient (K_{OW}) and aerobic soil half-life data were primarily compiled from U.S. and California pesticide registration data, but other sources were consulted in a few cases, including the USDA-ARS pesticide properties database (*27*) and the European Union Footprint database of pesticide properties (*28*). For the pyrethroids in particular, K_{OW} and aerobic half-life data were taken from the recent review by Laskowski (*13*) because much of the older historical data available elsewhere are unreliable. Finally, the *Daphnia magna* acute toxicity LC_{50} data were compiled from the U.S. Environmental Protection Agency Office of Pesticide Programs ecotoxicity database (*29*). *Daphnia magna* were chosen as a relative measure of aquatic toxicity because of

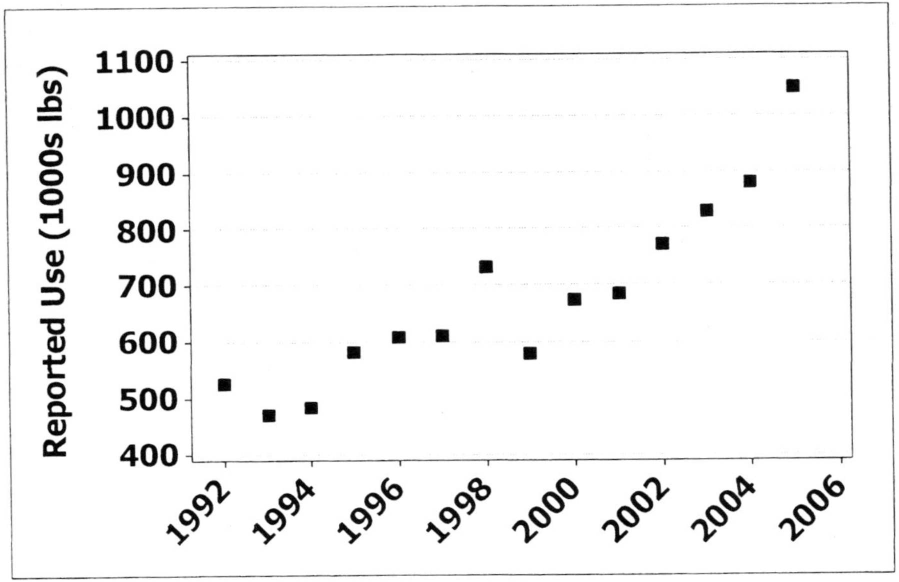

2005 TOP 7	**lbs**
permethrin	550k
cypermethrin	208k
bifenthrin	63k
cyfluthrin	50k
fenpropathrin	40k
λ-cyhalothrin	37k
esfenvalerate	33k

Figure 2. Overall trend in California reported use of 25 synthetic pyrethroids (20). Includes agricultural, commercial structura,l and landscape maintenance applications.

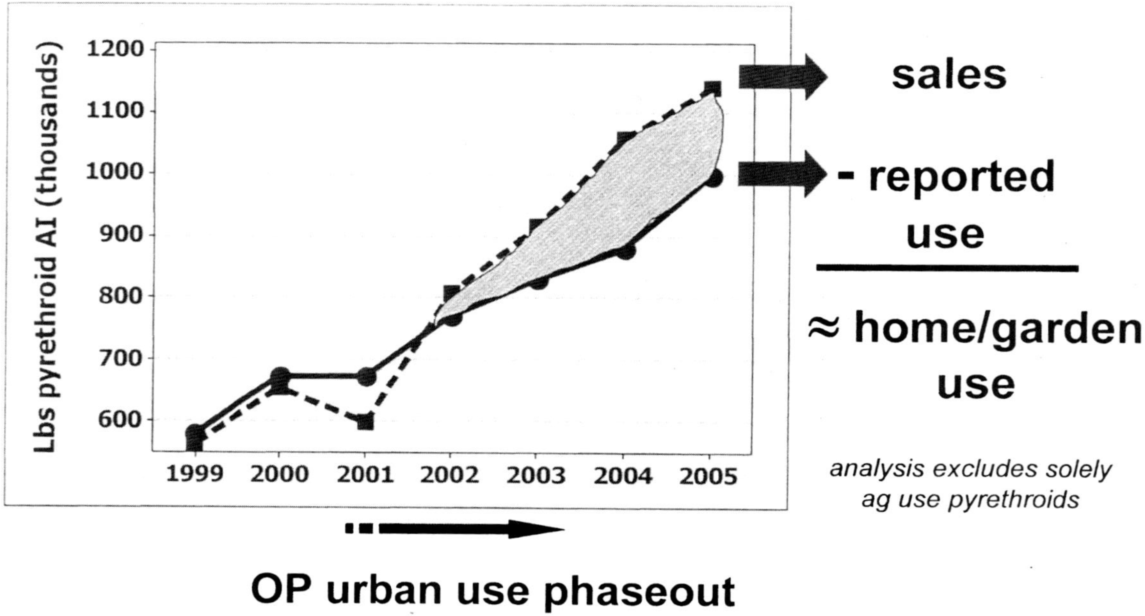

Figure 3. Approximate trend in residential home and garden pyrethroid use. Estimated by difference between sales(15) and reported use (16).

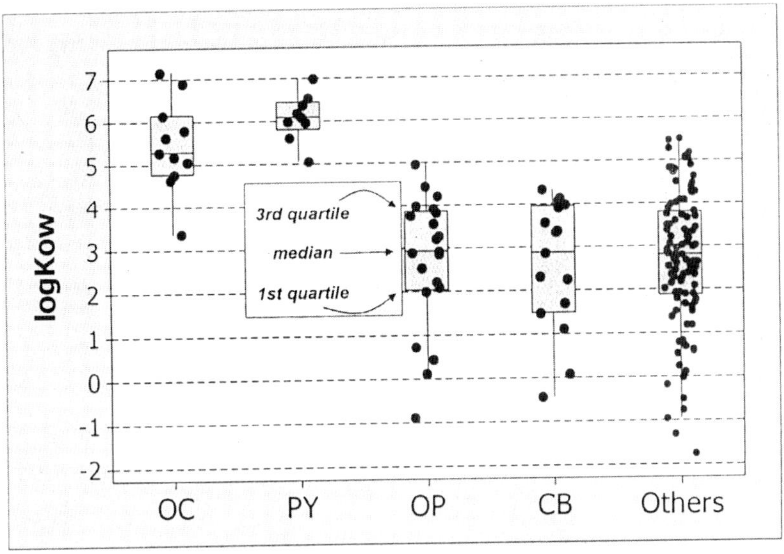

Figure 4. Comparison of log octanol/water partition coefficients for organochlorines (OC), pyrethroids (PY), organophosphates (OP), carbamates (CB) and other miscellaneous pesticides.

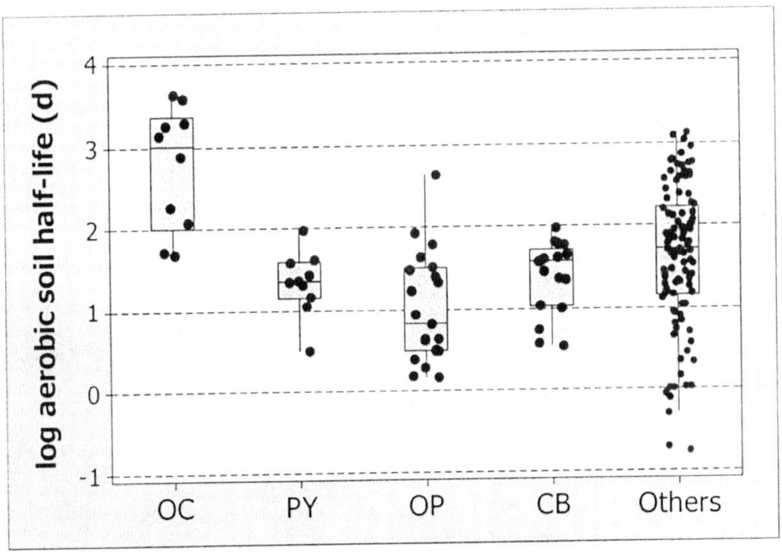

Figure 5. Comparison of laboratory aerobic soil half-lives for organochlorines (OC), pyrethroids (PY), organophosphates (OP), carbamates (CB) and other miscellaneous pesticides.

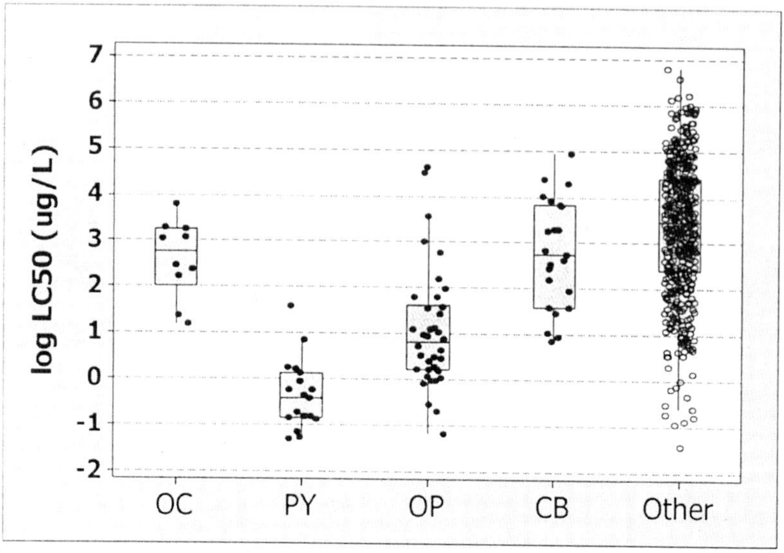

Figure 6. Comparison of Dapnia magna median lethal concentrations (LC50) for organochlorines (OC), pyrethroids (PY), organophosphates (OP), carbamates (CB, and other miscellaneous pesticides. Data from laboratory aqueous studies.

the large amount of data available and because arthropods are among the most sensitive aquatic organisms to insecticides and pesticides in general.

Hydrophobicity

It is evident from Figure 4 that the pyrethroids possess hydrophobicities comparable to those OCs selected for comparison. The median K_{OW} for the OC pesticides shown here is 2.0 x 10^5, while the pyrethroid median is 1.2 x 10^6. The difference is not significant and hydrophobicity as measured by K_{OW} clearly distinguishes the two groups from the rest of the pesticide groups. The median K_{OW}s for OPs and CBs are more than two orders of magnitude lower, essentially equal to that of the universe of "other" pesticides (~ 600-700). Extreme hydrophobicity is one of the key distinguishing characteristics of the pyrethroids, influencing their environmental transport and bioavailability via partitioning to sediment and dissolved organic matter.

Persistence

Although the PYs are similar to OCs in terms of hydrophobicity, they display much lower persistence (Figure 5). Organochlorine soil half-lives are often on the order of 1000s of days (i.e., years), but the median PY aerobic soil half-life of approximately 23 days is roughly comparable to that for CBs (37d) as well as the general universe of "other" pesticides (50d). The OPs have lower aerobic soil half-lives than the other groups, most likely due to the hydrolytic tendencies of many OPs. Interestingly, several OPs are among the most common pesticidal surface water contaminants in California in spite of their relatively low persistence. It should also be noted that, although PY persistence data in sediment are sparse, one study reported PY laboratory sediment half-lives of several months to greater than a year (*30*).

Aquatic Toxicity

Finally, the potential for pyrethroid acute toxicity to aquatic life is evident from the *D. magna* LC_{50} data comparison (Figure 6). The median LC_{50} for the pyrethroids of 0.4 ug L^{-1} is more than an order of magnitude lower than the OPs (6.4 ug L^{-1}) and 3–4 orders of magnitude lower than the remaining groups of OC, CB, and other pesticides. It is critical to note that the *D. magna* LC_{50} data in Figure 6 were all determined in aqueous solutions in the laboratory, so that the known reduction in bioavailability due to dissolved organic matter or sediments present in natural waters is not reflected in these data (*31, 32*).

Aquatic Risk Assessment Issues

Recent Detections in Freshwater Sediments

Recent findings of widespread pyrethroid detections in California sediments and associated toxicity in bioassays has led to intense interest in the fate and transport of synthetic pyrethroids in urban and agricultural environments. While pyrethroids have long been recognized as very highly toxic to fish and aquatic invertebrates, conventional wisdom held that they should have minimal bioavailability to aquatic organisms in waterbodies that receive runoff water from treated areas due to their extreme hydrophobicity and sorptive characteristics. Assessments to verify this assumption, however, were limited by the ability of analytical laboratories to quantify pyrethroids in environmental samples. Toxicity tests using the U.S. Environmental Protection Agency's standard bioassay organisms were unreliable because pyrethroids, if present, would presumably be associated with sediment, not in the water column. Additionally, the inherent tendencies of pyrethroids to adsorb to collection and test vessels compromised the ability to expose water column bioassay organisms to pyrethroid concentrations that reflect *in situ* exposures.

Weston et al. (*3*) conducted the first large-scale California pyrethroid study and used a sensitive analytical method and sediment bioassays to investigate the relationship between the presence of pyrethroids in sediment and toxicity. They demonstrated sediment samples collected from California waterways dominated by agricultural runoff were frequently toxic to the sediment-dwelling amphipod *Hyalella azteca* and toxicity was correlated with pyrethroid concentrations. Subsequently, Weston et al. (*4*) found similar correlations in sediment in a watershed receiving urban runoff; Amweg et al. (*5*) confirmed that pyrethroid-associated toxicity in sediment is common in urbanized watersheds throughout Northern California.

Response Strategies

In California, impairments of the aquatic environment caused by pesticides are addressed primarily by two public agencies. The Department of Pesticide Regulation (DPR) is mandated by California state law to protect the environment from environmentally harmful pesticides by prohibiting, regulating, or ensuring proper stewardship of those pesticides. The State Water Resources Control Board (State Water Board) and its nine affiliated Regional Water Quality Control Boards (Regional Water Boards) are the lead agencies for coordination and control of water quality in California. Generally, DPR and the State and Regional Water Boards collaborate on pesticide and water quality issues. As these agencies develop response strategies, they attempt to take advantage of the array of existing information on particular pesticides' environmental fate and behavior, toxicity, and potential sources. In the case of pyrethroids, these evaluations identified several areas where additional information is needed to fully assess the significance of pyrethroids detections in sediment as well as mitigate their off-site movement. Some of these issues, such as the significance of synergist co-occurrence, are more-or-less unique to the pyrethroids. Other questions, such as sources and transport mechanisms in urban environments, highlight a basic shortcoming in traditional pesticide regulation where agricultural usage has been the primary focus. Some of the scientific or technical issues that been raised include:

Sources

DPR's pesticide use reporting system (PUR) records the time, place, and type of site where pesticide products are applied and is an indispensable tool for investigating, for example, the history of pyrethroid applications in agricultural watersheds affected by pyrethroid contamination. Obviously, however, it cannot relate all site specific information, such as soil characteristics or slope, that may affect pesticide transport away from the site of application. Without this information, it is difficult to identify sites with high runoff potential that should be targeted with mitigation measures. It is even more difficult to investigate sources in urbanized watersheds because the PUR does not include detailed information on commercial, structural, or many other urban pesticide uses, and homeowner-applied uses are not reported at all.

Dispersal in the Environment

Transport in streams and rivers of sediment and colloidal material—and hydrophobic molecules like pyrethroids that may be adsorbed to them—is not well understood. Better knowledge of conditions that affect sediment movement and deposition and pyrethroid persistence in bed sediments can lead to an understanding of the types of streams and watersheds that are most vulnerable to toxicity in bed sediments.

Goals for Environmental Concentration

A goal for many projects aimed at reducing pollutant loading in affected water bodies is to develop water quality criteria—conditions, including concentrations of pollutants and durations of exposure, that will preclude detrimental toxic effects on sensitive aquatic organisms. Such criteria can act as performance targets for mitigation measures. The standard methodology for deriving criteria is for waterborne pollutants only (*33*); there is no standard methodology for sediment-bound pollutants. Those seeking environmental quality criteria for pyrethroids would not only need to identify toxicity values for sediment-dwelling organisms, they will need to address basic questions about the pyrethroids' bioavailability. For example, under what conditions are pyrethroid concentrations in interstitial water high enough to affect resident arthropods? Can ingestion of pyrethroid-laden particles cause toxicity? Could ingestion similarly be a route of exposure for organisms that inhabit the water column? Pyrethroids frequently co-occur in sediment with other pyrethroids. Is an additive toxicity model normalized to each pyrethroid's relative toxicity reasonable to predict toxicity in sediment? Another concern is how the toxicity of pyrethroids may be enhanced by the presence of synergists found in formulated pyrethroid products. Is it realistic to expect the co-occurrence of a pyrethroid active ingredient and its synergists in sediments off of the site of application? And at a more basic level, chronic toxicity studies with pyrethroids and arthropods have not been performed. What are the effects of chronic low-level exposures to sensitive sediment-dwelling organisms? Answers to such questions are critical for understanding the potential for pyrethroids to cause toxicity in ambient sediments and for ultimately developing appropriate mitigation strategies.

DPR, via its reevaluation (i.e., a data call-in for pesticide products like pyrethroids found to be a hazard to the environment) of over six hundred pyrethroid products, requires those who register pyrethroid products for sale in California to submit specific data that will help address key topic areas (*34*). The data requirements include aerobic and anaerobic aquatic sediment dissipation half-lives, acute and chronic sediment toxicity data for the amphipod *Hyallela azteca* and the midge *Chironomus tentans,* data identifying the processes by which pyrethroids move off-site from application sites to aquatic sediments, and data identifying management practices that can reduce or eliminate movement from application sites to aquatic sediments.

In summary, synthetic pyrethroids are a large group of insecticides whose use has been increasing for several decades. Pyrethroids are distinguished by three general characteristics: extreme hydrophobicity, rich stereochemistry, and broad spectrum high level insecticidal activity. The older photolabile pyrethroids are primarily limited to pressurized sprays and fogger products and represent about 10% of California sales. These products are often formulated with synergists such as piperonyl butoxide. In contrast, the photostable pyrethroids are widely used in essentially every type of situation where insecticides are applied in California, including agriculture, commercial structural applications, landscape maintenance, and a variety of home and garden applications. Few photostable pyrethroid products contain synergists. Synthetic photostable pyrethroids possess several desireable characteristics, including low potential to contaminate ground water, and low mammalian and avian toxicities. However, they have also been detected off-site in California aquatic sediments at concentrations sufficient to cause toxicity to sediment-dwelling organisms. In spite of a large body of data on pyrethroid environmental fate and chemistry, several questions remain from both a regulatory and scientific standpoint. These include questions about how pyrethroids are used in urban areas, urban fate and transport mechanisms, bioavailability in different matrices, and sediment persistence under different environmental conditions.

References

1. Croplife Foundation, National Pesticide Use Database: 2002, URL http://www.croplifefoundation.org/cpri_npud2002.htm
2. Weston, D. P.; J. You and M.J. Lydy. 2004. Distribution and toxicity of sediment-associated pesticides in agriculture-dominated water bodies of California's Central Valley. Environ. Sci. Technol. 38: 2752-2759.
3. Weston, D. P., R.J. Holmes, J. You and M.J. Lydy. 2005. Aquatic toxicity due to residential use of pyrethroid insecticides. Environ. Sci. Technol. 39: 9778-9784.
4. Amweg, E.L., D.P. Weston and N.M. Ureda. 2005. Use and toxicity of pyrethroid pesticides in the Central Valley, California, USA.
5. Starner, K., J. White, F. Spurlock and K. Kelley. Assessment of pyrethroid contamination of streams in high pyrethroid-use agricultural regions of California. Chapter X, This volume.
6. Davies, J.H. 1985. The Pyrethroids: An Historical Introduction. In: *The Pyrethroid Insecticides*; Leahy, J.P., Ed., Taylor and Francis, Philadelphia, PA, 1985, pp 1-41.
7. Schecter, M. S., Green, N. & Laforge, F. B. (1949). Constituents of pyrethrum flowers. XXIII. Cinerolone and the synthesis of related cylopenlenolones. J. Amer. Chem. Soc. 71 : 3165-3173.
8. Isobe, N, M. Matsuo and J. Miyamoto. 1984. Novel photoproducts of allethrin. Tetrahedron Letters, 25, 861-864.
9. Chen, Y. and J. E. Casida. 1969. Photodecomposition of pyrethrin I, allethrin, phthalthrin, and dimethrin. Modifications in the acid moiety. J. Agric. Food Chem. 17: 208-215.
10. Ando, T., L. Ruzo, J. L. Engel, and J.E. Casida. 1983. 3-(3,3-Dihalo-2-propenyl) analogues of allethrin and related pyrethroids: synthesis, biological activity, and photostability. J. Agric. Food Chem. 31:250-253.
11. Elliot, M., A.W. Farnham, N.F. Janes, P.H. Needham, D.A. Pulman and J.H. Stevenson. 1973. A photostable pyrethroid. Nature 246, 169-170.
12. Holmstead, R.L., J.E. Casida, L.O. Ruzo and D.G. Fulmer. 1978. Pyrethroid photodecomposition: permethrin. J. Agric. Food Chem. 26:590-595.
13. Laskowski, D.A. 2002. Physical and chemical properties of pyrethroids. Rev. Environ. Contam.Toxicol. 174: 49-170.
14. Litchfield, M.H. 1985. Toxicity to Mammals. In: *The Pyrethroid Insecticides*; Leahy, J.P., Ed., Taylor and Francis, Philadelphia, PA, 1985, pp 99-150.
15. Liu, W. Gan, J. and S. Qin. 2005. Separation and aquatic toxicity of enantiomers of synthetic pyrethroid insecticides. Chirality 17:S127:S133.
16. Liu, W., J. Gan, S. Lee and I. Werner. Isomer selectivity in aquatic toxicity

and biodegradation of bifenthrin and permethrin. Environ. Toxicol. Chem. 24:1861-1866.

17. Liu, W., Gan, J., Schlenk, D and W. A. Jury. 2005. Enantioselectivity in environmental safety of current chiral insecticides. Proc. Natl. Acad. Sci. 102:701:706.
18. Garrison, A.W. 2006. Probing the enantioselectivity of chiral pesticides. Environ. Sci. Technol. 40:16-23.
19. California Department of Pesticide Regulation Pesticide Sales Database, URL http://www.cdpr.ca.gov/docs/mlassess/nopdsold.htm
20. California Department of Pesticide Regulation Pesticide Use Database, URL http://www.cdpr.ca.gov/docs/pur/purmain.htm
21. Memorandum of Agreement Between the Environmental Protection Agency and Signatory Registrants Regarding the Registration of Pesticide Products Containing Chlorpyrifos, URL http://www.epa.gov/oppsrrd1/op/chlorpyrifos/moa62000.pdf
22. Central Valley Regional Water Quality Control Board Total Maximum Daily Loads (TMDLs) and Impaired Water Bodies 303(d) List, URL http://www.waterboards.ca.gov/centralvalley/programs/tmdl/index.htm
23. Wilen, C.A. 2001. Survey of residential pesticide use and sales in the San Diego Creek Watershed of Orange County, California. University of California Statewide IPM project. URL http://www.cdpr.ca.gov/docs/sw/contracts/sdcrk.pdf
24. Wilen, C.A. 2002. Survey of residential pesticide use in the Chollas Creek area of San Diego County and Delhi Channel of Orange County, California. University of California Statewide IPM project. URL http://www.cdpr.ca.gov/docs/sw/contracts/chollas_survey.pdf
25. Flint, M. 2003. A report of surveys taken in the Sacramento (Arcade Creek), Stockton (Five-Mile Slough) and San Francisco Bay Areas with comparisons to the San Diego Creek Watershed of Orange County, California. University of California Statewide IPM project. URL http://www.cdpr.ca.gov/docs/sw/contracts/ncalifsurvey_1.pdf
26. Kreidich, N., M.L. Flint, C.A. Wilen, M. Zhang. 2005. Tracking non-residential pesticide use in urban areas of California. University of California Statewide IPM project. URL http://www.cdpr.ca.gov/docs/sw/contracts/ucdavisrep.pdf
27. USDA-ARS Pesticide Properties Database, URL http://www.ars.usda.gov/Services/docs.htm?docid=14147
28. Footprint, the European Pesticides Properties Database, URL http://www.herts.ac.uk/aeru/footprint/
29. USEPA Office of Pesticide Programs Ecotoxicity Database, URL http://www.epa.gov/oppefed1/general/databasesdescription.htm#ecotoxicity
30. Gan, J., S. J. Lee, W. P. Liu, D. L. Haver, and J. N. Kabashima. 2005.

Distribution and persistence of synthetic pyrethroids in runoff sediments. J. Environ. Qual. 34:836-841.

31. Yang, W. , F. Spurlock, W. Liu, and J. Gan. 2006. Inhibition of aquatic toxicity of pyrethroid insecticides by suspended sediment. J. Environ. Tox. Chem. 25(7) 1913-1919.
32. Yang, W., F. Spurlock, W. Liu and J. Gan. 2006. Effects of dissolved organic matter on permethrin bioavailability to Daphnia species. J. Agric. Food Chem., 54(11) 3967-3972.
33. USEPA. 1985. Guidelines for deriving numerical national water quality criteria for the protection of aquatic organisms and their uses. PB-85-227049. U.S. Environmental Protection Agency, National Technical Information Service, Springfield, VA, USA.
34. Department of Pesticide Regulation. 2006. Notice of decision to begin reevaluation of certain pesticide products containing pyrethroids. Notice 2006-13. URL http://www.cdpr.ca.gov/docs/canot/2006/ca2006-13.pdf

Chapter 2

Sediment Toxicity in Agricultural Areas of California and the Role of Hydrophobic Pesticides

Donald P. Weston[1], Jing You[2], Erin L. Amweg[1,3], and Michael J. Lydy[2]

[1]Department of Integrative Biology, University of California, 3060 Valley Life Sciences Building, Berkeley, CA 94720-3140
[2]Fisheries and Illinois Aquaculture Center and Department of Zoology, Southern Illinois University, 171 Life Sciences II Building, Carbondale, IL 62901
[3]Arcadis U.S., 155 Montgomery Street, Suite 1500, San Francisco, CA 94104

To assess the impact of agricultural pesticides on sediment quality, 200 sediment samples were collected in California's Central Valley and tested for acute toxicity to the amphipod *Hyalella azteca*. Twenty-seven percent of the samples caused acute toxicity, and in 67% of these instances, the analytes were in sufficient concentration to explain it. Pyrethroids, most notably bifenthrin and lambda-cyhalothrin, reached concentrations associated with *H. azteca* toxicity in 55% of the toxic samples, or 61% if toxicity of compounds within the class is assumed to be additive. Chlorpyrifos reached acutely toxic concentrations in 20% of the samples. Organochlorines rarely, if ever, contributed to acute toxicity. While toxicity was sometimes observed in creeks and major rivers, small agricultural drains nearest the points of pesticide use were more affected than the water bodies to which they flow. The pesticides appear to be entering these drains largely by way of irrigation runoff during the summer months.

Introduction

About 40% of California is located in the Central Valley, a region of highly productive agricultural land. The area produces an exceptionally wide variety of crops, with the dominant commodities including alfalfa, hay, corn, rice, tomatoes, lettuce, citrus fruits, peaches, plums, almonds, nuts and grapes. The Central Valley provides nearly all commercial U.S. production of almonds, walnuts, figs, kiwifruit, nectarines, olives, pistachios, prunes, and raisins.

Maintaining the 7,000,000 acres of irrigated agricultural land within the Central Valley requires a complex network of watercourses to supply water and carry irrigation runoff (known as tailwater) to the region's major rivers. A total of 32,000 km of channels have been constructed, and there is an additional 2,400 km of natural creeks and rivers, many of which have been modified and are heavily used for agricultural supply water or drainage (*1*).

Habitat quality within this network of constructed and natural water courses is affected by the 57 million kg of pesticides used in the Valley annually (2004 usage data: www.cdpr.ca.gov/docs/pur/pur04rep/04_pur.htm). Since the early 1990s, rivers and other Central Valley water bodies receiving tailwater have been frequently found to be acutely toxic to a standard testing species, the cladoceran Ceriodaphnia dubia (2,3). In most cases when the causative agent could be identified, it was found to be one of the organophosphate pesticides, often diazinon or chlorpyrifos. These findings led to extensive water quality monitoring for the organophosphates, development of management practices to reduce organophosphate runoff, and widespread grower education efforts. Agricultural organophosphate use is now less than half of what it was in the early 1990s, and the frequency of C. dubia toxicity has decreased significantly.

Sediment quality in the Central Valley, on the other hand, has until recently, received little attention. An assessment done in 2002 and 2003 indicated widespread toxicity to the amphipod Hyalella azteca, and less frequently, to the midge, Chironomus dilutus (formerly C. tentans) (4). About 28% of the sediment samples collected were acutely toxic to H. azteca. Based on a comparison of the pyrethroid concentrations occurring in the samples to estimates of the concentrations in sediment likely to cause toxicity, pyrethroids were believed to contribute to the toxicity in the majority of the cases.

The Central Valley of California is unique in the amount of environmental data available on pyrethroid pesticides in sediments. The compounds are widely used, yet there has been little or no monitoring for the compounds in other agricultural areas throughout the world, and only limited data are available from other locations within California (*5,6*). Thus, a close assessment of the Central Valley data provides the opportunity to determine if environmental residues of pyrethroids are present in aquatic habitats, whether they are present at toxic levels, and which specific compounds within the class most often contribute to toxicity. The present study is intended to re-evaluate and build upon the findings of Weston et al. (*4*). In the intervening years there have been several improvements in our ability to assess sediment quality in the Central Valley.

First, the database is considerably larger. While Weston et al. (*4*) was based on 70 samples, the current study is based on an additional 130, and the Central Valley counties with available data have increased from 10 to 17 (of the 19 within the Valley). Improvements in analytical methods have also doubled the number of pyrethroid analytes quantified in recent samples. Secondly, in the former study it was necessary to estimate the concentrations of pyrethroids that would be acutely toxic to *H. azteca*, since precise measurement had not been made for most members of the class. However, such data are now available (*7*) making it possible to more confidently establish when pyrethroids may be contributing to observed toxicity. Finally, in addition to comparing observed concentrations to known toxic levels, several other approaches are now available to help identify the causative agent of toxicity.

Materials and Methods

Overall Sampling Design

The available data represents 133 sites throughout the Central Valley, 117 of which have complete toxicity and chemistry data (the remainder with only toxicity data). Most of these sites (81) were sampled on one occasion, 42 were sampled twice, six were sampled three times, and four were sampled four to five times. This effort yielded 200 samples, 180 of which have complete chemistry and toxicity data. The 70 samples of Weston et al. (4) are included within this total.

About two-thirds of the samples were provided by a study of sediment quality in waterways throughout the Valley receiving agricultural tailwater. Most of these waterways were of moderate size and intended to be representative of regional inputs rather than one or a few farms. Sites were selected to obtain even geographic coverage across the Valley, without regard to crop type or pesticides use in the vicinity. The remaining third of the samples were located in areas of high pyrethroid use. For this latter subset of samples, sites were established in 10 of the Central Valley counties with the highest annual agricultural pyrethroid use, as determined by mapping of pesticide use data from the California Department of Pesticide Regulation's Pesticide Use Reporting (PUR) database (www.cdpr.ca.gov/docs/pur/purmain.htm). These sites were roughly evenly divided between the counties in the northern half of the Valley (Sacramento River watershed) and the southern counties (San Joaquin River watershed).

Samples were collected from July 2002 through April 2006. Most samples were collected either at the end of the rainy season before agricultural irrigation begins, when sediment quality would be expected to be influenced by winter stormwater runoff (March-April), or the end of the irrigation season when tailwater return provides the primary route for transport of hydrophobic pesticides (mostly August samples, with a few in July and September).

Sampling Methodology

Sampling efforts were focused on fine-grained sediments given their higher organic carbon content, and therefore higher affinity for hydrophobic pesticides (8). Even in those waterways dominated by gravel or hardpan clay, it was usually possible to find soft sediment deposits from which to collect the samples. When possible, samples were composited over a reach of at least 30 m, though often a shorter segment was sampled because of limited soft sediment availability or access difficulties.

Sediments were collected from the bank or by wading into shallow water, using a stainless steel scoop to skim the upper 1-2 cm of the sediment column. A sample consisted of a dozen or more such scoops, composited in a solvent-cleaned 4 L glass jar. The sediment was held on ice until returned to the laboratory, where it was homogenized by hand mixing in a stainless steel bowl. Approximately 4% of the samples contained gravel, vegetation or other debris requiring removal by sieving on a 1 mm screen to obtain homogeneous material. Subsamples were taken from the mixing bowl for pesticide and total organic carbon analysis (both held at -20°C until analysis) and toxicity testing and grain size analysis (both held at 4°C).

Toxicity Testing

Sediments were tested for toxicity using the amphipod, *H. azteca*, using standard protocols (*9*). The only significant departures from these protocols were use of a slightly smaller amount of sediment (75 ml) and use of only the mortality endpoint (growth was measured in some samples but data are not presented). Briefly, 400 ml beakers were filled with 75 ml sediment and 250 ml moderately hard water, reconstituted by addition of salts to Milli-Q purified deionized water (Millipore Corp., Billerica, MA, USA). Each batch of test sediments was accompanied by a control sediment, the source of which varied over the four years that the samples were collected. In any given test, control sediment may have come from the American River at Folsom Lake, CA, San Pablo Dam Reservoir, Orinda, CA, or Lake Anza, Berkeley, CA. Ten individuals of *H. azteca*, 7-12 d in age, were added to eight replicates for each sediment, or five replicates in about 18% of the samples. Tests were conducted for 10 d, at 23°C, with a 16: 8 hr light:dark cycle, and feeding 1 ml YCT (yeast, cerophyll, trout food) per beaker per day. A few of the samples required gentle aeration to keep dissolved oxygen levels within test limits. Water was changed at the rate of two volume additions daily (total 500 ml) by an automatic water delivery system. Ammonia, hardness, alkalinity, conductivity, and pH were measured at the start and end of the test; temperature and dissolved oxygen were monitored regularly throughout the test. Water quality data are not presented but temperature was always within 1°C of nominal and dissolved oxygen was always above 2.5 mg/L as required by the standard protocol (*9*). After 10 d, the sediment was sieved on a 425 μm screen and the surviving animals enumerated.

Some samples exhibiting high mortality were tested in a dilution series using control sediment as the diluent, concentration steps of a factor of two (e.g., 50%, 25%, 12%, 6%), and three replicates per concentration. Control sediment and test sediment were thoroughly mixed by hand, and the test initiated 24 h later.

Test data were analyzed using ToxCalc software (Tidepool Scientific Software, McKinleyville, CA, USA). Test sediments were compared to control using Dunnett's procedure when parametric assumptions were met, with arcsine squareroot transformation when necessary. Steel's Many-One Rank test was used when parametric assumptions were not met. Median lethal concentrations (LC50) were determined in the sediment dilution series by maximum likelihood regression using probit transformation.

Chemical Analysis

All the sediment samples were analyzed for four pyrethroids: bifenthrin, esfenvalerate, lambda-cyhalothrin and permethrin. Three additional pyrethroids, cyfluthrin, cypermethrin, and deltamethrin, were added to the analyte list later, and were analyzed for in two-thirds of the samples. The pyrethroid fenpropathrin was quantified in only a single sample. Organochlorine pesticides analyzed included alpha-, beta-, delta-, and gamma-BHC, heptachlor, heptachlor epoxide, alpha- and gamma-chlordane, alpha- and beta-endosulfan, endosulfan sulfate, *p,p'*- DDE, *p,p'*- DDD, *p,p'*- DDT, aldrin, dieldrin, endrin, endrin aldehyde, endrin ketone, and methoxychlor. Chlorpyrifos was the only organophosphate insecticide quantified.

Analysis was performed on an Agilent 6890 series gas chromatograph equipped with an Agilent 7683 autosampler and an electron capture detector (Agilent Technologies, Palo Alto, CA, USA). Two columns from Agilent, a HP-5MS (30m x 0.25mm; 0.25μm film thickness) and a DB-608 (30m x 0.25mm; 0.25μm film thickness) were used. Five external standards solutions ranged from 5 to 250 ng/ml were used for calibration. The calibration curves were linear within this concentration range. Qualitative identity was established using a retention window of 1% with confirmation on a second column.

Prior to analysis, frozen sediment was thawed, centrifuged to remove excess water and homogenized. The extraction and cleanup methods were developed and validated in an earlier study (10). Two surrogates, 4,4'-dibromo-octafluorobiphenyl and decachlorobiphenyl, were added to the sediment prior to the extraction to verify extraction and cleanup efficiency. Approximately 20 g of sediment (wet weight) was mixed with anhydrous $MgSO_4$ and sonicated with 50 ml of 50:50 acetone:methylene chloride (v/v) for 3 minutes using a high intensity ultrasonic processor (Sonics and Materials Model VCX 400, Newtown,

CT, USA). The extract was centrifuged, decanted, and filtered. This procedure was repeated twice more. Extracts were combined, solvent exchanged with hexane and the volume reduced to 2 ml. Adsorption chromatography with Florisil, deactivated by mixing with distilled water (6% w/v), was used for extract cleanup. The pesticides were eluted from the column with 50 ml of 30% diethyl ether in hexane (v/v). The eluent was evaporated, redissolved in 2 ml of hexane and analyzed on the gas chromatograph. Additional dilution steps were needed for some field-collected samples due to elevated pesticide concentrations. With method detection limits of 0.22-0.85 ng/g dry weight, the method reporting limits were set at 1 ng/g for all the analytes.

Quality control measures included re-analysis of pyrethroid pesticides in five samples by an independent laboratory (California Department of Fish and Game, Rancho Cordova, CA) using gas chromatography-mass spectroscopy (GC-MS) and blind analysis of 13 spiked sediment samples. Qualitative agreement with the GC-MS analysis was excellent, with the GC-MS confirming GC-ECD-derived compound identity in all cases. Quantitative agreement was good, with GC-MS and GC-ECD quantitations having a median relative percent difference of 25%, and equal instances of the GC-MS yielding higher and lower values than the GC-ECD. Analysis of the 13 blind samples produced pyrethroid recoveries that were nearly always 50-120% of nominal values (median = 79%). Chlorpyrifos recoveries in the blind spikes were 33-104% (median = 66%), with the lower values possibly due to loss of the compound to overlying water. Organochlorine recoveries from the blind spikes were usually 60-100% (median = 74%).

Total organic carbon was measured using a CE-440 elemental analyzer (Exeter Analytical, Chelmsford, MA, USA) after acid vapor treatment to remove inorganic carbon. Grain size was determined by wet sieving, with silt and clay combined in the <64 μm fraction.

Results and Discussion

Sediment Properties

The sediments sampled were deliberately chosen to represent the finest-grained material available at each site, in which hydrophobic pesticides would be more likely to be present at measurable levels. The percentage of silt and clay particles within the samples ranged from 7-97% by weight, with a median of 42%. Eighty percent of the sites contained >25% silt and clay. Total organic carbon of the sediment samples ranged from 0.1-7.4%, with a median value of 1.1%. Seventy percent of the samples fell within the range of 0.5-2.5% organic carbon.

Toxicity Testing

Control survival was acceptable, with a median value of 94% across all tests, and was never below 86%. Test sediments, however, frequently caused acute mortality to *H. azteca*. Overall, a total of 53 out of the 200 samples (27%) exhibited toxicity, and 39 out of the 133 sites (29%) were toxic on at least one sampling occasion. The later statistic, while accurate, is somewhat distorted by those sites that were sampled on multiple occasions, when they may have only been toxic once. If the percentage of sites exhibiting toxicity is calculated only on the basis of the first sampling event at each site, regardless of findings in later events, (a more reasonable approximation of the frequency of toxicity if measured at a single point in time) the percentage of toxic sites decreases to 23%.

Toxicity, however, was not uniformly distributed throughout the Central Valley (Figure 1). Sites in the southern half of the Valley within the San Joaquin River watershed were twice as likely to show toxicity as those in the northern half within the Sacramento River watershed (37% of the southern sites vs. 19% in the north). In particular, the northwestern portion of the San Joaquin watershed, comprised of portions of San Joaquin and Stanislaus Counties, was an area of frequent sediment toxicity, with 15 out of 34 sites (44%) in these two counties causing toxicity. Fresno County, at the southern end of the Valley, also had a high frequency of toxicity, with 47% of the sites toxic on at least one occasion.

Of the various water body types within the Valley, unnamed drains showed the most frequent toxicity (Table I). These drains are entirely constructed water bodies, and because they serve a relatively small number of farms, are unnamed and do not appear on regional maps. Forty-one percent of the sampling sites in these drains showed toxicity, the highest of any water body type, as might be expected given the close proximity of these drains to the points of pesticide application, and the fact that water flow to the drains consists entirely of field runoff. The frequency of toxicity is reduced by nearly half in named drains (e.g., Island Field Drain, Colusa Drain, Button Ditch), water bodies serving larger watersheds and more critical to regional irrigation systems. Creeks showed a surprisingly high frequency of toxicity, with 40% of the creek sites toxic on at least one occasion. The creeks generally originate around the periphery of the Central Valley, with their headwaters in the surrounding mountains, and are natural waterbodies though their flow is highly managed for irrigation purposes in their agricultural reaches. The high frequency of toxicity in creeks is, however, somewhat distorted by the numerous creeks in western Stanislaus and San Joaquin counties that have consistently been found to contain toxic sediments. Excluding this region to obtain a more representative picture of the Central Valley, creek toxicity is reduced to 26% of the sites. Three of 11 river sites (27%) showed sediment toxicity (Calaveras River, Kaweah River, and the San Joaquin River near the town of Vernalis, CA).

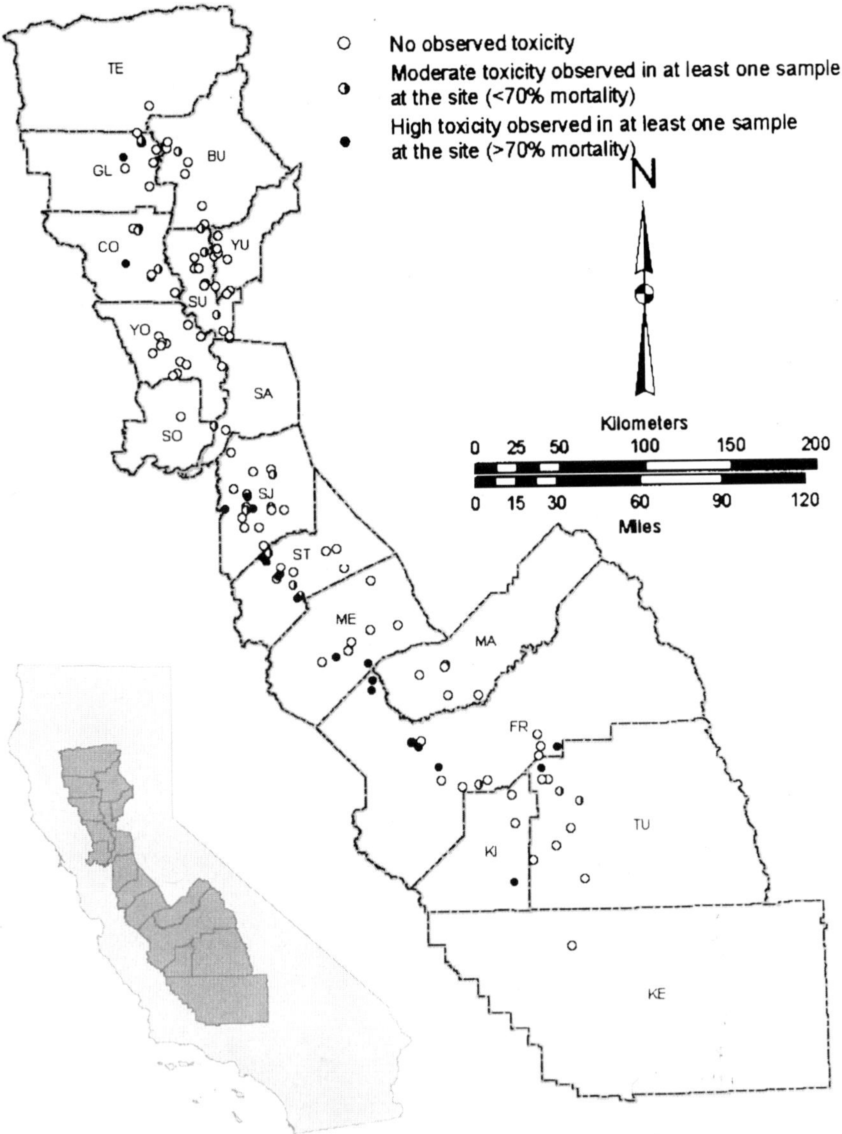

Figure 1. Distribution of H. azteca toxicity among the Central Valley sediment sampling sites. County names are abbreviated as: BU=Butte, CO=Colusa, FR=Fresno, GL=Glenn, KE=Kern, KI=Kings, MA=Madera, ME=Merced, SA=Sacramento, SJ=San Joaquin, SO=Solano, ST=Stanislaus, SU=Sutter, TE=Tehama, TU=Tulare, YO=Yolo, YU=Yuba. The Central Valley counties of Shasta and Placer are not shown as there were no samples taken in those locations. The inset map shows the location of the study area within California.

Table I. Frequency of *H. azteca* sediment toxicity in various water body types within the Central Valley

Water body type	*Number of sites*	*Percentage of sites*
Unnamed drains	34	41
Named drains	17	24
Canals	8	13
Sloughs	28	11
Creeks	35	40
Creeks excluding westside[a]	27	26
Rivers	11	27

[a] "Westside" is a local designation for the area on the west side of the San Joaquin River in portions of Stanislaus and San Joaquin Counties. It includes sampling sites in Hospital, Ingram, Del Puerto, and Orestimba Creeks, all of which consistently have had sediments toxic to *H. azteca*.

Nearly all sites were sampled either at the end of the winter rainy season (March/April) or late in the summer irrigation season (July/August/September). The timing of the sampling did not have a great effect on the frequency of sediment toxicity observed. In the late winter 29% of the samples showed acute toxicity, whereas in the late summer that proportion was 21%.

Contributors to Sediment Toxicity

One approach to identifying likely contributors to sediment toxicity is the use of toxic units (TU) normalized to sediment organic carbon (oc; Weston et al., 2004) defined as:

$$TU = \frac{\text{Actual sediment concentration of the analyte on oc basis}}{\text{Reported 10-d } LC_{50} \text{ concentration of the analyte on an oc basis}} \quad (1)$$

Application of the TU approach to explaining *H. azteca* toxicity requires that 10-d sediment LC50 values for the species be available for all toxicants of interest. These values have been published for all pyrethroids regularly analyzed in this study (cypermethrin 10-d LC50 = 0.38 µg/g oc, lambda-cyhalothrin = 0.45 µg/g oc, bifenthrin = 0.52 µg/g oc, deltamethrin = 0.79 µg/g oc, cyfluthrin = 1.08 µg/g oc, esfenvalerate = 1.54 µg/g oc, permethrin = 10.83 µg/g oc (*7,11*). The chlorpyrifos LC50 has been determined in three different sediments in our laboratory and averaged as 2.96 µg/g oc (*12*). The sediment LC50 of fenpropathrin to *H. azteca* has not been determined. However, in water-only exposures to aquatic life, the 5th percentile LC50 of fenpropathrin is 1.7

times that of permethrin (114 ng/L vs. 68 ng/L; *13*), and this ratio was applied to the permethrin sediment LC50 to derive an estimated fenpropathrin sediment LC50 of 18.2 µg/g oc.

Since *H. azteca* mortality of 50% would be expected at 1 TU, a value of 0.5 TU is herein used as a rough approximation of the concentration at which mortality would first appear, and a threshold above which the given toxicant is considered potentially responsible for mortality when observed (*4*). The 0.5 TU value is arbitrary, but suggests the compound is on the verge of reaching acutely toxic concentrations if not already surpassing them.

Pyrethroid pesticides are implicated by the TU analysis as a probable cause of the toxicity in the majority of cases. Since pyrethroids all have similar modes of neurotoxic action, the most probable interaction between members of the group is additivity. Assuming additivity of pyrethroid TUs, 61% of the acutely toxic samples (31 out of 51 samples; excluding two toxic samples with no chemistry data) contain at least 0.5 TU of total pyrethroids (Figure 2). The assumption of additivity, while reasonable, has not been proven specifically for pyrethroids, but it does not substantially affect the analysis. Even without the assumption of additivity, 55% of the acutely toxic samples contained at least 0.5 TU of at least one individual pyrethroid. Chlorpyrifos concentrations reached 0.5 TU in 20% of the toxic samples. Organochlorines almost never reached concentrations expected to be toxic to *H. azteca*. Estimated organochlorine LC50 values are available for gamma-BHC, endosulfan, DDE, DDT, DDD, dieldrin, endrin, and methoxychlor (*4*). Not a single toxic sample contained at least 0.5 TU of any organochlorine. There was only one instance of an organochlorine (endrin) present at 0.5 TU and this sample showed no acute toxicity.

After accounting for all analytes for which TU values could be calculated, there remained 33% of the toxic samples for which none of the measured analytes exceeded 0.5 TU, even if assuming additivity of TUs within the pesticide classes. Toxicity in these cases may have been due to sediment properties that enhanced bioavailability and toxicity of the measured analytes above that expected based on organic carbon normalization alone, or a result of other stressors including one of the many agricultural pesticides used in the Central Valley that are not among the analytes of this or any monitoring program.

The TU approach, even when considering pyrethroid TUs alone, was highly predictive of toxicity to *H. azteca* (Figure 3). Below about 0.5 TU of pyrethroids, toxicity was rarely seen, and when present tended to be fairly modest (<40% mortality). The three samples in the upper-left corner of Figure 3 (>80% mortality but less than 0.01 TU of pyrethroids) can be explained by the presence of chlorpyrifos that exceeded 0.8 TU in all three of these samples. Above 0.5 pyrethroid TUs, mortality rate climbed rapidly, as would be expected if pyrethroids were the primary causative agent. Above about 3 TU of pyrethroids, there was total or near total mortality in all samples. The seven

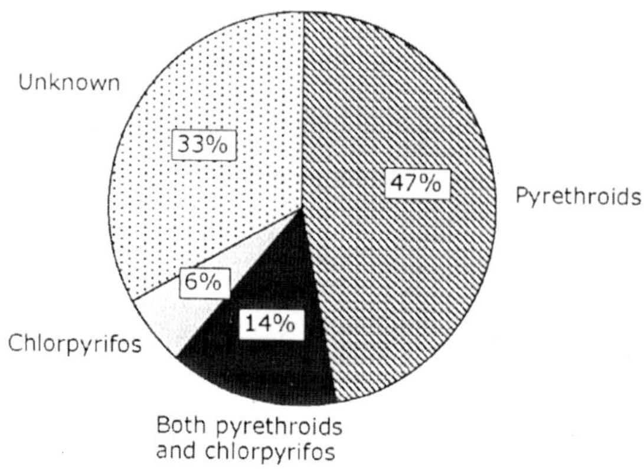

Figure 2. Proportions of the 51 toxic samples containing at least 0.5 toxic units of the indicated analytes, suggesting a potential causal relationship for the toxicity.

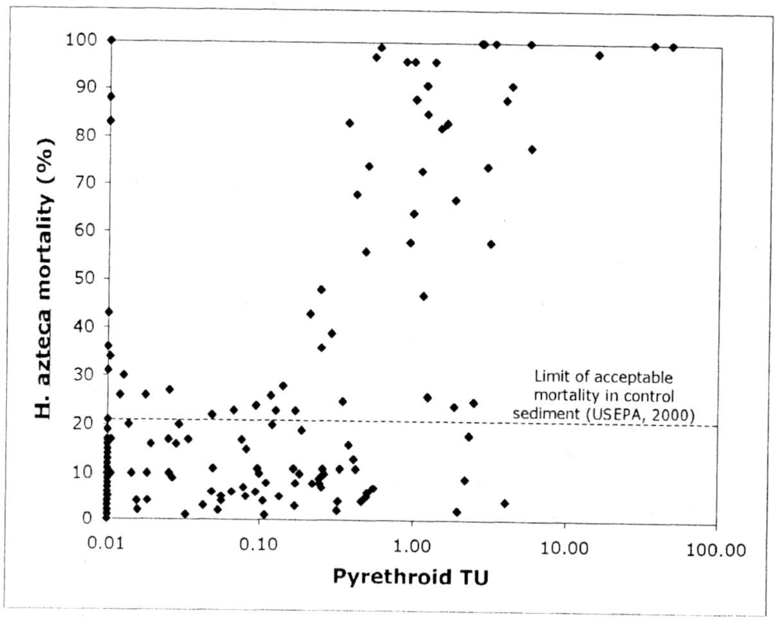

Figure 3. Relationship between total pyrethroid toxic units (TU) and H. azteca mortality in the same sediment samples. Data are from all 180 samples for which both chemistry and toxicity results are available. A TU value of 0.01 has been arbitrarily used for those samples in which pyrethroids were undetected. The threshold for acceptable mortality when testing control sediments following standard protocols (9) is shown to provide some benchmark against which to compare the level of mortality that may be indicative of sediment toxicity.

samples with surprisingly low toxicity (1-4 TU, but <30% mortality) tended to be from coarse sands: all but one of these samples were among the third of the samples with the lowest proportion of silt and clay. Previous work (*14*) has also reported an overestimate of pyrethroid toxicity by the TU approach in similar sediments. While there are a few data points that deviate slightly from the expected TU:mortality relationship of Figure 3, the relationship is remarkably good, and consistent with pyrethroids being the causative agent for much of the observed toxicity.

A second line of evidence to help infer causality for the toxicity is provided by dilutions of samples exhibiting high *H. azteca* mortality. A parameter referred to as "observed TUs" was calculated based on the toxicity test dilutions as:

$$\text{Observed TU} = 100/\text{Observed LC50 of test sediment determined by dilution} \quad (2)$$

The observed TU derived by toxicity testing could then be compared to the expected TU previously shown in equation 1, calculated based on chemical concentrations and literature-derived LC50 concentrations. Close agreement of the observed TU with the expected TU provides evidence that the compounds used to calculate the expected TU are indeed responsible for the toxicity.

This approach is commonly used in a Toxicity Identification Evaluation context for water samples (*15*), but a mathematical adjustment is necessary when applying it to hydrophobic toxicants in sediment. The bioavailability, and hence toxicity, of such materials is highly dependent upon the sediment organic carbon content (*8*), but rarely will the sediment used for dilution be of equal organic content to the test sediment. Dilution with a control sediment high in organic carbon will yield a higher LC50 estimate than if the control sediment diluent contained little organic carbon. If the organic carbon content of the test sediment and control sediment diluent are both known, it is possible to calculate the organic content of the diluted sediment when at its LC50 concentration, and then use this value to express the observed LC50 on an organic carbon adjusted basis as:

$$\text{Observed LC50oc} = \text{Observed LC50} \times \frac{\text{oc of undiluted test sediment}}{\text{oc of diluted sediment at the LC50}} \quad (3)$$

The observed LC50oc was then used in a manner analogous to equation 2 to obtain an observed TUoc. The approach assumes the LC50 is linearly related to the organic carbon content of the sediment, and expresses that LC50 as if the organic carbon at the LC50 concentration was equivalent to that of the original test sediment.

There were 11 test sediments that were tested in dilution series, and for which it was possible to compare the observed TUoc with the expected TU of either pyrethroids or chlorpyrifos (Table II). In only a couple cases was there precise agreement between the observed and expected TU, but in nearly every

Table II. Comparison of observed TU_{oc}, derived from toxicity testing of a dilution series, with the expected TU, derived from the sediment chemical data. The range in the observed TU_{oc} reflects the 95% confidence interval of the calculated LC50.

Sample site	*Sample date*	*Observed TU_{oc}*	*Expected pyrethroid TU (specific compound)*	*Expected chlorpyrifos TU*
SED11 (unnamed drain)	Aug. 28 2004	1.5-2.2	1.0 (esfenvalerate)	0
SED11 (unnamed drain)	Oct. 13 2004	1.1-1.6	1.1 (esfenvalerate)	0
NSJ18 (Orestimba Creek)	Aug. 12 2004	2.1-2.4	3.9 (lambda-cyhalothrin)	0.1
SED12 (Hospital Creek)	Oct. 13 2004	3.6-4.6	5.6 (bifenthrin, lambda-cyhalothrin)	0
SED15 (unnamed drain)	Mar. 24 2005	2.2-2.9	1.2 (lambda-cyhalothrin)	0.6
SED15 (unnamed drain)	Aug. 18 2005	19-136	36 (bifenthrin, lambda-cyhalothrin)	24
CS15 (Spring Creek)	Aug. 9 2005	1.6-2.2	0.9 (bifenthrin)	1.5
CS12 (unnamed drain)	Aug. 9 2005	1.9-2.5	0.6 (fenpropathrin)	0.8
FT19 (unnamed drain)	Aug 2 2005	2.8-3.7	0	1.6
FT19 (unnamed drain)	Aug. 19 2005	3.3-4.9	0	5.3
SED40 (Del Puerto Creek)	Dec. 7 2005	71-100	47 (bifenthrin)	0

case the two approaches to derive TUs agreed within a factor of two. Variation in LC50s of this magnitude are common when testing multiple sediments (*7,11*), and thus the expected TUs, derived using generalized LC50s, could easily incorporate a factor of two error when applied to specific sediments. Taking this potential variability into account, there was good agreement between observed and expected TUs, regardless of whether the putative toxicant was esfenvalerate, lambda-cyhalothrin, bifenthrin, fenpropathrin, or chlorpyrifos, further supporting the role of these compounds in causing the observed toxicity.

One complicating factor in this analysis is the nature of the toxicological interaction of pyrethroids and chlorpyrifos. For a few of the sediments in Table II, additivity of pyrethroid TUs was assumed, which is a reasonable assumption given the similar mode of action for all the pyrethroids. However, pyrethroid and chlorpyrifos TUs were calculated independently as the two groups have different modes of neurotoxicity. There are at least two studies indicating toxicity of pyrethroids and organophosphates is at least additive and potentially synergistic (*16,17*), but since such an interaction is not widely established, no implicit assumptions of their interaction was made here. Thus, for a few sediments when there could be substantial toxicity due to both pyrethroids and chlorpyrifos, such as CS15 and CS12 samples both from August 9, 2005, the expected TUs when including both pesticides remains uncertain.

Finally, the third line of evidence for causality comes from newly developed toxicity identification evaluation techniques for bulk sediments. Pyrethroids are atypical in that they become more toxic as the temperature decreases (*18*), whereas chlorpyrifos toxicity to *H. azteca* is temperature independent (Weston, unpub. data). Thus, simultaneous sediment toxicity test can be performed at the standard 23°C and at the reduced temperature of 18°C, and a doubling of toxicity at the lower temperature is suggestive of pyrethroids (Weston, unpub. data). Seven sediments from the current study were tested with *H. azteca* at reduced temperatures (Hospital, Del Puerto, and Spring Creeks, and the unnamed drains of SED11, CS12, FT19, and SED15). In every case the sediments were more toxic. Those samples in which chlorpyrifos was suspected to be contributing to the toxicity showed a statistically significant, but only slight temperature response, whereas those containing pyrethroids showed a strong temperature response, with typically an increase in toxicity of a factor of two or more (Weston, unpub. data).

Five sediments from the current study (Del Puerto and Spring Creeks, and the unnamed drains of CS12, FT19, and SED15) were also tested with piperonyl butoxide (PBO) in the overlying water. PBO enhances the toxicity of pyrethroids (*19,20*), but lessens the toxicity of chlorpyrifos (*21*). In four sediments in which pyrethroids were believed to be substantial contributors to toxicity based on TU calculations, PBO enhanced toxicity supporting the suspected role of pyrethroids (*12*). In one sediment sample containing only chlorpyrifos, toxicity was diminished by addition of PBO.

Two samples from the current study (Del Puerto and Hospital Creeks) were tested by addition of esterase to the overlying water. The enzyme cleaves the ester bond of pyrethroids, dramatically reducing toxicity (*22*). The procedure has been shown to reduce pyrethroid-related sediment toxicity to *H. azteca* when added to the overlying water (*23*). In the two samples tested, pyrethroids were suspected to be the cause of toxicity based on TU analysis, and in both cases addition of esterase substantially reduced that toxicity (*23*).

Patterns of Pesticide Use and Sediment Contamination

This study included analysis of 28 pesticides, the majority of which did not appear to play a significant role in determining toxicity to *H. azteca*. Only eight pesticides reached concentrations of at least half their estimated sediment LC50 to the species (Table III). Of these eight, three pesticides reached this 0.5 TU threshold at more than 5% of the sites: bifenthrin, lambda-cyhalothrin and chlorpyrifos.

The distribution of these three pesticides that most often exceeded the 0.5 TU threshold in Central Valley sediments is shown in Figures 4, 5, and 6. Their concentration in the sediment is shown on a ng/g basis, but with the concentration categories corresponding to the toxicity units of the respective compounds (undetected; less than 0.5 TU; 0.5 to 2 TU; greater than 2 TU – with all TUs calculated assuming a typical 1% sediment oc). To relate the concentrations to patterns of pesticide use, these maps also indicate the intensity of agricultural use of the given pesticide in each Central Valley county based on 2004 data from the California Department of Pesticide Regulation's PUR database (www.cdpr.ca.gov/docs/pur/purmain.htm). This database provides statistics on the mass of pesticide used in each county, however, the counties differ dramatically in size, the amount of their land area committed to agriculture, and the amount of agricultural land likely to have few or no pesticides applied (e.g., rangeland). Therefore, the usage in each county has been adjusted to the area of harvested cropland, using 2002 acreage figures from the California Department of Finance (www.dof.ca.gov/HTML/FS_DATA/STAT-ABS/tables/g8.pdf), and the maps indicate the amount of the given pesticide used annually in each county to produce a hectare of harvested crop. Thus, "high use" counties on these maps may not necessarily use a large amount of the pesticide on an absolute basis, but do use a relatively large amount within their land area of harvested cropland. Finally, these maps also illustrate the seasonal patterns of use for the given pesticide, as such information has ramifications for appropriate mitigation practices.

Table III. Proportion of sites (out of 117 total) with concentrations of one the measured analytes exceeding 0.5 TU. Those sites in bold type had *H. azteca* toxicity that statistically exceeded control.

Pesticide	*% sites at or above 0.5 TU*	*Water bodies with 0.5 TU exceedances*
Bifenthrin	16	3 rivers (San Joaquin, Tule, **Kaweah**)
		5 creeks (**Spring, Hospital, Del Puerto, Orestimba**, Root)
		2 sloughs (**Poso**, Elk Bayou)
		1 canal (**Stinson**)
		2 named drains (Crescent Ditch, **Boundary Drain**)
		4 unnamed drains (FS, **TL**, **MA**, **SED15**)
Lambda-cyhalothrin	9	1 river (San Joaquin)
		3 creeks (**Hospital, Del Puerto, Orestimba**)
		2 sloughs (**Murphy**, **Poso**)
		3 unnamed drains (FS, **MA**, **SED15**)
Chlorpyrifos	8	1 creek (**Spring**)
		1 slough (**Poso**)
		3 named drains (**Holland Drain**, Button Ditch, **Knestric Ditch**)
		4 unnamed drains (**FT19**, **CS12**, **SED15**, **AD2**)
Esfenvalerate	4	2 creeks (**Littlejohn**, **Del Puerto**)
		1 named drain (**Knestric Ditch**)
		2 unnamed drains (**AD6**, **SED11**)
Cypermethrin	3	1 named drain (**Knestric Ditch**)
		1 unnamed drain (**SED23**)
Permethrin	2	1 creek (Root)
		1 unnamed drain (**AD5**)
Fenpropathrin[a]	unknown	1 unnamed drain (**CS12**)
Endrin	1	1 named drain (TID#3)

[a]Fenpropathrin was only analyzed in a single sample, thus this table may underestimate its prevalence or contribution to toxicity.

Bifenthrin (Figure 4) was detected in 23% of the samples and reached the 0.5 TU level on at least one sampling occasion at 16% of the sites (i.e., one out of six Central Valley sites contained acutely toxic concentrations of bifenthrin in at least one sampling event). Concentrations of bifenthrin in sediments that would be potentially toxic to *H. azteca* are limited almost entirely to the southern Central Valley counties (San Joaquin County and southward).

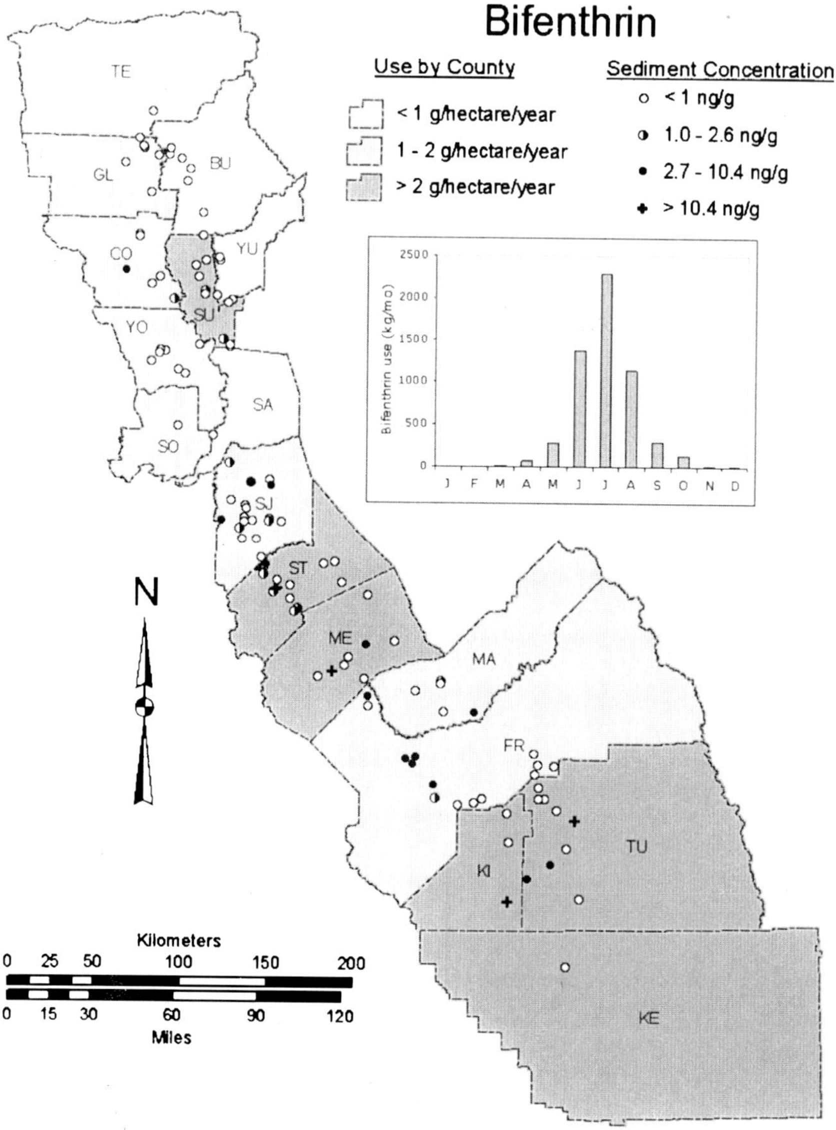

Figure 4. Concentration of bifenthrin at each of the study sites, with the highest concentration shown if the site was sampled on multiple occasions. The breakpoints between the four categories of concentration correspond to specific toxic unit (TU) thresholds assuming a generic organic carbon content of 1%: undetected, detectable but acute H. azteca toxicity unlikely (<0.5 TU), toxicity likely (0.5-2 TU), high toxicity likely (>2 TU). The figure also shows the annual usage of bifenthrin within each county, normalized to area of harvested cropland, and the monthly use of bifenthrin in Central Valley agriculture using 2004 data. County abbreviations defined in Figure 1.

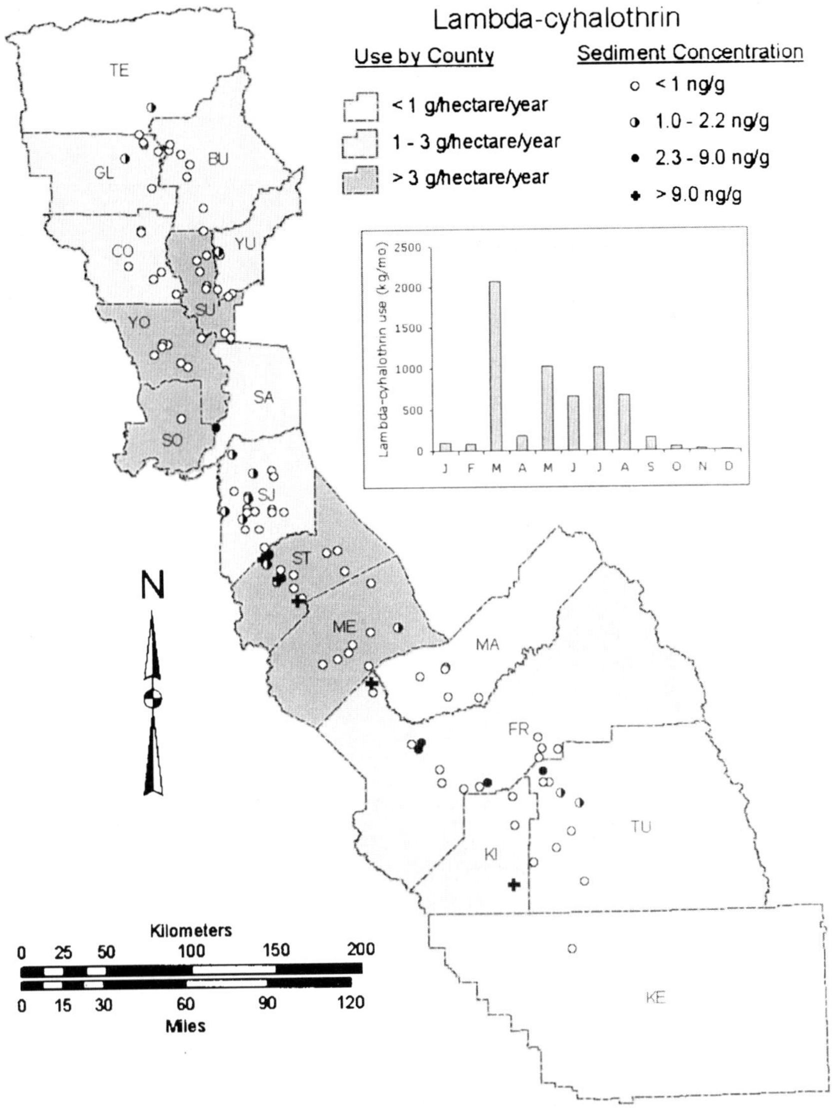

Figure 5. Concentration of lambda-cyhalothrin at each of the study sites, with the highest concentration shown if the site was sampled on multiple occasions. The breakpoints between the four categories of concentration correspond to specific toxic unit (TU) thresholds assuming a generic organic carbon content of 1%: undetected, detectable but acute H. azteca toxicity unlikely (<0.5 TU), toxicity likely (0.5-2 TU), high toxicity likely (>2 TU). The figure also shows the annual usage of lambda-cyhalothrin within each county, normalized to area of harvested cropland, and the monthly use of lambda-cyhalothrin in Central Valley agriculture using 2004 data. County abbreviations defined in Figure 1.

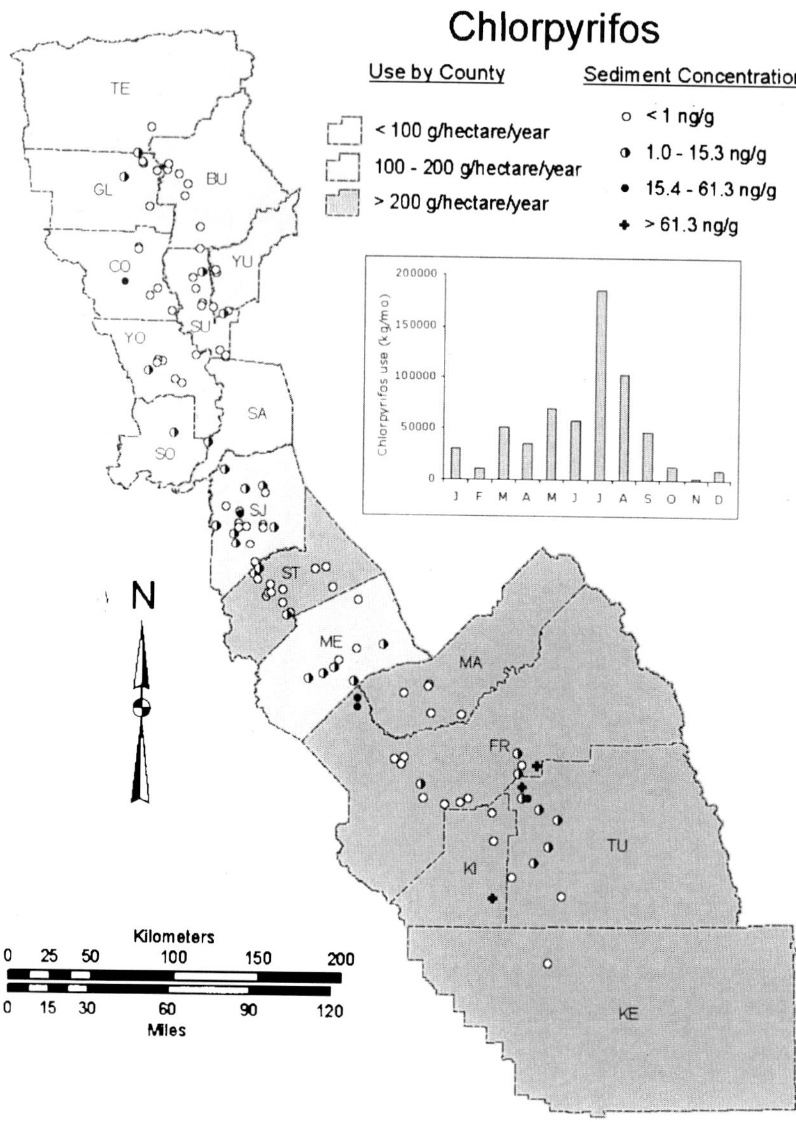

Figure 6. Concentration of chlorpyrifos at each of the study sites, with the highest concentration shown if the site was sampled on multiple occasions. The breakpoints between the four categories of concentration correspond to specific toxic unit (TU) thresholds assuming a generic organic carbon content of 1%: undetected, detectable but acute H. azteca toxicity unlikely (<0.5 TU), toxicity likely (0.5-2 TU), high toxicity likely (>2 TU). The figure also shows the annual usage of chlorpyrifos within each county, normalized to area of harvested cropland, and the monthly use of chlorpyrifos in Central Valley agriculture using 2004 data. County abbreviations defined in Figure 1.

Similarly, five of the eight southern counties use relatively large amounts of bifenthrin per unit cropland, and comparable amounts are applied in only a single northern county (Sutter). Every instance of very high bifenthrin concentrations in sediments (>10.4 ng/g, or >2 TUs assuming 1% oc) occurred in these southern high use counties. Bifenthrin use in the Central Valley is limited entirely to the summer months, with nearly all of the compound applied in June, July and August. Of those sites sampled in both summer and winter, sediment concentrations of bifenthrin were higher in the summer in 64% of the cases.

Lambda-cyhalothrin (Figure 5) exceeded 0.5 TU in 9% of the sampling sites. As was the case for bifenthrin, highest sediment concentrations of lambda-cyhalothrin was largely limited to the southern counties, with a high frequency of potentially toxic concentrations in Stanislaus and Fresno counties. Lambda-cyhalothrin use, however, is more equitably distributed between the northern and southern counties. Four of the six high use counties are to the north, indicating other factors besides use (e.g., soil type and potential for erosion, irrigation practices) play a significant role in determining the potential for contamination of surface water bodies. The greatest monthly use of lambda-cyhalothrin occurs in March when it is applied to alfalfa. Substantial quantities are also used in the May through August growing period on a variety of row crops.

Chlorpyrifos (Figure 6) approached or exceeded concentrations toxic to *H. azteca* in 8% of the sampling sites. Compared to bifenthrin and lambda-cyhalothrin, chlorpyrifos showed a greater frequency of detection (36%), but was often at non-toxic concentrations. Potentially toxic concentrations were scattered throughout San Joaquin, Fresno, Tulare and Kings Counties in the south, and Colusa County in the north. There is a small amount of chlorpyrifos used as a dormant spray on orchards in the winter months, but the vast majority is applied during the growing season, particularly in July and August.

Esfenvalerate approached or exceeded acutely toxic concentrations at 4% of the sites. Three of these sites were in San Joaquin County (Little John Creek and two unnamed drains), one in neighboring Stanislaus County (Del Puerto Creek), and one in Tulare County (Knestric Ditch). Use of esfenvalerate in the Central Valley is nearly equally split between winter and summer months. Fifty-five percent of the annual use of esfenvalerate occurs in April through October, with the remainder applied during winter months, largely on almond and stone fruit orchards (e.g., plums and peaches). Sediment concentrations of esfenvalerate, however, were higher far more often in the summer (78% of the cases) than in the winter for those sites that were sampled on both occasions, suggesting greater off-site transport from the crops with summer esfenvalerate applications.

Permethrin is the most heavily used of the pyrethroids in Central Valley agriculture, and was often detected in the sediment samples (40% frequency of detection). However, it is one of the least toxic of the pyrethroids to aquatic life (*13*), and approached toxic concentrations at only two sites; Root Creek in Madera County in an area dominated by pistachio orchards and an unnamed drain in San Joaquin County. Cypermethrin use in urban areas as a termiticide is far greater than its use in agriculture in California. It was detected in only 3% of the samples and reached the 0.5 TU threshhold in only two agricultural drains in the southern Central Valley (Fresno and Tulare Counties). Fenpropathrin was not among the analytes typically measured in this study. It was analyzed in only one sample when toxicity identification evaluation procedures suggested the presence of a pyrethroid, though none of the regularly quantified pyrethroid analytes were present. Further analysis of this single sample (unnamed drain in Glenn County) indicated the presence of 52 ng/g fenpropathrin.

Of all the pesticides measured, the most frequently detected were DDT and its degradates, DDE and DDD (frequency of detection 91%, 73%, and 34%, respectively). Maximum concentrations observed were 177 ng/g, 225 ng/g, and 15 ng/g, respectively. Even at these highest concentrations, these compounds were unlikely to significantly contribute to the observed *H. azteca* toxicity. The highest concentration of DDT corresponded to only 0.1 TU, and no other sample exceeded 0.02 TU of DDT. While DDT and its degradates may be of concern in the watershed for other reasons (e.g., bioaccumulation and trophic transfer), the concentrations now prevailing in Central Valley sediments appear to have little potential for acute toxicity, at least to *H. azteca*, and, based on more limited data, to *Chironomus dilutus* (*4*).

For the remainder of the pesticide analytes, detections were infrequent for most compounds, and when present were at low concentrations not expected to contribute to the observed toxicity based on estimated toxicity thresholds (*4*), though measured thresholds are lacking for many of the compounds. The frequency of detection (at 1 ng/g reporting limit) and maximum concentration observed were: alpha-BHC (3%, 37 ng/g), beta-BHC (7%, 7 ng/g), gamma-BHC (1%, 2 ng/g), delta-BHC (3%, 4 ng/g), heptachlor (4%, 3 ng/g), heptachlor epoxide (1%, 1 ng/g), aldrin (1%, 6 ng/g), alpha-chlordane (5%, 3 ng/g), gamma-chlordane (4%, 2 ng/g), dieldrin (22%, 374 ng/g), endrin (11%, 813 ng/g), endosulfan I (7%, 35 ng/g), endosulfan II (9%, 23 ng/g), endosulfan sulfate (11%, 14 ng/g), endrin aldehyde (4%, 11 ng/g), endrin ketone (5%, 138 ng/g), methoxychlor (16%, 190 ng/g).

Persistence of Toxicity and Sediment-associated Pesticides

As this study was intended to assess toxicity across California's Central Valley, the emphasis of the design was on broad geographic distribution of sampling sites, and the majority of sites were sampled on only one or two occasions. However, a few sites were sampled repeatedly over several years, providing an opportunity to assess the persistence of sediment toxicity and the pesticide responsible for it (Figure 7). Such a field-based approach to studying persistence provides no control over frequency of application on surrounding agricultural lands or sediment transport. Thus, it should be recognized that a decrease in chemical concentration may be the result of chemical degradation,

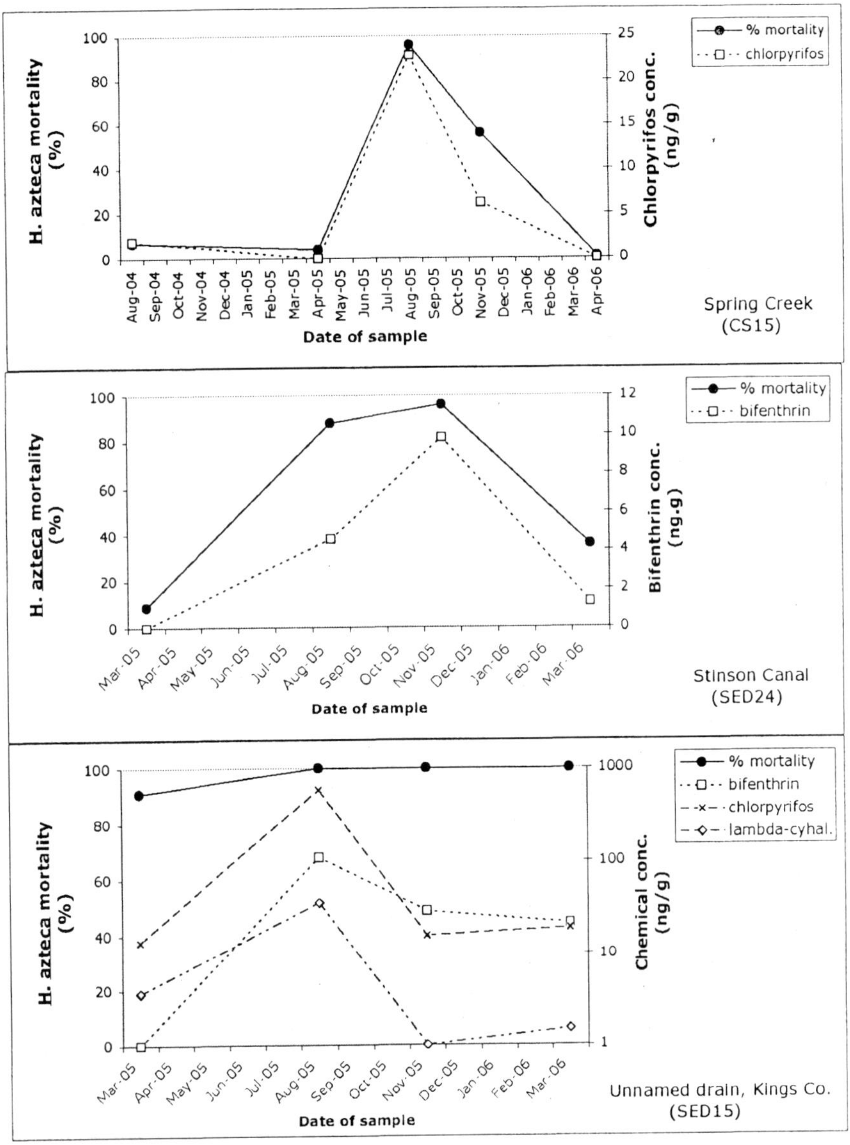

Figure 7. Persistence of toxicity and the pesticides likely contributing to it (based on toxic unit calculations) at three sampling sites at which sediments were repeatedly sampled.

burial of contaminated material beneath cleaner sediments, or transport of the contaminated material to more downstream sites by irrigation or storm-derived flow. The patterns discussed below, however, are not likely to be due to spatial heterogeneity, as triplicate samples were collected on one occasion from three sites discussed below (Sites CS15, SED24, SED15), and negligible differences were found in chemical concentration among the triplicates.

Site CS15, located in Spring Creek, Colusa County, CA, exhibited a dramatic increase in sediment toxicity in August 2005, and an accompanying increase in sediment chlorpyrifos concentration, presumably due to summertime use of the compound. However, there was a 73% reduction in chlorpyrifos concentration by November 2005, suggestive of degradation in place as August to November was a period of minimal irrigation flow and rainfall. By the following spring, the site contained no chlorpyrifos and was no longer toxic, at least in part due to sediment transport during the winter rains, including particularly heavy rainfall in March 2006.

Toxicity appeared in Stinson Canal (Site SED24) in the summer of 2005, most likely due to the presence of bifenthrin. The high toxicity and elevated bifenthrin concentrations persisted through November 2005, but were substantially diminished by the following spring. This site contained water only during the summer irrigation season, and thus persistence patterns observed in the winter months may not be indicative of aquatic systems.

SED15, an unnamed drain in Kings County, consistently showed high toxicity on every sampling occasion, due to a variety of pesticides including bifenthrin, chlorpyrifos and lambda-cyhalothrin. All these pesticides reached peak concentrations in August 2005, and chlorpyrifos in particular had been applied to an adjacent field within a couple weeks prior to the August sampling. By November 2005, bifenthrin concentrations had declined by 74%, chlorpyrifos by 97%, and lambda-cyhalothrin by 100%, though based on TUs the concentration of bifenthrin was still high enough even after the decrease to account for mortality of *H. azteca*. In the intervening three months between the August and November samples there was minimal irrigation flow and no storm runoff capable of eroding sediments (<0.5 cm rainfall from August to November, 2005 in nearby Visalia, CA; cdec.water.ca.gov/cgi-progs/queryMonthly?VSL), suggesting contaminant degradation as the reason for the reduction in concentrations.

The San Joaquin River near Vernalis was sampled every spring or summer from 2002 to 2005 (not shown in Figure 7). Toxicity was observed in 2002, possibly due to esfenvalerate, but no toxicity was observed in subsequent years. In 2004 there was 1.2 TU of bifenthrin in the sediments at this site, though without observed toxicity.

Overall, there were 31 instances when toxicity was seen and another sample taken at the same location in a subsequent sampling event 2-12 months later. Among these 31 instances when toxicity was seen in the first occasion, there was a 45% chance of observing it in the subsequent sample. In the complete study dataset of 200 samples, there was a 27% frequency of toxicity, thus there was a substantially greater tendency to find toxicity at a site if it has historically shown toxicity. Persistence of toxicity may indicate slow contaminant degradation, minimal sediment transport, or on-going pesticide inputs from surrounding farmland. However, for slightly more than half the sites exhibiting sediment toxicity, toxicity is relatively short-lived and not observed if tested a few months later.

The field data gives the overall impression of less environmental persistence of the pyrethroids than is indicated by the very limited published data. Laboratory-based studies are available only for permethrin and bifenthrin, but indicate half-lives in aquatic sediments usually in the range of 6 months to a couple years (*24*). Comparison between the field and lab persistence estimates is complicated by the difficulties noted above in deriving persistence from field data, but the data certainly indicate that further study of pyrethroid persistence in sediments is warranted.

Conclusions

Sediment toxicity is widespread in agricultural areas of the Central Valley, occurring in 29% of the sites. As would be expected, it is far more common in constructed agricultural drains than in natural creeks and rivers. These findings have important implications for the management of water quality in agriculture-affected waterbodies. Decisions will need to be made as to whether constructed drains, which are nearer the points of pesticide use and in which sediment toxicity is more common, are to be held to the same environmental quality standards as the natural water bodies to which they discharge. Nevertheless, even in the natural creeks and rivers, slightly over one-quarter of the sites showed acute sediment toxicity to *H. azteca*. Sediment toxicity in these water bodies might be expected to elicit a stronger response from environmental management agencies. However, in the Central Valley the flow and in some cases even the course of these natural creeks and rivers are routinely manipulated for purposes such as protection of water supplies, providing for tailwater return flow, flood control, or protection of fish or fisheries. Thus, the distinction between constructed and natural watercourses is blurred, and provides an uncertain distinction when interpreting these findings regarding sediment toxicity.

There is also the question of effects of pyrethroid pesticides on the resident invertebrate community. This study focused only on toxicity to *H. azteca*, and it is likely that effects on resident organisms would be more difficult to document. First, based on the limited data available, *H. azteca* appears to be one of the more sensitive invertebrates to pyrethroids of those species tested (*4*). Secondly, these same water bodies have received pesticides and other agricultural pollutants for many decades, and attempting to assess pyrethroid-related compositional changes in aquatic communities that may be already degraded is fraught with difficulties. Thus, this study assessed toxicity based on laboratory exposures to a standard sediment testing species (*9*), in part because it allowed better identification of causality for that toxicity, but the work did not address effects of pyrethroids or other toxicants on the resident organisms.

H. azteca is widely used throughout the United States for sediment toxicity assessment. The integration of sediment chemistry data with measured toxicity thresholds, as incorporated in the TU approach, proved highly effective in predicting which samples were likely to be toxic and identifying potential causative agents. Despite the necessary simplifying assumptions (the arbitrary 0.5 TU threshold for the onset of toxicity, the reasonable but untested assumption of the additivity of pyrethroid toxicity, the implied independence of pyrethroid and organophospate toxicity, the use of total organic carbon used as the sole normalizing factor for bioavailability), the TU approach when applied to pyrethroids and chlorpyrifos was successful in predicting toxicity with 84% accuracy; observing toxicity in samples with >0.5 TU of either pyrethroids or chlorpyrifos, and not observing it when below that threshold. In 9% of the samples the approach underestimated the likelihood of toxicity, with toxicity observed despite <0.5 TU pyrethroids or chlorpyrifos, possibly due in part to unmeasured contaminants in the sediments. In 7% of the samples the approach overestimated the potential for toxicity (non-toxic despite >0.5 TU), often in cases of very coarse, low organic carbon sediments in which unquantified factors appeared to influence bioavailability.

There is strong evidence that pyrethroids, most notably bifenthrin and lambda-cyhalothrin (and secondarily, esfenvalerate), are responsible for much of the observed toxicity. Their role was implicated by the TU analysis, the dilution series data, and three toxicity identification evaluation procedures (temperature manipulation, PBO and esterase). Approximately one out of four Central Valley sediment samples contained bifenthrin, and it was acutely toxic in one out of six sites. One out of six sediments contained lambda-cyhalothrin, and it was acutely toxic in one out of 12 sites. These compounds are not the most used pyrethroids in California agriculture, falling in 6th and 5th place, respectively, on a statewide basis (led by, in decreasing order, permethrin, fenpropathrin, esfenvalerate, and zeta-cypermethrin). Their contribution to *H. azteca* toxicity is attributable to a high sensitivity of the species to these pyrethroids (*7*), or could be a consequence of greater environmental persistence leading to sediment concentrations out of proportion to their use.

While this study focused on agriculture-dominated water bodies, it is important to recognize that surface waters in the Central Valley and elsewhere can also be affected by urban pyrethroid use. Non-agricultural bifenthrin use in California (primarily for pest control around homes and other structures) is twice that of agricultural use, and the compound is a frequent contributor to *H. azteca* toxicity in urban creeks (*14,25*). The amount of lambda-cyhalothrin used for agricultural purposes in California is about 50% greater than its non-agricultural use.

The only non-pyrethroid found to be a significant contributor to *H. azteca* toxicity was the organophosphate, chlorpyrifos. It contributed to toxicity about half as often as bifenthrin, and with comparable frequency as lambda-cyhalothrin. However, the use of chlorpyrifos in Central Valley agriculture is 100-fold greater than for these pyrethroids. The comparatively low incidence of sediment toxicity (in proportion to its use) is probably due in part to a lower hydrophobicity of the compound, allowing for greater dispersal and dilution of dissolved phase residues.

Historically, water quality concerns related to agricultural pesticides in the Central Valley have been greatest in the winter months, when organophosphates are applied to orchards, and heavy rains wash the residues in to surface water bodies (*2,3*). However, for the pesticides that appear to present a threat to sediment quality, the summer months are of greater concern. The compounds that contributed most to *H. azteca* toxicity were used largely (lambda-cyhalothrin ,esfenvalerate, chlorpyrifos) or entirely (bifenthrin) in the summer months. The sediments of the water bodies studied had higher concentrations of all these compounds in the summer months (56-78% of the cases, depending on the pesticide). At least for the pyrethroids, half-lives for residues in aerobic soils, as in farm fields, are on the order of 1-2 months (*26*). Thus, in the five months from peak use (July) to the first heavy winter rains (usually December), it is likely that much of the pesticide will have been degraded.

The greater summer use of the contaminants of concern, the higher concentrations in sediment usually observed during summer, and the relatively short persistence in farm soils all suggest that summer irrigation return flows, rather than winter storms, are likely to be the more important mechanism for transporting contaminated soils to the drains and creeks on which this study focused. Winter rains, and the accompanying high flows, may play a significant role in further downstream transport, moving the contaminated sediments in to the major rivers. If irrigation runoff is in fact the principal mechanism for transport to aquatic systems, this finding has important implications for management practices, since control of irrigation runoff and its associated eroded soil is more feasible than control of winter storm runoff. Measures developed for agricultural erosion control (e.g. polyacrylamide addition to irrigation water (*27*), vegetated ditches or filter strips (*28*)), particularly if focused on the finer particle sizes, ought to be effective in controlling entry of the pesticides of concern in to surface water bodies.

Acknowledgments

This study was supported by the Central Valley Regional Water Quality Control Board through the Irrigated Lands Program and the PRISM Program, and by the California Bay-Delta Authority. We thank Irrigated Lands Program staff (W. Croyle, M. Lopez-Read, J. Swanson, and S. Azimi-Gaylon) for their support and encouragement throughout the study. We also thank A. King for assistance in preparing some of the figures, and are grateful for the assistance of M. Johnson and other University of California, Davis staff in collecting some of the sediments.

References

1. Central Valley Regional Water Quality Control Board. Considerations of Water Body Designations to Comply with the Provisions of the Water Quality Control Plan for ISWP. Rancho Cordova, CA. 1992.
2. Kuivila, K.M.; Foe, C.G. Concentrations, transport and biological effects of dormant spray pesticides in the San Francisco Estuary, California. Environ. Toxicol. Chem. **1995**, *14*, 1141-1150.
3. Werner, I.; Deanovic, L.A.; Connor, V.; de Vlaming, V.; Bailey, H.C.; Hinton, D.E. Insecticide-caused toxicity to Ceriodaphnia dubia (Cladocera) in the Sacramento-San Joaquin River Delta, California, USA. Environ. Toxicol. Chem. **2000**, *19*, 215-227.
4. Weston, D.P.; You, J.; Lydy, M.J. Distribution and toxicity of sediment-associated pesticides in agriculture-dominated water bodies of California's Central Valley. Environ. Sci. Technol. **2004**, *38*, 2752-2759.
5. Anderson, B.S.; Phillips, B.M.; Hunt, J.W.; Worcester, K.; Adams, M.; Kapellas, N.; Tjeerdema, R.S. Evidence of pesticide impacts in the Santa Maria River watershed, California, USA. Environ. Toxicol. Chem. **2006**, *25*, 1160-1170.
6. Phillips B.M,; Anderson, B.S.; Hunt, J.W.; Huntley, S.; Tjeerdema, R.S.; Kapellas, N.; Worcester, K. Solid-phase sediment toxicity identification evaluation in an agricultural stream. Environ Toxicol Chem. **2006**, *25*:1671-1676.
7. Amweg, E.L.; Weston, D.P.; Ureda, N.M. Use and toxicity of pyrethroid pesticides in the Central Valley, California, USA. Environ. Toxicol. Chem. **2005**, *24*, 966-972 (correction in 24, 1300-1301).
8. Di Toro, D.M.; Zarba, C.S.; Hansen, D.J.; Berry, W.J.; Swartz, R.C.; Cowan, C.E.; Pavlou, S.P.; Allen, H.E.; Thomas, N.A.; Paquin, P.R. Technical basis for establishing sediment quality criteria for nonionic organic chemicals using equilibrium partitioning. Environ. Toxicol. Chem. **1991**, *10*, 1541-1583.
9. U.S. Environmental Protection Agency. Methods for measuring the toxicity and bioaccumulation of sediment-associated contaminants with freshwater invertebrates. EPA/600/R-99/064, Office of Research and Development, Washington, D.C., 2000.
10. You, J.; Weston, D.P.; Lydy, M.J. A sonication extraction method for the analysis of pyrethroid, organophosphate, and organochlorine pesticides from sediment by gas chromatography with electron-capture detection. Arch. Environ. Contam. Toxicol. **2004**, *47*:141-147.
11. Maund, S.J.; Hamer, M.J.; Lane, M.C.G.; Farrelly, E.; Rapley, J.H.; Goggin, U.M.; Gentle, W.E. Partitioning, bioavailability, and toxicity of the pyrethroid insecticide cypermethrin in sediments. Environ. Toxicol. Chem. **2002**, *21*, 9-15.

12. Amweg, E.L.; Weston, D.P. Whole sediment toxicity identification evaluation tools for pyrethroid insecticides: II. Piperonyl butoxide addition. Environ. Toxicol. Chem. (In press)
13. Solomon, K.R.; Giddings, J.M.; Maund, S.J. Probabalistic risk assessment of cotton pyrethroids: I. Distributional analyses of laboratory aquatic toxicity data. Environ. Toxicol. Chem. **2001**, *20*, 652-659
14. Amweg, E.L.; Weston, D.P; You, J.; Lydy, M.J. Pyrethroid insecticides and sediment toxicity in urban creeks from California and Tennessee. Environ. Sci. Technol. **2006b**, *40*, 1700-1706.
15. U.S.Environmental Protection Agency. Methods for aquatic toxicity identification evaluations: Phase III toxicity confirmation procedures. EPA 600/3-88-036, Environmental Research Laboratory, Duluth, MN., 1989.
16. Denton, D.L.; Wheelock, C.E.; Murray, S.A.; Deanovic, L.A.; Hammock, B.D.; Hinton, D.E. Joint acute toxicity of esfenvalerate and diazinon to larval fathead minnows (Pimephales promelas). Environ. Toxicol. Chem. **2003**, *22*, 336-341.
17. Belden, J.B.; Lydy, M.J. Joint toxicity of chlorpyrifos and esfenvalerate to fathead minnows and midge larvae. Environ. Toxicol. Chem. **2006**, *25*, 623-629.
18. Wadleigh, R.W.; Koehler, P.G.; Preisler, H.K.; Patterson, R.S.; Robertson, J.L. Effect of temperature on the toxicities of ten pyrethroids to German cockroach (Dictyoptera: Blattellidae). J. Econ. Entomol. **1991**, 84, 1433-1436.
19. Amweg, E.L.; Weston, D.P.; Johnson, C.S.; You, J.; Lydy, M.J. Effect of piperonyl butoxide on permethrin toxicity in Hyalella azteca. Environ. Toxicol. Chem. **2006a**, *25*, 1817-1825.
20. Weston, D.P.; Amweg, E.L.; Mekebri, A.; Ogle, S.; Lydy, M.J. Aquatic effects of aerial spraying for mosquito control over an urban area. Environ. Sci. Technol. **2006**, *40*, 5817-5822.
21. Bailey, H.C.; Digiorgio, C.; Kroll, K.; Miller, J.L.; Hinton, D.E.; Starrett, G. Development of procedures for identifying pesticide toxicity in ambient waters: carbofuran, diazinon, chlorpyrifos. Environ. Toxicol. Chem. **1996**, *15*, 387-845.
22. Wheelock, C,E,; Miller, J.L.; Miller, M.J.; Phillips, B.M.; Huntley, S.A.; Gee, S.J.; Tjeerdema, R.S.; Hammock, B.D. Use of carboxylesterase activity to remove pyrethroid-associated toxicity to Ceriodaphnia dubia and Hyalella azteca in Toxicity Identification Evaluations (TIEs). Environ. Toxicol. Chem. **2006**, *25*:973-984.
23. Weston, D.P.; Amweg, E.L. Whole sediment toxicity identification evaluation tools for pyrethroid insecticides: I. Esterase addition. Environ. Toxciol. Chem. (In press).
24. Gan, J.; Lee, S.J.; Liu, W.P.; Haver, D.L.; Kabashima, J.N. Distribution and persistence of pyrethroids in runoff sediments. J. Environ. Qual. **2005**, *34*, 836-841.

25. Weston, D.P.; Holmes, R.W.; You, J.; Lydy, M.J. Aquatic toxicity due to residential use of pyrethroid insecticides. *Environ. Sci. Technol.* **2005**, *39*, 9778-9784.
26. Laskowski, D.A. Physical and chemical properties of pyrethroids. *Rev. Environ. Contam. Toxicol.* **2002**, *174*, 49-170.
27. Sojka, R.E.; Lentz, R.D. Polyacrylamide for furrow-irrigation erosion control. *Irrigation Journal* **1996**, *64*:8-11.
28. Moore, M.T.; Bennett, E.R.; Cooper, C.M.; Smith, S.; Shields, F.D.; Milam, C.D.; Farris, J.L. Transport and fate of atrazine and lambda-cyhalothrin in an agricultural drainage ditch in the Mississippi Delta, USA. *Agric. Ecosys. Environ.* **2001**, *87*:309-314.

Chapter 3

Occurrence of Pyrethroids in Bed and Suspended Sediments in California

Michelle L. Hladik and Kathryn M. Kuivila*

U.S. Geological Survey, 6000 J Street, Sacramento, CA 95819

Pyrethroids have been detected in sediments collected over the past decade from a variety of California locations. While most published studies sampled in close proximity to pyrethroid application, the studies included in this chapter focused on watersheds located farther from pyrethroid sources. The four watersheds in this chapter varied considerably in size (1-100,000 km^2) and included both agricultural and urban land use. Four pyrethroids (bifenthrin, λ-cyhalothrin, τ-fluvalinate and permethrin) were detected in bed and suspended sediments. Bifenthrin and permethrin were detected the most frequently and at the highest concentrations, with maximum concentrations of 24 and 70 ng/g dry weight, respectively. Occurrences of individual pyrethroids were correlated to higher use in some but not all cases.

Introduction

The Pesticide Fate Research Group of the U.S. Geological Survey (USGS) has been measuring pyrethroid insecticides in bed and suspended sediments in California for years and detected pyrethroids in samples as early as 1997. While the objective of the studies described herein was to understand the transport and fate of many current-use pesticides, pyrethroids were included in the suite of analytes.

Pyrethroids are detected mainly in the sediment because these compounds are hydrophobic (log K_{oc} >5; ref *1*) and tend to partition onto particles rather than remain in the dissolved phase. Pyrethroids are highly toxic to aquatic organisms, especially those that live in the sediment; 10-day LC_{50} values for *Hyalella azteca* have been found to range from 0.52 µg/g organic carbon for bifenthrin to 11 µg/g organic carbon for permethrin (*2*). Sediment toxicity is likely to be of greatest concern close to the pyrethroid application sites and in depositional zones just downstream of those sites. Typically, dilution of sediment-bound pyrethroids occurs as non-contaminated sediments enter from other downstream sources; however, sediment sorting has been shown to cause downstream pyrethroid enrichment in a single-source drainage canal (*3*).

This chapter describes the occurrence of pyrethroids in bed or suspended sediments at four locations in California (Carpinteria Marsh, Mallard Island, Yolo Bypass and Salton Sea) from 1997 to 2005 (see Figure 1 for general locations). The Carpinteria Marsh study focused on bed sediments to monitor coastal marsh contamination from urban inputs. The Salton Sea study combined investigations of both bed and suspended sediments in a watershed that has

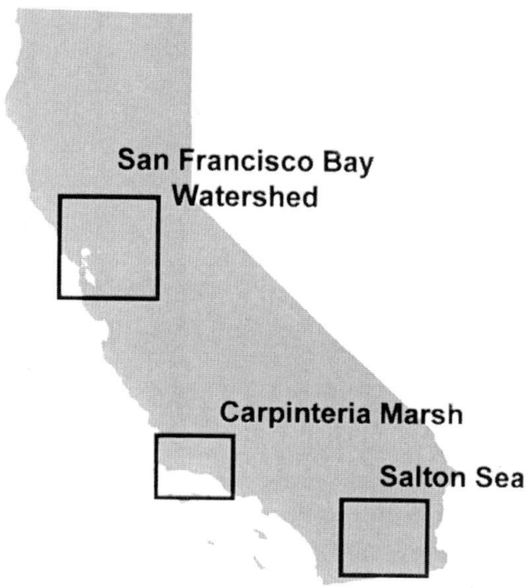

Figure 1. Areas in California where pyrethroids were detected. San Francisco Bay includes Mallard Island and Yolo Bypass.

intensive agricultural activity, is home to endangered species and an important ecosystem for the Pacific Flyway. The Yolo Bypass and Mallard Island studies (both located within the San Francisco Bay watershed) focused on suspended sediments that are transported into San Francisco Bay, one of the most economically and ecologically important estuaries in the United States. Sampling, detections and use for each location are summarized in Table I. Other studies have measured pyrethroids in sediments in California, but most have been close to agricultural (*4*) or urban (*5*) sources. To achieve a greater understanding of pyrethroid transport and toxicity, watersheds farther from sources need to be monitored; these include larger rivers and estuaries.

Table I. Summary of locations included in this chapter.

Location	*Year(s)*	*Sediment Matrix*	*Watershed Size (km²)*	*Pyrethroids Detected*	*Environmental Setting*
Carpinteria Marsh	2002, 2003	bed	0.9	bifenthrin, permethrin	coastal marsh contamination
Salton Sea	2001, 2002	bed and suspended	5,359[a]	bifenthrin, λ-cyhalothrin, permethrin	ecosystem for Pacific Flyway
Yolo Bypass	2004, 2005	suspended	5,036[b]	bifenthrin, τ-fluvalinate	fish habitat
Mallard Island	1997	suspended	99,587	bifenthrin	flows to San Francisco Bay

[a] Alamo and New Rivers: 2,428 km^2; Whitewater River: 3,931 km^2

[b] Willow Slough: 697 km^2; Knights Landing Ridge Cut: 4,339 km^2

Sampling and Analysis Techniques

Sample Collection

Bed sediment samples were collected with either a stainless steel spoon from the top 2 cm of depositional zone, a 9-inch (22.9 cm) Ekman grab sampler, or a 2-inch (5.1 cm) diameter, Teflon-barreled hand corer. In some cases multiple grabs were required to obtain approximately 0.5 L for each site.

Suspended sediment was isolated from large-volume water samples (100-1000 L). At each site, water was collected with a peristaltic pump into 20-L stainless steel soda kegs and then pumped through a Westfalia continuous-flow centrifuge at 2 L/min within 6 hours of collection.

Extraction and Quantitation

Sediment samples were homogenized before microwave-assisted solvent extraction. Sediment samples were extracted wet, at approximately 50% moisture. Matrix was removed from the extract using either a carbon cartridge or stacked carbon and alumina cartridges and sulfur was removed with gel permeation chromatography. Samples were analyzed on a Varian Saturn ion-trap gas chromatography/mass spectrometry in MS and MS/MS mode. Further details for both pyrethroid methods and the procedure to measure percent organic carbon can be found elsewhere (*6,7*). The specific pyrethroids analyzed varied by study, as more pyrethroids were added over time to reflect new compounds of interest. All studies included bifenthrin, cyfluthrin, cypermethrin, λ-cyhalothrin, esfenvalerate and permethrin. The Yolo Bypass study also included deltamethrin, fenpropathrin, sumithrin and τ-fluvalinate. All concentrations were quantified on a sediment dry weight basis. Method detection limits, based on the standard deviation of seven replicate samples, were 1-2 ng/g for the pyrethroids measured (*6,7*); however, the calculated limit of detection for bifenthrin using the MS/MS mode method for the Mallard Island samples was 0.2 ng/g.

Carpinteria Marsh

Carpinteria Marsh is a restored estuarine wetland located east of Santa Barbara, California, along the coast of the Pacific Ocean. This 0.9 km^2 marsh receives pesticide inputs from applications on nurseries, greenhouses, orchards, row crops and residential areas within the watershed. A creek channel on the west side of the marsh was sampled along a transect beginning at the edge of the marsh (Figure 2). Four stations (A-D) were sampled along the transect in July of 2002 and August of 2003. At each station, bed sediments were collected at different channel depths to account for tidal inundation cycles: the creek channel (low), creek bank (mid), and the vegetated marsh edge (high) (Figure 2). Typically, the creek channel was saturated throughout the tidal cycle, the intertidal bank was flooded intermittently, and the marsh edge was rarely flooded.

Two pyrethroids, bifenthrin and permethrin, were detected in the bed sediments. Bifenthrin was detected at stations A and B in 2002 (Figure 3a) and at all stations (A-D) in 2003 (Figure 3b). In 2002, similar concentrations were found at stations A and B (14 to 22 ng/g) and the distribution was throughout the

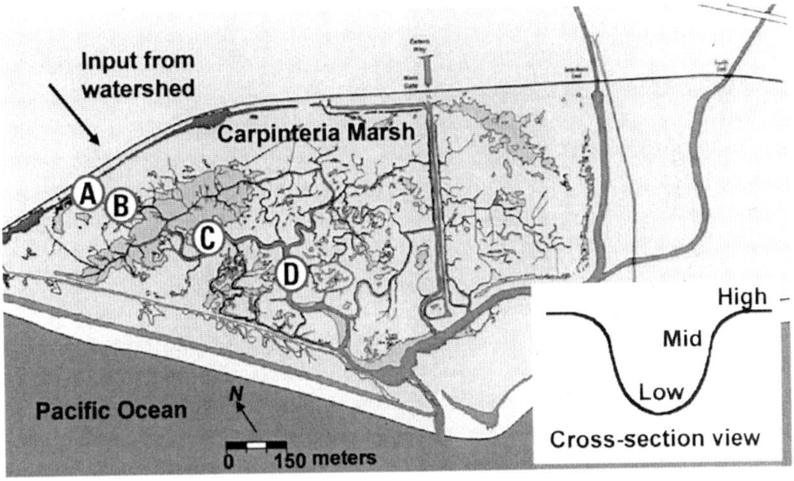

Figure 2. Diagram of sampling at Carpinteria Marsh. Stations were located along a transect that increased in distance from the input. Additionally each station had a cross-section sampled.

channel but no bifenthrin was detected downstream of station B. In 2003, maximum concentrations decreased along the transect from 24 ng/g at station A to 1.3 ng/g at station D, with distributions once again being found in varying portions of the channel. Bifenthrin concentrations did not show a clear elevation pattern along the transect of the channel in either year. Percent organic carbon for sediments containing bifenthrin ranged from 0.8 to 4.1 percent. Calculated LC50's for Hyalella azteca in sediments with similar organic carbon concentrations (using 0.52 µg/g organic carbon; ref 2) ranged from 4 to 21 ng/g. Of the eleven bifenthrin detections, six exceeded the corresponding LC50.

Permethrin was only detected at station A in 2002 (Figure 3a) and stations A and B (Figure 3b) in 2003, with concentrations ranging from 20 to 69 ng/g. Permethrin was detected in the creek channel and bank, but not on the marsh edge. Percent organic carbon in the sediment for the permethrin detections ranged from 0.6 to 2.8 percent; therefore, calculated LC_{50}'s for *Hyalella azteca* in these sediments (using 11 µg/g organic carbon; ref 2) were 60 to 300 ng/g. None of the permethrin detections exceeded the LC_{50}.

Overall, bifenthrin and permethrin show different patterns of detection by elevation suggesting different transport during high and low flows, possibly dependent on timing or location of application. Differences in degradation rates on the marsh edge versus the channel could also influence their fate. Bifenthrin is more stable than permethrin; half-lives for bifenthrin range from 277 to 770 days versus 99 to 141 days for permethrin (8).

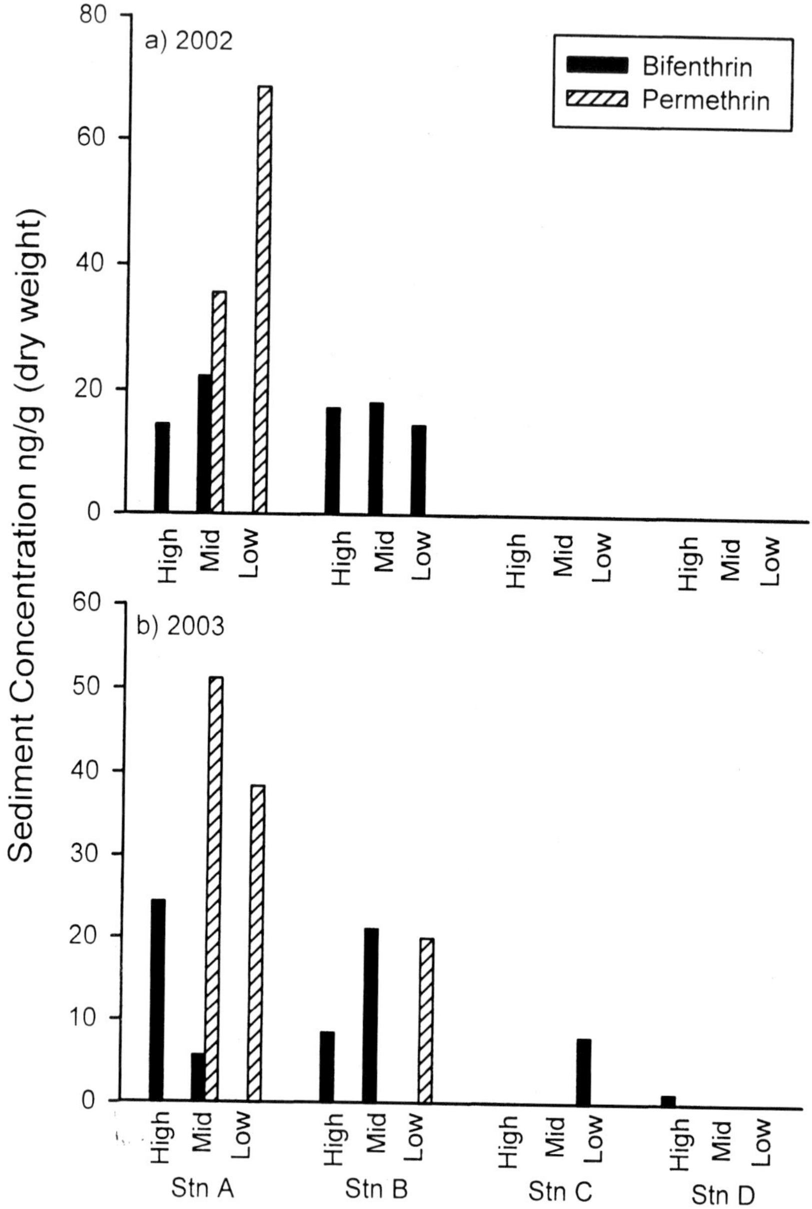

Figure 3. Concentrations of bifenthrin and permethrin at Carpinteria Marsh in a) 2002 and b) 2003. Four stations were sampled along a transect and within stations a cross-section was sampled at the high, mid and low portions of the channel.

Salton Sea

Located in Southern California, the Salton Sea is a closed-basin lake that receives runoff from the surrounding areas. Three rivers drain into the Salton Sea: the Alamo and New Rivers from the south and the Whitewater River from the north (Figure 4). Within the Imperial Valley, the Alamo and New Rivers drain an area of 2,428 km^2 that is primarily agricultural. In contrast, the Whitewater River watershed, 3,931 km^2, is a mixture of agriculture, urban and undeveloped upland areas.

Pesticides in water, suspended sediments, and bed sediments were measured at the three river outlets (Figure 4). Samples were collected in the fall of 2001 and the spring and fall of 2002. Details of the sampling times and locations can be found elsewhere (7).

Three pyrethroids, bifenthrin, λ-cyhalothrin, and permethrin, were detected in the suspended and bed sediments. For each sampling period, the concentrations for λ-cyhalothrin and permethrin are plotted in Figure 5. Bifenthrin had only one detection of 7.5 ng/g in bed sediments from the Whitewater River in the fall of 2001. The Alamo River had the highest number of detections

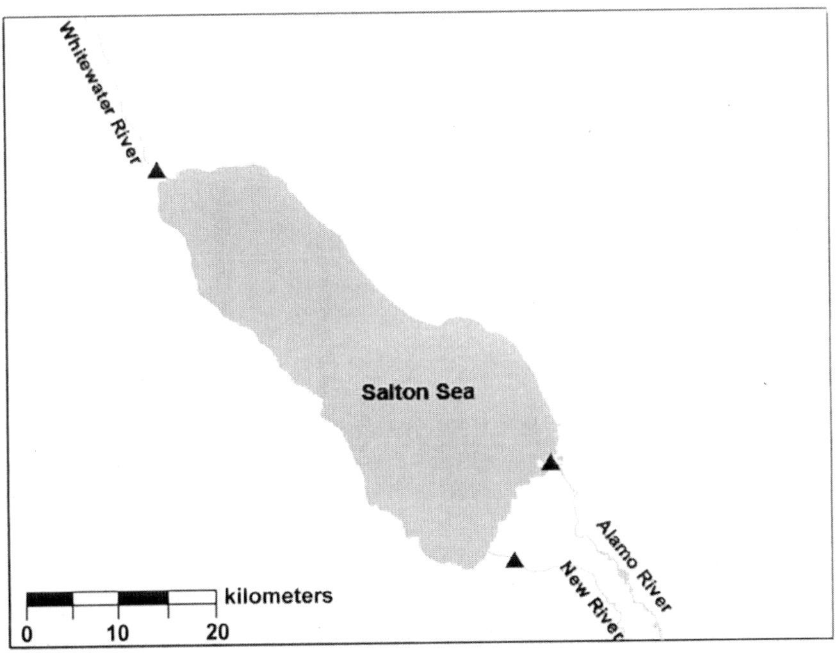

Figure 4. Locations of sampling sites for the Salton Sea.

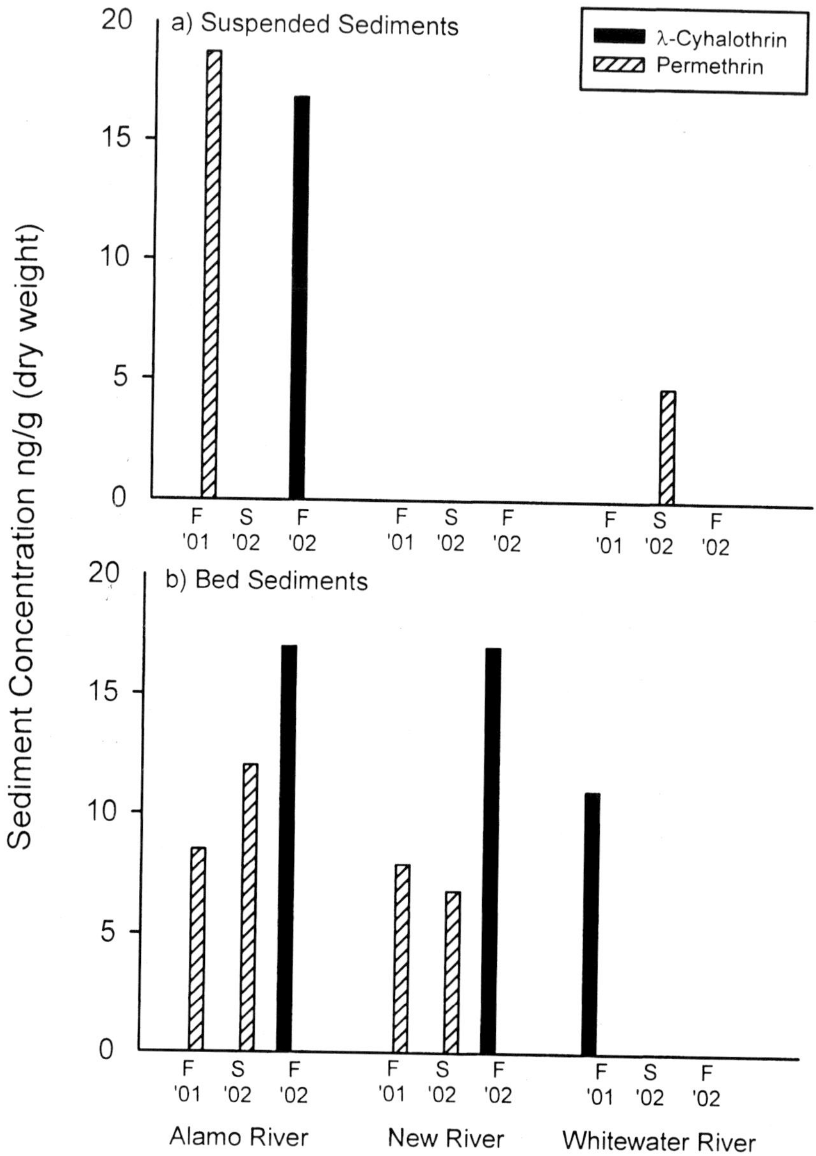

Figure 5. Concentrations of λ-cyhalothrin and permethrin measured in the Salton Sea in a) suspended sediments and b) bed sediments in fall of 2001, spring of 2002 and fall of 2002. Data from ref 7.

and the highest concentrations (8-19 ng/g) followed by the New River (4-17 ng/g) and the Whitewater River (5-11 ng/g).

Permethrin and λ-cyhalothrin were applied for agricultural use in the highest amounts (DPR PUR database; ref *9*) to the Salton Sea watershed and were detected most frequently and at the highest concentrations in the sediments. These areas are assumed to be primarily agricultural and other pesticide applications are assumed to be negligible. Concentrations were greater in the Alamo and New Rivers than in the Whitewater River, reflecting the intensive agriculture of the Imperial Valley. The number of detections and concentrations were higher in the fall than in the spring and the use of each pesticide was slightly higher in the summer (preceding the fall sampling). Pesticide use in the Salton Sea watershed (for 2001) in the summer (April through October) was 661 and 2,955 kg for λ-cyhalothrin and permethrin, respectively, and winter use (November through March) was 449 and 2,028 kg (*7,9*). Bifenthrin use was much lower overall, with an average of 182 kg applied in the summer and 11 kg applied in the winter.

For bed sediments, the concentrations can be compared to LC_{50} values corrected for percent organic carbon concentrations of 0.3 to 1.2 percent. The one bifenthrin detection and all the λ-cyhalothrin detections exceeded the calculated LC_{50}'s for *Hyalella azteca* (2.6 and 1.3 to 2.6 ng/g, respectively; ref *2*). In contrast, none of the permethrin concentrations exceeded the LC_{50} (43 to 130 ng/g; ref *2*). Toxicity has not been determined for pelagic organisms exposed to contaminants on suspended sediments but for illustrative purposes the suspended sediment concentrations can be compared to LC_{50}'s for *Hyalella azteca*. The percent organic carbon concentrations ranged from 0.9 to 5.5 percent for the suspended sediment samples. The toxicity pattern for the suspended sediments is the same as the bed sediment detections; the one λ-cyhalothrin detection exceeded the LC_{50} (5.4 ng/g) while the permethrin detections did not exceed the LC_{50} (98-600 ng/g).

Yolo Bypass

Yolo Bypass is a flood-control area for the lower Sacramento River near Sacramento (Figure 6). The Bypass floods in about 60% of years with peak flows occurring in the winter or spring, typically between January and March (*10*). The Bypass receives water from agricultural areas that are potential sources of pesticides. Pesticide concentrations were measured in several different inputs to the Bypass to determine sources of pesticides and potential impacts to critical life stages of native fish (*6*).

Two source watersheds were sampled: Willow Slough and Knight's Landing Ridge Cut (KLRC) with watershed sizes of 697 and 4,339 km^2, respectively.

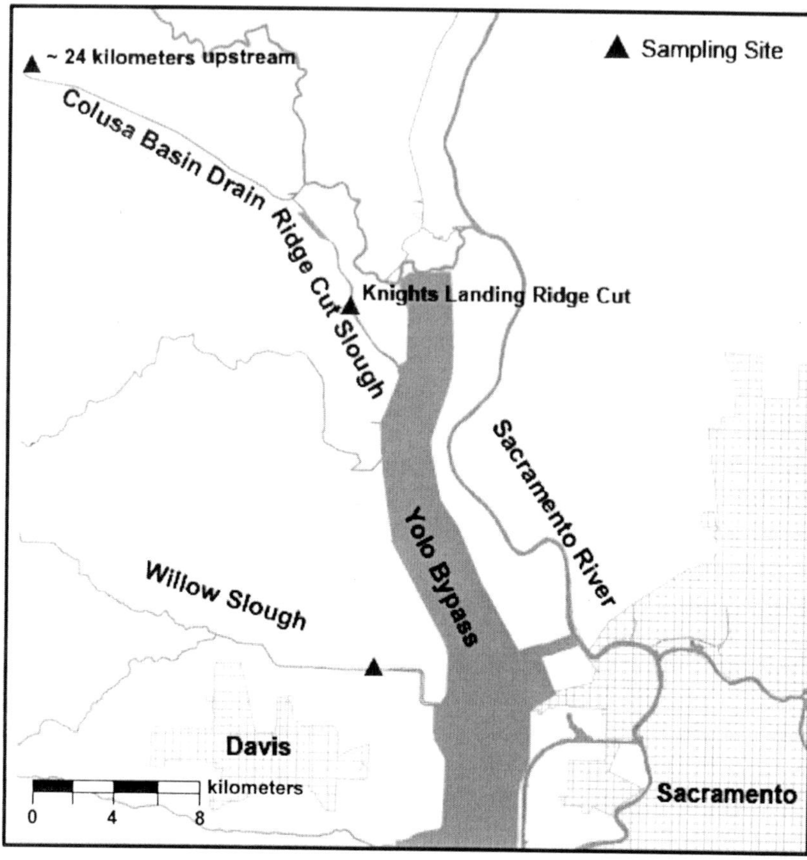

Figure 6. Location of Yolo Bypass and sampling sites.

Water and suspended sediment samples were collected in water years 2004, 2005, and 2006 (a water year is defined as the 12-month period from October to September and is designated as the calendar year in which it ends). Details for the sampling times and locations can be found elsewhere (6). The Colusa Basin Drain site is located within the KLRC watershed and was sampled only in water year 2006 to further characterize the pesticide sources.

Two pyrethroids were detected in the Yolo Bypass suspended sediments. Bifenthrin was detected at three sites in water years 2005 and 2006 and τ-fluvalinate was detected at two sites in water years 2004 and 2005 (Table II). Concentrations for both compounds were relatively low. Bifenthrin concentrations were all below 2 ng/g, while τ-fluvalinate concentrations were slightly higher at 1 to 11 ng/g. There are no LC_{50} values reported for τ-fluvalinate

Table II. Concentrations of bifenthrin and τ-fluvalinate on suspended sediments detected in the Yolo Bypass during water years 2004, 2005 and 2006 (Data from refs *11,12*); nd = not detected.

Site	*Water Year*	*Bifenthrin (ng/g, dry weight)*	*τ-Fluvalinate (ng/g, dry weight)*
Colusa Basin Drain	2006	0.41	nd
Knights Landing Ridge Cut	2004	nd	1.4
	2005	nd	11
	2006	0.80	nd
Willow Slough	2005	1.7	2.7
	2006	0.47	nd

in bed sediments to compare with the suspended sediment concentrations. All bifenthrin suspended sediment concentrations measured were below bed sediment LC_{50} toxicity values for *Hyalella azteca* (7-11 ng/g at 1.3 to 2.2 percent organic carbon; ref *2*).

Agricultural use of both bifenthrin and τ-fluvalinate was low in the watershed so the relation between known pyrethroid use and detections is unclear. In 2004, only 470 and 50 kg of bifenthrin were applied to the KLRC (including Colusa Basin Drain) and Willow Slough watersheds, respectively (*9,12*). The only watershed in the Yolo Bypass with agricultural use of τ-fluvalinate was KLRC with 30 kg applied in 2004 (*9,12*). In addition, since this is an area with very little urban input, pyrethroid use by non-professionals (urban applications by a professional is included in the DPR database) is unlikely to be important.

Mallard Island

Mallard Island is located downstream of the Sacramento-San Joaquin River Delta at the eastern edge of Suisun Bay (which is the most landward area of San Francisco Bay; see Figure 7). This drainage area encompasses 99,587 km^2 and includes two major river systems, the Sacramento and the San Joaquin Rivers. Suspended sediments and associated pesticides are transported past Mallard Island into San Francisco Bay from the agricultural and urban areas in the Central Valley of California.

The sampling site is tidally influenced and the discharge cannot be measured using traditional methods. Instead the net daily flow, called Net Delta Outflow, is calculated using a computer program (DAYFLOW; ref *13*). During a high-

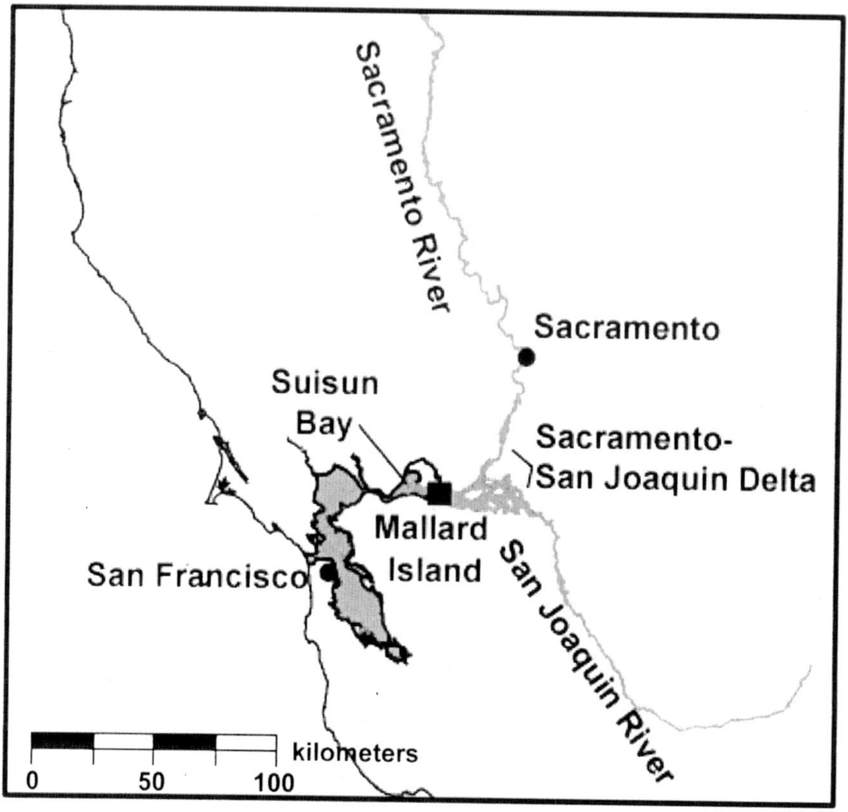

Figure 7. Location of Mallard Island.

flow runoff event in 1997, water samples containing suspended sediments were collected twice-daily at slack before and after ebb from January 2 through 17.

Suspended sediment concentrations and the net Delta outflow at Mallard Island versus time are shown in Figure 8. High-flow events transport elevated suspended sediment concentrations and, consequently, high loads of suspended sediments. Typically, the flow and suspended sediment peaks correspond to elevated concentrations of pesticides associated with the suspended sediments. In this region, the first significant rainfall of the year occurs between late December and January and transports the first flush of pesticides that have been applied since the last significant rainfall.

Bifenthrin was detected during the high-flow event in six of 30 samples collected during January. Concentrations ranged from 0.3 to 1.1 ng/g (Figure 8). The highest bifenthrin concentration coincided with the maximum suspended sediment concentration and co-occurred with maximum concentrations of

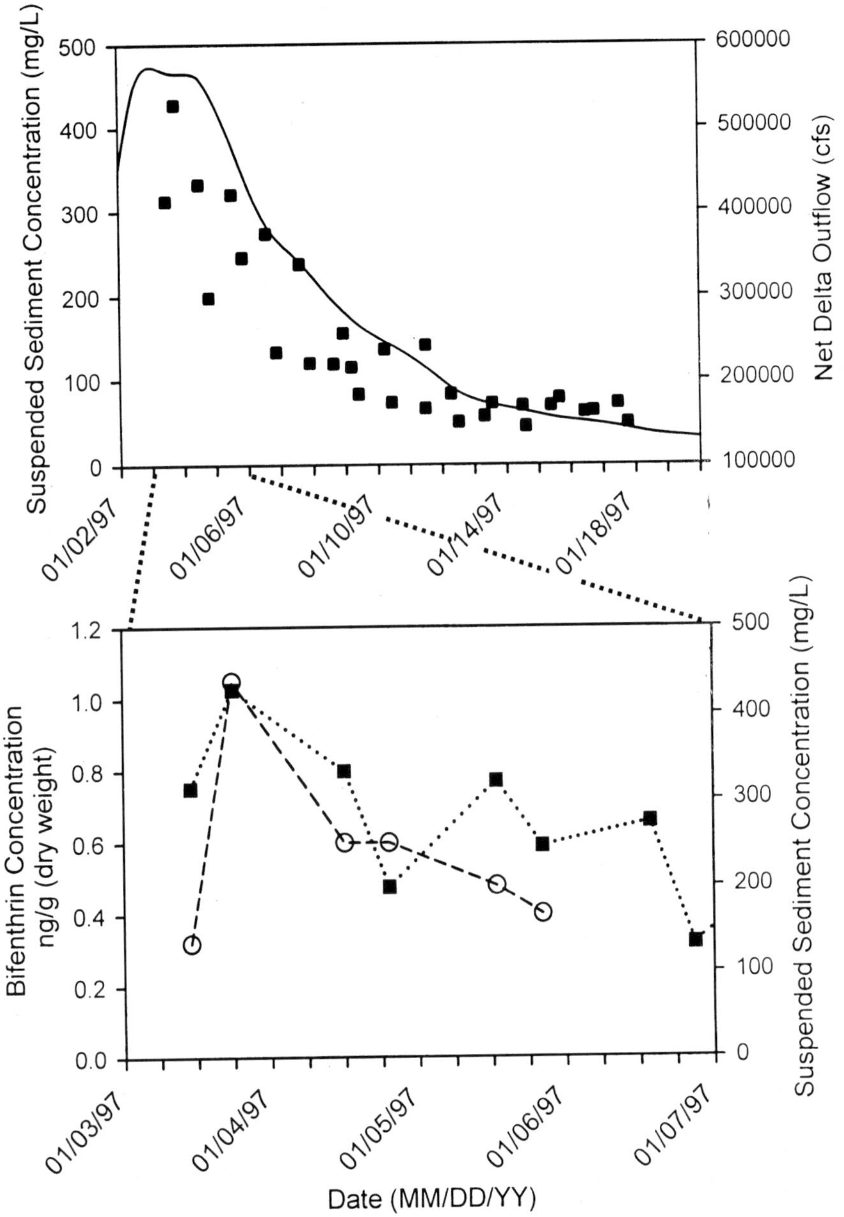

Figure 8. Suspended sediment concentrations (■) and flow (line) at Mallard Island are plotted against the date (mm/dd/yy) in the top graph. Bifenthrin concentrations (○) and suspended sediment concentrations (■) are plotted against the date (mm/dd/yy) in the bottom graph.

other current-use pesticides, such as chlorpyrifos, thiobencarb and trifluralin. The maximum concentration of bifenthrin was low (1 ng/g) relative to the other pesticides (20-50 ng/g; data not shown) but it is also more toxic. LC_{50} values for *Hyalella azteca* are 5 ng/g for bifenthrin (assuming one percent organic carbon; ref *2*) versus 400 ng/g for chlorpyrifos (*14*).

Agricultural and urban professional use of bifenthrin for the San Francisco Bay watershed in 1997 was primarily agricultural with 1,610 kg applied (DPR PUR database; ref *9*). Non-professional use was probably minimal since the increase in pyrethroid use did not begin until after 2000 with the phase-out of organophosphate pesticides (*15*). To our knowledge, this is the earliest known detection of a pyrethroid in California surface waters. Since 1997, the agricultural and urban professional use of bifenthrin has increase almost 7-fold, with 10,740 kg applied in 2005 (*9*).

Conclusions

Pyrethroids were detected across California, from small to large watersheds, in both suspended and bed sediments, and from different types of land uses. The four locations sampled were downstream of where applications occurred and represent watersheds that vary by 5 orders of magnitude in size – from 1 km^2 (Carpinteria Marsh) to nearly 100,000 km^2 (Mallard Island). Sampling occurred during both high-flow and low-flow conditions. The land use in the different watersheds varied from primarily agricultural (Salton Sea and Mallard Island) to mixed urban and agriculture (Yolo Bypass and Carpinteria Marsh).

Bifenthrin was detected at all locations and frequently at concentrations greater than the LC_{50} value for *Hyalella azteca*. In contrast, permethrin was detected at two of the four locations and always at concentrations less than the LC_{50} value. The two other pyrethroids, λ-cyhalothrin and τ-fluvalinate, were only detected at one location each. The concentrations for λ-cyhalothrin were greater than the LC_{50}, while the LC_{50} value for τ-fluvalinate is unknown.

Carpinteria Marsh had the highest concentrations of all pyrethroids measured and the highest concentrations of bifenthrin (24 ng/g) and permethrin (70 ng/g). The pattern of occurrence of the two pyrethroids differed by elevation with bifenthrin detected both on the marsh edge and in the channel and permethrin detected only in the channel; therefore, organisms may experience different exposures, depending on their habitat. Understanding the fate of contaminants in coastal salt marshes such as Carpinteria is critical to maintaining healthy coastal ecosystems.

The riverine inputs to the Salton Sea had the highest number of pyrethroids detected (three) in both bed and suspended sediments. The source of these pyrethroids is most likely from the intense agricultural activity of the surrounding area. The high loads of suspended sediment (and associated pesticides) from the rivers are deposited in broad shallow deltas which harbor large numbers of fish and birds and are adjacent to federal and state wildlife refuges.

Yolo Bypass is a medium sized watershed that carries a large amount of suspended sediments during high-flow conditions. Two pyrethroids, bifenthrin and τ-fluvalinate were detected. Most noticeable was the relatively low registered use of these pyrethroids, suggesting that it may be difficult to predict occurrence from application amounts. One of the pyrethroids, τ-fluvalinate, has not been analyzed by other researchers in California and little is known about its occurrence and toxicity. Yolo Bypass is an important habitat to 42 fish species (*10*), many which use it as a migration corridor or as a rearing/ spawning ground during the winter and early spring when it floods.

Mallard Island represents a very large watershed that is the major freshwater input to the ecologically-important San Francisco Bay-Estuary. During periods of high-flow, large amounts of suspended sediments are transported into the Bay. Bifenthrin was the only pyrethroid detected but even its low concentrations might not be predicted from its low registered use in such a large watershed.

The results from this series of studies demonstrate the occurrence of pyrethroids in a variety of California surface waters and suggest a strategy for the design of pesticide studies. First, it is important to sample sites close to the application location and sites farther downstream to gain a better understanding of pesticide transport. Sampling watersheds of varying sizes adds information on the importance of spatial scale. Second, measuring pesticides in different environmental compartments provides an overall understanding of the fate and effects of pesticides in the environment. Analysis of bed sediments allows characterization of habitat for benthic organisms while analysis of suspended sediments contributes information on the transport and fate of pyrethroids. Finally, it is useful to analyze all the pesticides applied in the watershed, not just the high-use ones. While high-use pesticides are important to monitor, other pesticides may have high application rates relative to their toxicity or be relatively persistent in the environment, and should not be overlooked.

Acknowledgements

Funding for the Carpinteria Marsh projects was provided by an EPA STAR grant (R82867601). Funding for the Mallard Island was provided by the USGS Toxic Substances Hydrology Program. Funding for the Yolo Bypass study was provided by the California Resources Agency/California Bay-Delta Authority (Contract # 460002755) and the USGS Federal–State Cooperative Program. Funding for the Salton Sea study was provided by the State Water Resources Control Board and the USGS Federal–State Cooperative Program.

References

1. Laskowski, D. A. Physical and chemical properties of pyrethroids. *Reviews of Environmental Contamination and Toxicology,* **2002**, *174*, 49-170.
2. Amweg, E. L.; Weston, D. P.; Ureda, N. M. Use and toxicity of pyrethroid pesticides in the Central Valley, California, USA. *Environmental Toxicology and Chemistry,* **2005**, *24*, 966-972.
3. Gan, J.; Lee, S. J.; Liu, W. P.; Haver, D. L.; Kabashima, J. N. Distribution and persistence of pyrethroids in runoff sediments. *Journal of Environmental Quality,* **2005**, *34*, 836-841.
4. Weston, D. P.; You, J.; Lydy, M. J. Distribution and toxicity of sediment-associated pesticides in agriculture-dominated water bodies of California's Central Valley. *Environmental Science & Technology,* **2004**, *38*, 2752-2759.
5. Weston, D. P.; Holmes, R. W.; You, J.; Lydy, M. J. Aquatic toxicity due to residential use of pyrethroid insecticides. *Environmental Science & Technology,* **2005**, *39*, 9778-9784.
6. Smalling, K. L.; Orlando, J. L.; Kuivila, K. M. Analysis of Pesticides in Surface Water and Sediment from Yolo Bypass, California, 2004–2005, 2005. U.S. Geological Survey Scientific Investigations Report 2005-5220. p 20.
7. LeBlanc, L. A.; Schroeder, R. A.; Orlando, J. L.; Kuivila, K. M. Occurrence, Distribution and Transport of Pesticides, Trace Elements and Selected Inorganic Constituents into the Salton Sea Basin, California, 2001-2002, 2004. U.S. Geological Survey Scientific Investigations Report 2004-5117. p 40.
8. Qin, S. R.; Budd, R.; Bondarenko, S.; W., L.; Gan, J. Enantioselective degradation and chiral stability of pyrethroids in soil and sediment. *J. Agric. Food Chem,* **2006**, *54*, 5040-5045.
9. ______ Pesticide Use Reporting (PUR). California Department of Pesticide Regulation. http://www.cdpr.ca.gov/docs/pur/purmain.htm.
10. Sommer, T.; Harrell, B.; Nobriga, M.; Brown, R.; Moyle, P.; Kimmerer, W.; Schemel, L. California's Yolo Bypass: Evidence that flood control can be compatible with fisheries, wetlands, wildlife, and agriculture. *Fisheries,* **2002**, *26*, 6-16.
11. Smalling, K. L. unpublished data, 2006. USGS.
12. Smalling, K. L.; Orlando, J. L.; Kuivila, K. M. Occurrence of pesticides in water, sediment, and soil from the Yolo Bypass, California. *San Francisco Estuary and Watershed Science,* **2006**, *5*, 17.
13. ______ DAYFLOW. California Department of Water Resources. http://www.iep.ca.gov/dayflow/index.html.
14. Brown, R. P.; Landre, A. M.; Miller, J. A.; Kirk, H. D.; Hugo, J. M. Toxicity of sediment-associated chlorpyrifos with the freshwater invertebrates *Hyalella azteca* (amphipod) and *Chironomus tentans* (midge).

1997. DECO-ES-3036 Health and Environmental Research Laboratories, Dow Chemical, Midland, MI, USA.
15. ______ Pesticides in Urban Surface Water: Urban Pesticide Use Trends Report 2005. TDC Environmental. http://www.up3project.org/documents/UP3_Use_Report_2005.pdf.

Chapter 4

Assessment of Pyrethroid Contamination of Streams in High-Use Agricultural Regions of California

Keith Starner[1], Jane White[2], Frank Spurlock[1], and Kevin Kelley[1]

California Departments of [1]Pesticide Regulation and [2]Food and Agriculture, 1001 I Street, Sacramento, CA 95812

Over 100 surface water and bed sediment samples were collected from four agricultural regions within the state of California and analyzed for a suite of pyrethroid insecticides. Total organic carbon (TOC) was determined for sediment samples from each sampling site, and a toxicity unit (TU) analysis was performed to identify sediment concentrations that could potentially result in toxicity to *Hyallela azteca*. Overall, 60% of samples had detectable pyrethroids in either water or sediment, and 30% of sediment samples contained > 1 TU. The results highlight the need for the development of methods to reduce or eliminate offsite movement of pyrethroid insecticides.

Introduction

Synthetic pyrethroid insecticides are applied to a variety of crops in California throughout the year. In 2005, over 140,000 kg of pyrethroid insecticide active ingredients were applied to agricultural fields throughout the state. The primary pyrethroids used in California agriculture, in order of decreasing amount applied in 2005, were permethrin, fenpropathrin, esfenvalerate, cypermethrin, lambda-cyhalothrin, bifenthrin and cyfluthrin (*1*).

Due to the toxicity of the pyrethroids to aquatic organisms *(2-3)*, (Table 1), offsite movement of these compounds into surface water is of concern. Recent monitoring studies conducted in agricultural regions of California have shown pyrethroid contamination of both surface water and stream bed sediment *(4-9)*. Considering their high and potentially increasing use in California, reliable information regarding the environmental fate of these compounds is increasingly important.

Table 1. Pyrethroid Sediment Median Lethal Concentrations (LC50)[a]

Compound	*LC50*
Bifenthrin	0.52
Cyfluthrin	1.08
Cypermethrin	0.38
Esfenvalerate	1.54
Lambda-cyahlothrin	0.45
Permethrin	10.83

[a] LC50 are average 10-day for *H. azteca* (ug/g organic carbon) *(2-3)*.

The pyrethroid insecticides are extremely hydrophobic, with high adsorption coefficients and very low water solubility *(10)*. Due to their adsorption properties, determining the impact of pyrethroids on aquatic systems requires monitoring of suspended and bed sediments in addition to monitoring surface waters.

To assess the extent of pyrethroid contamination of aquatic environments in California, both surface waters and stream bed sediments were sampled at numerous sites throughout the state and analyzed for pyrethroid insecticides.

Materials and Methods

Site Descriptions

Using data from the California Department of Pesticide Regulation's (DPR) Pesticide Use Reporting (PUR) database *(1)*, several regions of the state were identified which historically have at least one period per year, lasting two or more months, of relatively high pyrethroid use. These regions include the Salinas Valley (Monterey County), the Imperial Valley (Imperial County), the Feather River region in the Sacramento Valley, and several areas of the San Joaquin Valley (Figure 1).

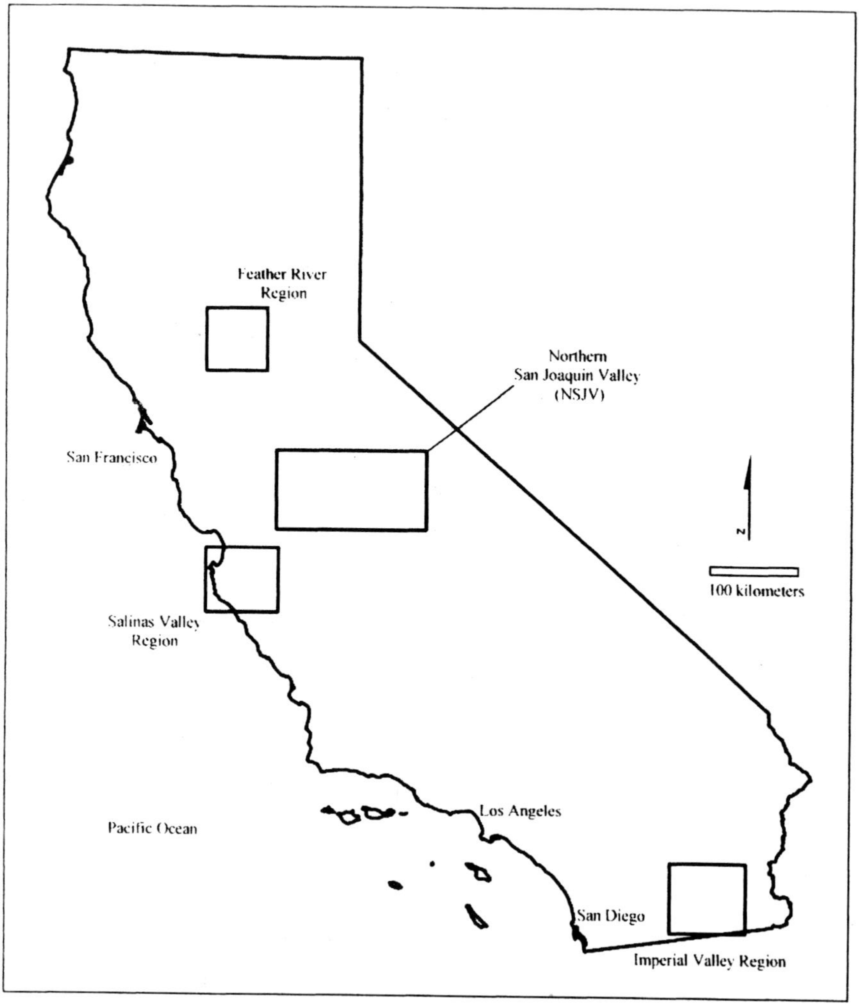

Figure 1. State of California with the four study areas indicated.

The four regions represent a variety of climates, pyrethroid use patterns, and agricultural practices, factors which influence the potential for offsite movement of pyrethroids.

Salinas Valley, Monterey County

Monterey county is located on the central coast of California (Figure 1). Over 18,000 kg of pyrethroid insecticide active ingredients were applied to agricultural crops in Monterey County in 2005 (*1*). Of this amount, most was applied within the Salinas Valley. The valley is approximately 25 km wide and 110 km long and extends from the city of Castroville in the north to King City in the south. Much of this area is cultivated year-round, with associated intensive use of pyrethroid insecticides and other pesticides.

The primary high-use period in the Salinas Valley is from April through September; 80% of all pyrethroid use in the region occurs during these months. During this period in 2005, over 14,000 kg of pyrethroid active ingredients were applied, with nearly 60% of this amount applied to lettuce. Use on lettuce, spinach, celery, artichokes and broccoli accounted for 80% of the amount applied in this period.

Imperial Valley, Imperial County

Imperial county is located in southeastern California (Figure 1). Pyrethroid insecticides are applied throughout the year there, with over 10,000 kg applied in 2005 (*1*). Virtually all agricultural use of pyrethroid insecticides in Imperial County takes place within the Imperial Valley (Figure 1). From the southeastern shoreline of the Salton Sea, the high use region extends east approximately 40 km and south to the US/Mexico border. In 2005, over 10,000 kg of pyrethroids were applied to agricultural crops within the Imperial Valley.

Two distinct periods of relatively high pyrethroid use occur within the Imperial Valley region. The highest use occurs during the fall (October and November), with applications made primarily to vegetable crops, including lettuce, onions and sugarbeets. A second period of high use occurs in late winter (February and March), when applications are made primarily to alfalfa. These two high use periods together account for over 60% of the annual pyrethroid use in the region.

Feather River/Sacramento Valley

Pyrethroid insecticides are applied throughout the Sacramento Valley (Butte, Colusa, Glenn, Sacramento, Solano, Sutter, Tehama, Yolo and Yuba counties) on a variety of crops, including fruits, nuts, vegetables and rice. There are several areas within the Sacramento Valley where pyrethroids are applied in significant amounts. However, the Feather River region, approximately 70 km north of the city of Sacramento (Figure 1), is one of the regions of heaviest pyrethroid use. This region includes portions of Sutter, Yuba and Butte counties. From May to August of 2005, approximately 3,000 kg of pyrethroid active ingredients were applied to agricultural crops (primarily peaches) in this region; the use during this period accounts for 80% of the annual pyrethroid use in the region.

Northern San Joaquin Valley

Pyrethroid insecticides are applied throughout the eight county region of the San Joaquin Valley (Fresno, Kern, Kings, Madera, Merced, San Joaquin, Stanislaus and Tulare counties) on a variety of crops, including nuts, fruits, corn, cotton and alfalfa. Due to the high use adjacent to nearby rivers and streams, the greatest potential for movement of pyrethroids to surface water bodies occurs in the northern San Joaquin Valley (NSJV) (Figure 1). This area includes parts of San Joaquin, Stanislaus and Merced counties.

In this region, the primary pyrethroid use period generally occurs between May and August. In 2005, over 15,000 kg of pyrethroid active ingredients were applied in the region from May through August, with applications to almonds and pistachios accounting for nearly 40 percent of the total use during the period.

Sampling Procedures

Samples were collected between July 2004 and June 2006, primarily during the dry season. During the first 12 months of the 24-month study (Phase A), each region was sampled three times. In the second half of the study (Phase B) samples were collected only from the Salinas and Imperial Valley regions. At each site, surface water and bed sediment were collected for pyrethroid analysis. Sites included tributary streams and mainstem rivers.

For sediment, a steel trowel was used to collect 100 g samples by gently scraping the top layer of the sediment column. The top 2-3 cm of the sediment column was collected. Sediment samples were collected into glass sample jars for pyrethroid analysis; an additional sediment sample was collected for TOC analysis. Sediment samples were transported on wet ice and transferred as soon as practical to frozen storage at 0° C until extraction for chemical analysis.

Surface water grab samples were collected directly into 1-liter amber glass bottles. Grab samples were collected from just below the surface, using a grab pole consisting of a glass bottle at the end of an extendable pole. Samples were not transferred from the original sample bottles until extraction at the laboratory.

Amber bottles were sealed with Teflon-lined lids and transported and stored on wet ice or refrigerated at 4°C until extraction for chemical analysis.

Analytical Procedures

Chemical analysis of all samples was performed by the California Department of Food and Agriculture's Center for Analytical Chemistry. Analytical method reporting limits are given in Tables II and III.

During the first half (Phase A) of the 24-month study, all samples were analyzed using analytical Method A. During Phase B of the study, an improved analytical method with additional analytes and lower reporting limits (Method B) was adopted for all sample analyses (Tables II and III).

For all pyrethroid water analyses, the whole sample, including any suspended sediment, was extracted in the sample bottle (*in toto*) and the pyrethroid residues reported on a whole sample basis (water plus suspended sediment).

Analytical Method A

Sediment samples were homogenized, followed by extraction with acetonitrile using an orbital shaker. Samples were concentrated and solvent exchanged to hexane. Extracts were analyzed with a gas chromatograph equipped with an electron capture detector. Confirmation of residues was completed using gas chromatography/mass spectrometry (GC/MS).

Whole water samples were extracted using liquid-liquid extraction with methylene chloride. Samples were concentrated and solvent exchanged to hexane. The extracts were analyzed by GC/MS.

Analytical Method B

Sediment samples were homogenized and copper powder was added to eliminate elemental sulfur. The samples were extracted with 1:1 acetone:hexane using an orbital shaker. The extracts were concentrated and then cleaned using a Florisil® column and analyzed with a gas chromatograph equipped with an electron capture detector. Confirmation or residues was completed by GC/MS when concentrations were within the sensitivity range of that instrument.

Whole water samples were extracted using liquid-liquid extraction with hexane. Sample extracts were concentrated and cleaned using a Florisil® column and analyzed using a gas chromatogram equipped with an electron capture detector. Confirmation of residues was completed by GC/MS when concentrations were within the sensitivity range of that instrument.

Table II. Reporting Limits for Whole Water Analytical Methods

Compound	*Reporting Limit (RL) in ug/L*	
	Method A	*Method B*
Bifentrhin	0.005	0.005
Cyfluthrin	0.08	0.015
Cypermethrin	0.08	0.015
Deltamethrin	NI[a]	0.015
Esfenvalerate	0.05	0.015
Fenpropathrin	NI	0.015
Lambda-cyhalothrin	0.02	0.015
Permethrin	0.05	0.015
Resmethrin	NI	0.015

[a] NI = not included.

Table III. Reporting Limits for Sediment Analytical Methods

Compound	*Reporting Limit (RL) in ug/g*	
	Method A	*Method B*
Bifentrhin	0.01	0.0010
Cyfluthrin	0.01	0.0010
Cypermethrin	0.01	0.0010
Deltamethrin	NI	0.0010
Esfenvalerate	0.01	0.0010
Fenpropathrin	NI	0.0010
Lambda-cyhalothrin	0.01	0.0010
Permethrin	0.01	0.0010
Resmethrin	NI	0.0015

[a] NI = not included.

TOC Analysis

Representative sediment samples from each sampling location were analyzed for TOC. TOC was determined using a DC-85A Total Organic Carbon Analyzer from Automated Custom Systems Inc., following acid treatment to remove inorganic carbon.

Results and Discussion

Pyrethroids were detected in three of the four regions, with an overall detection frequency of 61% (Table IV). Detection frequency was highest in the Salinas Valley region (85%), and was ca. 25% in the Imperial and NSJV regions. No pyrethroids were detected in the Feather River region.

For all regions, most detections were in bed sediment; there were relatively few detections in whole water samples (Tables V and VI), likely due to their low solubility. There were no detections of cyfluthrin, deltamethrin or resmethrin in any of the four regions. Many sediment samples had detections of multiple pyrethroid active ingredients. This was particularly true for the Salinas region, where a variety of vegetable crops are grown year-around.

A toxicity unit (TU) analysis identified sediment concentrations that could potentially result in toxicity to *H. azteca*. TU was calculated by dividing the organic carbon normalized concentration of the detected pyrethroid by its associated LC50 value (Table I). Trace detections and nondetected concentrations were assumed to be zero. At the time of this analysis, sediment toxicity data for fenpropathrin were not available; consequently, fenpropathrin was not included in the TU analysis. Pyrethroid toxicity was assumed to be additive; when multiple pyrethroid active ingredients were detected in a single sediment sample, their individual TUs were added together. A summary of the results of the TU analysis are shown in Table VII.

Overall, 30% of sediment samples had > 1 pyrethroid TU (Table VII), indicating that those sediments would be expected to be acutely toxic to *H. azteca*. Amweg *et al.* (*2*) showed that significant pyrethroid toxicity occurs in sediment at about 0.5 TU; the 1 TU benchmark used here is then a relatively conservative one. Approximately 45% of all sediment samples had > 0.5 TU.

The highest frequency of detection (85%) and exceedance of the 1 TU benchmark (42%) both occurred in the Salinas region (Tables IV and VII). Even considering only the earlier (Phase A) data, utilizing the less sensitive analytical method A, the Salinas samples still contained detectable concentrations of pyrethroids 60% of the time (Table IV).

Table IV. Summary of Pyrethroid Detections in Water and Sediment Samples from Four Regions of California, 2004-2006

Summary data	*Region*			
	Imperial	*Salinas*	*NSJV*	*Feather*
Sampling Sites	6 (5) [c]	14 (5)	4	4
No. of Samples [a]	21 (15)	76 (15)	11	12
Samples with Detections [b]	5 (4)	65 (9)	3	0
Overall Detection Frequency (%)	24 (27)	85 (60)	27	0

[a] No. of samples is each, water and sediment.

[b] Samples with detections is for at least one active ingredient, water or sediment.

[c] Values in parentheses are Phase A data only.

Table V. Pyrethroid Concentrations in Whole Water[a]

Compound	*Region*			
	Imperial	*Salinas*	*NSJV*	*Feather*
Bifenthrin	ND [b]	trace	ND	ND
Cypermethrin	ND	0.055	ND	ND
Esfenvalerate	ND	trace	ND	ND
Lambda-cyahlothrin	0.0274	ND	0.11- 014	
Permethrin	trace	trace - 0.08	ND	ND

[a] Concentrations in ug/L.

[b] ND = not detected.

Table VI. Pyrethroid Concentrations in Bed Sediment [a]

Compound	*Region*			
	Imperial	*Salinas*	*NSJV*	*Feather*
Bifenthrin	ND	0.0013 - 0.0790	trace	ND
Cyfluthrin	ND	ND	ND	ND
Cypermethrin	ND	0.0020 - 0.0118	ND	ND
Esfenvalerate	trace - 0.02	0.002 - 0.06	ND	ND
Fenpropathrin	ND	0.0017 - 0.0094	ND	ND
Lambda-cyahlothrin	0.04 - 0.31	0.0018 - 0.1441	trace - 0.02	ND
Permethrin	trace	0.00167 - 0.1441	ND	ND

[a] Concentrations in ug/g, dry sediment.

Table VII. Toxicity Unit Calculation, Sediment Samples

Summary data	*Region*			
	Imperial	*Salinas*	*NSJV*	*Feather*
Sampling Sites	6 (5) [b]	14 (5)	4	4
No. of Samples	21 (15)	76 (15)	11	12
Samples with est. > 1 TU	4 (3)	32 (3)	1	0
Percent Samples with est. > 1 TU	19 (20)	42 (20)	9	0
Source of estimated toxicity [a]	LAM	ESF, BIF	LAM	none

[a] LAM = lambda cyhalothrin, ESF = esfenvalerate, BIF = bifenthrin.

[b] Values in parentheses are Phase A only data.

The higher detection frequency in Salinas sediment samples relative to the other regions is likely due at least partially to the higher organic carbon content of the bed sediments in that region. The percent TOC measured in bed sediments from the Salinas Valley sites ranged from about 2.0 to 3.5; in the other three regions, the TOC was significantly lower, generally less than 1.0 (Table VIII). Due to the hydrophobic nature of the pyrethroids, accumulation in sediment organic carbon is expected.

Additionally, high pyrethroid use in the Salinas region likely played a role in the observed differences in detection frequency. The overall rate of pyrethroid use is substantially higher in the Salinas region than in the other regions (Table VIII).

Table VIII. Region Characteristics

Characteristic	*Region*			
	Imperial	*Salinas*	*NSJV*	*Feather*
Sediment TOC (%)	< 1.0	2.0 - 3.5	< 1.0	0.5 - 1.5
Pyrethroid Use/Area	3.3	11	1	1.9
Primary Use Season	Mar / Oct	Apr - Sep	May - Aug	May - Aug
Primary Crops	alfalfa/lettuce	lettuce, spinach	almonds	peaches

[A] Use per unit area units are kg/hectare, over the entire use region and normalized to NSJV Region use (not an application rate).

The differences in detection frequencies could also be partly due to differences between regions in the rate of off-site movement of the pyrethroids. In general, pesticides move off-site from agricultural fields into surface waters in runoff or drainage induced by either rain or irrigation (*11*). The four primary factors that affect pesticide transport in runoff are climate (amount and intensity of rainfall, as well as the timing of rainfall with respect to pesticide applications), soil characteristics (soil texture, organic matter content, surface crusting and compaction, and slope and topography of the field), agricultural management practices (irrigation practices, erosion control efforts, pesticide formulation and application rate), and the chemical and physical properties of the specific pesticides applied (*11, 12*). Some or all of these factors could vary between the four regions, and potentially contribute to the observed differences in detection frequencies. Further investigation into these regional factors may provide additional insight into the observed differences in detection frequencies.

The data presented here demonstrate the potential of the pyrethroid insecticides to accumulate in stream bed sediments to concentrations capable of causing acute toxicity to aquatic organisms. Streams in areas with high and consistent use of pyrethroids appear to be particularly vulnerable to pyrethroid contamination. It is therefore imperative to quantify their transport into water bodies, assess their impact on aquatic ecosystems, and determine methods to reduce or eliminate the offsite movement of the pyrethroid insecticides.

Acknowledgements

We would like to thank Paul Lee, Teresa Woroncieka and Suzanne Matsumoto of the California Department of Food and Agriculture, Center for Analytical Chemistry, and Kean S. Goh, Carissa Ganapathy, Jesse Ybarra, Heidi Dietrich and Amrith Gunasekara of the Department of Pesticide Regulation, Environmental Monitoring Branch.

References

1. California Department of Pesticide Regulation Pesticide Information Portal. http://calpip.cdpr.ca.gov/cfdocs/calpip/prod/main.cfm (accessed April 17, 2007).
2. Amweg, E. L.; Weston, D. P.; Ureda, N. M. *Environ. Toxicol. Chem.* **2005**, *24*, 966-972; *erratum 24*, 1300-1301.
3. Maund, S. J.; Hamer, M. J.; Lane, M. C. G.; Farrelly, E.; Rapley, J. H.; Goggin, U. M.; Gentle, W. E. *Environ. Toxicol. Chem.* **2002**, *21*, 9-15.
4. Anderson, B. S.; Phillips, B. M.; Hunt, J. W.; Worcester, K.; Adams, M.; Kapellas, N.; Tjeerdema, R. S. *Environ. Toxicol. Chem.* **2006**, *25*, 1160-1170.
5. Bacey, J.; Spurlock, F.; Starner, K.; Feng, H.; Hsu, J.; White, J.; Tran, D. M. *Bull. Environ. Contam. Toxicol.* **2005**, *74*, 864-871.
6. Gill, S.; Spurlock, F. Technical Memorandum, California Department of Pesticide Regulation, Sacramento, California. **2004**, 14pp.

7. Kelley, K.; Starner, K. Technical Memorandum, California Department of Pesticide Regulation, Sacramento, California. **2004**, 41pp
8. Weston, D. P.; You, J. C.; Lydy, M. J. *Environ. Sci. Technol.* **2004**, *38*, 2752-2759.
9. Walters, J; Kim, D.; Goh, K. S. Technical Memorandum, California Department of Pesticide Regulation, Sacramento, California. **2002**, 8pp.
10. Laskowski, D. A. *Rev. Environ. Contam. Toxicol.* **2002**, *174*, 49-170.
11. Larson, J.; Capel, P. D.; Majewski, M. S. *Pesticides in Surface Waters: Distribution, Trends, and Governing Factors*; Pesticides in the Hydrologic System; Ann Arbor Press, Inc.: Chelsea, MI, 1991; Vol. 3, pp 217-219.
12. Leonard, R. A. In *Pesticides in the soil environment: Processes, impacts, and modeling*; Cheng, H. H., Ed.; Soil Science Society of America Book Series; Soil Science Society of America, Inc.: Madison, WI, 1990; No. 2, pp. 303-349.

Analytical Methodology

Chapter 5

Quantification of Pyrethroid Insecticides at Sub-ppb Levels in Sediment Using Matrix-Dispersive Accelerated Solvent Extraction with Tandem SPE Cleanup

Jing You[1,3], Donald P. Weston[2], and Michael J. Lydy[1,*]

[1]Fisheries and Illinois Aquaculture Center and Department of Zoology, Southern Illinois University, 171 Life Science II Building, Carbondale, IL 62901
[2]Department of Integrative Biology, University of California, 3060 Valley Life Sciences Building, Berkeley, CA 94720
[3]Current address: Department of Biochemistry, Chemistry and Physics, University of Central Missouri, Warrensburg, MO 64093
*Corresponding author: email: mlydy@siu.edu; telephone: 618-453-4091; fax: 618-453-6095

A rapid and sensitive analytical method was developed to quantify pyrethroid insecticide concentrations in sediment with method detection limits (MDL) at biologically relevant concentrations (sub-ppb levels). Pyrethroids were isolated from the sediment by matrix-dispersive accelerated solvent extraction, and the extracts cleaned with tandem solid phase cartridges packed with graphite carbon black and primary/secondary amine. Method detection limits, relative recoveries and precision were determined for two type I and six type II pyrethroids at several different concentrations using four different sediments. Additional fractionation with a Florisil column further improved method sensitivity. The method was successful at analyzing pyrethroids with MDLs of 0.11-0.85 ppb (dry sediment), and recoveries were 73.7-151.7% for sediment spiked with pyrethroids at 0.2 ppb dry weight.

Introduction

The dominance of organophosphate insecticides (OPs) over the past several decades has led to a focus of environmental monitoring on dissolved phase pesticides and their toxicity (*1,2*). However, pyrethroids are becoming increasingly important in agriculture, and the use of OPs was drastically curtailed by the recent withdrawal of nearly all products for residential use containing chlorpyrifos or diazinon. As the major replacement, use of pyrethroids, a group of synthetic insecticides derived from naturally occurring pyrethrin, has increased dramatically in recent years. As a result there is an emerging need for a better understanding of the environmental fate and effects of pyrethroid residues in aquatic systems. Pyrethroids are highly hydrophobic and strongly bind to sediment particles when entering aquatic systems, so monitoring suspended or bedded sediments is more appropriate for these compounds.

Recently, field monitoring showed a strong linkage between the presence of pyrethroid residues in sediments and benthic invertebrate toxicity from both agricultural and urban areas in California (*3-5*). Agricultural use of pyrethroids has resulted in appearance of pesticide residues in sediments of creeks receiving return flow from irrigated fields, and over 60% of the observed sediment toxicity to a sensitive invertebrate, *Hyalella azteca,* can be explained by pyrethroid residues (*3*). On the other hand, the vast majority of insecticides sold for consumer use now contain pyrethroids as their active ingredients, and they are widely used around homes by professional pest control applicators as well. Non-agricultural use of pyrethroids has increased dramatically, and commercial non-agricultural use in California (324,000 kg) in 2005 was seven times greater than in the early 1990s (*California Department of Pesticide Regulation data; (http://www.cdpr.ca.gov/docs/pur/purmain.htm*). It has been reported that the use of pyrethroids in urban areas in California, and particularly the use of bifenthrin, has led to substantial contamination of nearby aquatic systems with these compounds (*4,5*).

Though some environmental monitoring of pyrethroids has been conducted, the efforts lag far behind the transition from OPs to pyrethroids in the marketplace. Part of the reason for a lack of data is that analytical methods to detect pyrethroids in sediment have not been broadly available or standardized. Methods to assess pyrethroids in sediment are still under development and the reported method detection limits (MDL) for pyrethroids in dry sediments are in the range of 1 to 25 μg/kg (*6-9*). To achieve the sensitivity requirement, gas chromatography (GC) coupled with an electron capture detetor (ECD) (*8,9*) or mass spectrometer (MS)(*10,11*) have been generally used as detection techniques. However, prior to GC quantification, a time-consuming sample preparation process is required to isolate pyrethroid residues from sediment by solvent extraction, and to remove the co-extracted interferences by cleanup due to the strong binding of pyrethroids to sediment and the complexity of sediment matrices. Additionally, co-extracted matrix components may inhibit the isomerization of pyrethroids during GC analysis and enhance GC response, which can introduce error in the quantification of these compounds (*12*).

Although pyrethroids have low mammalian and avian toxicity, they are extremely toxic to fish and aquatic invertebrates (*13,14*). As a result, toxicity of pyrethroids to sensitive species may exist at field sites where pyrethroids concentrations are barely detectable or not detected at all using current analytical techniques. As shown in Table I, in sediment with about 1% total organic carbon (TOC), most pyrethroids, excluding permethrin, will cause 50% mortality to the aquatic amphipod *H. azteca* in a 10-d exposure at concentrations in the range of 4 to 10 µg/kg dry weight (dw) (LC50). While these values represent acute LC50s, it is important to recognize that the onset of toxicity, as well as growth impairment, will occur at about half of these concentrations (*13, 14*).

Table I. Pyrethroid sediment 10-d median lethal concentration (LC50, µg/kg dry weight (dw)), growth lowest-observable-effect concentration (LOEC, µg/kg dw) for invertebrate *Hyalella azteca* and target method detection limit (MDL, µg/kg dw) for analyzing pyrethroid in sediment with 1% organic carbon. Target MDL based on one-ninth the LOEC to allow for pyrethroid toxicity at lower temperatures and the presence of multiple pyrethroids with additive toxicity (see text).

Pyrethroid	LC50 [a]	LOEC [a]	Target MDL
Bifenthrin	5.2	3.4	0.4
λ-Cyhalothrin	4.5	1.8	0.2
Cyfluthrin	10.8	6.2	0.7
Cypermethrin [b]	3.6	0.77 (NOEC [c])	0.1
Deltamethrin	7.9	2.0	0.2
Esfenvalerate	15.4	6.1	0.7
Permethrin	108	83	9.2

[a] Amweg et al. (*13*)

[b] Based on median of three values in Maund et al. (*14*)

[c] NOEC = no-observed-effect concentration

Yet in field situations, these lab derived toxicity estimates most likely underestimate the potential toxicity. Pyrethroids are more toxic at cooler temperatures (*15*), and while *H. azteca* LC50 values are commonly measured at 23 °C, at cooler, but environmentally realistic temperatures, their toxicity is considerably greater. Pyrethroid toxicity to *H. azteca* increases by three-fold at 13 °C relative to 23 °C (*Weston, unpub. data*). Also, environmental samples have commonly been shown to contain 2-4 pyrethroids (*3-5*), and if their toxicity is presumed to be additive, each individual analyte would need to be quantifiable at about one-third the concentration of that considered independently. Thus, taking the effects of temperature and the co-occurrence of multiple pyrethroids into consideration, lab-derived, single compound estimates of toxicity could easily be under-protective by a factor of nine or more. While *H. azteca* may be among the more sensitive species to pyrethroids, it is also one that is widely used nationwide in sediment monitoring, and it is clear that in order to quantify pyrethroids at levels toxic to this species, detection limits in the range of 0.1-0.7 μg/kg are required (Table I). Therefore, development of a sensitive and rapid method for quantifying pyrethroid insecticides in sediment at environment-relevant concentrations (sub-ppb levels) with minimal matrix effects is of critical need due to the rapid increase in pyrethroid use and the extremely high toxicity of pyrethroids to aquatic organisms.

The objective of the current study was to develop and validate an analytical method to detect pyrethroids in sediment with expected MDL at sub-ppb levels.

Materials and Methods

Chemicals

Pyrethroids analyzed in this study included two type-I pyrethroids, bifenthrin and permethrin, and six type-II pyrethroids, *lambda*-cyhalothrin, cyfluthrin, cypermethrin, deltamethrin, esfenvalerate and fenpropathrin (Figure 1). The eight pyrethroids were selected because of their heavy usage and potentially high toxicity to benthic invertebrates. Pyrethroid standards were purchased from ChemService Inc. (West Chester, PA, USA) and stock solutions of 1 mg/ml were prepared in hexane.

Diatomaceous earth (DE), clean sea sand, anhydrous Na_2SO_4, anhydrous $MgSO_4$, copper powder, normal phase adsorbents (alumina, silica and Florisil, 60~100 mesh) and various solvents were purchased from Fisher Scientific (Pittsburgh, PA, USA). All solvents used in the current study were pesticide grade. Prior to use, copper powder was treated with diluted HNO_3 to remove oxides, rinsed with distilled water and methanol, and then dried under nitrogen. Anhydrous Na_2SO_4 and $MgSO_4$ were baked at 400 °C for 4 h and the normal phase adsorbents were baked at 130 °C over night. Solid phase extraction (SPE) cartridges packed with Supelclean™ primary/ secondary amine (PSA) and Supelclean ENVI-Carb graphite carbon black (GCB) were purchased from Supelco (Bellefonte, PA, USA). Bulk PSA absorbent was obtained from Varian (Harbor City, CA, USA). Two surrogate standards, 4,4'-dibromooctafluoro-biphenyl (DBOFB) and decachlorobiphenyl (DCBP) (Supelco) were added to the sediment prior to extraction to verify the performance of the extraction and cleanup processes. A gel permeation chromatography (GPC) calibration standard, including 25 mg/ml corn oil, 1.0 mg/ml bis (2-ethylhexyl) phthalate,

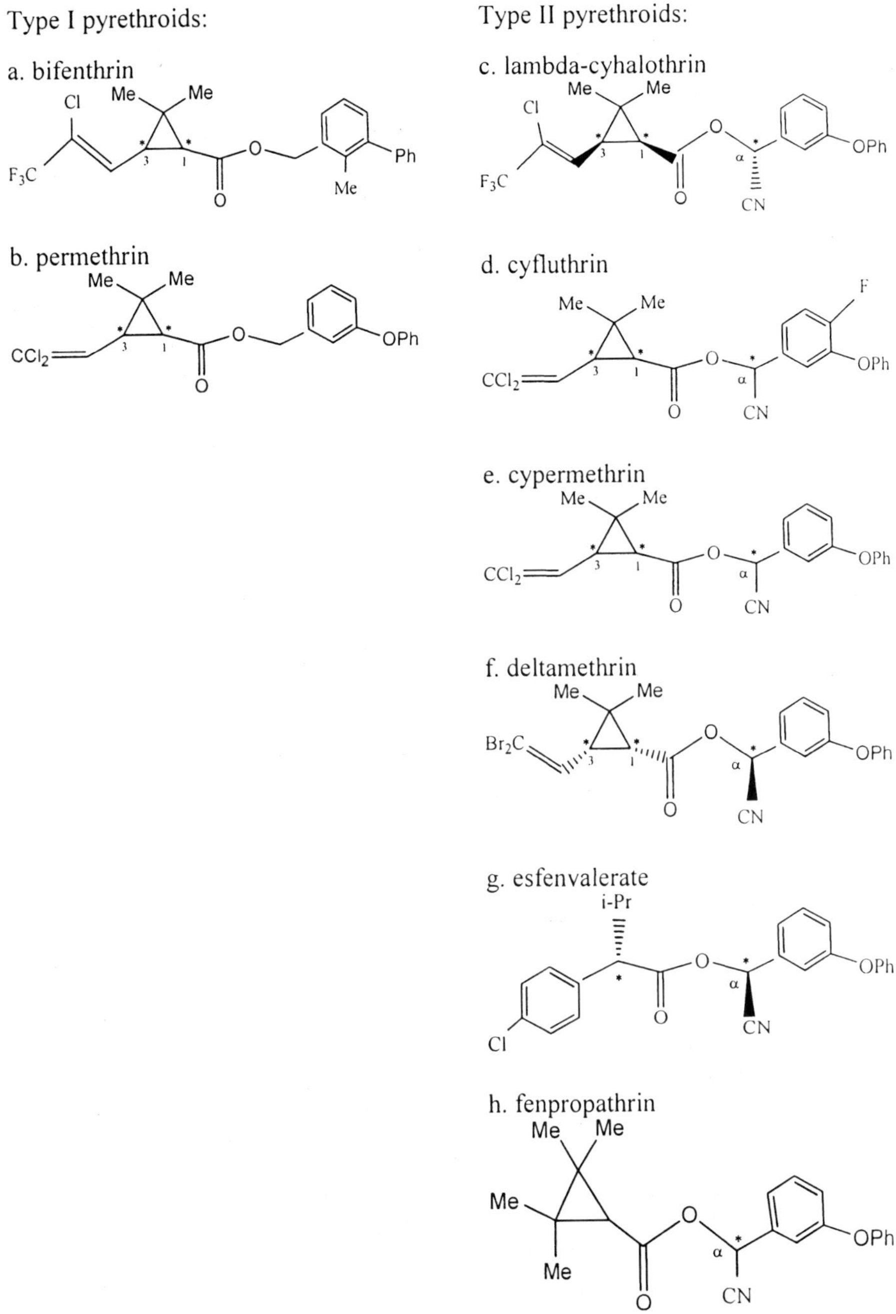

Figure 1. Structures of the type I and type II pyrethroids used in the current study. The stars (*) indicate chiral centers in the molecule.

0.2 mg/ml methoxychlor, 0.02 mg/ml perylene and 0.08 mg/ml sulfur, was purchased from Fisher Scientific.

Sediments spiking

Four reference sediments with different characteristics were collected from the American River, California (AR), a farm pond in Wichita, Kansas (KS), Pacheco Creek near Hollister, California (CA), and Bearskin Lake in Grand Marais, Minnesota (MN). Sediments were spiked with 0.2, 0.5, 1, 5 or 10 μg/kg dw of tested pyrethroids to develop and validate extraction, cleanup and fractioning methods. The AR and CA sediments were previously used for pyrethroid toxicity tests with *H. azetca,* and their TOC contents were 1.4 and 6.5%, respectively (*13*). An EA 1110 CHN analyzer (CE Instruments, Thermoquest Italia, Milan, Italy) was used to measure TOC in the other two sediments after removing carbonates by treating with 3 mol/L of HCl, and the TOC content of these sediments were 1.31 ± 0.20, and 7.85 ± 0.20% for the KS and MN sediments, respectively.

No target pyrethroids were detected in the reference sediments. Sediments were spiked with appropriate quantities of pyrethroids in acetone. After spiking, sediments were thoroughly mixed using a stainless steel paddle driven by an overhead motor for over 1 h, and then stored at 4 °C overnight prior to use.

Sediment extraction

Matrix-dispersive accelerate solvent extraction

A matrix-dispersive accelerated solvent extraction (ASE) method was developed using a Dionex 200 ASE with 33 ml stainless steel cells and 60 ml glass collection vials (Dionex, Sunnyvale, CA, USA). Before extraction, frozen sediment was thawed, centrifuged to remove excess water and homogenized. After adding the surrogates, DBOFB and DCBP, approximately 10 g of sediment (wet weight (ww)) was mixed with appropriate amounts of the drying agent diatomaceous earth (DE), dispersion absorbent, and 2 g of copper powder, and transferred into 33 ml of stainless steel cells. Normal phase adsorbent was used as the dispersion agent for in-line cleanup of the extracts by trapping polar interference in the ASE cell. The amounts of drying and dispersion agents were optimized to lower the amount of co-extracted interference and improve pyrethroid recoveries. After loading the cells onto the ASE extractor, the samples were extracted with methlyene chloride and acetone (1:1, v/v) at an elevated temperature and pressure. The static extraction time was set at 5 min, and the influence of extraction cycles on extraction efficiency was studied.

Optimization parameters for the matrix-dispersion ASE method are summarized in Table II.

Table II. Optimization of matrix-dispersion accelerated solvent extraction.

Parameter	Values
Sediment : drying agent (diatomaceous earth) ratio	3:1, 2:1, 1:1, 1: 1.5
Extraction temperature (°C)	60, 80, 100, 120, 140
Extraction pressure (psi)	1000, 1500, 2000
Static extraction cycle	1, 2, 3
Type of dispersion absorbent	Silica, Florisil, Alumina-N
Amount of dispersion absorbent (g)	1, 2, 3

The extracts in the collection vials were dried with 12 g of anhydrous Na_2SO_4 and concentrated to 5 ml under a stream of N_2 at 50 °C and 15 psi using a TurboVap II evaporator (Zymark, Hopkinton, MA, USA). After solvent exchange, the extract was then evaporated to 1 ml under a gentle stream of N_2 using a Pierce Model 1878 Reactivap™ (Rockford, IL, USA) prior to further cleanup.

Sonication-assisted solvent extraction

For comparison purposes, sediments were also extracted followed our previously established sonication-assisted solvent extraction procedure (*8*). In brief, anhydrous $MgSO_4$ dried sediment (20 g ww) was sonicated three times with a solvent mixture of methylene chloride and acetone (1:1, v/v). The extracts were decanted, filtered and combined. After solvent-exchange, the extract was evaporated to 1 ml for further cleanup.

Sediment extract cleanup

Three commonly used cleanup techniques were compared, including Florisil adsorption column, high performance gel permeation chromatography (HPGPC) and solid phase extraction (SPE). Copper powder was introduced to sediment before extraction to remove the sulfur interference when Florisil columns and SPEs were used as cleanup methods, while no copper was required when HPGPC was employed for cleanup due to its ability to separate sulfur from the target pesticides (*9*).

Florisil adsorption column cleanup was performed following our previously developed method *(8)*. Briefly, the concentrated extracts were solvent-exchanged to hexane and loaded onto an adsorption column packed with 10 g of Florisil deactivated with 6% water and pre-conditioned with 20 ml of hexane. The target pyrethroids were eluted with a 50 ml ethyl ether and hexane mixture (3:7, v/v), and the polar matrix components were retained on the Florisil column. The eluent was concentrated and solvent-exchanged to acidified (0.1% acetic acid) hexane prior to gas chromatography analysis.

The HPGPC cleanup was conducted with an Agilent 1100 high performance liquid chromatograph (Agilent Technologies, Palo Alto, CA, USA) coupled with a Foxy Jr. ™ Fraction collector (ISCO, Inc. Lincoln, NE, USA). The separation was completed on a 300 mm × 19 mm Envirogel™ GPC cleanup column with a 5 mm × 19 mm precolumn (Waters, Milford, MA, USA) and the flow rate of the mobile phase (methylene chloride) was set at 5 ml/min. The extracts were solvent exchanged to methylene chloride, filtered through a 0.2 µm filter (Whatman Inc., Florham Park, NJ, USA), and injected into the GPC system through a Rheodyne 7225 manual injector with a 0.5 ml loop (Cotati, CA, USA). The elution profile was monitored with an ultraviolet (UV) detector set at a wavelength of 254 nm, and fractions were collected every 12 s. The collected fractions were solvent-exchanged to acidified hexane and analyzed using gas chromatography. The GPC calibration standard was analyzed daily prior to sample processing to verify GPC performance.

A primary/secondary amine (PSA) cartridge was connected to the bottom of a graphite carbon black (GCB) cartridge, and 1cm of anhydrous Na_2SO_4 was used to cap the GCB adsorbent bed. The purpose for the anhydrous Na_2SO_4, GCB and PSA was to remove residual water, planar pigments, and polar interferences, respectively. Prior to use, the tandem SPE cartridges were conditioned with 3 ml of hexane. The sample was then loaded onto the cartridges and the sample tube rinsed with 1 ml of hexane to remove any residue compound from the tube. This rinse was also loaded onto the cartridges. A three ml mixture of toluene/ hexane or methylene chloride/ hexane at appropriate ratios was used as elution solutions at a flow rate of approximately 1 drop/s to elute the pyrethroids off of the cartridges. The collected eluent was solvent-exchanged to acidified hexane for gas chromatography analysis or to hexane for further fractionating.

Pesticide fractionating by Florisil column

For sediments containing extremely low concentrations of pyrethroids, an additional fractionating step was included following the cleanup to separate pyrethroids from other chlorine-containing chemicals, such as polychlorinated biphenyl (PCBs) and organochlorine pesticides (OCs). Three grams of Florisil, which was activated by heating at 90 °C overnight and partially deactivated by mixing with 6% (w/v) distilled water, was used to pack the column for the fractionating procedure. The deactivated Florisil column was conditioned with 5 ml of hexane, and the cleaned extract was transferred onto the column. The column was first eluted with 20 ml of hexane and the eluent was discharged, and then 15 ml of 10% ethyl ether in hexane (v/v) was used to elute the pyrethroids off of the column. The second fraction was collected, evaporated, solvent-exchanged to 0.1 ml of acidified hexane and analyzed using gas chromatography.

Gas chromatography quantification

Analysis of the final extracts was performed on an Agilent 6890 series gas chromatograph equipped with an Agilent 7683 autosampler and a micro-electron capture detector (GC-ECD, Agilent Technologies, Palo Alto, CA). Two columns, a HP-5MS (30m × 0.25mm × 0.25μm film thickness, Agilent Technologies) and a DB-608 (30m × 0.32 mm × 0.50μm film thickness, Agilent Technologies) were used to confirm the analytical results. Helium and nitrogen were employed as the carrier and makeup gas, respectively. The flow rates of carrier gas were 3.8 ml/min and 1.8 ml/min for the HP-5MS and DB-608 columns, respectively. A 2 μl sample was injected into the GC using a pulsed split-less mode. When the separation was conducted with the HP-5MS column, the oven was set at 100 °C, heated to 180 °C at 10 °C/min, then to 205 °C at 3°C/min, and held at 205 °C for 4 min, then heated to 280°C at 20°C/min and held at this temperature for 10 min. When the separation was conducted with the DB-608 column, the oven was set at 100 °C, heated to 250 °C at 10 °C/min, then to 280 °C at 3 °C/min, and held at 280 °C for 15 min. Calibration was based on area using six external standards. The standard solutions were made by dissolving 5, 10, 25, 50, 100, 250 or 500 μg/L of each pyrethroid and surrogate in acidified hexane. These solutions were analyzed using the GC-ECD methods detailed above. The calibration curves were linear within this concentration range. Qualitative identity was established using a retention window of 0.5% with confirmation on a second column.

Measurement of matrix components

Matrix components left in the sediment extracts after each cleanup and fractionation procedure were assessed following the spectrophotometer methods of van Handel (*16*). In brief, 0.1 ml of extract was digested with concentrated H_2SO_4 and quantified spectrometrically with a vanillin-H_3PO_4 reagent. The efficiency of each cleanup and fractionation procedure to remove the matrix component was calculated by dividing the amounts of matrix removed with the cleanup procedure by the amounts of matrix in the original extracts.

Results and Discussion

Method development

Optimization of the extraction procedure

A matrix-dispersion accelerated solvent extraction (ASE) method was developed to isolate pyrethroid residues from sediment. Because of the strong binding between pyrethroids and sediment organic carbon, it generally requires an exhaustive extraction method to separate matrix components from analytes. Traditional extraction methods, such as sonication-assisted solvent extraction (*8*) and Soxhlet extraction (*10*) have been successfully used to extract pyrethroids from sediment; however, these methods are very labor, time and solvent consuming.

Since Dionex commercialized it in 1994, ASE has been reported as a rapid and effective alternative to traditional extraction methods for soils (*17-19*), and U.S. Environmental Protection Agency (EPA) also recommended it for semi-volatile organic compounds extraction from solid matrices (*20*). As an automated extraction technique, ASE extensively reduces the labor and time needed for sample preparation. For example, the extraction time for a sediment sample can be reduced from 16 h for a Soxhlet extraction to 15 min per sample using ASE. In the current study, matrix dispersion absorbents were introduced into the ASE extraction cell to facilitate in-line cleanup. Several parameters for the ASE extraction were optimized to improve the extraction efficiencies for pyrethroids, while reducing the amount of co-extracted interference (Table II).

A mixture of methylene chloride and acetone was used as extraction solvents because of their high solubility for pyrethroids (*8*), and diatomaceous earth (DE) was chosen as a drying agent to avoid possible clogging of the ASE cell which can occur when commonly used drying agents such as anhydrous $MgSO_4$ and Na_2SO_4 are used (*21*). The ratio of DE to sediment was evaluated in order to maximize the sediment amount in the ASE cell. There was no significant difference in pyrethroid recoveries with different DE: sediment ratios, therefore the lowest DE-sediment ratio (2:1), which could produce a free flow DE-sediment mix was used. This is consistent with the study by Lehotay and Lee (*21*) where the amount of drying agent used did not significantly affect recoveries of non-polar pesticides.

Pyrethroids were effectively extracted from sediment in a short period of time compared to the traditional extraction techniques due to the higher temperatures and pressures used with the ASE procedure. Solvents at elevated temperature have a lower viscosity, higher diffusion rates and are therefore more capable of solublizing the analytes. Similarly, elevated pressures help keep the solvent in the liquid phase, allowing the solvent to better penetrate the sample matrix, thereby increasing solvent-analyte interactions. Results showed that pyrethroids could be extracted well when temperatures reached 80 °C, and that no significant differences were observed when sediments were extracted at 80, 100, 120 and 140 °C. Two pressures, 1500 and 2000 psi, were tested with the 100 °C extraction temperature, and pyrethroid recoveries dropped slightly from 1.0 - 1.5 times at 2000 psi in comparison to 1500 psi. Therefore, 100 °C and 1500 psi were selected as the extraction temperature and pressure. In each static extraction cycle, analytes equilibrated between the extraction solvent and the matrix, therefore, an increase in the number of extraction cycles could potentially increase extraction efficiency; however, the extraction time and solvent usage would increase as well. Results showed that two cycles of 5 min static extractions provided the best quantitative extraction of pyrethroids from sediment.

Normal phase adsorbents, including activated Florisil, silica gel, and alumina, were added into the ASE cells for in-line removal of polar interferences. A preliminary study was conducted to select the best normal phase adsorbent. Pyrethroids were eluted through columns packed with single adsorbents and pyrethroids recoveries ranged from 55.8-103.8, 74.0-110.4 and 0-22.5% for Florisil, silica and alumina columns, respectively. Silica had the best recoveries and was selected as the dispersive adsorbent. Different amounts of silica (0, 1, 2, and 3 g) were dispersed with DE dried sediment prior to extraction, and results showed the in-line cleanup in all cases, demonstrated clean chromatograms without loss of target analytes. Sediment dispersed with 3 g silica provided the cleanest chromatogram. In summary, the optimized ASE conditions were to use 10 g wet sediment mixed with 2 g of copper power, 5 g of DE, and 1 g of silica which was transferred into a ASE cell packed with 2 g silica, and extracted at 100 °C, 1500 psi for two cycles of 5 min using static extraction with a flush volume of 60% of cell volume.

The newly developed ASE method was compared to the previously used sonication-assisted solvent extraction (*8*) (Figure 2). As shown in Figure 2, extraction efficiencies of the two methods were similar, while ASE provided less interference peaks, especially in the later portion of the chromatogram where pyrethroids were found. Additionally, as an automatic technique, ASE was capable of running 24 samples per day compared to 6 samples per day with the sonication method, and solvent usage was reduced from 150 ml using the sonication method to 60 ml per sample for the ASE method.

Optimization of cleanup methods

Two cleanup methods, namely high performance gel permeation chromatography (HPGPC) and tandem solid phase extraction (SPE) with primary/secondary amine (PSA) and graphite carbon black (GCB) cartridges, have been developed in the current study and the two methods were compared to our previously established Florisil adsorption column method to evaluate the effectiveness of reducing matrix interference (*8*).

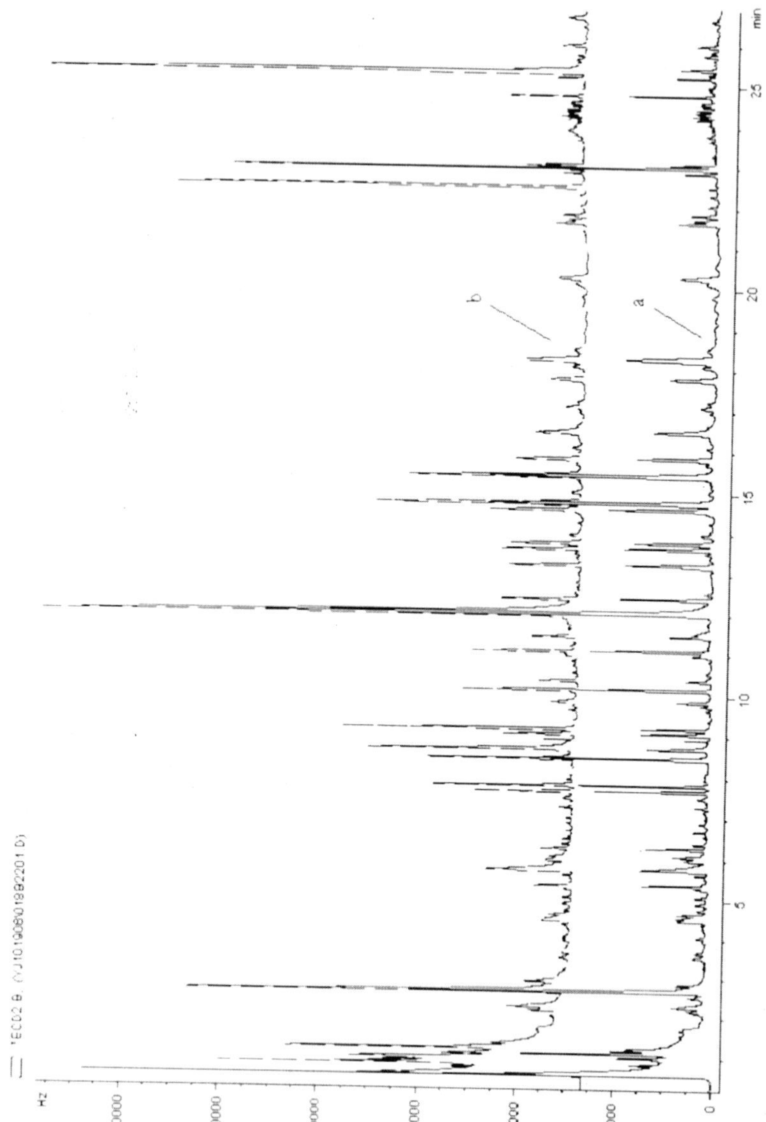

Figure 2. Comparison of matrix-dispersive accelerated solvent extraction (a) and sonication-assisted solvent extraction (b) for pyrethroids in sediment.

Gel permeation chromatography has been widely used to remove large molecular weight matrix components, such as pigments and lipids, from the extracts by size exclusion. The HPGPC method, which uses smaller size particles in its packing material and higher pressure, extensively reduced analytical time and solvent usage compared to traditional GPC. Elution profiles for pyrethroid, organochlorine (OC), and organophosphate insecticides (OP) and poly-chlorinated biphenyls (PCBs) are listed in Figure 3. As shown in the figure, HPGPC not only removed the higher molecule weight matrix components and sulfur from the extracts, but also achieved separation of pyrethroids from the OC and OP insecticides and PCBs. The retention times for the corn oil standard and sulfur were 6.3 and 14.5 min, respectively. Pyrethroids were the first pesticides to elute (7.5 - 8.5 min), while the OC and OP insecticides eluted at 8.5-12 min, and the PCBs eluted at 11-12 min. *Lambda*-cyhalothrin was the first pyrethroid to elute and permethrin was the last to elute. Over 96% of the eight target pyrethroids were recovered in the selected fraction of 7.5-8.5 min, however, the pyrethroid fraction was visibly yellow possibly due to the presence of un-separated pigment residues.

Though HPGPC can be effectively used to clean sediment extracts for pyrethroid analysis, not all analytical laboratories have this type of equipment. The tandem SPE cleanup method that uses SPE cartridges filled with PSA and GCB packing materials were then investigated to provide a simpler alternative cleanup method. Black carbon is a strong adsorbent for planar compounds, thereby removing pigments and sterols from the sediment extract. The GCB cleanup step provided a dramatic visible improvement in the color of the extract with a dark green extract being introduced into the GCB cartridge and a clear solution exiting the cartridge. However, no significant difference was noted in the chromatograms of sediment extracts with and without GCB SPE cleanup. Therefore, another SPE cartridge (PSA) was connected after the GCB cartridge for removing the matrix interference. The PSA is a silica gel coated with a polymerically bonded ethylenediamine-N-propyl phase that contains both primary and secondary amines. With both normal phase and anion-exchange retention mechanisms, PSA showed a strong affinity for polar compounds, such as fatty acids and phenols, which are detectable by the GC detector. Schenck et al. (*22*) recommend PSA and aminopropyl cartridges for cleaning fruit and vegetable extracts for pesticide residue analysis after comparing the cleanup efficiency of GCB, C18, strong anion exchange, aminopropyl and PSA cartridges. Even though PSA has been successfully used for trace pesticide analysis for agricultural products (*22-24*), its application with sediment extracts has not yet been reported. Our preliminary test used 25 mg of PSA absorbent which was shaken with 1 ml of sediment extract showed that matrix interference, especially in the front portion of the chromatogram, was dramatically reduced, while acceptable recoveries (86.9 to 115.5%) of the target pyrethroids spiked at a concentration of 5 μg/kg dw were achieved with relative standard deviations (RSD) of 3.7-15.0%.

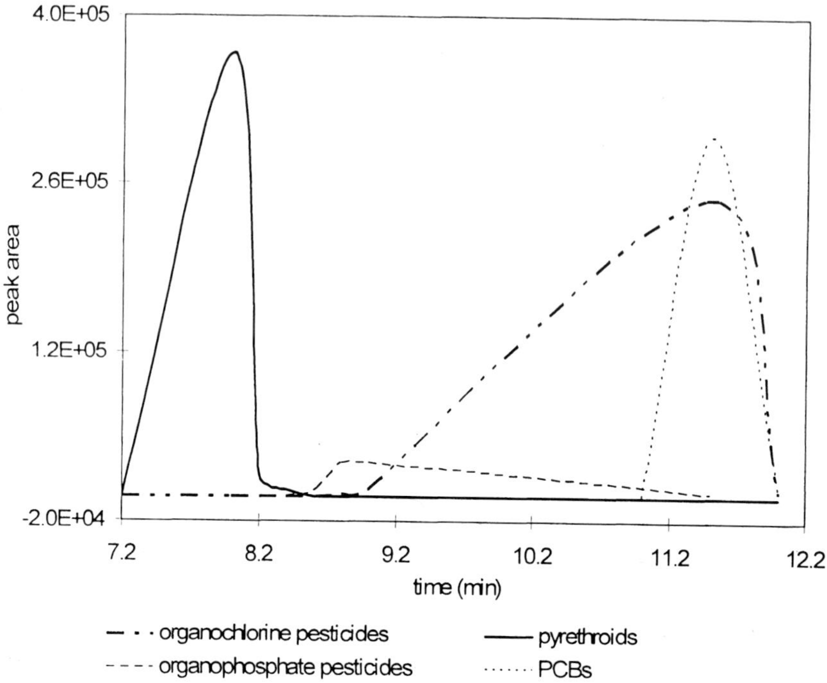

Figure 3. Elution profiles for pyrethroid, organochlorine and organophosphate insecticides, and PCBs determined using high performance gel permeation chromatography

Elution solvents with different elution strength were studied in order to achieve the highest recovery of target pyrethroids with the lowest matrix interference. Mixtures of differing percentages of toluene (1, 5, and 10%) and methylene chloride (5, 10, 20, 30, 50, 80, and 100%) in hexane were tested. Deltamethrin could not be eluted from the cartridge with 1 and 5% toluene in hexane. Six ml of a 10% toluene in hexane solution quantitatively recovered all of the pyrethroids, however, co-eluted pigments (light green extract) were obviously higher than that eluted with the methylene chloride-hexane mixture. This may be explained in part by the higher solubility of planar chemicals in toluene. Therefore, a methylene chloride-hexane mixture was used and the effect of the percentage of methylene chloride in the hexane on pyrethroid recovery was studied (Figure 4). Deltamethrin was the last pyrethroid eluted and it could not be quantitatively recovered until the methylene chloride was increased to 30%, therefore, 3 ml of a 30% methylene chloride in hexane solution was selected and recovery of pyrethroids from the tandem cartridges ranged from 94.5 to 102.6% (Figure 4).

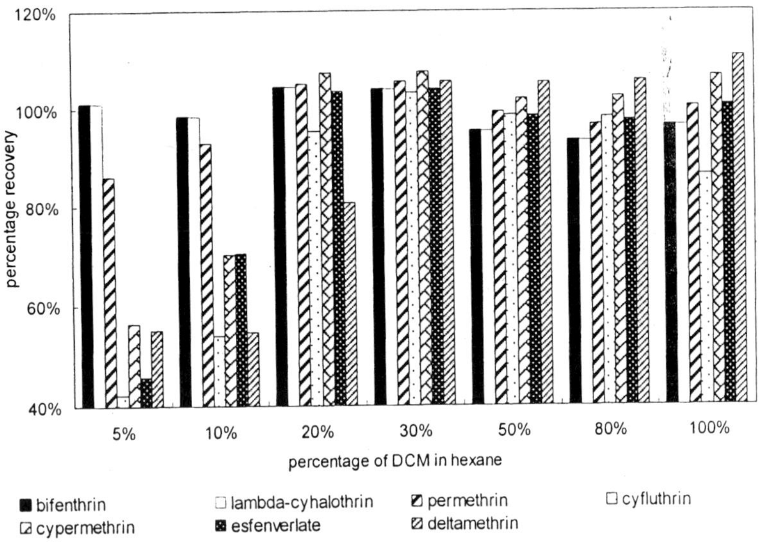

Figure 4. The influence of elution solvents on pyrethroid recoveries from tandem solid phase extraction with graphite carbon black (GCB) and primary/secondary amine (PSA) cartridges.

The effectiveness of the three methods, Florisil adsorbent column, HPGPC and tandem SPE, were evaluated by comparing the amount of matrix interference in the final solution and the cleanness of the GC chromatograms for various sediments (Figure 5). The matrix interference in sediment extracts without cleanup from 10 g (ww) of KS, CA, or MN sediments were in the range of 1250-2500 μg, while the Florisil, HPGPC, and tandem SPE cleanup removed 25.7-47.7%, 35.6-84.7% or 28.4-70.0% of matrix components from the sediment extracts, respectively. Extracts from a field-contaminated sediment collected from California were cleaned with the three methods, and the chromatograms are presented in Figure 5. Though removal of matrix interference was similar among the three methods, the chromatograms showed that the Florisil column removed the least interference, while the other two techniques were much better and showed good promise at reducing GC detectable interference (Figure 5).

Fractionating by Florisil adsorption column

In order to achieve lower MDLs, it is not only desirable to remove the larger matrix components, but also common sediment-associated contaminants which are detected by ECD, such as OCs and PCBs. Therefore, additional fractionating was used to separate pyrethroids from other sediment-associated contaminants.

The cleaned extract obtained from the tandem SPE was then passed through a column packed with 3 g Florisil deactivated with 6% of distilled water. Twenty ml of hexane was used as the first elution solvent and this fraction contained PCBs and some OCs, including *alpha*- and *gamma*-BHC, heptachlor, aldrin, chlordanes and the DDT group. Sulfur residues, which were not fully removed by the copper powder, were also present in this fraction. Pyrethroids were eluted from the column next using 15 ml of 10% ethyl ether in hexane (v/v). Chlorpyrifos and several OCs including *beta*- and *delta*-BHC, heptachlor epoxide, endosulfan I, dieldrin and endrin were co-eluted with pyrethroids in this fraction, however they did not interfere with detection of the pyrethroids due to their different retention times. The Florisil column removed additional matrix components by retaining polar components that were not removed during the tandem SPE cleanup step. The cleanup efficiency of the matrix components from sediment extracts increased from 28.4-70.0% for the tandem SPE method alone to 87.5-97.2% when tandem SPE was coupled with Florisil fractionating. Due to the low matrix interference in the final solution, the final volume of the pyrethroid-containing extract was concentrated 10 times to 100 μl without violating the requirement of maximum injection amount of matrix components of 6 μg. The greater GC response found for the more concentrated extracts resulted in lower MDLs. Figure 6 shows the chromatograms for the KS sediment extracts spiked with 0.5 μg/kg of each pyrethroid, OC and chloropyrifos with tandem SPE cleanup (Figure 6A), and then after Florisil fractionation (Figure 6B). The fractioning step extensively increased the signal / noise ratio for the target pyrethroids. Therefore, tandem SPE with PSA and GCB was chosen as the best cleanup procedure, while additional Florisil fractioning could be used if lower MDLs are desired. A flowchart for the cleanup and fractionating procedures is illustrated in Figure 7.

GC-ECD quantification

As shown in Figure 1, multiple chiral centers exist in pyrethroid molecules and multiple peaks observed due to the separation of the diasteroisomers, so individual pyrethroids are quantified as the sum of the observed isomers. Isomerization of pyrethroids in the GC inlet has been reported (*25,26*). Our previous study (*12*) showed that co-extracted sediment matrix components could mask active sites in the GC inlet, which resulted in an inhibition of the isomerization process along with an enhancement of chromatographic response for pyrethroids. Acetic acid (0.1%, v/v) was used as an analyte protectant to remove the difference in responses noted for pyrethroids in matrix-free standard solutions and matrix-containing sediment extracts.

Method Validation

The MDL is an important parameter to validate an analytical method, and it is defined as the minimum concentration of a substance that can be measured and reported with 99% confidence that the analyte concentration is greater than zero (*27*). The protocol to determine the MDL requires a minimum of seven replicates of a given spiking concentration in a range of one to five times that of the projected lowest concentration that the detector in the analytical method can measure (*28*). The MDL is then calculated as follows: MDL = s $t_{(0.99,\ n-1)}$ where s is a standard deviation of the seven replicate measurements and $t_{(0.99,\ n-1)} = 3.14$ is a *t*-distribution value taken at a confidence level of 0.99 and degrees of freedom of six. The MDL was estimated for the eight pyrethroids extracted from sediment following tandem SPE cleanup with or without Florisil fractionation. Sediment type may affect MDLs due to varying matrix components; so four sediments (AR, KS, CA, and MN) with organic carbon content ranging from 1.1 to 7.9% were spiked with 1 μg/kg dw of each pyrethroid, and seven replicates were analyzed with the procedure without fractionation (Table III). Different MDLs were observed for the four sediments and cleaner chromatograms (better baseline and less interference) were inconsistent with lower MDLs which ranged from 0.07 to 0.62 and 0.07 to 0.35 μg/kg dw for KS and CA sediments, respectively. The MDLs for the pyrethroids from the AR sediment (0.23-0.50 μg/kg dw) were slightly higher than the other sediments due to the existence of more interference on the chromatograms. The effectiveness of Florisil fractionation on MDLs were evaluated by spiking the AR sediment with 0.2 μg/kg dw of each pyrethroid, and analyzing the sediment in seven replicates with the procedure including Florisil fractionating. The MDLs of 0.11 to 0.85 μg/kg dw showed the potential of additional fractionation to measure pyrethroid at lower concentrations in sediment. The KS sediment spiked with 0.2 μg/kg dw of each pyrethroid and a blank KS sediment sample were analyzed simultaneously with tandem SPE cleanup followed by Florisil fractionation and the chromatogram of the spiked pyrethroids is shown in Figure 8 after subtracting the sediment blank.

Recovery and precision were evaluated for pyrethroids spiked at different concentrations in the four sediments (*27*). As shown in Table IV, with a concentration as low as 0.5 μg/kg dw, pyrethroids with the exception of permethrin, were readily detected in the four sediments with recoveries of 81.4-136.4, 68.6-118.1, 74.7-133.0 and 82.4-112.5% for the KS, AR, CA and MN sediments, respectively. Precision was good with RSD of 5.7-9.1, 5.8-10.3 and 8.1-14.9% for the KS, CA and MN sediments, respectively. Permethrin was the exception and its recovery (121.5-278.9%) was unexpectedly high at this concentration. As discussed previously, the AR sediment contained more co-eluting interference; therefore the RSD for that sediment were higher (20.9-

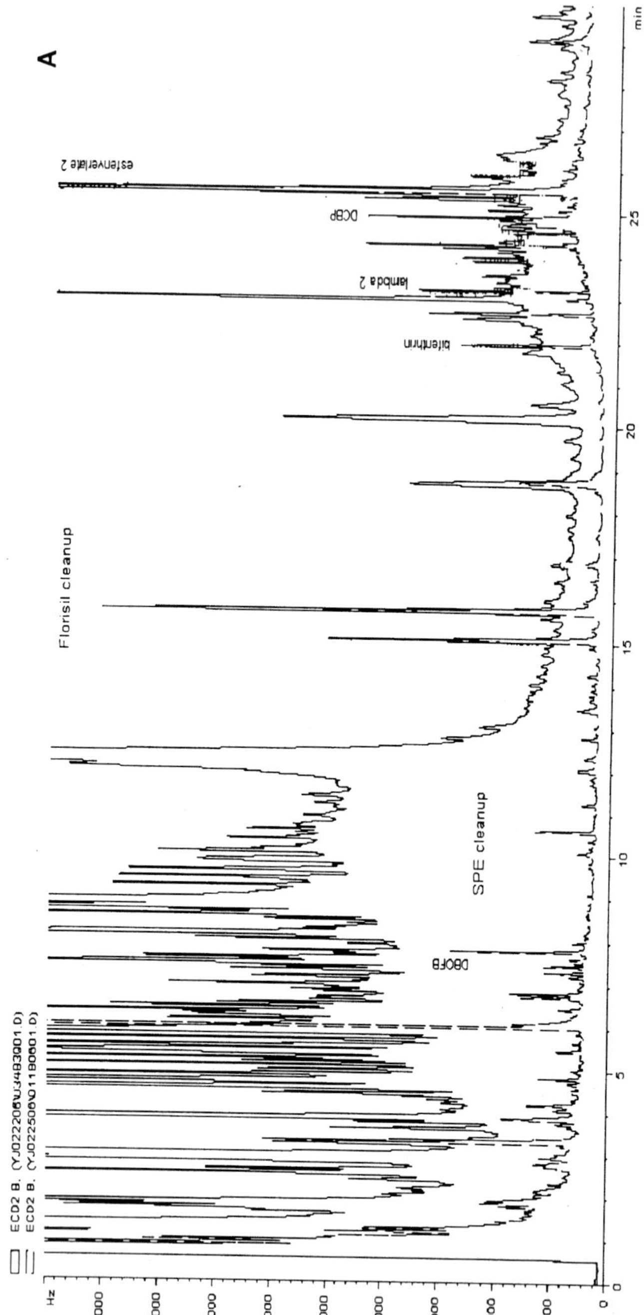
A
Florisil cleanup
SPE cleanup
esfenverlate 2
DCBP
lambda 2
bifenthrin
DBOFB
Hz
min

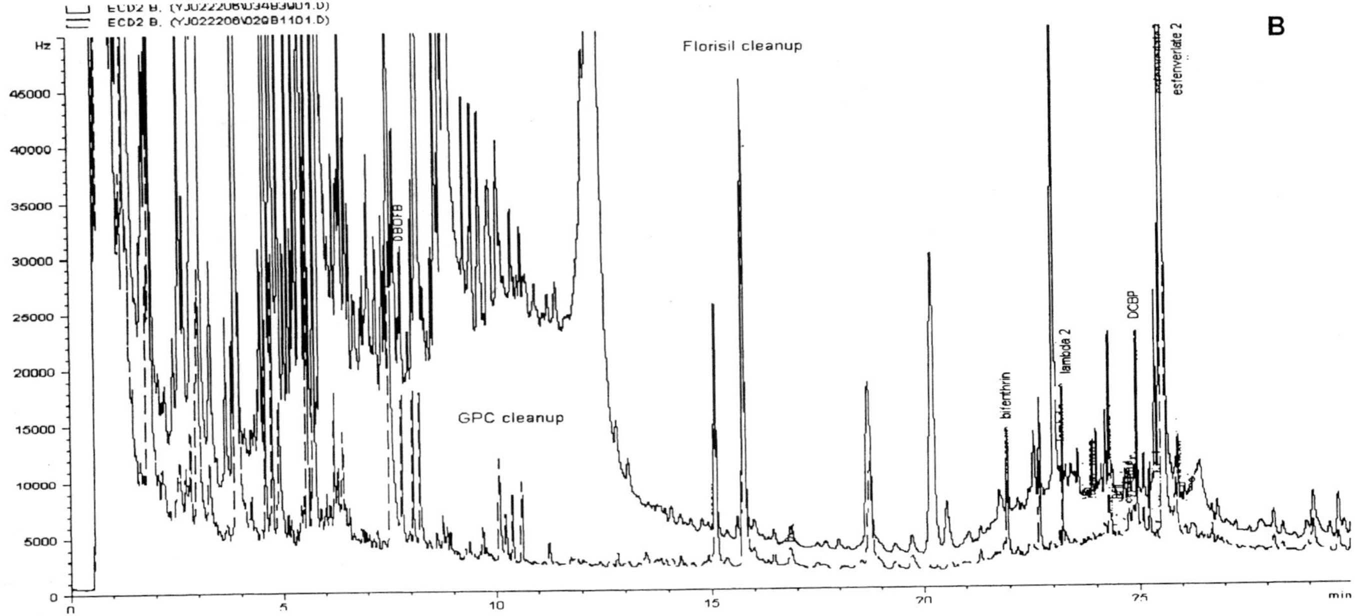

Figure 5. Chromatograms of an extract for a field-collected sediment treated with different cleanup methods: A. Comparison of Florisil adsorption cleanup and tandem solid phase extraction (SPE) cleanup B. Comparison of Florisil adsorption cleanup and high performance gel permeation chromatography (HPGPC) cleanup

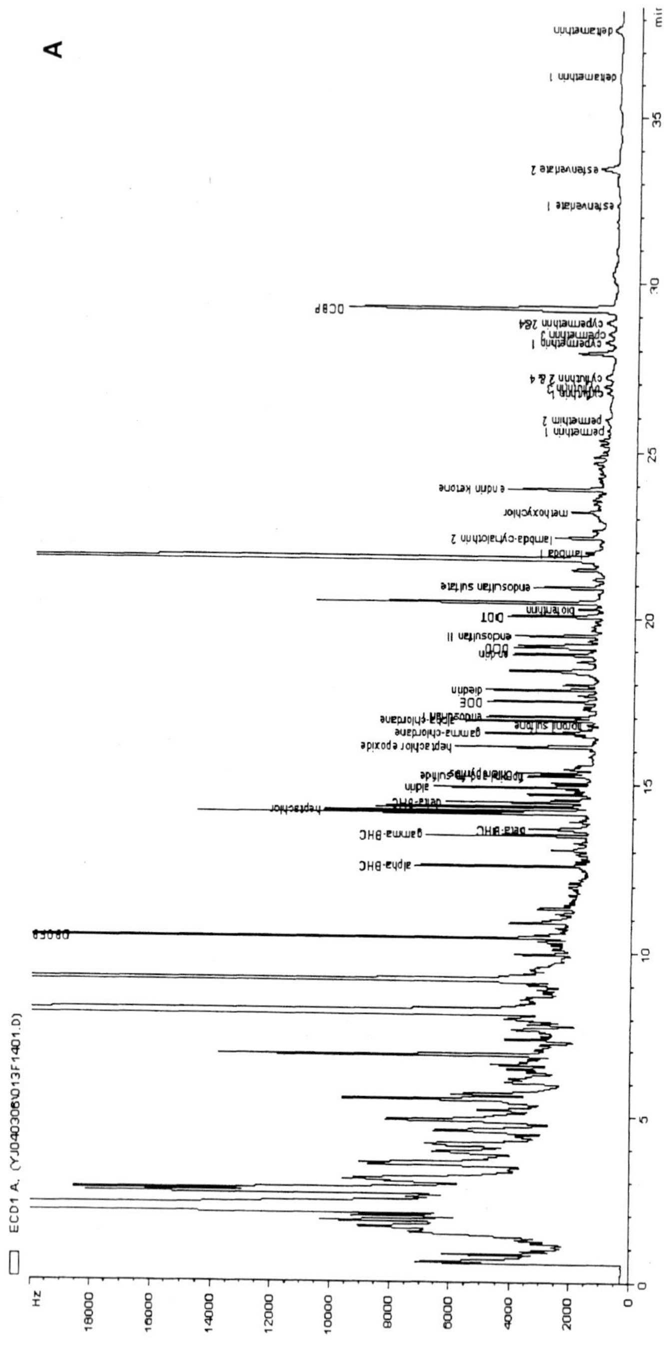
A
ECD1 A, (YJ040306\013F1401.D)
Hz
18000
16000
14000
12000
10000
8000
6000
4000
2000
0
0
5
10
15
20
25
30
35
min
DBOFB
alpha-BHC
gamma-BHC
beta-BHC
heptachlor
delta-BHC
aldrin
heptachlor epoxide
gamma-chlordane
alpha-chlordane
endosulfan I
DDE
dieldrin
endrin
DDD
endosulfan II
DDT
bifenthrin
endosulfan sulfate
lambda-cyhalothrin 2
methoxychlor
endrin ketone
permethrin 1
permethrin 2
cypermethrin 1
cypermethrin 2&4
DCBP
esfenvelrate 1
esfenvelrate 2
deltamethrin 1
deltamethrin

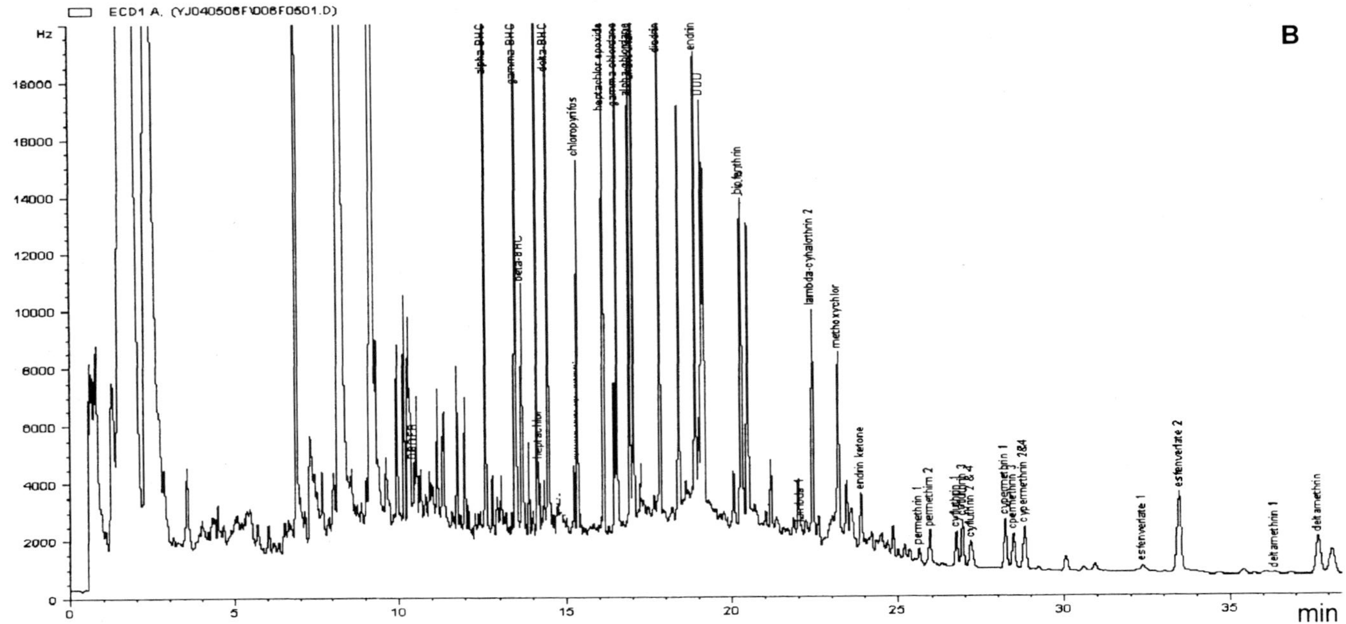

Figure 6. Chromatograms of laboratory-spiked sediment collected from Wichita, KS at a concentration of 0.5 µg/kg dry weight. (A) Extract was cleaned with tandem solid phase extraction alone. (B) Extract was cleaned with tandem solid phase extraction with Florisil fractioning.

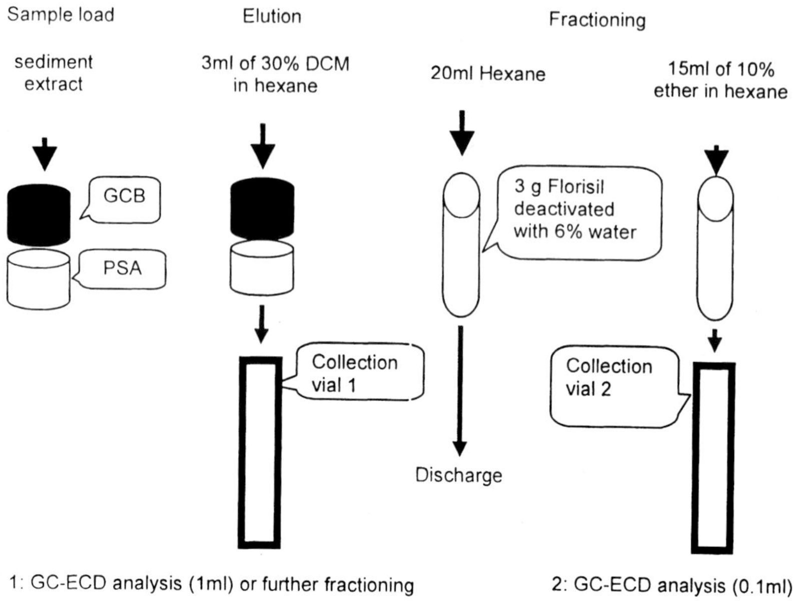

Figure 7. Flowchart for the proposed cleanup method.

Table III. Method detection limit (MDL, µg/kg dry weight) for the target pyrethroids in four sediments collected from American River, CA (AR), Wichita, KS (KS), Pacheco Creek near Hollister, CA (CA) and Bear-Skin Lake, MN (MN) processed by two cleanup methods.

Pyrethroid	MDL for Method 1[a]					Method 2[b]
	KS	AR	CA	MN	Method[c]	AR
Bifenthrin	0.18	0.26	0.07	0.22	0.26	0.13
λ-Cyhalothrin	0.15	0.30	0.11	0.15	0.30	0.11
Cyfluthrin	0.07	0.23	0.10	0.14	0.23	0.17
Cypermethrin	0.11	0.35	0.16	0.19	0.35	0.29
Deltamethrin	0.10	0.50	0.08	0.19	0.50	0.13
Esfenvalerate	0.10	0.43	0.18	0.14	0.43	0.15
Permethrin	0.62	0.23	0.35	0.42	0.62	0.85
Fenpropathrin [d]	NA	0.45	NA	NA	0.45	0.16

[a] Sediment extracts were cleaned with tandem solid phase extraction and spiking concentration of pyrethroids in sediment was 1 µg/kg dry weight.

[b] Sediment extracts were cleaned with tandem solid phase extraction followed by Florisil fractionation and spiking concentration of pyrethroids in sediment was 0.2 µg/kg dry weight.

[c] Method MDL was the highest of all four sediments.

[d] only spiked in AR sediment

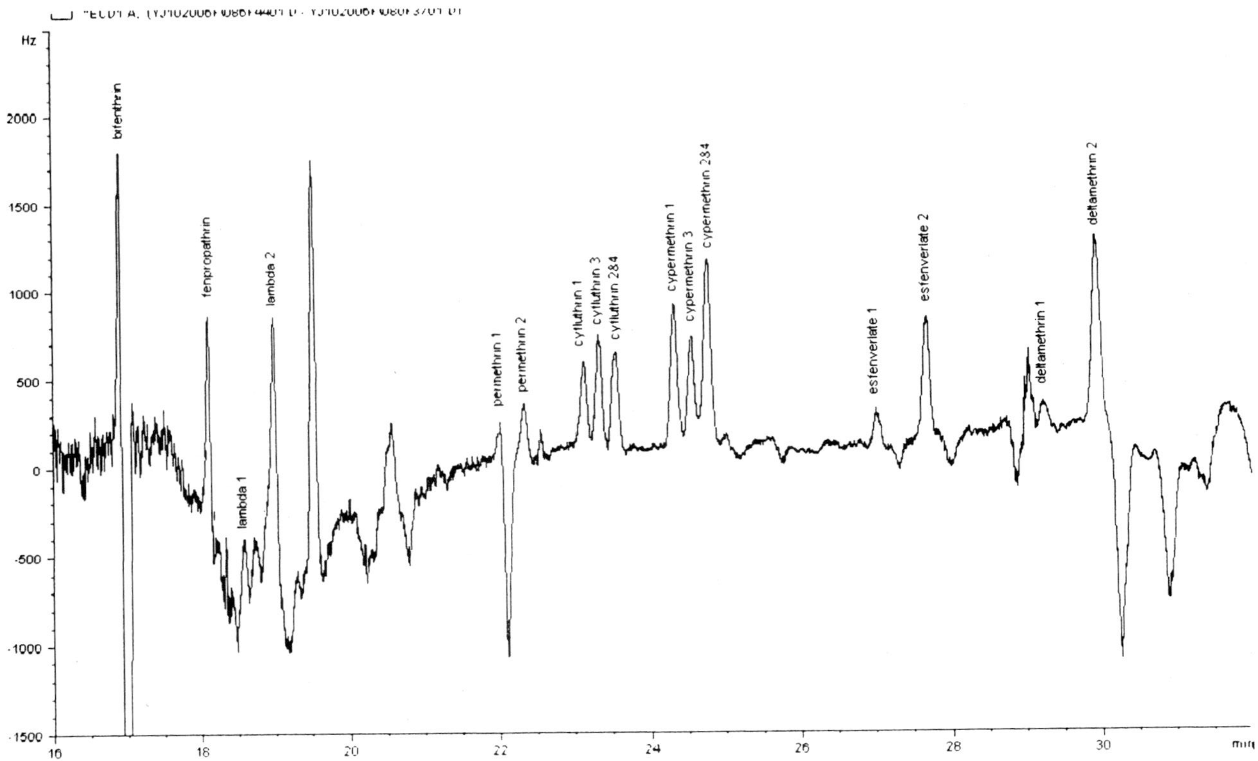

Figure 8. Chromatogram of laboratory-spiked sediment collected from Wichita, KS at a concentration of 0.2 μg/kg dry weight.

43.6%) than for the other sediments. The overall high recoveries and RSD for permethrin at the low concentration may be attributed to its relatively low response on the GC-ECD which resulted in a higher incidence of quantification errors caused by matrix-induced chromatographic response enhancement and possible coeluted interference. This might also explained why the MDL for permethrin increased when the additional fractionation step was added (Table III). Different concentrations of pyrethroids were also spiked into KS and AR sediments to evaluate recovery and precision of the proposed method (Table V). Results showed that extraction efficiencies were comparable for all spiking levels across the sediments; however, precision was better at the higher concentration. As shown in Table V, when the permethrin concentration exceeded 1 μg/kg dw, acceptable recoveries (100.5-119.4% for 1-10 μg/kg dw) were achieved and this concentration is much lower than the expected method detection limit of 9 μg/kg dw for permethrin (Table I).

Table IV. Percentage recovery (PR) and relative standard deviation (RSD) for the target pyrethroids spiked at 0.5 μg/kg dry weight in four sediments collected from American River, CA (AR), Wichita, KS (KS), Pacheco Creek near Hollister, CA (CA) and Bear-Skin Lake, MN (MN).

Pyrethroid	PR (RSD) (%) (n=3)			
	KS	AR	CA	MN
Bifenthrin	136.4 (8.2)	118.1 (28.2)	74.7 (5.8)	95.8 (14.6)
λ-Cyhalothrin	104.1 (9.1)	90.0 (43.6)	102.1 (6.8)	93.1 (10.2)
Cyfluthrin	81.7 (5.7)	75.4 (32.1)	98.7 (6.3)	88.9 (10.3)
Cypermethrin	108.5 (6.7)	68.6 (24.0)	133.0 (7.6)	105.4 (11.3)
Deltamethrin	81.4 (7.8)	77.1 (24.7)	85.9 (6.0)	82.4 (14.9)
Esfenvalerate	102.7 (7.5)	84.4 (20.9)	108.4 (10.3)	112.5 (8.1)
Permethrin	163.1 (24.1)	121.5 (46.8)	158.7 (14.2)	278.9 (9.6)

Conclusion

A sensitive and rapid analytical method was developed for analyzing pyrethroid insecticides at sub-ppb levels in sediment. If needed, OC and OP insecticides could also be detected simultaneously using this sediment extraction method without the Florisil fractionation step. The combination of matrix-dispersive ASE and SPE techniques allowed for sensitive analysis of pyrethroid insecticides, while saving analytical time and reducing solvent usage. With MDLs of 0.23-0.45 and 0.11-0.29 μg/kg dw for methods with and without additional fractionation, the proposed method fulfilled the requirement of expected MDLs of 0.1-0.7 μg/kg dw for highly toxic pyrethroids. The attainment of these detection limits will allow quantification of pyrethroids at about the concentration they would begin to elicit toxicity to *H. azteca*, allowing for worst case assumptions of low temperature and the simultaneous presence of several pyrethroids. Yet even these detection limits provide little margin of safety (i.e., ability to quantify the compounds at less than toxic levels), presenting a continued challenge to develop even more sensitive analytical methods capable of quantifying pyrethroids at toxicologically relevant concentrations.

Table V. Percentage recovery (PR) and relative standard deviation (RSD) for the target pyrethroids in sediments collected from Wichita, KS (KS) and American River, CA (AR) spiked at different concentrations (C_s, µg/kg dry weight).

Pyrethroid	Recovery (RSD) (%) (n=3)					
	KS sediment, C_s (µg/kg)			AR sediment, C_s (µg/kg)		
	0.2	1	5	1	5	10
Bifenthrin	103.2 (28.7)	88.8 (8.0)	91.7 (16.6)	97.9 (8.4)	106.6 (1.4)	73.9 (0.1)
λ-Cyhalothrin	75.6 (35.4)	86.8 (11.2)	72.1 (7.4)	81.4 (9.5)	109.5 (8.7)	99.6 (6.9)
Cyfluthrin	93.5 (28.5)	108.8 (0.4)	101.4 (13.2)	89.4 (7.2)	100.2 (10.0)	117.1 (2.1)
Cypermethrin	132.6 (34.5)	84.8 (4.1)	114.6 (11.7)	102.7 (11.1)	96.2 (7.8)	121.6 (0.1)
Deltamethrin	151.7 (13.7)	143.7 (5.8)	86.0 (3.6)	83.7 (13.8)	94.8 (4.1)	82.3 (8.0)
Esfenvalerate	74.4 (31.7)	101.5 (4.1)	91.4 (6.3)	78.1 (16.1)	90.6 (9.3)	106.2 (1.2)
Permethrin	146.9 (92.5)	115.7 (4.9)	119.4 (24.9)	100.5 (7.3)	109.5 (8.7)	111.1 (6.6)
Fenpropathrin	73.7 (35.4)	115.2 (11.6)	85.9 (7.0)	90.8 (14.2)	NA [a]	91.2 (7.7)

[a] Not spiked into the sediments

Acknowledgments

This research was funded in part by the Agricultural Water Quality Grant Program of the California State Water Resources Control Board.

References

1. Kuivila, K. M.; Foe, C. G. *Environ. Toxicol. Chem.* **1995**, *14*, 1141-1150.
2. Werner, I.; Deanovic, L. A.; Connor, V.; Vlaming, V. de; Bailey, H. C.; Hinton, D. E. *Environ. Toxicol. Chem.* **2000**, *19*, 215-227.
3. Weston, D. P.; You, J.; Lydy, M. J. *Environ. Sci. Technol.* **2004**, *38*, 2752-2759.
4. Weston, D. P.; Holmes, R.W.; You, J.; Lydy, M. J. *Environ. Sci. Technol.* **2005**, *39*, 9778-9784
5. Amweg, E. L.; Weston, D. P.; You, J.; Lydy, M. J. *Environ. Sci. Technol.* **2006**, *40*, 1700-1706.
6. Castro, J.; Sanchez-Brunete, C.; Tadeo, J. L. *J. Chromatogr.* A, **2001**, *918*, 371-380.
7. Esteve-Turrillas, F. A.; Aman, C. S.; Pastor, A.; Gurdia, M. de la. *Anal. Chim. Acta.* **2004**, *522*, 73-78
8. You, J.; Weston, D. P.;Lydy, M. J. *Arch. Environ. Contam. Toxicol.* **2004**, *47*, 141-147.
9. You, J.; Lydy, M. J. *Arch. Environ. Contam. Toxicol.* **2004**, *47*, 148-153.
10. Woudneh, M. B.; Oros, D. R. *J. Chromatogr.* A, **2006**, *1135*, 71-77
11. Yasin, M.; Baugh, P. J.; Bonwick, G. A.; Davies, D. H.; Hancock, P.; Leinoudi, M. *J. Chromatogr.* A, **1996**, *754*, 235-243
12. You, J.; Lydy, M. J. *J. Chromatogr.* A, **2007**, in review
13. Amweg, E. L.; Weston, D. P.; Ureda, N. M. *Environ. Toxicol. Chem.* **2005**, *24*, 966-972. Erratum *24*, 1300-1301.
14. Maund, S. J.; Hamer, M. J.; Lane, M. C. G.; Farrelly, E.; Rapley, J. H.; Goggin, U. M.; Gentle, W. E. *Environ. Toxicol. Chem.* **2002**, *21*, 9-15.
15. Wadleigh, R. W.; Koehler, P. G.; Preisler, H. K.; Patterson, R. S.; Robertson, J. L. *J. Econ. Entomol.* **1991**, *84*, 1433-1436.
16. van Handel, E. *J. Am. Mosq. Control. Assoc.* **1985**, *1*, 302-304.
17. Fisher, J. A.; Scarlett, M. J.; Stott, A. D. *Environ. Sci. Technol.* **1997**, *31*, 1120-1127.
18. Gan, J.; Papiernik, S. K.; Koskinen, W. C.; Yates, S. R. *Environ. Sci. Technol.* **1999**, *33*, 3249-3253.
19. Zuloaga, O.; Etxebarria, N.; Fernández, L. A.; Madariage, J. M. *Fresenius' J. Anal. Chem.* **2000**, *367*, 733-737.
20. US Environmental Protection Agency. Pressurized fluid extraction (PFE), EPA method 3545. US Environmental Protection Agency, Washington, DC, **1996**.
21. Lehotay, S. J.; Lee, C. H. *J. Chromatogr. A,* **1997**, *785*, 313-327.
22. Schenck, F. J.; Lehotay, S. J.; Vega, V. *J. Sep Sci.* **2002**, *25*, 883-890.
23. Okihashi, M.; Kitagawa, Y.; Akutsu, K.; Obana, H.; Tanaka, Y. *J. Pestic. Sci.* **2005**, *30*, 368-377.

24. Sharif, Z.; Man, Y. B. C.; Hamid, N. S. A.; Keat, C. C. *J. Chromatogr. A,* **2006**, *1127*, 254-261.
25. Liu, W.; Qin, S.; Gan, J. *J. Agric. Food Chem.* **2005**, *53*, 3814-3820.
26. Maštovská, K.; Lehotay, S. J. *J. Chromatogr. A,* **2004**, *1040*, 259-272.
27. Collins, G. Environmental monitoring and assessment program: chemical method evaluation guidance, EPA 620/R-96/001. Environmental Monitoring and Assessment Program: US Environmental Protection Agency, Cincinnati, OH, **1996**.
28. Code of Federal Regulations (CFR) 40 Part 136, Appendix B,. U.S. Government, Washington D.C, **1993**; pp 554-555.

Chapter 6

Analysis of Pyrethroid Pesticides in Sediment and Waters by EPA Method 8270 Gas Chromatography/Mass Spectrometer (GC/MS) Narrow-Range Scan Selected Ion Monitoring

Richard L. Heines and Peter W. Halpin

Caltest Analytical Laboratory, 1885 North Kelly Road, Napa, CA 94558

An existing commonly used United States Environmental Protection Agency (EPA) Gas Chromatography Mass Spectrometry (GC/MS) method with quadrupole mass spectrometer and conventional electron ionization was modified by adding 15 pyrethroid and 6 pyrethrins analytes plus piperonyl butoxide (PBO) and using the option stated in the method of selected ion monitoring. Per method recommendations three ions are used per analyte for confidence in analyte confirmation. This approach was adapted from the United States Geological Survey (USGS) and the California Department of Food and Agriculture's (CDFA) GCMS narrow range scan techniques and yields reporting limits of 330 to 660 ppt in sediment and reporting limits of 5–10 ppt in waters.

Environmental Need for Pyrethroid Pesticide Analysis

Pyrethroid pesticides appear to be a significant source of sediment toxicity in urban and agriculturally dominated streams [*1*]. The toxicity appears to be widespread in the Sacramento and San Francisco California Bay areas [*2*], and occurs at low part per billion concentrations in sediment.

Recent sediment toxicity studies indicate thresholds of acute toxicity to arthropods at low µg/Kg levels (as low as 2-10 µg/Kg for bifenthrin, lambda-cyhalothrin and deltamethrin). [*3*] Environmentally relevant reporting limits for sediments should start at 1 µg/Kg dry weight [*4*], and lower if possible with a target of one tenth the LC-50 (The concentration that causes mortality in 50% of the test organisms). Reporting limits for water should at least meet the State of California's Central Valley Regional Water Quality Control Board's (CVRWQCB) Irrigated Lands Program requirement of 0.05 µg/L, and would ideally be in the low ng/L range to meet the lowest listed water quality goals. Table VI lists the 2003 CVRWQCB water quality goals for the pyrethroids, including the August 2007 update. [*5*] This application of the EPA GCMS method includes all the pyrethroids of current regulatory and use-based research interest [*5, 6, 7, 8, and 9*] in California monitoring programs. The EPA method 8270 [*10*] as described meets all monitoring criteria for sediments, and also meets the concentration related monitoring goals [*5*] for 9 of 10 pyrethroids in fresh water. The exception in fresh water is for cypermethrin, which has a State of California CVRWQCB water quality goal of 0.002 µg/L. This method application as described has a reporting limit of 0.005 µg/L for cypermethrin.

The analytes added to the method include the following synthetic pyrethroids: allethrin, bifenthrin, cyfluthrin, cyhalothrin, cypermethrin, esfenvalerate/fenvalerate, fluvalinate, fenpropathrin, permethrin, phenothrin, resmethrin, tralomethrin/deltamethrin and tetramethrin. Those listed together (separated only by a '/') are unable to be separated by GC/MS analysis. Tralomethrin converts in the injection port to Deltamethrin [*11, 12*] and therefore these two analytes are reported together. Fenvalerate and Esfenvalerate are also reported together [*12, 13*]. More recently we added the plant-derived pyrethrins (pyrethrin I, cinerin I and jasmolin I; and pyrethrin II, cinerin II, and jasmolin II) which are the six insecticidal components of the pyrethrum technical mix. Piperonyl butoxide (PBO) was also added due to it frequently being used with the pyrethrins. Tables IV and V list calibration levels and Method Detection Limits. Spike recoveries and laboratory derived control limits are listed for the synthetic pyrethroids. The existing data on the pyrethrins and PBO spike recoveries is insufficient to calculate standard control limits (where n = 20) for this application. Average recoveries and control limits from a limited data set are provided for the aqueous samples.

Due to the extremely high K_{OC} values of these compounds they tend to be associated with particles [*14, 15*]. Many argue that the primary data collection efforts for pyrethroids should be in the sediment matrix due to the high affinity of these compounds for solids, and the observed acute toxicity in sediments. Concern for the possibility of short duration pulses of pyrethroids in water and dissolved organic matter has led to demand for analysis of run-off, and stormwater as well, [*16*] and is addressed in this method.

Preference of EPA Source Method

In California some of the demand for analysis of pyrethroids is in a regulatory context, and an EPA source method is preferred if feasible. EPA environmental methods are widely used, well understood, and commonly available. When the chemical characteristics of environmental analytes of interest are suitable for addition to an existing EPA method it is generally beneficial to adapt the existing method [*14*]. Minimal modifications should be made, and should be proven to not reduce the quality of data obtained. In this application of EPA method 8270, all of the method QC monitoring parameters are kept, and run with the analytes of interest. A copy of the full EPA method 8270 is available on-line at http://www.epa.gov/sw-846/8_series.htm. The primary difference between a standard application of method 8270 and this application is the addition of the pyrethroid, pyrethrins and PBO analytes; the volume of sample injected; and the mode of scanning employed by the mass spectrometer.

Narrow Range Scan Selected Ion Monitoring

In this application of method 8270 the mass spectrometer scanning function is used in a narrow-range scan instead of full scan. In full scan mode, the mass spectrometer is constantly scanning the range of 35 to 500 amu. In the selected ion mode the mass spectrometer is only scanning for the predetermined three most abundant ions per analyte [*13, 18, 19*] per timed event. Scanning for fewer ions provides much greater equivalent sensitivity compared to full range scan data. This approach is consistently applied to all target and method quality control analytes for all samples for this application. Analyte identification is based on chromatographic relative retention time, accompanied by measurement of the most abundant ion, in the proper ratio to the second most abundant ion, and with the presence of a third characteristic ion. 'Narrow Range Scan-Selected Ion Monitoring' is not the commonly misconstrued 'single ion monitoring'. In the narrow range scan, the software is set to run the method as a series of timed events. In this application there are 13 timed events corresponding to the retention times of a few analytes at a time. In each timed event, the mass spectrometer is scanning for three ions per analyte. There are 20 target analytes, but 42 compounds total due to the extensive QC parameters required by the EPA 8270 method. In each timed event the mass spectrometer is scanning for 8 to 20 ions, with one timed event including 32 ions. As the run time progresses, the scan is looking for a rotating list of ions, dropping the ones pertaining to the earlier retention time analytes, and adding the ions for the next retention time analytes, so at any given point in the method the mass spectrometer is scanning for the least number of relevant ions. In this way sensitivity is maximized, yet each analyte is still confirmed by three characteristic ions.

Since this method was first employed, we have made a separate analytical run for the pyrethrins and PBO. The method approach using 8270 is the same as for the synthetic pyrethroids, but simplifies the number of ions required per timed event, and eliminates the possibility of interference from the pyrethrin standards themselves, as the pyrethrins are sold as a technical mix, including natural waxes, pigments and other non-target matrix.

Methods and Materials

Standards: Initially, individual pyrethroid standards were purchased followed by custom pyrethroid standards mixes from Accustandard and Restek. Pyrethrin and PBO standards were purchased from Chem Service and Accustandard. See Table I for part numbers, concentrations and solvent.

Individual standards were run in full scan mode to determine correct retention times and evaluation of ion spectra for characteristic ions with the most abundant masses, and ratios of primary to secondary ions. See Table II for retention times, Table III for the list of ions. Column bleed ions were considered in choosing representative analyte ions that would not be influenced by column bleed at low levels of quantification. Timed events were then programmed into the software to correspond to optimal scan criteria per rolling retention time frame.

EPA method 8270 QC criteria are used to evaluate method performance. Each batch of 20 or fewer samples is accompanied by a Laboratory Control Standard (LCS) spiked in mid-range of calibration; and sample Matrix Spike and Matrix Spike Duplicate (MS/MSD) with precision reported; and a Method Blank. Extraction Surrogates are added to each sample prior to extraction and are reported, but per EPA method criteria, the results are not normalized to the extraction surrogate recoveries. Determination of Method Detection Limit (MDL) is per the criteria set forth in the Code of Federal Regulations (CFR), Title 40 part 136, appendix B [*20*]. Both the lowest acceptable standard concentration and the MDL must be considered in determining the reporting level in an EPA method. The Reporting Limit must be at or above the MDL and no lower than the lowest acceptable calibration standard, conforming to the EPA definition of Minimum Level for reporting. In some cases we have been able to routinely calibrate lower than our calculated MDLs, in which case the reporting limit is set at or above the MDL, not simply at the level of the lowest calibration standard. MDLs are calculated annually, and Report Limits in Tables IV and V reflect a rounding up to be reasonably sure that subsequent MDL studies will still be supportive (less than or equal to) of the designated Reporting Limit.

Instrumentation and Analysis

This method was initially demonstrated on an Agilent model 6890 GC with a model 5973 quadrapole mass spectrometer, using a diffusion vacuum pump. Instrumentation was from 1997 and yielded calibration/reporting limits of 0.005 µg/L to 0.02 µg/L in waters and 0.33 µg/Kg in sediment. The current instrumentation that is referenced in the rest of this work is an Agilent 6890 GC with a model 5975 mass spectrometer using a performance turbo vacuum pump. The instrumentation is from 2005. The newer instrument has faster scan time, more sensitivity and greater vacuum; however either era instrument will achieve the current sediment reporting goals, and similar aqueous reporting limits.

System inertness is believed to be important to optimizing this technique for the low concentrations of interest. The GC injection port weldament is silco-steel coated as is the seal. Double gooseneck silanized liners were used. The split/splitless injection port was run in a splitless pressure pulse mode with the goal of decreasing residence time in the injection port. An autosampler is used to inject 5 µL of the concentrated sample as opposed to the method default 1-2 µL. The method was developed so that isomers of target analytes elute without complete resolution. A ramped flow rate is used to control the elution of the isomers so that they give approximately 80% resolution. The isomers are summed and quantified as 'total'. Some of the target analytes have up to 4 diastereomers. To identify these compounds a midpoint retention time is used, and the entire retention time of the analyte is bracketed +/- 0.5 minutes on each side of the multiple isomers in the method software. An internal standard is added to the sample extract at the instrument. Instrument tune is checked every 12 hours. An experienced mass spectra analyst opens each sample data file and reviews each analyte to assure proper identification and integration.

The injection port is run with a high pressure pulsed program. Injection port temperature is 250°C, with a starting pressure of 15 psi, pulsing to 80 psi for 0.75 minutes. Column is a J&W DB-5MS 30 meter capillary column with a 0.25 mm diameter and a film thickness of 0.25 µm. Helium is used as the carrier gas, with flow at 1.5 mL/min.

The GC oven was set at 85°C and heated 25°/min to 140°C, then 12°C/min to 260°C, then 2°C/min to 280°C and then 18°C/min to 310°C. Run time is 26.87 min. Analytes eluted from the capillary column are introduced into the mass spectrometer. Identification of target analytes is accomplished by comparing their selective mass ion ratios with the selective mass ion ratios of authentic standards. Quantitation is accomplished by comparing the response of a major (quantitation) ion relative to an internal standard using a five-point calibration curve. Second Source calibration check standards are employed. Control limits are established based on laboratory in-house data.

Extraction in Sediment

Sediment is collected into a glass jar with Teflon lined lid. At the laboratory, the entire sample is homogenized with a stainless steel spatula in a baked Kimball porcelain dish. From the thoroughly mixed sample, a 30 g +/- 0.1 g portion is placed in a 40 mL VOA vial and centrifuged at approximately 1000 rpm for 5 minutes. The supernatant water is decanted to waste. The remaining solids are transferred to a 250 mL wide mouth jar. Sodium sulfate is added and mixed with the sample until the sample has no free liquid and has a sandy texture. The sample and sodium sulfate mix is then spiked with 200 μL of surrogate solution. The extraction procedure consists of adding 100 mL of a 1:1 mix of methylene chloride and acetone [*21*] and sonicating in a chilled sonic bath at 3-6°C temp for 20 minutes. The extraction process is carried out three times sequentially. After each extraction, the supernatant solvent is poured off through a pre-washed filter paper (Whatman # 144125) containing roasted sodium sulfate to remove moisture. All extraction solvent is concentrated to a final volume of 2 mL using a Turbovap water bath/nitrogen blow down. Final concentration factor is 30:2 (15x). Separate from the extraction process, an aliquot of the sediment is analyzed for percent moisture in order to convert final results to dry weight.

Extraction in Water by EPA Method 3510 Separatory Funnel Liquid-Liquid Extraction [22]

For water, a one-Liter sample is collected in glass with Teflon-lined lid. The entire sample is poured into a Teflon 2-liter separatory funnel. Many have reported that pyrethroids may be lost to container walls, to varying degrees depending on the presence or absence of solids and organic materials in the sample. To try to recover all analyte possible from the sample container walls 60 mL. of Methylene chloride is added to the empty sample container, the lid screwed on tightly, and the sample bottle shaken with the solvent for approximately 30 seconds (this is standard protocol in EPA pesticide extraction of waters). The sample bottle solvent rinse is added to the separatory funnel. The water sample is extracted three times with 60 mL. of methylene chloride and shaken ten minutes each time on an auto-shaker. The combined solvents from the bottle rinse and all three extractions are then concentrated to 1mL final volume with a Turbovap heated water bath with nitrogen blow down. Final concentration factor is 1,000:1. Pyrethroids have a strong affinity for solids, and rapidly adsorb to sediment particles [*14,15*]. Pyrethroids can be found on the suspended solids, in the dissolved organic matter, and to a limited extent in the dissolved phase [*14*]. Any pre-filtering prior extraction may cause a reduction in total pyrethroids measured due to loss of analyte adsorbed to particles filtered out. Because of this, environmental water analyses for pyrethroids is typically based on 'whole water', that is water with all sediments/suspended solids included. Alternatively total 'whole water' and filtered/centrifuged fractions can be collected. When whole water samples are analyzed by liquid-liquid extraction, results will include pyrethroids bound to the suspended solids and other organic matter as well as any dissolved fraction present.

Hold Times

USGS [*19*] and CDFA [*13*] sample storage-hold time studies in water indicated lower recovery of spiked samples in times as short as 3 days to 13 days depending on the analyte. The shortest recommended hold times from the USGS and CDFA studies were cyhalothrin and permethrin: 3 days; and bifenthrin, cyfluthrin, cypermethrin, and esfenvalerate: 13 days. Short hold times require coordination with the lab to extract the samples in time. Based on the studies referenced, using the EPA method 8270 default hold time for waters (7 days) would be inadequate when cyhalothrin, esfenvalerate or permethrin are target analytes.

State of California CVRWQCB Irrigated Lands criteria is to freeze sediment samples until extraction. Default sediment/soil extraction times in method 8270 are 14 days. Method 8270 hold times are based on storage at 4°C. We freeze sediments until analyses.

Acknowledgment

The authors acknowledge the work of the USGS and the California Department of Food and Agriculture for their application of the GC/MS narrow range scan selected ion monitoring approach and for their assistance in helping us with this specific application. Don Weston of UC Berkeley was helpful in providing spiked sediments for analyses, and valuable commentary based on his pyrethroid work conducted with Michael Lydy and JC You. At Caltest valuable contributions were made by Christine Horn, Estrellita Manuel, and Raquel DeGuzman.

Table I. Pyrethroid & Pyrethrin Standards

Standard	*Source*	*Part Number*	*Concentrat.*	*Solvent*
Custom Pyrethroid Mix	AccuStandard	S12458-R3	1000 ppm	Methanol
Custom Pyrethroid Mix	Restek	559092	50 µg/ml	Methanol
CCC Check Compounds	AccuStandard	CLP-0110	2000 ppm	DCM
SPCC	AccuStandard	CLP-010-10x	2000 ppm	DCM
Base Neutral Surrogates	AccuStandard	CLP-BNS	1000 ppm	DCM
Internal Standard Mix	AccuStandard	Z-014JPAK	4000 ppm	DCM
Pyrethrins	Chem Service	PS097	Neat	
PBO	Chem Service	45626	Neat	
Pyrethrins	Accustandard	P187S	200 ppm	Methanol
PBO	Accustandard	P-3485	100 ppm	Methanol

Accustandard custom pyrethroid mix is comprised of: resmethrin, danitol (fenpropathrin), bifenthrin, l-cyhalothrin, permethrin, cyfluthrin, cypermethrin, fenvalerate, tau-fluvalinate, tralomethrin, deltamethrin, tetramethrin, allethrin, sumithrin, esfenvalerate.

Restek custom pyrethroid mix is comprised of: bifenthrin, cyfluthrin, cypermethrin, deltamethrin, esfenvalerate, fenpropathrin, fenvalerate, l-cyhalothrin, permethrin (cis & trans), resmethrin, tau-fluvalinate, tralomethrin.

Explanation of EPA Method 8270 Quality Control acronyms: CCC is Calibration Check Compounds; SPCC is System Performance Check Compounds

Table II. Pyrethroid and Pyrethrin Retention Times

Analyte	*CAS #*	*Retention Time*	*Diastereomeric Pairs**
(CCC)1,4-Dichlorobenzene	106-46-7	7.5	
(IS)1,4-Dichlorobenzene-d4	3855-82-1	7.5	
(SPCC)N-Nitroso-di-n-propylamine	621-64-7	7.9	
(SS)Nitrobenzene-d5	4165-60-0	8.1	
(IS)Naphthalene-d8	1146-65-2	8.8	
(CCC)Hexachlorobutadiene	87-68-3	8.8	
(SPCC)Hexachlorocyclopentadiene	77-47-4	9.6	
(SS)2-Fluorobiphenyl	321-60-8	9.9	
(IS)Acenaphthene-d10	15067-26-2	10.7	
(CCC)Acenaphthene	83-32-9	10.7	
(SPCC)4-Nitrophenol	100-02-7	10.8	
(SPCC)2,4-Dinitrophenol	51-28-5	11.1	
(CCC)N-Nitrosodiphenylamine	86-30-6	11.5	
(IS)Phenanthrene-d10	1517-22-2	13.1	
Allethrin	584-79-2	15.2	4
(CCC)Fluoranthene	206-44-0	16.4	
Cinerin I	25402-06-6	16.9	
(SS)p-Terphenyl-d14	1718-51-0	17.5	
Jasmolin I	4466-14-2	17.9	
Pyrethrin I	121-21-1	18.0	
Piperonyl Butoxide(PBO)	51-03-6	18.5	
(CCC)1,4-Dichlorobenzene	106-46-7	7.5	
(IS)1,4-Dichlorobenzene-d4	3855-82-1	7.5	
Resmethrin	10453-86-8	19.6	2
Bifenthrin	82657-04-3	20.4	1
Tetramethrin	7696-12-0	20.9	2
Fenpropathrin(Danitol)	39515-41-8	20.9	1
Phenothrin(Sumithrin)	26002-80-2	21.4	2
Cinerin II	121-20-0	21.7	
(IS)Chrysene-d12	1719-03-5	21.9	
L-Cyhalothrin	68085-85-8	22.2	1
Jasmolin II	1172-63-0	22.8	
Pyrethrin II	121-29-9	22.8	
(CCC)Di-n-octylphthalate	117-84-0	23.9	
Permethrin	52645-53-1	24.0	2
Cyfluthrin	68359-37-5	25.1	4

Table II. ***Continued.***

Analyte	*CAS #*	*Retention Time*	*Diastereomeric Pairs**
Cypermethrin	52315-07-8	25.5	4
(IS)Perylene-d12	1520-96-3	27.0	
Fenvalerate/Esfenvalerate (!)	51630-58-1	26.7	2
Fluvalinate	69409-94-5	26.6	2
(CCC)Benzo(a)pyrene	50-32-8	26.8	
Deltamethrin/Tralomethrin (#)	66841-25-6	27.2	2

*achiral column (DB-5MS) cannot separate individual enantiomers, but can separate diastereomers. (Each diastereomeric pair consists of two co-eluting enantiomers) This table column represents the number of peaks generated, which are then quantitated as one peak representing all enantiomers. Results are then expressed as total pesticide residue.

Explanation of EPA Method 8270 Quality Control acronyms: CCC is Calibration Check Compounds; IS is Internal Standard; SPCC is System Performance Check Compounds; SS is Surrogate Standard,

Table III. Pyrethroid and Pyrethrin Characteristic Ions

Analyte	*Quantitation Ion*	*Qualifying Ion*	*Monitoring Ion*
(CCC)1,4-Dichlorobenzene	146	148	111
(IS)1,4-Dichlorobenzene-d4	152	150	115
(SPCC)N-Nitroso-di-n-propylamine	43	70	130
(SS)Nitrobenzene-d5	82	54	128
(IS)Naphthalene-d8	136	68	108
(CCC)Hexachlorobutadiene	225	223	227
(SPCC)Hexachlorocyclopentadiene	237	235	239
(SS)2-Fluorobiphenyl	172	171	170
(IS)Acenaphthene-d10	164	162	160
(CCC)Acenaphthene	153	154	152
(SPCC)4-Nitrophenol	65	139	109
(SPCC)2,4-Dinitrophenol	184	63	154
(CCC)N-Nitrosodiphenylamine	168	167	169
(IS)Phenanthrene-d10	188	187	189
Allethrin	123	79	91
(CCC)Fluoranthene	202	203	200
Cinerin I	123	93	150
(SS)p-Terphenyl-d14	244	243	245
Jasmolin I	123	81	164
Pyrethrin I	123	81	162
Piperonyl Butoxide(PBO)	176	177	149
Resmethrin	123	171	128
Bifenthrin	181	165	166
Tetramethrin	164	123	79
Fenpropathrin(Danitol)	97	181	265
Phenothrin(Sumithrin)	123	183	81
Cinerin II	107	121	167
(IS)Chrysene-d12	240	120	236
L-Cyhalothrin	181	197	208
Jasmolin II	107	135	167
Pyrethrin II	107	91	133
(CCC)Di-n-octylphthalate	149	279	150
Permethrin	183	165	91
Cyfluthrin	206	226	165
Cypermethrin	181	91	165
(IS)Perylene-d12	264	260	265
Fenvalerate/Esfenvalerate	167	181	225

Table III. *Continued.*

Analyte	*Quantitation Ion*	*Qualifying Ion*	*Monitoring Ion*
Fluvalinate	250	181	252
(CCC)Benzo(a)pyrene	252	253	250
Deltamethrin/Tralomethrin	181	253	93
(CCC)1,4-Dichlorobenzene	146	148	111
(IS)1,4-Dichlorobenzene-d4	152	150	115

Table IV. Aqueous: Low Calibration Standard, MDL, Reporting Limit, Average Recovery & Lab Water Spike Recovery Control Limits

Analyte	*Lowest Std. µg/L*	*MDL ng/L*	*RL ng/L*	*Avg. Recov.*	*Control Limits**	*n*
Bifenthrin	0.5	4.2	5.0	80.4	58-121	20
Cyfluthrin	2.5	3.7	5.0	81.6	54-109	20
Cyhalothrin	0.5	4.7	5.0	83.8	58-107	20
Cypermethrin	5.0	3.1	5.0	79.1	52-103	20
Fenvalerate/Esfenvalerate	10.0	6.1	10.0	82.3	61-113	20
Fluvalinate	2.5	3.8	5.0	67.0	45-96	20
Fenpropathrin(Danitol)	2.5	4.1	5.0	89.1	75-109	20
Permethrin	2.5	4.9	5.0	81.1	55-116	20
Resmethrin	2.5	6.3	10.0	74.2	36-126	20
Deltamethrin/Tralomethrin	10.0	6.2	10.0	80.8	28-135	20
Allethrin	2.5	3.5	5.0	89.6	79-100	6
Phenothrin	0.5	3.2	5.0	81.3	74-88	6
Tetramethrin	2.5	3.7	5.0	94.4	83-106	6
Cinerin I	4.4	20.0	50.0	84.1	56-112	4
Jasmolin I	1.5	20.0	50.0	87.0	54-120	4
Pyrethrin I	14.9	60.0	100.0	118.5	33-204	4
Cinerin II	2.8	8.0	50.0	73.8	40-107	4
Jasmolin II	1.1	4.0	50.0	80.0	49-111	4
Pyrethrin II	11.0	110.0	500.0	124.8	10-240	4
Piperonyl Butoxide(PBO)	1.0	0.3	1.0	89.5	54-125	4
2-Fluorobiphenyl(SS)		NA	NA	38.3	12-64	20
p-Terphenyl-d14(SS)		NA	NA	81.3	47-115	20

Reporting Limits are at the lowest calibration standard (or higher) as adjusted by the extraction concentration factor, and must be at or above the Method Detection Limit (MDL) and may be rounded up above the highest of the MDL and low calibration standard.

*Control Limits based on +/- three standard deviations from the mean recovery of spike samples. Twenty samples are usually used to generate laboratory in-house control limits.

Table V. Sediments: Low Calibration Standard, MDL, Reporting Limit, Average Recovery and Spike Recovery Control Limits

Analyte	*Lowest Std. µg/L*	*MDL µg/Kg*	*RL µg/Kg*	*Average Recovery*	*Control Limits**	*n*
Bifenthrin	0.5	0.27	0.33	84.8	68-101	10
Cyfluthrin	2.5	0.18	0.33	103.8	69-139	10
Cyhalothrin	0.5	0.25	0.33	90.8	71-110	10
Cypermethrin	5.0	0.19	0.33	102.6	72-132	10
Fenvalerate/Esfenvalerate	10.0	0.4	0.66	97.4	70-125	10
Fluvalinate	2.5	0.2	0.33	87.9	38-137	10
Fenpropathrin(Danitol)	2.5	0.17	0.33	91.3	71-110	10
Permethrin	2.5	0.22	0.33	93.8	49-138	10
Resmethrin	2.5	0.21	0.33	75.2	62-88	10
Deltamethrin/Tralomethrin	10.0	0.39	0.66	103.6	66-141	10
Allethrin	2.5	0.24	0.33	92.9	53-133	6
Phenothrin	0.5	0.31	0.33	96.2	16-176	6
Tetramethrin	2.5	0.27	0.33	85.2	71-99	6
Cinerin I	4.4	0.7	2.0			
Jasmolin I	1.5	1.1	2.0			
Pyrethrin I	14.9	2.6	5.0			
Cinerin II	2.8	0.4	2.0			
Jasmolin II	1.1	0.5	2.0			
Pyrethrin II	11.0	13.0	20.0			
Piperonyl Butoxide(PBO)	1.0	0.02	0.07			
2-Fluorobiphenyl(SS)		NA	NA	69.7	0-144	20
p-Terphenyl-d14(SS)		NA	NA	92.6	59-126	20

Reporting Limits are at the lowest calibration standard (or higher) as adjusted by the extraction concentration factor, and must be at or above the Method Detection Limit (MDL) and may be rounded up above the highest of the MDL and low calibration standard.

Pyrethrins (cinerin I&II, jasmolin I&II, pyrethrin I&II) and PBO sediment spike data not available

* Control Limits based on +/- three standard deviations from the mean recovery of spike samples. Twenty samples are usually used to generate laboratory in-house control limits.

Table VI. State of California Central Valley Regional Water Quality Control Board Water Quality Goals for Pyrethroids

Pyrethroid Analyte	*CVRWQCB lowest Water Quality Goal µg/L*
Bifenthrin	110
Cyfluthrin	180
Cyhalothrin	35
Cypermethrin	0.002
Esfenvalerate/ Fenvalerate	175
Fluvalinate	70
Fenpropathrin	180
Permethrin	0.03 freshwater 0.001 saltwater
Resmethrin	210
Tralomethrin/ Deltamethrin	53

State of California CVRWQCB Irrigated Lands Program draft monitoring criteria requires a minimum reporting limit of 0.05µg/L for bifenthrin, cyfluthrin, cypermethrin, esfenvalerate, l-cyhalothrin, permethrin, and fenpropathrin. The application of the EPA method 8270 as described meets the Irrigated Lands Program draft monitoring reporting limit criteria and meets all other CVRWQCB listed water quality goals except cypermethrin, where this method's reporting limit is currently 0.005 µg/L, and permethrin in saltwater, where this method's reporting limit is also currently 0.005 µg/L.

References

1. Weston DP, You JC, Lydy M.J. 2004. Distribution and Toxicity of Sediment-associated Pesticides in Agriculture-dominated Water Bodies of California's Central Valley. Environ Sci Technol 38(10):2752-2759.
2. Johnson B. 2005. Diazinon and Pesticide Related Toxicity in Bay Area Urban Creeks, Water Quality Attainment Strategy and Total Maximum Daily Load (TMDL), Proposed Basin Plan Amendment and Staff Report, August 2005. San Francisco Bay Regional Water Quality Control Board, Oakland, CA. USA
3. Amweg E, Weston DP, Ureda NM. 2005: Use and Toxicity of Pyrethroid Pesticides in the Central Valley, California, USA, Environ Toxicol Chem; Vol. 24 (4); 966-972
4. 'TENTATIVE' California Regional Water Quality Control Board Central Valley Region Monitoring and Reporting Program Order No. R5-2005-_____ For Individual Dischargers Under Order No. R5-2005-

_____Individual Conditional Waiver of Waste Discharge Requirements For Discharges from Irrigated Lands November 2005

5. Marshack J. 2003. A Compilation of Water Quality Goals. California Central Valley Water Quality Control Board, 11020 Sun Center Drive #200 Rancho Cordova, CA 95670 USA
6. Working Draft, California Regional Water Quality Control Board, Central Valley Region, Monitoring and Reporting Program Order No. R5-2007-____ For Coalition Groups Under Amended Order No. R5-2006-0053 Coalition Group Conditional Waiver of Waste Discharge Requirements for Discharges From Irrigated Lands. Revision 29 March 2007
7. Moran K. 2005. Pesticides in Urban Surface Water, Urban Pesticides Use Trends Annual Report. San Francisco Estuary Project, Oakland, CA, USA
8. Kelley, K. 2003. Monitoring Surface Waters and Sediments of the Salinas and San Joaquin River Basins for Synthetic Pyrethroid Pesticides. Study 219. California Department of Pesticide Regulation, Environmental Monitoring Branch, 1001 I Street, Sacramento, California 95812 USA
9. Starner, K. 2004. A Preliminary Assessment of Pyrethroid Contamination of Surface Waters and Bed Sediments in High Pyrethroid-Use Regions of California. Study 224. California Department of Pesticide Regulation, Environmental Monitoring Branch, 1001 I Street, Sacramento, California 95812 USA
10 U.S. EPA. 2005. SW-846 Manual, Test Methods for Evaluating Solid Waste, Physical/Chemical Methods. Method 8270. U.S. Government Printing Office, Superintendent of Documents, Washington, DC 20402 USA.
11. Valverde A, 2001. What Are We Determining Using Gas Chromatographic Multiresidue Methods: Tralomethrin or Deltamethrin? Journal of Chromatography A, 943 (2001) 101-111.
12. SGE International 2003 Technical Article TA-118-C
13. White J, Feng H, Cooper C, Rivera L, Jackson T. 2003. Determination of Pyrethroids in Sediment Water. EMON-SM-52-7. California Dept. of Food and Agriculture, Center for Analytical Chemistry, Environmental Monitoring Section, Sacramento, CA USA
14. Spurlock F, 2003. Probabilistic Estimation of Dissolved Phase Pyrethroid Concentrations from Whole Water Analytical Data. EH03-06. Environmental Monitoring Branch California Department of Pesticide Regulation P.O. Box 4015 Sacramento CA 95812 USA
15. Liu W, Gan JJ, Lee S, and Kabashima JN. 2004 Phase Distribution of Synthetic Pyrethroids in Runoff and Stream Water. Environmental Toxicology and Chemistry, Vol. 23, No. 1, pp. 7–11, 2004
16. Oros, DR. and Werner I. 2005. Pyrethroid Insecticides: An Analysis of Use Patterns, Distributions, Potential Toxicity and Fate in the Sacramento-San Joaquin Delta and Central Valley. White Paper for the Interagency

Ecological Program. SFEI Contribution 415. San Francisco Estuary Institute, Oakland, CA, USA.

17. U.S. EPA. 2005. SW-846 Manual, Test Methods for Evaluating Solid Waste, Physical/Chemical Methods. Section 2.2 U.S. Government Printing Office, Superintendent of Documents, Washington, DC 20402 USA.
18. Zaugg SD, Sandstrom MW, Smith SG, Fehlberg KM. 1995 Methods of Analysis by the U.S. Geological Survey National Water Quality Laboratory – Determination of Pesticides in Water by C-18 Solid-Phase Extraction and Capillary-Column Gas Chromatography/Mass Spectrometry with Selected-Ion Monitoring. Open File Report 95-181, USGS Denver, Colorado, USA
19. Sandstrom MW, Stroppel ME., Foreman WT, Schroeder MP. 2002. Methods of Analysis by the U.S. Geological Survey National Water Quality Laboratory-Determination of Moderate-Use Pesticides and Selected Degredates in Water by C-18 Solid Phase Extraction and Gas Chromatography/Mass Spectrometry. Water Resources Investigations Report 01-4098, U.S. Geological Survey Method O-2002-01, USGS Denver, Colorado, USA.
20. United States Government Printing Office, 732 N. Capitol Street NW, Washington D.C. 20401
21. You J, Weston D, Lydy M.J. 2004. A Sonication Extraction Method for the Analysis of Pyrethroid, Organophosphate, and Organochlorine Pesticides from Sediment by Gas Chromatography with Electron-Capture Detection. Arch Environ Contam Toxicol 47, 141-147
22. U.S. EPA. 2005. SW-846 Manual, Test Methods for Evaluating Solid Waste, Physical/Chemical Methods. Method 3510. U.S. Government Printing Office, Superintendent of Documents, Washington, DC 20402 USA.

Chapter 7

Solid-Phase Microextraction (SPME) Methods to Measure Bioavailable Concentrations in Surface Waters

Wesley Hunter[1], Weichun Yang[1], Frank Spurlock[2], and Jay Gan[1]

[1]Department of Environmental Sciences, University of California Riverside, 2258 Geology Building, Riverside, CA 92521
[2]California Department of Pesticide Regulation, Environmental Monitoring, 1001 I Street, Sacramento, CA 95812

Due to strong affinity for the solid phase, pyrethroid bioavailability is reduced considerably in surface water containing dissolved organic matter or suspended solids. The bioavailable fraction for water column invertebrates is often considered to be the freely dissolved fraction, because it is that fraction which can cross the cellular membrane. Thus, selective methods that measure the freely dissolved concentration are good indicators of the bioavailability of pyrethroids in surface waters. Solid phase microextraction (SPME) was used in several studies as a selective sampling approach to evaluate the bioavailability of pyrethroids in surface water. The SPME measurements correlated well with the pyrethroid bioaccumulation by *Daphnia magna*. The LC50 estimated by SPME also correlated well with the observed LC50 to *Ceriodaphnia dubia*. These results indicate the usefulness of SPME as a tool for evaluating the bio-accumulation and toxicity of pyrethroids to water column invertebrates in the presence of dissolved organic matter or suspended solids. Results from these studies suggest that the presence of dissolved organic carbon and suspended solids should be taken into account when evaluating organism exposure to pyrethroids, especially when monitoring runoff or irrigation effluent which usually contain high levels of dissolved organic matter and suspended solids.

Introduction

Pyrethroid insecticides are used throughout the world for pest control in both urban and agricultural settings. Pyrethroids belong to the class of non-ionic hydrophobic organic contaminants (HOCs) with log K_{OC} in the range of 5-7 (*1, 2*). Because of their high hydrophobicity, pyrethroids have been considered to be nearly immobile in the environment. However, pyrethroids may move into surface waters via surface runoff or erosion of soil particles, especially during storm events (*3-8*).

Pyrethroids exhibit high acute toxicity to aquatic organisms including water-column invertebrates (*9*). For instance, the LC50s of permethrin and bifenthrin to the invertebrate *Ceriodaphnia dubia* are in the sub-ppb range, at 0.5 μg L^{-1} and 0.078 μg L^{-1}, respectively (*10*). Concern about the potential toxic effect of pyrethroids in surface waters has increased as pyrethroid contamination has been detected in bed sediments of many streams and rivers (*7, 8*).

Because pyrethroids are extremely hydrophobic they tend to bind to dissolved organic matter (DOM) and suspended solids (SS) in surface waters. Studies have shown that the HOC bioavailability and observed toxicity decrease in the presence of DOM (*11-13*) and SS (*4, 9, 14, 15*). This decrease in the availability of HOCs to organisms is attributed to the formation of large and possibly very polar HOC-DOM complexes that are unable to pass through the cellular membrane. Therefore, the total chemical concentration obtained by the conventional liquid-liquid extraction method (LLE) (which releases "bound" pyrethroid residues from DOM or SS) overestimates the toxicity because it does not take into account the reduced bioavailability of HOCs due to sorption. The freely dissolved concentration (C_{free}) is the fraction that is considered bioavailable because it is free to cross the cellular membrane.

To better understand the bioavailability of HOCs in surface waters, a number of methods have been developed to selectively measure C_{free} in aqueous solutions. These methods have various limitations. For instance, dialysis membranes are time consuming and may adsorb lipophilic compounds, fluorescence quenching is chemical specific, and Tenax extraction disturbs the equilibrium partitioning of the chemical. Recently, two different forms of solid-phase microextraction (SPME) have been used to measure C_{free} in aqueous samples (*16*). In the first type of SPME, a thin fiber coated with a non-polar polymer such as polydimethylsiloxane (PDMS) is fitted into a syringe-type sampler, and is used manually (and sometimes automatically) to sample an aqueous solution. The fiber is directly inserted into a GC for elution and analysis. A variation of the conventional SPME method involves the use of disposable PDMS fibers. Disposable PDMS fibers have been termed a biomimetic tool because they have been shown to "mimic" the bioaccumulation of HOCs by organisms when exposed along with the test organism (*16-19*). For instance, studies have demonstrated that PDMS fibers accurately predicted the bioavailability of PAHs in soils and sediments (*16, 20*).

PDMS fibers, used in the injector-type SPME mode or disposable mode, can be used to measure C_{free} in aqueous matrices (*16, 17*) because the sampling process is selective, allowing only the freely dissolved molecules to diffuse into the coated polymer phase. C_{free} is related to the concentration in the fiber C_{PDMS} by:

$$C_{free} = C_{PDMS} / K_{PDMS} \quad (1)$$

where K_{PDMS} is the solute PDMS -water partition coefficient. In this review, we will summarize the use of disposable PDMS fibers and injector-type SPME in evaluating the role of DOM and SS in inhibiting the uptake and toxicity of pyrethroids to water-column invertebrates.

Method Development

Disposable PDMS fibers (35 μm PDMS coating, Polymicro Technologies Inc., Phoenix, AZ, USA) were used in several studies to investigate the bioavailability of pyrethroids in water containing DOM and SS (*21-23*). The use of disposable PDMS fibers have the advantage that they can be exposed simultaneously with test organisms in the same test system, thus reducing variables that might affect measurements. ^{14}C-labeled pyrethroids were used in these studies with PDMS fibers. Pesticide accumulation in the PDMS fibers (C_{PDMS}) was determined by direct liquid scintillation counting (LSC), while the organism uptake was measured after sample combustion, which was then followed by LSC.

One requirement for using SPME techniques to measure C_{free} is that the fiber must absorb a negligible amount of the analyte from the test system so that the solute partition equilibria is not affected. In the following studies using

PDMS fibers, it was found that <5% of the total pyrethroid concentration in the test system was absorbed into the fiber (*21, 22*), and thus the phase equilibrium in the samples was not significantly disturbed (*16, 17*).

In order to understand the uptake kinetics of pyrethroids into PDMS fibers, pyrethroid accumulation in PDMS fibers was plotted against time and fit to an empirical first-order rate equation:

$$C_{PDMS} = C_m(1 - e^{-kt}) \tag{2}$$

where t is the exposure time and C_m is the maximum (equilibrium) concentration in the fiber. In the instance of permethrin, regression showed that good correlation ($R^2 > 0.90$) and equilibrium (95% of C_m) was reached after 230 h. The time to reach equilibrium was too long to use the fibers at a steady state, so a 24-h exposure interval was chosen to coincide with the time interval for the organism bioaccumulation tests. C_{free} can be determined from the non-steady state measurements of C_{PDMS} using eq. 1 providing that the exposure interval is identical to that for which the non-steady state K_{PDMS} is determined..

In one study (*23*), SPME (30-μm PDMS coating; 1.0-cm length) was used to measure C_{free} via a previously published method (*24*). Briefly, before sampling, the SPME fiber was activated for 3 min at 260°C in the GC inlet. The activated SPME fiber was then immersed into the sample solution 2 cm from the surface and sampling time was 15 min. The sample solution was stirred with a small bar at 600 rpm. The analyte was then desorbed from the SPME fiber by injection into the GC inlet for 3 min and analyzed by a GC equipped with an electron-capture detector. External standards prepared in deionized water at known concentrations were sampled and analyzed under the same conditions for quantitation of C_{free}.

Effect of DOM on Bioavailability

DOM is ubiquitous in natural surface water environments. Previous studies have shown that DOM can significantly reduce the bioavailability, and hence the toxicity, of HOCs, including pyrethroids (*11-13*). We used permethrin as a model pyrethroid compound to evaluate the effect of DOM on the uptake and toxicity of water column invertebrates using disposable PDMS fibers as a biomimetic tool (*21*). Three DOM sources were used in this study: lake water, pond water, and an extract from a commercial compost. The study consisted of bioaccumulation experiments and acute toxicity (LC50) assays involving water column invertebrates.

Bioaccumulation experiments

The influence of DOM on the bioavailability of permethrin was evaluated by performing 24-h bioaccumulation tests using *D. magna* as the test species. Different DOC concentrations were prepared by diluting the lake water and the aqueous compost extract (0, 1.0, 5.0, 10, 20, and 30 mg L^{-1}) and the pond water (0, 0.5, 1, 2, 5, and 10 mg L^{-1}). Each test jar was then spiked with the same amount of ^{14}C-permethrin. After allowing the test systems to equilibrate, PDMS fibers and *D. magna* (7-14 day old) were simultaneously introduced into each test jar. After the 24-h exposure period, the permethrin concentrations in the PDMS fibers (C_{PDMS}) and permethrin body residue (*BR*) (based on dry biomass) in *D. magna* were determined. C_{PDMS} was further used to obtain C_{free} from eq. 1.

In the presence of DOM, C_{free} is related to C_t (nominal spiked concentration) to account for phase distribution:

$$C_{free} = \frac{C_t}{1 + K_{DOC}[DOC]} \tag{3}$$

where $[DOC]$ is the dissolved organic carbon concentration in the aqueous sample, and K_{DOC} is the permethrin DOC-water partition coefficient.

The bioaccumulation factor (BAF) (*BR* divided by the total chemical concentration in the sample) is related to C_t in the presence of DOM by a similar relationship:

$$BAF = \frac{BAF_0}{1 + \alpha K_{DOC}[DOC]} \tag{4}$$

where BAF_0 is the *BAF* measured in the DOM-free control, and α is an empirical factor indicating the availability of DOM-adsorbed pyrethroids for uptake into the organism. The α value is determined by comparing K_{DOC} from eq. 3 and αK_{DOC} from eq. 4. If $\alpha \approx 1$, the fraction of pyrethroid adsorbed to DOM is not bioavailable. If $\alpha < 1$, then the pyrethroid adsorbed to DOM is partially available to the organism.

The effect of DOM on the bioaccumulation of permethrin to *D. magna* was evaluated by comparing the *BAF* values from each DOC source. It was found that *BAF* values decreased with increasing $[DOC]$ in all samples (Figure 1). C_{PDMS} values followed a similar decreasing trend, demonstrating that the PDMS fibers mimicked the effect on permethrin uptake by *D. magna* (Figure 2).

The relationship between C_{PDMS}, *BAF* and $[DOC]$ was well described by eq. 3 and eq. 4, respectively, for the three DOM sources. Both the C_{PDMS} (Figure 1) and *BAF* values (Figure 2) showed a good correlation for all DOM sources (R^2 = 0.90-0.95 and R^2 = 0.86-0.99, respectively). These results show that increasing $[DOC]$ decreased the bioavailability of permethrin and illustrate the use of PDMS fibers for directly measuring the bioavailable permethrin in the presence of DOM.

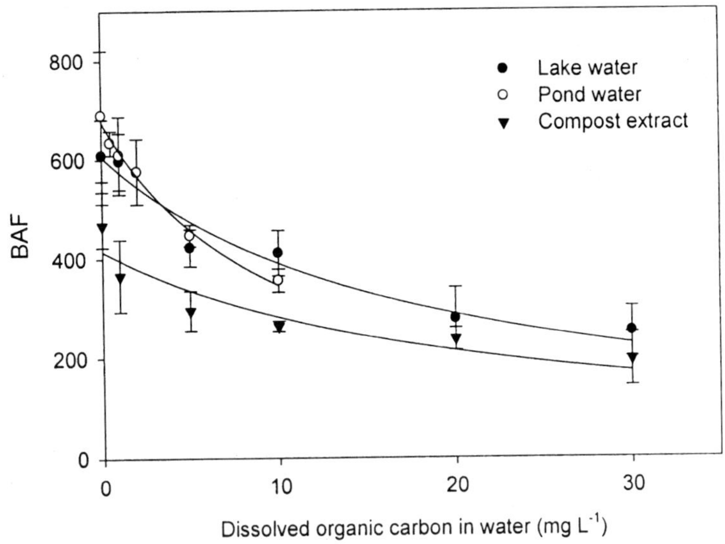

Figure 1. BAF of permethrin for D. magna *exposed to aqueous solutions with DOM from different sources. Error bars are standard deviations (4 replicates) (Reproduced with permission from reference 21. Copyright 2006 Amercan Chemical Society.)*

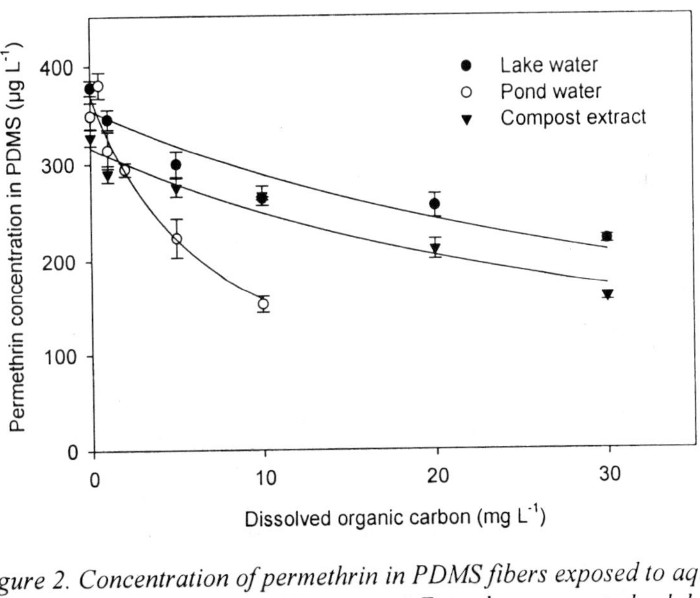

Figure 2. Concentration of permethrin in PDMS fibers exposed to aqueous solutions with DOM from different sources. Error bars are standard deviations (4 replicates). (Reproduced with permission from reference 21. Copyright 2006 Amercan Chemical Society.)

The bioavailability of the permethrin adsorbed to the DOM was evaluated by calculating the α value (Table 1). A paired *t*-test between K_{DOC} and αK_{DOC} showed no statistically significant difference between the two values for any of the DOM sources at the $\alpha = 0.05$ level of significance (Table 1). This means that there is insufficient evidence in these data to conclude that α is not equal to 1. Consequently, we concluded that DOM-adsorbed permethrin in these tests was not bioavailable to *D. magna*.

Table 1. K_{DOC} (×10^4) and αK_{DOC} (×10^4) derived by regression of C_{PDMS}, *BAF*, and LC50 data from water containing DOM.

DOM source	*Estimation Method*		
	PDMS	*BAF*	*LC50*
Lake water	4.74 ± 2.84	5.52 ± 0.89	3.4 ± 0.5
Pond water	11.21 ± 1.38	9.51 ± 0.66	9.2 ± 0.7
Compost extract	3.10 ± 0.69	4.69 ± 1.39	3.1 ± 0.3

SOURCE: Reproduced with permission from reference 21.

To further evaluate whether the PDMS fibers were detecting the bioavailable fraction, C_{PDMS} was plotted against *BR* measurements from all three DOM sources. A positive linear correlation was found ($R^2 = 0.51$, $P < 0.0001$) (Figure 3), indicating that the PDMS fibers used were good indicators of permethrin bioavailability to *D. magna*. In another study, a moderately good linear correlation between *BR* and C_{free} was also found for cyfluthrin ($R^2=0.40$, $P < 0.0001$) (Figure 4) (*25*). Together, these results further support the concept that only the freely dissolved chemical is available for uptake by *D. magna*.

Acute toxicity experiments

Another way of looking at the bioavailability of pyrethroids in the presence of DOM is by evaluating the effect on the observed toxicity to aquatic invertebrates. We performed 96-h static *C. dubia* LC50 assays in water samples containing DOM using U.S. Environmental Protection Agency guidelines for effluent toxicity (*26*). The DOC concentrations for lake water, pond water and compost extract solutions were the same as in the above bioaccumulation experiments. Permethrin was spiked into each sample to give a range of initial concentrations (0 to 4.8 μg L^{-1}). Five *C. dubia* (< 24 h old) were added to each test container and the number of surviving organisms were counted after 96 h of exposure. The LC50 for permethrin was estimated by Probit analysis.

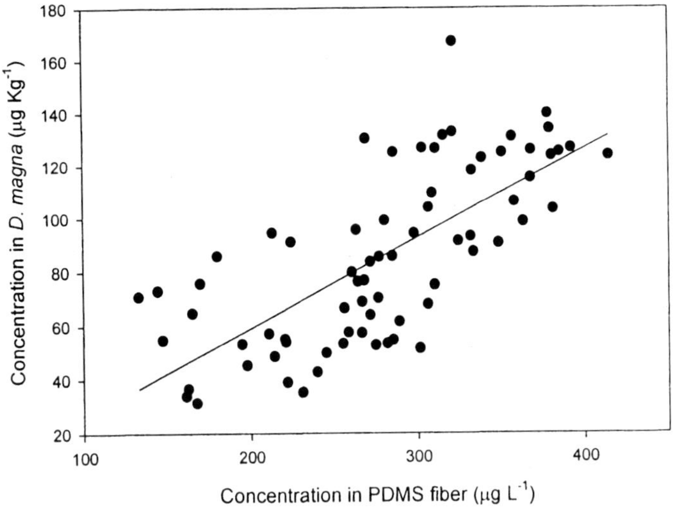

Figure 3. Correlation between body residues of permethrin in D. magna *and permethrin accumulation in PDMS fibers exposed to the same aqueous samples containing DOM. (Reproduced with permission from reference 21. Copyright 2006 Amercan Chemical Society.)*

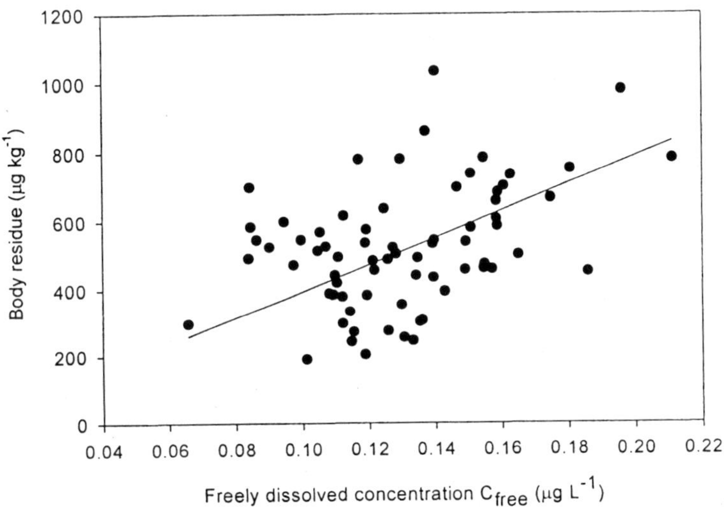

Figure 4. Correlation between body residue of cyfluthrin in D. magna *and freely dissolved concentrations measured by disposable PDMS fibers. (R eproduced with permission from reference 25. Copyright 2007 American Society of Agronomy, Crop Science Society of America, and Soil Science Society of America.)*

The LC50 for permethrin in the DOM-free control was similar to previously published values (0.48-0.56 μg L^{-1}) (*10*). In water samples containing DOM, the LC50 of permethrin consistently increased with increasing [*DOC*] (Table 2). A statistically significant (α = 0.05) increase in LC50 was observed in pond water at [*DOC*] ≥ 5 mg L^{-1} and in lake water at [*DOC*] ≥ 10 mg L^{-1} (Table 2).

Table 2. Effect of [*DOC*] on LC50 of permethrin to *C. dubia* from three DOM sources (mean and confidence intervals)

[DOC] (mg L⁻¹)	*LC50 (μg L⁻¹)*		*[DOC] (mg L⁻¹)*	*LC50 (μg L⁻¹)*
	Lake Water	*Compost Extract*		*Pond Water*
0	0.52 (0.38-0.65)	0.48 (0.39-0.58)	*0*	0.56 (0.41-0.68)
1	0.57 (0.42-0.69)	0.52 (0.39-0.63)	*0.5*	0.51 (0.38-0.62)
5	0.54 (0.43-0.66)	0.49 (0.388-0.60)	*1*	0.59 (0.48-0.72)
10	0.74 (0.57-0.95)	0.59 (0.42-0.74)	*2*	0.66 (0.49-0.81)
20	0.78 (0.63-0.95)	0.73 (0.52-0.90)	*5*	0.76 (0.57-0.95)
30	1.09 (0.81-1.39)	0.92 (0.71-1.19)	*10*	1.03 (0.81-1.32)

SOURCE: Reproduced from reference 21. Copyright 2006 Amercan Chemical Society.

In aqueous DOM solutions, the relationship between LC50 and [*DOC*] is similar to that for bioaccumulation:

$$LC50 = (1 + \alpha K_{DOC}[DOC]) \times LC50_{(0)} \tag{5}$$

where $LC50_{(0)}$ is the LC50 measured in the DOM-free control. The LC50 values and [*DOC*] were fit to eq. 5 for each DOM source and regression analysis showed significant linear correlations in each case (R^2 = 0.92-0.98, p < 0.005) and included the pooled data from all three DOM sources (R^2 = 0.92, p < 0.0001). These results demonstrate that the DOM in water effectively decreased the observed acute toxicity of permethrin to *C. dubia*. The degree to which the pyrethroids adsorbed to the DOM was available to cause toxicity to *C. dubia* was evaluated by determining the α value (Table 1). The αK_{DOC} values determined by this method were compared with independently measured K_{DOC} and no significant difference was found, indicating that $\alpha \approx 1$ in these systems. This finding suggests that the DOM-associated permethrin was also not available to cause acute toxicity to *C. dubia*.

To evaluate the effectiveness of using PDMS fibers to estimate LC50, the K_{DOC}s determined by C_{PDMS} were used along with the $LC50_{(0)}$ measured in DOM-free water to predict LC50s for *C. dubia* in DOM solutions (*21*). A highly significant linear correlation between observed and predicted LC50 was found with a slope near 1.0 (Figure 5). The good correlation further validated the use of PDMS fibers for measuring bioavailable pyrethroid concentrations DOM-containing in water samples with known DOC concentrations.

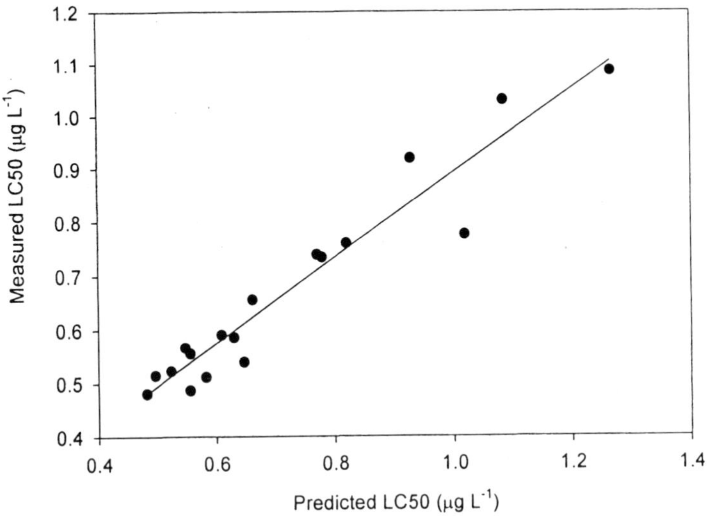

Figure 5. Measured LC50 vs. predicted LC50 of permethrin for C. dubia *in water containing DOM from different sources. (Reproduced from reference 21. Copyright 2006 Amercan Chemical Society.)*

Effects of DOM properties on bioavailability

It should be noted that not only does the quantity of DOM affect the bioavailability of pyrethroids, but studies have also shown that the properties of DOM can also have an effect. For instance, a study by Day (*11*) showed that Aldrich humic acid had a greater influence on decreasing *BAF* values for pyrethroids and *D. magna* at concentrations comparable to the above study than DOM from sources in the above study. The uptake of deltamethrin and fenvalerate by *D. magna* decreased by 81-88% in the presence of Aldrich humic acid at a DOC concentration of 13.1 mg L^{-1}. LC50 values for fenvalerate to *D. magna* in Day's study (*11*) increased by a factor of 17. In another study, Hodge et al. (*12*) did not find a significant reduction of acute toxicity of fenvalerate to *D. magna* when in the presence of DOM from algal sources at 1.4-4.8 mg L^{-1}. Other studies have shown that quality characteristics of DOM (molecular weight, specific UV absorptivity, and aromaticity, functional group content) may influence the effect of DOM on reducing the bioavailability of HOCs such as PAHs and PCBs (*13, 27-30*).

In a separate study (*25*), we examined the effect of selected DOM properties on pyrethroid partitioning and bioavailability using a large number of natural surface water samples that were collected from various sites in Southern California. K_{DOC} values were determined by the following relationship:

$$K_{DOC} = \frac{(C_t - C_{free})/[DOC]}{C_{free}} \quad (6)$$

where C_t is the original spiked concentration. The K_{DOC} values determined by eq. 6 were correlated with various DOM properties including ABS_{254} (UV absorption at 254 nm), carboxylic and phenolic acid content, and the E4/E6 ratio (degree of aromaticity). Of these, only the carboxylic acid content showed a significant correlation with K_{DOC} ($R^2 = 0.72$, $P = 0.002$) (Figure 6).

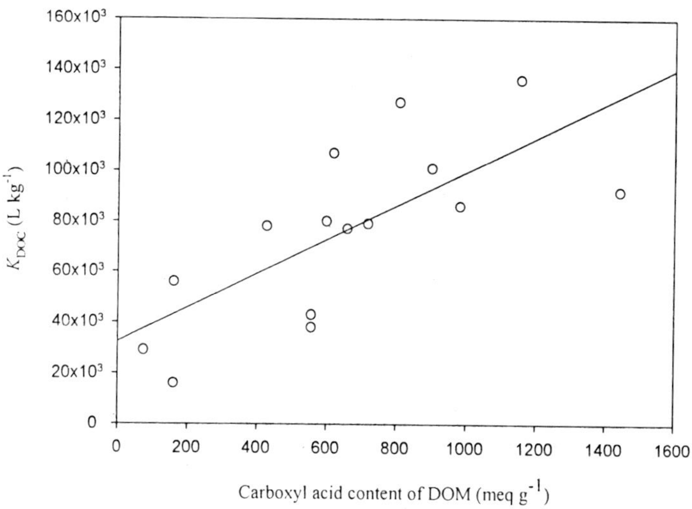

Figure 6. K_{DOC} values for cyfluthrin and permethrin, plotted against carboxylic acid content of DOM in test samples ($R^2 = 0.72$, $P = 0.002$). (Reproduced with permission from reference 25. Copyright 2007 American Society of Agronomy, Crop Science Society of America, and Soil Science Society of America.)

Effect of suspended solids on bioavailability

Bioaccumulation experiments

Natural surface waters usually contain suspended solids (SS), and SS levels can be particularly high in rain-induced runoff, irrigation runoff, and waterbodies receiving large amounts of such runoff effluents. Similar to the role of DOM, the presence of SS in water also decreases the pyrethroid C_{free}, which can greatly alter the whole effluent toxicity. The C_{free} of a pyrethroid in SS-containing water is related to total chemical concentration via eq. 7 (*31*):

$$C_{free} = \frac{C_t}{1 + K_{SS}[SS] + K_{DOC}[DOC]} \quad (7)$$

where [*SS*] is the suspended solid concentration, and K_{SS} is the pyrtheroid adsorption coefficient for SS. When sorption to DOC and SS is considered as a lumped process, eq. 7 simplifies to:

$$C_{free} = \frac{C_t}{1 + K_d[SS]} \quad (8)$$

where K_d is the apparent adsorption coefficient for the combined SS and DOM phases.

When PDMS fibers are used to measure C_{free} in systems with SS, the following relationship may be used to describe the effect of [*SS*] on C_{PDMS}:

$$C_{PDMS} = \frac{C_{PDMS(0)}}{1 + K_d[SS]} \quad (9)$$

where $C_{PDMS(0)}$ is the pyrethroid concentration in a PDMS fiber exposed in the SS-free control. The *BAF* is similarly related to [*SS*]:

$$BAF = \frac{BAF_{(0)}}{1 + \alpha K_d[SS]} \quad (10)$$

As with the case of water containing DOM, $\alpha \approx 1$ suggests that the SS-adsorbed pyrethroid is not available for uptake by the organism. If $\alpha < 1$, some of the adsorbed pesticide is bioavailable.

To evaluate the influence of SS on the bioavailability of pyrethroids, static bioaccumulation tests (24-h) were conducted using *D. magna* as the test species (*22*). Four different sediments were used to prepare the SS suspensions. The sediments included San Diego Creek sediment (SDC), Salinas River sediment (SR), Miles Creek sediment (MC) and a field sediment from a strawberry field furrows (FF). Disposable PDMS fibers were used to measure C_{free} in the SS solutions. Suspended solid concentrations in aqueous solutions were prepared at 0, 25, 50, 100, or 200 mg L^{-1}. Each test jar was spiked with ^{14}C-permethrin or ^{14}C-bifenthrin at a nominal concentration below the LC50 for that specific pyrethroid. Disposable PDMS fibers (3.0 cm long) and six *D. magna* (7-14 d old) were simultaneously introduced into the test chambers and removed after 24-h of exposure.

Results from the above experiments showed that the *BR* and *BAF* values consistently decreased with increasing [*SS*] for each of the SS sources. The decrease was substantial for the higher SS levels. For instance, the *BAF* for permethrin and bifenthrin decreased by 41-63% in samples containing SS at 100 mg L^{-1} when compared to the SS-free control. When the *BAF* values were fit to eq. 10, good correlations were found for both bifenthrin and permethrin (R^2 = 0.85-0.99). The C_{PDMS} decreased similarly with increasing [*SS*] for both permethrin and bifenthrin, and a good correlation was found when the data was fit to eq. 9 (R^2 = 0.70-0.99). These results support the use of PDMS fibers as a measure of the freely dissolved form of pyrethroids and not the fraction adsorbed to the SS or DOM.

To evaluate the degree of availability of the pyrethroids adsorbed to the SS, the α value was determined by comparing K_d and αK_d values from regression analysis (Table 3). It was found that α was consistently ≥ 1 for all pyrethroids and sediments (Table 3). This finding suggests that the pesticide adsorbed to SS was not available for uptake by *D. magna* under the test conditions.

The effectiveness of using PDMS fibers as a biomimetic tool was evaluated by plotting the *BR* of *D. magna* and C_{PDMS}. The linear relationships were significant for both bifenthrin (*R*2 = 0.41; $p < 0.0001$; slope = 2.40 ± 0.24) and permethrin (R^2 = 0.70; $p < 0.0001$; slope = 2.47 ± 0.14) (Figure 7). The slopes for bifenthrin and permethrin were essentially identical, suggesting that the pyrethroids may act similarly in water samples containing SS. The *BR* in *D. magna* in this test could be estimated for both bifenthrin and permethrin by multiplying C_{PDMS} by a value of approximately 2.4. Thus, the PDMS fibers may be effective biomimetic tools for estimating pyrethroid bioaccumulation by *D. magna* in SS-containing water samples.

Toxicity inhibition experiments

To further evaluate the bioavailability of pyrethroids in water containing SS, we conducted a study to evaluate the effect of SS on the acute toxicity of the pyrethroids in surface water to *C. dubia* (*23*). Manual SPME was used to measure C_{free} in the samples. The relationship between the LC50 and SS can be described by the following relationship:

$$LC50 = (1 + \alpha K_d[SS]) \times LC50_{(0)} \tag{11}$$

In this study, four pyrethroids were tested: bifenthrin, permethrin, cypermethrin, and esfenvalerate. The same sediments (SDC, SR, MC, and FF) used in the above bioaccumulation study were used to prepare the SS suspensions for this study. The [*SS*] ranged from 0 to 200 mg L^{-1}. Static 96-h acute toxicity (LC50) tests were conducted using *C. dubia*, from which the LC50s were calculated using probit analysis.

Table 3. Apparent partition coefficient K_d for bifenthrin and permethrin in suspended solid solutions estimated from C_{PDMS} and *BAF* measurements

	Bifenthrin			
Source sediment[a]	*Unwashed* ($\times 10^3$)		*Washed* ($\times 10^3$)[b]	
	PDMS	*BAF*	*PDMS*	*BAF*
SDC	5.7±2.2	12.4±1.2	4.2±1.5	6.5±1.0
FF	7.1±1.5	15.1±2.2	4.0±1.2	9.0±1.8
SR	4.5±1.0	6.1±1.5	1.4±0.5	5.6±1.4
MC	5.8±0.8	13.1±2.1	7.9±1.1	7.1±1.9
	Permethrin			
Source sediment[a]	*Unwashed* ($\times 10^3$)		*Washed* ($\times 10^3$)[b]	
	PDMS	*BAF*	*PDMS*	*BAF*
SDC	7.0±0.8	7.4±1.0	4.9±0.2	4.9±0.8
FF	12.3±1.2	15.0±1.7	9.7±0.8	10.1±1.6
SR	2.7±0.6	5.9±1.2	1.8±0.6	5.5±1.3
MC	12.2±1.0	14.8±1.4	7.8±0.3	12.5±1.9

[a] SDC = San Diego Creek sediment; FF = field furrow sediment; SR = Salinas River sediment; MC = Miles Creek sediment.

[b] Sediment was washed to remove some of the organic carbon content.

(Reproduced with permission from reference 22. Copyright 2006 Society of Environmental Toxicology and Chemistry.)

For each pyrethroid, as the [*SS*] increased, the LC50 values generally increased, shifting the toxicity curve to the right (Figure 8). For instance, at the highest [*SS*] levels (100 or 200 mg L^{-1} SS), the LC50 values were about 2-13 times greater than the corresponding $LC50_{(0)}$ values. The LC50 and [*SS*] data were fit to eq. 11, and showed good linear correlation (R^2 = 0.94-0.99) (Figure 9), suggesting that the injector-type SPME was measuring only the freely dissolved chemical concentration.

To determine the degree to which these pyrethroids were available to cause toxicity to *C. dubia* in the presence of SS, the α value from eq. 11 was determined. The α values for this study ranged from 0.28-2.39 (mean = 1.07 ± 0.35; 95% confidence interval). These values were somewhat more variable than found in previous studies. The variability may be partially due to experimental error associated with LC50 and K_d values. However, an overall mean of $\alpha \approx 1$ shows that the freely dissolved fraction of the pyrethroids was responsible for most of the toxicity to *C. dubia*, while the pesticides adsorbed to SS accounted for little, if any, of the toxicity.

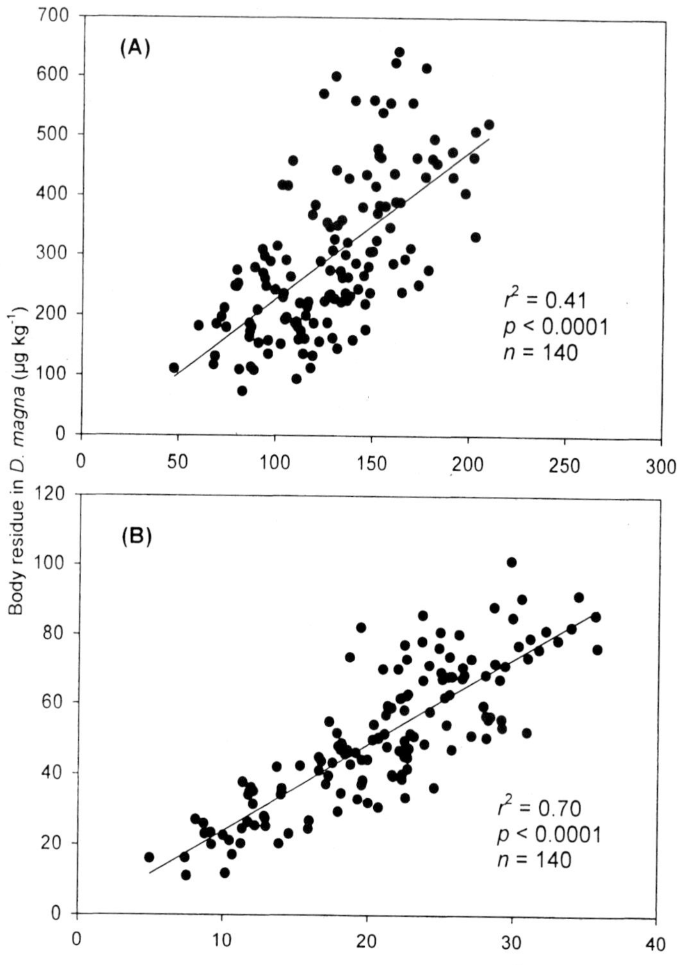

Figure 7. Correlation between (A) bifenthrin and (B) permethrin body residues in D. magna *and pesticide concentration in PDMS fibers in water containing suspended solids (n=140). (Reproduced with permission from reference 22. Copyright 2006 Society of Environmental Toxicology and Chemistry.)*

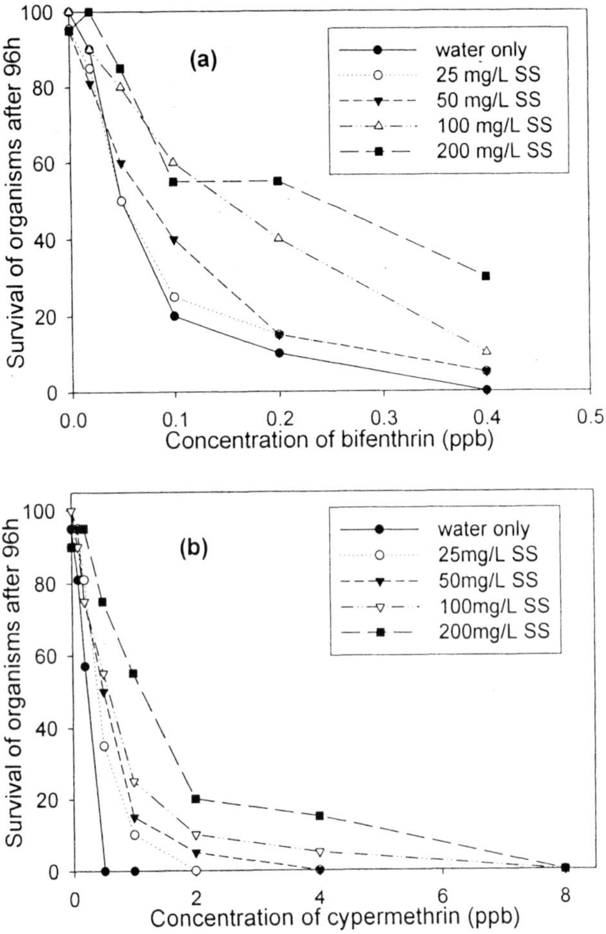

Figure 8. Toxicity curves for C. dubia *in water containing different levels of suspended solids from San Diego Creek sediment. (A) Bifenthrin; and (B) Cypermethrin. (Reproduced with permission from reference 23. Copyright 2006 Society of Environmental Toxicology and Chemistry.)*

The applicability of using manual SPME as a tool to predict pyrethroid bioavailability in water containing SS was investigated by comparing SPME-estimated LC50s to observed LC50 values. SPME-measured K_d and $LC50_{(0)}$ were used to estimate the LC50s of pyrethroids in the various SS-treatments by fitting to eq. 11. It was found that 95% of the estimated LC50s came within a factor of two of the measured LC50s. These results indicate that the acute toxicity of pyrethroids in SS-containing water may be predicted with reasonable accuracy by using K_d measurements obtained by SPME and $LC50_{(0)}$ (Figure 10).

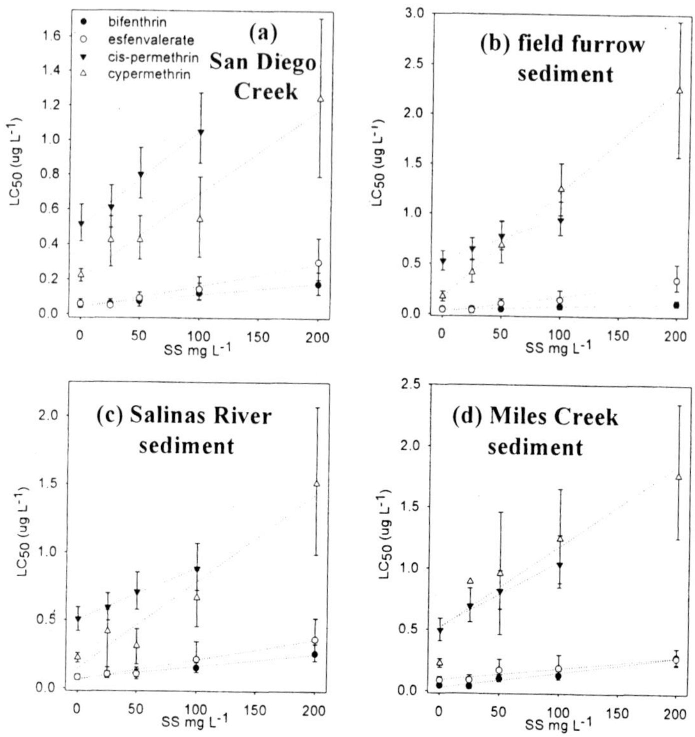

Figure 9. Correlation between C. dubia *LC50 values and suspended sediment (SS) concentrations. (Reproduced with permission from reference 23. Copyright 2006 Society of Environmental Toxicology and Chemistry.)*

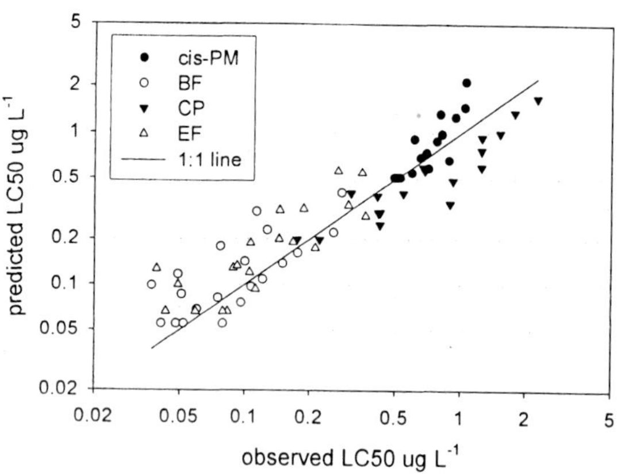

Figure 10. Correlation between predicted and observed LC50 values. (Reproduced with permission from reference 23. Copyright 2006 Society of Environmental Toxicology and Chemistry.)

Conclusions

Due to the high hydrophobicity of pyrethroids, the presence of DOM and SS in surface water decreases the freely dissolved pyrethroid concentration, and thus, the bioavailability of pyrethroids to water-column invertebrates. Studies using PDMS fibers as a biomimetic tool showed that DOM- or SS-sorbed pyrethroids were not available for uptake by *D. magna* or for causing toxicity to *C. dubia*. The uptake of pyrethroids by *D. magna* consistently decreased as the level of DOC or SS increased. Pyrethroid acute toxicity to *C. dubia* consistently decreased in waters containing DOM or SS. Changes in pesticide uptake or toxicity, as well as the influence of specific DOM properties on bioavailability, can be estimated using phase partition information derived from disposable PDMS fibers or injector-type SPME sampling. The reduction in the bioavailability of pyrethroids due to the presence of SS and DOM should be considered when evaluating the exposures of water-column invertebrates and in the water quality monitoring of runoff effluents and surface water. SPME sampling techniques are simple, inexpensive, and may serve as an effective tool for assaying the bioavailability of pyrethroids in runoff and water columns.

References

1. Laskowski, D.A. *Rev. Environ. Contam. Toxicol.* **2002**, *174*, 49-170.
2. Bondarenko, S.; Putt, A.; Kavanaugh S.; Poletika, N.; Gan, J.Y. *Environ. Toxicol. Chem.* **2006**, *25*, 3148-3154.
3. House, W.A.; Long, J.L.A.; Rae, J.E.; Parker, A.; Orr, D.R. *Pest Manage. Sci.* **2000**, *56*, 597-606.
4. Schulz, R.; Liess, M. *Environ. Toxicol. Chem.* **2001**, *20*, 185-190.
5. Kabashima, J.N.; Lee, S.J.; Haver, D.L.; Goh, K.S.; Wu, L.S.; Gan, J. In *Pesticide Decontamination and Detoxification;* Gan, J.J.; Zhu, P.C.; Aust, S.D.; Lemley, A.T. ACS Symposium Series 863; American Chemical Society: Washington, D.C., MA, 2004. pp 213-230.
6. Gan, J.; Lee, S.J.; Liu, W.P.; Haver, D.L.; Kabashima, J.N. *J. Environ. Qual.* **2005**, *34*, 836-841.
7. Weston, D.P.; You, J.; Lydy, M.J.; *Environ. Sci. Technol.* **2004**, *38*, 2752-2759.
8. Weston, D.P.; Holmes, R.W.; You, J.; Lydy, M.J. *Environ. Sci. Technol.* **2005**, *39*, 9778-9784.
9. Hill, I.R. *Pestic. Sci.* **1989**, *27*, 429-465.
10. Mokry, L.E.; Hoagland, K.D. *Environ. Toxicol. Chem.* **1990**, *9*, 1045-1051.
11. Day, K.E. *Environ. Toxicol. Chem.* **1991**, *10*, 91-101.
12. Hodge, V.A.; Fan, G.T.; Solomon, K.R.; Kaushik, N.K.; Leppard, G.G.; Burnison, B.K. *Environ. Toxicol. Chem..* **1993**, *12*, 167-176.

13. Haitzer, M.; Abbt-Braun, G.; Traunspurger, W.; Steinberg, C.E.W. *Environ. Toxicol. Chem.* **1999**, *18*, 2782-2788.
14. Hall, W.S.; Dickson, K.L.; Saleh, F.Y.; Rodgers, J.H. *Arch. Environ. Contam. Toxicol.* **1986**, *15*, 529-534.
15. Ekelund, R.; Granmo, A.; Berggren, M.; Renberg, L.; Wahlberg, C. *Bull. Environ. Contam. Toxicol.* **1897**, *38*, 500-508.
16. Mayer, P.; Vaes, W.H.J.; Wijnker, F.; Legierse, K.C.H.M.; Kraaij, R.H.; Tolls, J.; Hermens, J.L.M. *Environ. Sci. Technol.* **2000**, *34*, 5177-5183.
17. Ramos, E.U.; Meijer, S.N.; Vaes, W.H.J.; Verhaar, H.J.M.; Hermens, J.L.M. *Environ. Sci. Technol.* **1998**, *32*, 3430-3435.
18. Vaes, W.H.J.; Ramos, E.U.; Verhaar, H.J.M.; Seinen, W.; Hermens, J.L.M. *Anal. Chem.* **1996**, *68*, 4463-4467.
19. Leslie, H.A.; Oosthoek, A.J.P.; Busser, F.J.M.; Kraak, M.H.S.; Hermens, J.L.M. *Environ. Toxicol. Chem.* **2002**, *21*, 229-234.
20. van der Wal, L.; Jager, T.; Fleuren, R.H.L.J.; Barendregt, A.; Sinnige, T.L.; Van Gestel, C.A.M.; Hermens, J.L.M. *Environ. Sci. Technol.* **2004**, *38*, 4842-4848.
21. Yang, W.C., Spurlock, F.; Liu, W.P.; Gan, J.Y. *J. Agric. Food Chem.* **2006**, *54*, 3967-3972.
22. Yang, W.C.; Gan, J.Y.; Hunter, W.; Spurlock, F. *Environ. Toxicol. Chem.* **2006**, *25*, 1585-1591.
23. Yang, W.C.; Spurlock, F.; Liu, W.P.; Gan, J.Y. *Environ. Toxicol. Chem.* **2006**, *25*, 1913-1919.
24. Liu, W.P.; Gan, J.J.; Lee, S.; Kabashima, J.N. *Environ. Toxicol. Chem.* **2004**, *23*, 7-11.
25. Yang, W.C.; Hunter, W.; Spurlock, F.; Gan, J. *J. Environ. Qual.* **2007**, *36*, 1678-1685.
26. *Methods for measuring the acute toxicity of effluents and receiving waters to freshwater and marine organisms,* URL http://www.epa.gov/WET/disk2/atx.pdf
27. Gauthier, T.D.; Seitz, W.R.; Grant, C.L. *Environ. Sci. Technol.* **1987**, *21*, 243-248.
28. Muir, D.C.G.; Hobden, B.R.; Servos, M.R. *Aquat. Toxicol.* **1994**, *29*, 223-240.
29. Chin, Y.P.; Aiken, G.R.; Danielsen, K.M. *Environ. Sci. Technol.* **1997**, *31*, 1630-1635.
30. Akkanen, J.; Vogt, R.D.; Kukkonen, J.V.K. *Aquat. Sci.* **2004**, *66*, 171-177.
31. Chin, Y.P.; Gschwend, P.M. *Environ. Sci. Technol.* **1992**, *26*, 1621-1626.

Chapter 8

Solid-Phase Microextraction (SPME) Methods to Measure Bioavailable Concentrations in Sediment

S. Bondarenko[1], J. Gan[1], and F. Spurlock[2]

[1]Department of Environmental Sciences, University of California, 2258 Geology Building, Riverside, CA 92521
[2]California Department of Pesticide Regulation, 1001 I Street, Sacramento, CA 95812

The freely dissolved concentration (C_{free}) in porewater can be used to improve prediction of sediment toxicity by pyrethroids. We used solid-phase microextraction (SPME) to analyze C_{free} of eight pyrethroids in sediment porewater. External calibration was applied to obtain C_{free} of chemicals, whereas internal calibration with ^{13}C-*cis*-permethrin was used to determine total concentration (C_w). Total porewater concentration measured by using the isotopic-SPME method was well correlated with data obtained by exhaustive liquid-liquid extraction (LLE). Method detection limits (MDLs) of the SPME methods were lower than the 10th percentile of the reported LC50s for aquatic invertebrates, with relative standard deviation < 20%. The SPME method was further used on field contaminated samples. Measuring C_{free} by SPME may represent a good alternative to the estimation of total or OC normalized sediment concentrations for predicting sediment toxicity from pyrethroid contamination.

Introduction

Hydrophobic organic compounds (HOCs) such as organochlorine insecticides, polychlorinated biphenyls, polycyclic aromatic hydrocarbons, and pyrethroids, tend to accumulate in the bed sediment due to their high organic carbon partitioning coefficient (K_{OC}). Consequently potential toxicity to benthic organisms is an important concern. Fate and transport of HOCs in sediment depend on many factors, including sediment characteristics, sediment-chemical contact time and properties of HOCs. Exposure of sediment dwelling invertebrates to HOCs occurs primarily through transport of freely dissolved molecules of pollutants across cell membranes or direct ingestion of contaminated food particles. For soil invertebrates uptake of chemicals depends on physical characteristics of the organisms (soft and hard bodied) and physiology of the gut. In contrast, sediment invertebrate uptake mechanisms are not yet well understood (*1, 2*). Several studies have demonstrated that uptake through porewater is likely the dominant route for sediment exposure (*3, 4*).

Equilibrium partitioning theory (EqP) is widely used to describe bioavailability of HOCs in sediment-water systems. EqP is based on the assumption of HOC partition equilibrium between porewater phase and sediment phase, leading to the conclusion that sediment toxicity can be estimated either from porewater or from organic carbon (OC)-normalized sediment concentration, regardless of the exposure route (*3*). OC-normalization of sediment concentrations has been often used for estimating sediment toxicity caused by HOCs, including that by pyrethroids (*5-8*). The preference is mainly due to the availability of relatively robust methods for measuring sediment concentration that are exclusively based on exhaustive extraction techniques.

The bioavailability of HOCs under field conditions often differs from that predicted by using a constant K_{OC} for different types of sediments along with EqP (*2, 9*). Numerous studies showed variable K_{OC} values for the same compound due to sediment OC characteristics such as aromaticity, lipid content, black carbon content, and environmental factors, such as the contact time between the sediment and the contaminant, i.e., aging (*10-13*). Thus, the use of OC-normalization along with constant K_{OC} is likely to yield inaccurate estimates of HOC sediment toxicity in many cases.

The use of porewater concentration also has complications because the total porewater concentration C_w is the sum of both the freely dissolved concentration C_{free} and the concentration complexed with dissolved organic carbon (DOC). The presence of DOC in porewater complicates the measurement of C_w for HOCs with $\log K_{ow} > 5$ (*14*). Lately, a wide variety of extraction techniques have been tested for quantifying C_{free}. These methods include equilibrium dialysis, ultracentrifugation, reversed phase separation, size exclusion chromatography, fluorescence quenching, headspace, semipermeable membrane devices, and solid-phase microextraction (SPME). Overview of these techniques and their limitations was given in detail elsewhere (*14, 15*).

Solid-phase microextraction (SPME) was introduced by Pawliszyn and colleagues (*16*) and has been successfully used to measure C_{free} of HOCs in porewater (*17-20*). This technique is based on either "matrix-SPME" that uses the whole sediment for sampling or "negligible-depletion SPME" that uses a small amount of porewater under negligibly depletive conditions. For instance, SPME was used to quantify C_{free} of polyaromatic hydrocarbons (PAHs) in sediment porewater, and those SPME-determined concentrations were closely correlated with bioaccumulation of PAHs by benthic organisms (*20*). In recent studies, we applied SPME in analyzing pyrethroids in runoff effluents and surface water samples (*21-23*), as reviewed in a separate chapter in this book by Hunter et al. Those studies showed that both uptake of pyrethroids by *Daphnia magna* and acute toxicity to *Ceriodaphnia dubia* demonstrated a higher correlation to the SPME-detected concentrations than to C_w obtained with LLE.

In this chapter, we review the development of SPME methods for analysis of pyrethroids in sediment porewater. More details can be found in our published studies (*24, 25*). A significant amount of new information is also included, especially on the use of ^{13}C-*cis*-permethrin in GC-MS-MS analysis. The later application allows for the simultaneous determination of C_{free} and C_w in the same run. The proposed methods may be used for screening sediments for potential toxicity from pyrethroid contamination.

Materials and Methods

Sediment and Porewater Preparation

Several sediments were used in this study. The sediments were collected at the surface (0-5 cm) with a hand shovel into plastic containers and transported to the laboratory. Prior to use and characterization, sediments were wet sieved through a 2-mm screen to remove large particles. The OC content was determined by high temperature combustion of acidified sediments and subsequent analysis of the evolved CO_2. Fresh sediment porewater was prepared by centrifuging 200 g (wet weight) of the sieved sediment in a 250-mL polyethylene centrifuge bottle at 10,000 rpm for 30 min. The supernatant was carefully pipetted from multiple replicates into a 250-mL glass bottle for use in the following experiments. The DOC level in porewater samples was measured on an Apollo 9000 Carbon Analyzer (Teledyne Instruments, Mason, OH), using a high temperature combustion method.

SPME Analysis

The manual and automatic injector-type PDMS fibers were purchased from Supelco (Bellefonte, PA). Before use, new fibers were conditioned by heating at 320 °C in the GC inlet for 2 h, while reused fibers were cleaned and activated by heating at 260 °C for 3 min. For manual SPME, 10 mL of sample in 20-mL glass scintillation vials was used. The fiber immersion depth in the sample solution was fixed at 2 cm from the surface, and the solution was stirred at 600 rpm with a disposable magnetic bar made of 12 × 0.12 mm (diameter) rust-resistant steel wire. After exposure, the fiber was manually injected into the GC for analysis. For automatic SPME analysis, 9 mL of sample in 10-mL amber glass vial with PTFE-coated septa was placed on a CTC Combi-PAL autosampler (Varian, Palo Alto, CA). Agitation of sample was performed by an automatic device at 250 rpm. To avoid carry-over between samples, the fiber was desorbed for an additional 5 min at 320 °C between consecutive sample runs. External calibration standards for SPME analysis were prepared in deionized water and analyzed under the same conditions on the same day of analysis at six different concentrations (1000, 500, 100, 40, 10, 5, and 1 ng L^{-1}) for GC-ECD and GC-MS-MS analyses. For internal standard calibration using ^{13}C-*cis*-permethrin on GC-MS-MS a concentration of 1 ng L^{-1} was used.

Measuring SPME Water Partition Coefficients (K_{SPME}), Uptake (k_1) and Elimination (k_2) Rates

Pesticide uptake was determined for both the 7- and 30-µm polydimethylsiloxane (PDMS) fibers. Eight common pyrethroids were selected in this study: bifethrin, fenpropathrin, *cis*-and *trans*-permethrin, lambda-cyhalothrin, cyfluthrin, cypermethrin, esfenvalerate, and deltamethrin. Sediment porewater was spiked with pyrethroids at 40 ng L^{-1}, and then shaken for 10 min. The amount of acetone in each sample was < 0.01% (v/v). Exposure time was 10, 20, 30, 40, 60, 90, 120, 180, and/or 240 min. For each analysis a new vial was used and analysis was performed in three or four replicates. After exposure, the fiber was injected into the GC for analysis. The amount of pesticide desorbed from the fiber was calibrated by injecting pyrethroid standards in hexane under the same chromatographic conditions as for SPME analysis.

The volume of PDMS on the fiber was 0.028 µL and 0.132 µL for 7- and 30-µm PDMS fiber, respectively. The data obtained from manual and automated SPME analysis was fit to one-compartment model (*26*) using GraphPad Prism (v.4.03, GraphPad Software, San Diego, CA). From the curve fitting, uptake rate (k_1), elimination (k_2) rate, SPME water partition coefficient (K_{SPME}), and minimum sampling time (t_s) were determined.

Measuring Pyrethroid Concentrations in Porewater

Spiked porewater samples were analyzed in parallel by SPME with the 30-μm PDMS fiber (to measure C_{free}) and LLE (to measure C_w) to evaluate the capability of SPME for detecting C_{free}. Sediment porewater samples were spiked with a mixture of all eight pyrethroids at 500, 100 or 40 ng L^{-1}. The content of acetone in the spiked samples was < 0.01% (v/v). All spiked samples were vigorously shaken for 5 min and equilibrated for 30 min or more before analysis. Preliminary experiments showed that pyrethroids added to a porewater sample reached an apparent equilibrium within 10 min after pesticide addition. Four replicates of 10-mL porewater samples were used for each concentration level. The SPME sampling interval was fixed at 20 min, and the stirring speed was 600 rpm.

Field contaminated samples were analyzed using SPME to evaluate method performance. The samples were manually collected at two agricultural runoff sites in Orange County, CA. Wet sediment was weighed into a 250-ml polyethylene centrifuge bottle and the sample was centrifuged at 10,000 rpm to generate porewater. The porewater was analyzed using 30-μm PDMS fiber under the same conditions as described above.

To obtain C_w in porewater after SPME sampling, the same porewater sample was extracted with ethyl acetate. The water sample was transferred to a glass separatory funnel and vigorously mixed with 10 mL of ethyl acetate for 1 min, followed by the collection of the ethyl acetate phase into a 50-mL pear-shaped flask. The same extraction step was repeated for two additional times and solvent extracts were combined. The extract was dried by passing through 2 g of anhydrous sodium sulfate and then condensed to 0.5-1.0 mL on a vacuum rotary evaporator. A 1.0-μL aliquot was injected into the 6890 GC for analysis. Preliminary experiments showed that the recovery of the LLE procedure was 75-98% for the selected pyrethroids. Calibration standards were prepared in hexane-acetone (1:1) and analyzed on the same day as the samples.

Measuring C_{free} and C_w with SPME-Isotopic GC-MS-MS Analysis

Internal calibration with stable isotope labeled ^{13}C-*cis*-permethrin was used in SPME and GC-MS-MS analysis to estimate C_w. In the same analysis, external calibration was performed using deionized water with known concentrations of pyrethroids (1-1000 ng L^{-1}) to estimate C_{free}. The external calibration was found to be linear over the entire concentration range. The amount of acetone in each sample was < 0.01% (v/v). Fiber exposure time was 20 min, and other SPME sampling conditions were the same as given above.

GC-ECD and GC-MS-MS Analysis

Manual SPME and LLE analyses were carried out on an Agilent 6890 GC coupled with two electron capture detectors (ECD). After injection, the sample was split through a Y-connector into a DB-5MS column (30 m × 0.25 mm × 0.25 μm) and a DB-1701 column (30 m × 0.25 mm × 0.32 μm) (J&W Scientific, Folsom, CA). The dual columns were used to provide confirmation of the resolved peaks. The column temperature was held at 160 °C for 1 min, ramped to 300 °C at 10 °C min^{-1}, and then held at 300 °C for 6 min. The column flow rate was 1.5 mL min^{-1} (helium). The inlet temperature was 260 °C, and the detector temperature was 320 °C. The make-up gas flow rate was 60 mL min^{-1} (nitrogen). The injector port was used in pulsed splitless mode (50 psi at 3 min). External calibration was used for quantification. For compounds with multiple peaks, the sum of all peak areas was used for calibration and quantitation.

Automated SPME analyses were performed on a 3800 Varian GC system (Varian, Palo Alto, CA) equipped with a DB-5MS column (30 m × 0.25 mm × 0.25 μm) (J&W Scientific, Folsom, CA), a 1200 triple quadrupole mass spectrometer detector (Varian, Palo Alto, CA), and a Combi-PAL automated SPME sampler (Varian, Palo Alto, CA). The column program was the same as for manual SPME-GC-ECD analysis. The injector was used in the pulsed split mode (40 psi at 3 min). The injector, transfer line, and source temperature were 300 °C, 280 °C, and 170 °C, respectively. Electron impact mode at 70 eV and argon as collision gas were used. The following parent (precursor) ions were used for pyrethroids identification and quantitation: 181 (166) for bifenthrin, 181 (152) for fenpropathrin, lambda-cyhalothrin, and deltamethrin, 183 (153) for *cis*- and *trans*-permethrin, 189 (174) for ^{13}C-*cis*-permethrin, 163 (127) for cyfluthrin and cypermethrin, and 167(125) for esfenvalerate.

Method Validation

Method detection limits (MDLs) and method precision were determined using spiked sediment porewater samples. The MDL of each pyrethroid in a given matrix was determined by multiplying the one-sided 99% *t* statistic by the standard deviation obtained from four analyses of a matrix spike at 40 ng L^{-1}. The sampling and analytical conditions were the same as given above. The method precision was obtained by calculating the relative standard deviation (RSD) of pesticide concentrations from replicated analyses that were derived during the method development phase.

Theory

Concentration in the SPME fiber assuming one compartment system with a first-order kinetics can be described as (*26*)

$$C_{\mathrm{SPME},t} = \frac{k_1}{k_2} C_0 (1 - e^{-k_2 t}) \tag{1}$$

where C_0 (ng mL^{-1}) is initial analyte concentration at 0 time, $C_{\mathrm{SPME},t}$ (ng mL^{-1}) is analyte concentration in the fiber coating at time t (min), k_1 is uptake rate constant (min^{-1}), and k_2 is elimination rate constant (min^{-1}). This model can be used when depletion is negligible or when the following condition is met

$$\frac{k_1 V_{\mathrm{SPME}}}{k_2 V_{\mathrm{s}}} << 1 \tag{2}$$

where V_{SPME} (mL) is the volume of the fiber coating, and V_{s} (mL) is the volume of sample. At equilibrium, partition coefficient K_{SPME} can be measured as ratio of k_1 to k_2. In order to measure C_{free} in the kinetic state, the freely dissolved amount that is extracted from sample into the fiber coating should be negligibly small and matrix presented in the aqueous phase should not interfere with SPME measurement (*26*). The minimum SPME sampling time (t_{s}) can be calculated as (*26*)

$$t_{\mathrm{s}} \geq -\frac{1}{k_2} \ln\left(\frac{0.05}{k_2 + 0.05}\right) \tag{3}$$

Results and Discussion

Fiber Uptake Kinetics

Uptake curves for the 7-μm and 30-μm PDMS fibers are shown in Figure 1 for bifenthrin and Figure 2 for cypermethrin. Similar curves were also observed for the other pyrethroids. Since the depletion was less than 10 % (*27*) at all sampling time intervals, the first-order compartment model was used for

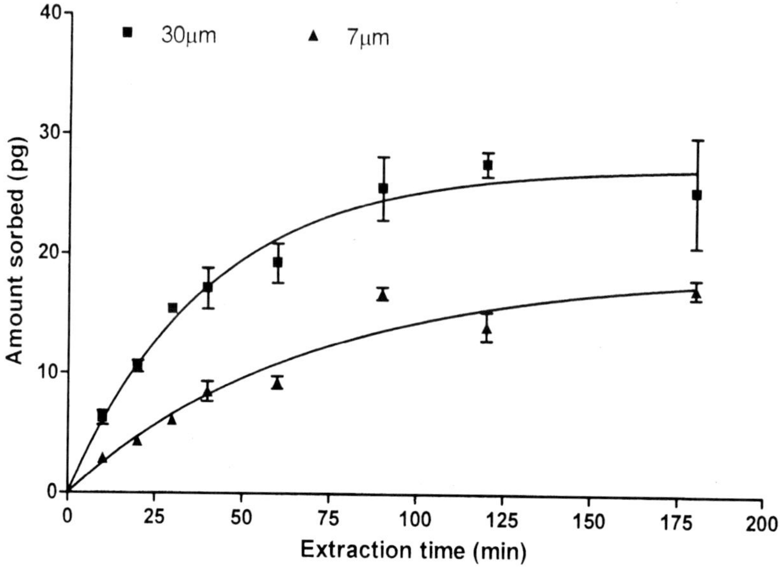

Figure 1. Uptake curves of bifenthrin into 7- and 30-μm PDMS-coated SPME fibers

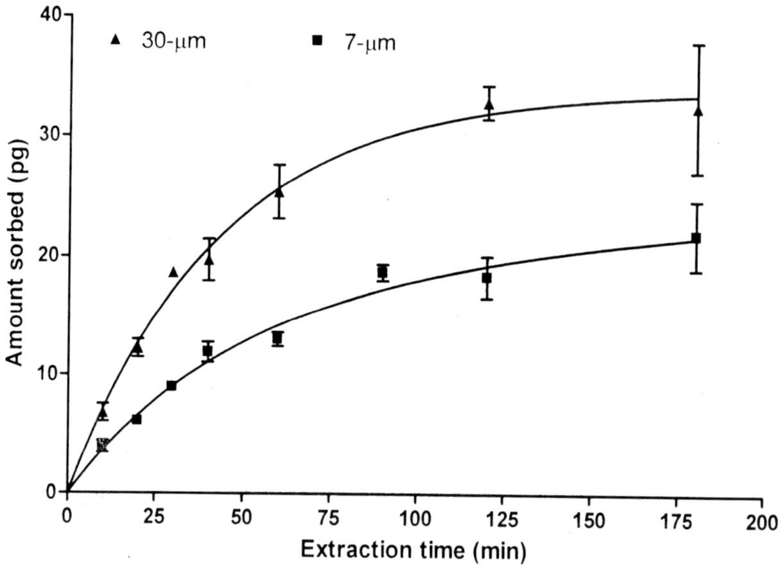

Figure 2. Uptake curves of cypermethrin into 7- and 30-μm PDMS-coated SPME fibers

describing the kinetics. The parameters derived from curve fitting are presented in Table 1 and Table 2. Uptake rate constants (k_1) in matrix-free solutions ranged from 194 to 417 min^{-1} for the 7-μm fiber, and from 74 to 274 min^{-1} for the 30-μm fiber, indicating slightly different diffusion kinetics of pyrethroids into fibers with different thicknesses of PDMS coating. Elimination rate constants (k_2) were relatively consistent, ranging from 0.010 to 0.037 min^{-1} for all pyrethroids and different coating thicknesses. It has been observed that for highly hydrophobic compounds, one of the limiting steps in kinetic phase is diffusion of free chemicals through the stagnant water layer surrounding the fiber coating (*28-31*). Extensive mixing of the sample, which may be achieved using automated agitation or sonication, can significantly reduce the unstirring water layer and improve kinetic uptake (*28, 32, 33*). Similar diffusion and elimination rates were obtained by using automated SPME coupled with an agitation device. These results suggest that under the conditions used here for both manual and automated SPME analyses, the effect of the stagnant layer may be negligible.

The ratio k_1V_{SPME}/k_2V_s was < 0.089 for the 7-μm PDMS fiber and <0.133 for the 30-μm PDMS. According to Vaes et al. (*26*), under the selected conditions the fiber uptake in the sample free of matrix and the sample containing matrix

Table 1. Uptake rate constant (k_1, min^{-1}), elimination rate constant (k_2, min^{-1}), minimum sampling time (t_s, min) and PDMS-water partition coefficient (K_{SPME}) for 7-μm PDMS fiber

Compound	k_1	k_2	*Log (K_{SPME})*	$\frac{k_1V_{SPME}}{k_2V_s}$	R^2	t_s	*dept.*[a] *(%)*
Bifenthrin	249	0.015	4.22	0.047	0.88	17.5	1.1
Fenpropathrin	417	0.015	4.44	0.078	0.95	17.4	2.0
λ-Cyhalothrin	243	0.017	4.15	0.040	0.91	17.2	1.0
c-Permethrin	315	0.01	4.50	0.088	0.95	18.2	1.6
t-Permethrin	412	0.013	4.50	0.089	0.95	17.8	1.9
Cyfluthrin	280	0.016	4.24	0.049	0.88	17.4	1.4
Cypermethrin	319	0.015	4.33	0.060	0.92	17.5	1.5
Esfenvalerate	209	0.018	4.06	0.032	0.76	17.1	1.1
Deltamethrin	194	0.014	4.14	0.039	0.71	17.6	1.1

[a] $dept.(\%) = \frac{C_{SPME}V_{SPME}}{C_0V_s}$

should be the same. Other studies also showed that no matrix effect occurred in fiber uptake in matrix-containing solutions (*15, 26, 28, 34, 35*).

Uptake of pyrethroids by the 7-μm PDMS fiber was more rapid than the 30-μm fiber. Since reaching equilibrium between the SPME fiber and the sample will take a long time (approximately 180 min), an estimated sampling time (min) and depletion (%) were calculated. In addition, the 30-μm PDMS fiber was selected over the 7-μm fiber due to its ability to absorb more chemicals under the same conditions, enabling a better sensitivity of detection. The following conditions were chosen for the subsequent experiments: 30-μm PDMS fiber, sampling time of 20 min, and 10 mL of sample volume for manual injection or 9 mL for automated SPME anlaysis. Because the sampling was carried out in the non-equilibrium state, sampling conditions including sampling interval, fiber immersion depth, and solution stirring speed were precisely controlled.

Table 2. Uptake rate constant (k_1, min^{-1}), elimination rate constant (k_2, min^{-1}), minimum sampling time (t_s, min) and PDMS-water partition coefficient (K_{SPME}) for 30-μm PDMS fiber

Compound	k_1	k_2	*Log (K_{SPME})*	$\frac{k_1 V_{SPME}}{k_2 V_s}$	R^2	t_s	*dept.*[a] *(%)*
Bifenthrin	123	0.022	3.75	0.074	0.84	16.6	2.6
Fenpropathrin	291	0.029	4.00	0.133	0.91	15.8	5.5
λ-Cyhalothrin	128	0.027	3.68	0.063	0.78	16.0	2.5
c-Permethrin	174	0.018	3.98	0.127	0.88	17.1	3.9
t-Permethrin	182	0.018	4.00	0.133	0.92	17.1	4.1
Cyfluthrin	119	0.02	3.77	0.078	0.85	16.8	2.5
Cypermethrin	145	0.018	3.91	0.106	0.80	17.1	3.1
Esfenvalerate	116	0.037	3.49	0.041	0.61	15.0	2.0
Deltamethrin	74	0.021	3.55	0.046	0.82	16.7	1.5

[a] $dept.(\%) = \frac{C_{SPME} V_{SPME}}{C_0 V_s}$

PDMS-water partioning coefficients (K_{SPME}) were further calculated for the various pyrethroid compounds in the apparent steady phase. Higher K_{SPME} values were obtained for the 7-μm PDMS fiber than for the 30-μm fiber. For instance, log (K_{SPME}) of pyrethroids for the 7-μm PDMS fiber ranged from 4.06 to 4.50 whereas for the 30-μm fiber, it ranged from 3.49 to 4.00. Assuming that depletion of analyte (amount of analyte on the fiber over the initial amount of analyte in the sample solution) should be < 10% to meet the negligible-depletion requirement, the minimum sample volume estimated for non-depletive SPME analysis should be 4.4 mL and 6.6 mL for 7- and 30-μm PDMS fiber, respectively, which are smaller than the volume (9 or 10 mL) used in all of our studies (*27*).

Linear Range, Detection Limits and Precision

The amount of an analyte sorbed on the SPME fiber under closely controlled and optimized conditions should be proportional to the concentration in the sample (*16, 33*). Under the conditions used, all pyrethroids displayed linear calibration ranges covering concentrations from 5 to 1000 ng L^{-1}. Full calibration curves were performed during the study and found to be linear with a consistent slope (RSD < 20%).

MDLs of pyrethroids ranged from 4 to 10 ng L^{-1} for SPME coupled with GC-ECD analysis, and 1-5 ng L^{-1} for SPME coupled with GC-MS-MS analysis. MDLs closely depended on the detection system as well as the sample matrix, and therefore can be further improved by using GC-MS analysis in the negative chemical ionization mode (NCI) or GC-MS-MS in the EI mode (Figure 3).

The precision of SPME calibration was examined over a two month period of time. The results show insignificant variation of the response factors of pyrethroids through external calibration (Figure 4). RSD was < 20% which is satisfactory for most routine analyses. Therefore, SPME analysis of pyrethroids can be done using calibration standard checks along with periodic full calibration plots.

Method precision was evaluated by calculating RSD of pyrethroid concentrations from replicated porewater analyses. The variation in replicated analyses (n = 4) was generally similar between LLE and SPME. The overall RSD for all analyses (n = 216) was 16.9 % for LLE, and 18.1% for SPME. Precision of the SPME method can be further improved by using automated SPME. For example, RSD of pyrethroid concentrations in replicated porewater analyses was found to be < 15% for repeated automated SPME analyses. Therefore, when sampling conditions such as fiber immersion depth, sampling time, and stirring speed are carefully controlled, SPME sampling of porewater for pyrethroids should be as reproducible as the conventional LLE method.

Determination of C_{free} in Porewater

Spiked sediment porewater samples were simultaneously analyzed by both SPME and LLE methods. Deionized water was spiked and analyzed by the two methods in each run for the purpose of quality control. Pesticide concentrations were analyzed by the two methods at three nominal concentrations (40, 100, and

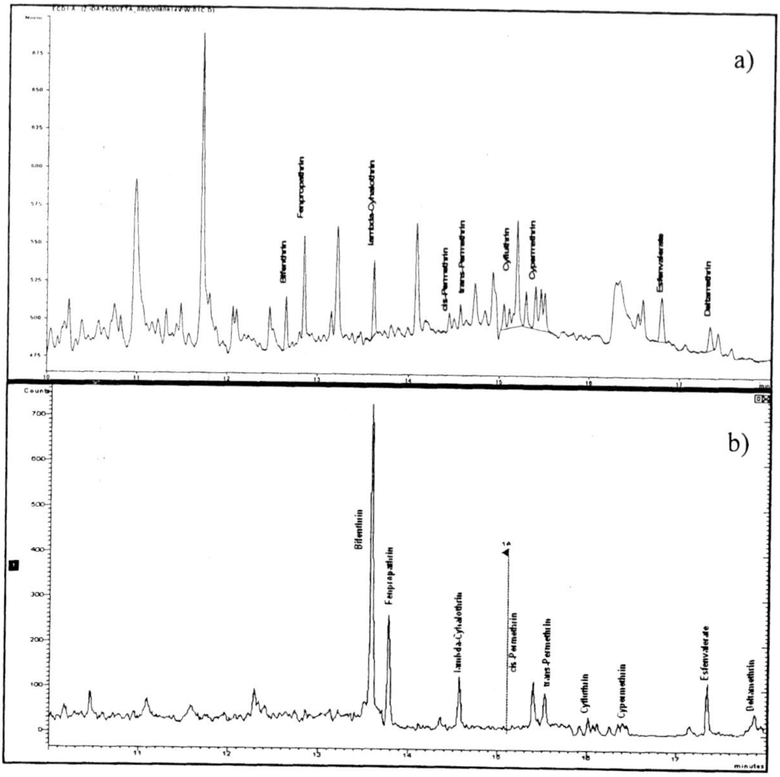

Figure 3. Chromatograms of pyrethroids in a porewater sample obtained with SPME coupled with (a) GC-ECD, and (b) GC-MS-MS

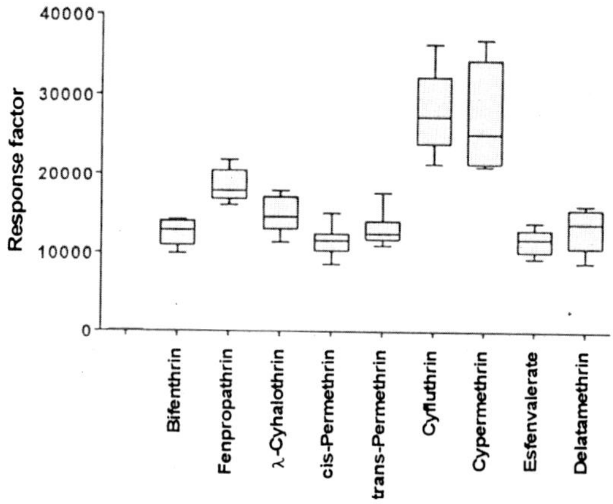

Figure 4. Variation of calibration curve response factors of pyrethroids over 2 months. Boxes represent highest and lowest values, horizontal lines represent mean values, and bars represents standard deviation.

500 ng L^{-1}) in deionized water, a freshwater sediment porewater (San Diego Creek, Orange, CA), and a marine sediment porewater (Little Harbor Beach, Wareham, MA). Results for the 100 ng L^{-1} are shown in Figure 5 for the marine sediment porewater and in Figure 6 for the freshwater sediment porewater. In the deionized water, the concentrations obtained by SPME and LLE were similar in most of the treatments based on two-tailed *t*-tests ($\alpha = 0.05$). This observation demonstrates that with the absence of an adsorbent phase (i.e., DOC), C_{free} was the same as C_w, and that the SPME method was capable of detecting C_{free}. Matrix and the stagnant layer effect when sampling in the kinetic state can potentially affect C_{free} measurement in porewater samples. However, the level of DOC in the porewater samples was always < 100 mg L^{-1}. Matrix effects, if any, are expected to be negligible (*15, 26, 28, 34, 35*).

In the spiked porewater samples, the concentrations determined by SPME were always only a fraction of those measured by LLE for the same samples (Figures 5 and 6). In the freshwater sediment porewater, the average fraction of C_{free} was about 37% of total concentration for fenpropathrin, 27.8% for *trans*-permethrin, 11.2-14.4% for bifenthrin, λ-cyhalothrin, *cis*-permethrin, cyfluthrin, and cypermethrin, and only 4.1-5.7% for esfenvalerate and deltamethrin. In the marine sediment porewater, the corresponding freely dissolved fractions were even smaller, ranging from 3.2 to 13.3%, with the freely dissolved fraction again being the largest for fenpropathrin, and the smallest for esfenvalerate and deltamethrin. The difference between the two analyses was attributed to sorption to DOC. We concluded that the dominant fraction of pyrethroids in the porewater samples was associated with the DOC phase, especially for the marine sediment.

The SPME method was further applied to the analysis of "aged" sediment samples collected from field sites in Orange County, CA, and C_{free} was compared against the total sediment concentration (Figure 7). While several pyrethroids were found at relatively high concentrations following exhaustive solvent extraction, the corresponding C_{free} values were very small. This observation was in agreement with reported studies showing that the total sediment concentrations were poorly correlated with bioavailability and yield inaccurate estimates of toxicity in aged sediment samples (*36*). In contrast, C_{free} measured by SPME provides a better prediction for sediment toxicity as it is related to contaminant bioavailability.

Determination of Porewater C_w using Isotopic-SPME

Another application of SPME for porewater analysis is the determination of total porewater concentration C_w. This application relies on the use of isotopically labeled internal standards in GC-MS analysis (*27, 37*). C_w of

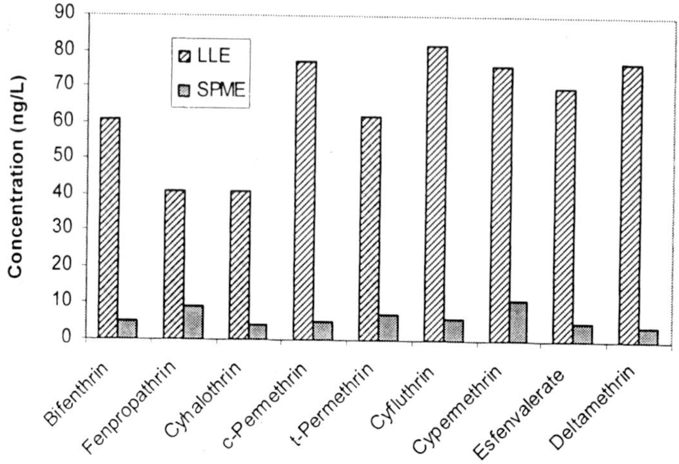

Figure 5. Aqueous concentrations of pyrethroids in a marine sediment porewater detected by LLE and SPME

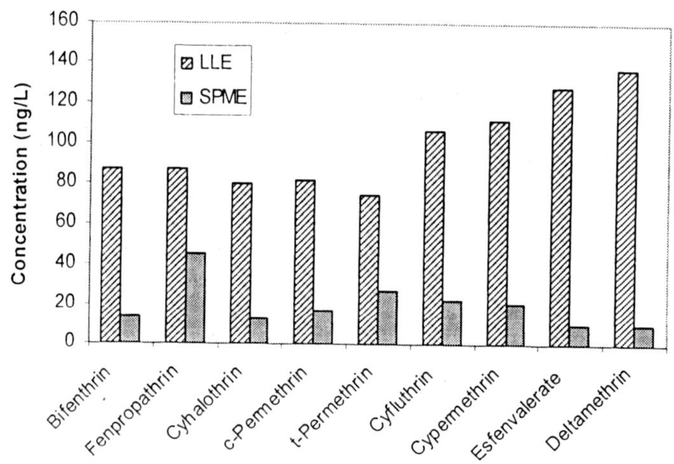

Figure 6. Aqueous concentrations of pyrethroids in a freshwater sediment porewater detected by LLE and SPME

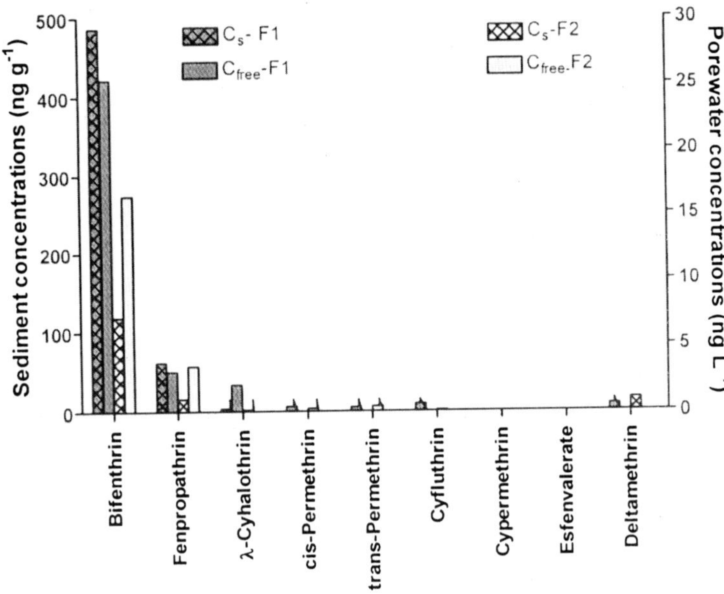

Figure 7. Pyrethroid concentrations in two field-contaminated sediments (F1 and F2) and the derived porewater samples

pyrethroids in porewater is the sum of C_{free} and DOC-complexed concentration and can be determined by SPME using isotopically labeled internal standard as

$$C_{w} = C_{i.s.} \frac{A_{w-SPME}}{A_{i.s.-SPME}} \tag{4}$$

where $A_{w\text{-SPME}}$ and $A_{i.s.\text{-SPME}}$ are absolute responses of nonlabeled and labeled compounds as given by SPME sampling, C_w is the total concentration of the non-labeled compound, and $C_{i.s.}$ is the total concentration of the labeled compound which is a known value. Identical partitioning coefficients for the labeled and nonlabeled forms of the same compound and rapid partitioning equilibrium between DOC and water phases are prerequisites for the validity of eq. 4. Ideally, the labeled compound should be the same as the analyte. However, for pyrethroids, there is a scarcity of ^{13}C-labeled standards and a complete lack of deuterium-labeled standards. C_w of a compound for which the labeled form is not available can be estimated by including a correction factor (K), which can be estimated from analyzing correlation of C_{free} between the labeled compound (i.e., ^{13}C-*cis*-permethrin) and a non-labeled pyrethroid (i.e., compounds other than *cis*-permethrin). Correction factors for pyrethroids estimated from using four different sediment porewater samples were 0.79 for bifenthrin, 0.41 for fenpropathrin, 0.98 for λ-cyhalothrin, 0.97 for *trans*-permethrin, 1.09 for cyfluthrin, 1.13 cypermethrin, 1.93 for esfenvalerate, and 1.88 for deltamethrin. Figure 8 shows the measured C_w values of pyrethroids using exhaustive LLE and isotopic-SPME method. Comparison using two-tailed paired *t*-test (α =0.05) showed no significant difference between the two methods in determining C_w. Thus, the SPME-isotopic GC-MS-MS approach is a good alternative to measuring both C_w and C_{free} of sediment porewater for pyrethroids. Compared to the conventional LLE approach, the isotopic-SPME procedure has several advantages. The isotopic-SPME method is solvent-less and much less laborious than the LLE approach as SPME sampling and analysis are fused into a single step. In isotopic-SPME analysis, both C_w and C_{free} can be obtained from the same run using the same sample. When coupled with an automated SPME sampler, the sample throughput is essentially determined only by the GC run time and can be greatly improved over the LLE method.

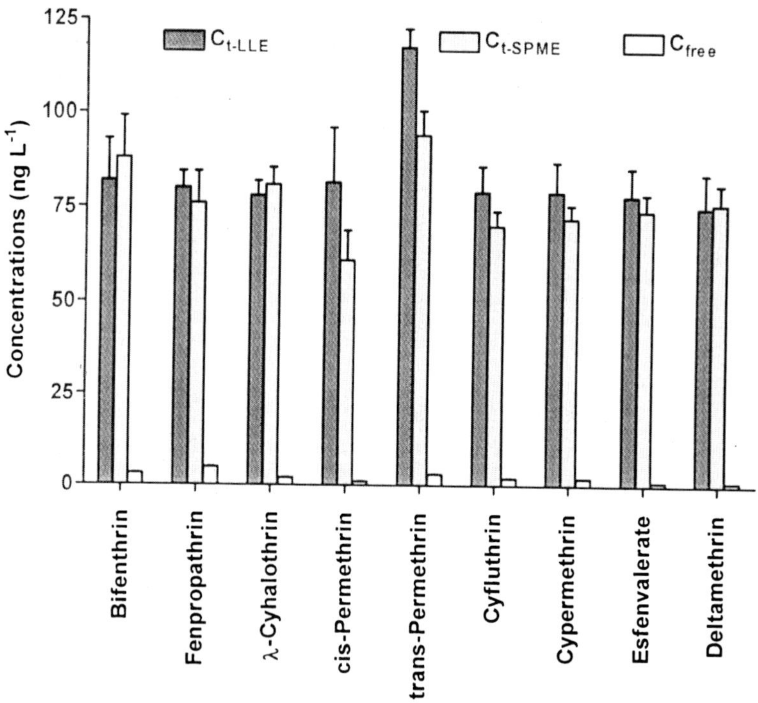

Figure 8. Total concentrations measured by LLE and SPME along with freely dissolved concentrations of pyrethroids in a marine sediment porewater

Distribution Coefficient K_{DOC}

Using SPME-measured C_{free} and C_w determined using either LLE or isotopic-SPME, the DOC-water partition coefficient K_{DOC} can be calculated as

$$K_{DOC} = \frac{(C_w - C_{free})/[DOC]}{C_{free}} \tag{5}$$

where $[DOC]$ is the level of DOC in the sample. Distribution coefficients K_{DOC} were calculated for various pyrethroids using the above approach. The averaged K_{DOC} values measured by the two different methods in the marine sediment porewater are listed in Table 3. K_{DOC}s measured by isotopic-SPME were in close agreement with those given by the LLE method. K_{DOC} was the smallest for fenpropathrin and the largest for esfenvalerate and deltamethrin, ranging 2.2-10.9 × 10^5 L kg^{-1} for the eight test pyrethroids. A similar pattern of relative K_{DOC}s was observed when several other sediments were used for deriving K_{DOC}s (Figure 9). However, the marine sediment porewater K_{DOC}s were found to be 2-10 fold greater than those derived using the other sediments. These findings together suggest that characteristics of DOC affected the phase distribution of pyrethroids in porewater. Estimated K_{DOC}s in this study correlated well with previously published results for the marine sediment (*24*).

Table 3. Partition coefficient K_{DOC} values in a marine sediment pore water estimated using isotopic-SPME and LLE methods[a]

Compound	*Isotopic-SPME*	*LLE*
Bifenthrin	4.33×10^5	4.01×10^5
Fenpropathrin	2.16×10^5	2.29×10^5
λ-Cyhalothrin	5.41×10^5	5.22×10^5
cis-Permethrin	5.78×10^5	7.68×10^5
trans-Permethrin	5.36×10^5	6.70×10^5
Cyfluthrin	6.04×10^5	6.86×10^5
Cypermethrin	6.25×10^5	6.88×10^5
Esfenvalerate	10.92×10^5	11.47×10^5
Deltamethrin	10.61×10^5	10.41×10^5

[a] Data are averages of four measurements and relative standard deviations of all data were below 11.2%

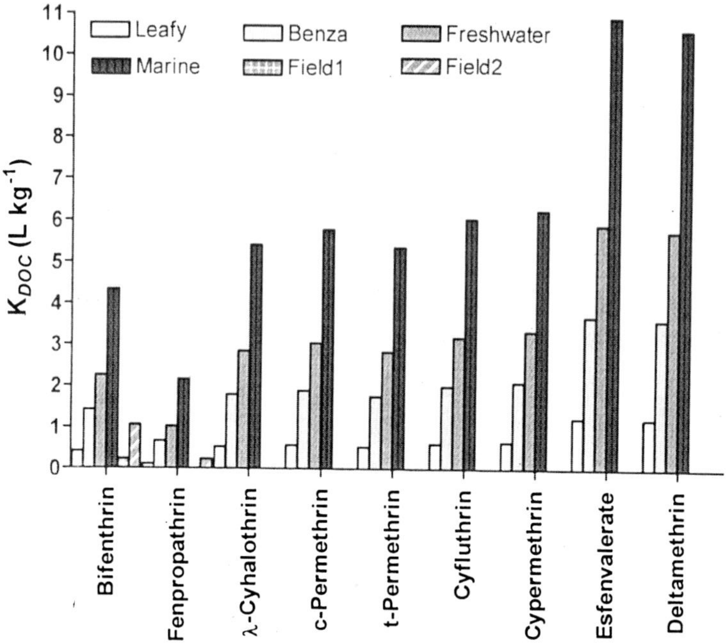

Figure 9. Averaged K_{DOC} values measured using SPME in porewater samples from various sediments. The data are averages of four measurements and relative standard deviations of all data were below 11.2%

Conclusions

Variability in K_{OC} and K_{DOC} complicates the use of OC normalization to accurately predict sediment toxicity from pyrethroid contamination. Solid-phase microextraction (SPME) provides a direct measure of pyrethroid C_{free} in sediment porewater, so is an improved approach to assessing sediment toxicity. The proposed SPME method allows the measurement of C_{free} when the sampling is coupled with GC-ECD. Moreover, when coupled with GC-MS analysis and ^{13}C-*cis*-permethrin as internal standard, SPME also gives the total porewater concentration C_w in the same analysis. These analyses may be further used to calculate distribution coefficients K_{DOC} that are otherwise difficult to quantify. The MDLs and method precisions of the developed SPME methods are satisfactory for analysis of pyrethroids. These parameters may be further improved by using GC-MS-NCI or GC-MS-MS with automated SPME sampling and injection. The proposed methods may be used for screening sediments for potential toxicity from pyrethroid contamination.

References

1. *Bioavailability of Contaminants in Soils and Sediments: Processes, Tools, and Applications*; Committee on Bioavailability of Contaminants in Soils and Sediments; The National Academies Press: Washington, D.C., 2003; pp.119-215.
2. Mcgroddy, S.E.; Farrington, J.W. *Environ. Sci. Technol.* **1995**, *29*, 1542-1550.
3. Di Toro, D.M.; Zarba, C.S.; Hansen, D.J.; Berry, W.J.; Swartz, R.C.; Cowan, C.E.; Pavlou, S.P.; Allen, H.E.; Thomas, N.A.; Paquin, P.R. *Environ. Toxicol. Chem.* **1991**, *10*, 1541-1583.
4. Landrum, P.F. *Environ. Sci. Technol.* **1989**, *23*, 588-595.
5. Weston, D.P.; Holmes, R.W.; You, J.; Lydy, M.J. *Environ. Sci. Technol.* **2005**, *39*, 9778-9784.
6. Weston, D.P.; You, J.; Lydy, M.J. *Environ. Sci. Technol.* **2004**, *38*, 2752-2759.
7. Amweg, E.L.; Weston, D.P.; You, J.; Lydy, M.J. *Environ. Sci. Technol.* **2006**, *40*, 1700-1706.
8. Maund, S.J.; Hamer, M.J.; Lane, M.C.G.; Farrelly, E.; Rapley, J.H. *Environ. Toxicol. Chem.* **2002**, *21*, 9-15.
9. Ronday, R.; Van Kammen-Polman, A.M.M.; Dekker, A.; Houx, N.W.H.; Leistra, M. *Environ. Toxicol. Chem.* **1997**, *16*, 601-607.
10. Kukkonen, J.V. K.; Landrum, P.F. *Aquatic Toxicology.* **1998**, *42*, 229-242.
11. Kukkonen, J.V.K.; Landrum, P.F.; Mitra, S.; Gossiaux, D.C.; Gunnarsson, J.; Weston, D. *Environ. Sci. Technol.* **2003**, *37*, 4656-4663.
12. Kukkonen, J.V.K.; Mitra, S.; Landrum, P.F.; Gossiaux, D.C.; Gunnarsson, J.; Weston, D. *Environ. Toxicol. Chem.* **2005**, *24*, 877-885.
13. Maruya, K.A.; Risebrough, R.W.; Horne, A. J. *Environ. Sci. Technol..* **1996**, *30*, 2942-2947.
14. Suffet, I.H.; Jafvert, C.; Kukkonen, J.; Servos, M.; Spacie, A.; Williams, L.; Noblet, J. In *Bioavailability: physical, chemical, and biological interactions;* Hamelink, J.; Landrum, P.; Bergman, H.; Benson, W., Eds.; SETAC Series, Lewis Publisher: Boca Raton, FL, 1994; pp. 93-108.
15. Heringa, M.B.; Pastor, D.; Algra, J.; Vaes, W.H.J.; Hermens, J.L.M. *Anal. Chem.* **2002**, *74*, 5993-5997.
16. Arthur, C.; Pawliszyn, J. *Anal. Chem.* **1990**, *62*, 2145-2148.
17. Mayer, P.; Vaes, W.H. J.; Wijnker, F.; Legierse, K.C.H.M.; Kraaij, R.H.; Tolls, J.; Hermens, J.L.M. *Environ. Sci. Technol.* **2000**, *34*, 5177-5183.
18. Vanderwal, L.; Jager, T.; Fleuren, R.H.L.J.; Barendregt, A.; Sinnige, T.L.; Vangestel, C.A.M.; Hermens, J.L.M. *Environ. Sci. Technol.* **2004**, *38*, 4842-4848.

19. Kraaij, R.; Mayer, P.; Busser, F.J.M.; Van Het Bolscher, M.; Seinen, W.; Tolls, J.; Belfroid, A.C. *Environ. Sci. Technol.* **2003**, *37*, 268-274.
20. Hawthorne, S.B.; Grabanski, C.B.; Miller, D.J.; Kreitinger, J.P. *Environ. Sci. Technol.* **2005**, *39*, 2795-2803.
21. Yang, W.C.; Gan, J.Y.; Hunter, W.; Spurlock, F. *Environ. Toxicol. Chem.* **2006**, *25*, 1585-1591.
22. Yang, W.C.; Spurlock, F.; Liu, W.P.; Gan, J.Y. *J. Agric. Food Chem.* **2006**, *54*, 3967-3972.
23. Yang, W.C.; Spurlock, F.; Liu, W.P.; Gan, J.Y. *Environ. Toxicol. Chem.* **2006**, *25*, 1913-1919.
24. Bondarenko, S.; Putt, A.; Kavanaugh, S.; Poletika, N.; Gan, J. *Environ. Toxicol. Chem.* **2006**, *25*, 3148-3154.
25. Bondarenko, S.; Spurlock, F.; Gan, J. *Environ. Toxicol. Chem.* **2007** (In press)
26. Vaes, W.H.J.; Ramos, E.; Verhaar, H.J.M.; Seinen, W.; Hermens, J.L.M. *Anal. Chem.* **1996**, *68*, 4463-4467.
27. Poerschmann, J.; Zhang, Z.; Kopinke, F.D.; Pawliszyn, J. *Anal. Chem.* **1997**, *69*, 597-600.
28. Ramos, E.; Meijer, S.N.; Vaes, W.H.J.; Verhaar, H.J.M.; Hermens, J.L.M. *Environ. Sci. Technol.* **1998**, *32*, 3430-3435.
29. Oomen, A.G.; Mayer, P.; Tolls, J. *Anal. Chem.* **2000**, *72*, 2802-2808.
30. Heringa, M.B.; Hermens, J.L.M. *Trac-Trends in Analytical Chemistry.* **2003**, *22*, 575-587.
31. Mayer, P.; Karlson, U.; Christensen, P.S.; Johnsen, A.R.; Trapp, S. *Environ. Sci. Technol.* **2005**, *39*, 6123-6129.
32. Zeng, E.Y.; Noblet, J.A. *Environ. Sci. Technol.* **2002**, *36*, 3385-3392.
33. Louch, D.; Motlagh, S.; Pawliszyn, J. *Anal. Chem.* **1992**, *64*, 1187-1199.
34. Poerschmann, J.; Kopinke, F.D.; Pawliszyn, J. *Environ. Sci. Technol.* **1997**, *31*, 3629-3636.
35. Schneider, A.R.; Paolicchi, A.; Baker, J.E. *International Journal of Environmental Analytical Chemistry.* **2006**, *86*, 789-803.
36. Alexander, M. *Environ. Sci. Technol..* **2000**, *34*, 4259-4265.
37. Poerschmann, J.; Kopinke, F.D.; and Pawliszyn, J. *J. Chromatog. A.* **1998**, *816*, 159-167.

Fate and Transport Processes

Chapter 9

Synthetic Pyrethroids in Agricultural Surface Waters: Exposure, Effects, and Risk Mitigation

Ralf Schulz and Sebastian Stehle

Institute for Environmental Sciences, University Koblenz-Landau, Campus Landau, Fortstrasse 7, D–76829 Landau, Germany

This book chapter gives an overview of synthetic pyrethroids in agricultural surface waters with particular emphasis on the aspects of exposure, effects and risk mitigation. Exposure from agricultural sources is largely short-term in nature and edge-of-field runoff appears to be a rather important source of entry for both aqueous-phase and particle-associated compounds. Concentrations of pyrethroids detected vary between a few nanograms in water up to several hundred micrograms in sediments or suspended particles. Three case studies on pyrethroid effects are included. Interspecific (predation) and intraspecific (cannibalism) interactions increased the susceptibility of model communities to pulse exposure with fenvalerate in artificial stream microcosms. Another stream microcosm study with cypermethrin that simulated agricultural runoff conditions with elevated turbidity levels indicated the importance of reduced bioavailability under field conditions in which there is naturally more suspended and dissolved material than in controlled laboratory studies. This study also suggested the behavioural reactions of mayfly nymphs as important, i.e. nymphs exposed to higher flow rates reduced their pyrethroid exposure by a flow avoidance behaviour. A third study used *in situ* bioassays with the amphipod species *Gammarus pulex* in a stream system receiving transient fenvalerate runoff. Here, an avoidance reaction was evident in that the amphipods actively migrated from a site contaminated with fenvalerate into an uncontaminated tributary during the runoff event. Hence, under field conditions, avoidance reactions of mobile non-target organisms appear to be a relevant process which, if not considered, may lead to overestimation of real-world toxicity e.g. in *in situ* studies while there may be no manifest responses at the in-stream population level. Due to the relatively low water solubility of pyrethroids, in stream mitigation measures using aquatic plants such as vegetated ditches or constructed wetlands appear to be a valuable management tool. A case study is presented below, which demonstrates that 300-m stretches of vegetated ditches effectively mitigated concentrations of 666 µg/L bifenthrin or 375 µg/L lambda-cyhalothrin experimentally-added to simulate a storm runoff event. It may be concluded, that by incorporating vegetated drainage ditches or constructed wetlands into a watershed management program for pyrethroids, that agriculture can continue to decrease the potential impact of nonpoint-sources to downstream aquatic receiving systems.

Introduction

Any unintended loss of pesticide is not only wasteful, but also represents a reduced efficiency and incurs increased costs to the user and the nontarget environment (*1*, *2*). Nonpoint-source pesticide pollution from agricultural areas is widely regarded as a significant source of contamination of surface waters (*3-5*). Various routes of nonpoint-source pesticide transport into surface waters have been addressed elsewhere (*6-8*).

Pyrethroids are synthetic derivatives of pyrethrins, which are natural insecticides that are produced by certain species of *Chrysanthemum*. Pyrethroids are neurotoxins and target insects' central nervous system (*9*, *10*). According to Oros and Werner (*11*), the pyrethroids of greatest interest to water quality include bifenthrin, cyfluthrin, cypermethrin, deltamethrin, esfenvalerate, lamba-cyhalothrin, and permethrin. These insecticides are applied in urban areas primarily for structural pest control, in agricultural areas on row crops (e.g., alfalfa, cotton, and lettuce) and orchards (e.g., almonds, pistachios, and peaches), and some are used in the home in pet sprays and shampoos.

Pyrethroids generally have low vapor pressures and Henry's Law constants which suggest that they are not easily volatilized into the atmosphere (*12*). They have high octanol/water partition coefficients (K_{OW}) so they tend to partition into lipids. They also have very high water/organic carbon (K_{OC}) partition coefficients, which suggests that the greatest risk to aquatic organisms would be through potential exposure to pyrethroid contaminated sediments (*13*). Although laboratory K_{ow} studies suggest that pyrethroids may bioconcentrate, depuration is rapid and bioaccumulation through the food web seems not to be a significant route of exposure (*14*). Pyrethroids readily bind to suspended particulate materials in the water column including clay, soils, sediment particles, and organic matter, which act as primary vectors for pyrethroid transport through aquatic systems while also reducing their bioavailability (*15*). Indeed, sorption to plants or sediments has been suggested as a method to mitigate the acute toxicity of pyrethroids by reducing their short-term bioavailability in the water column (*16*, *17*).

Most aquatic invertebrates and fish are highly susceptible to pyrethroid insecticides (*18*, *19*). Pyrethroid 96-h LC_{50}s for fish, aquatic insects and crustaceans are well below 1 µg/L, in contrast, molluscs are relatively insensitive to these chemicals (*18*). Overall, most aquatic invertebrates are more sensitive to pyrethroids than fish. Pyrethroids are several orders of magnitude more toxic to fish than the organophosphate pesticides they are replacing in many agricultural, commercial and residential applications.

The main objective of this chapter is to provide an overview of pyrethroids in agricultural surface waters with particular focus on exposure, effects and risk mitigation. This chapter contains a combination of a review of published literature and case studies covering various aspects of fate and effects. With regard to exposure, data from a recent review (*20*) were compiled and complemented with further literature screening. On the pyrethroid effects side, this chapter addresses three different case studies highlighting the importance of biological interactions, the interference between non-chemical abiotic factors and toxicity, as well as the ecological complexity and hence the difficulty in interpreting results obtained under field conditions in agricultural surface waters.

The third part of this chapter considers vegetation in surface waters, specifically in vegetated ditches or constructed wetlands, as a risk mitigation option for pyrethroid insecticides using a case study conducted in the Mississippi delta on bifenthrin and lambda-cyhalothrin. Since wetlands have a high ability to retain and process material, it seems reasonable that constructed wetlands, acting as buffer strips between agricultural areas and receiving surface waters, could mitigate the impact of pyrethroids. The effectiveness of wetlands for reduction of hydrophobic chemicals, e.g. most pyrethroids, should be as high as for suspended particles and phosphorus, since these chemicals enter aquatic ecosystems mainly in particle-associated form following surface runoff (*21*, *22*).

Exposure

This literature overview on pyrethroid exposure in surface waters from agricultural nonpoint-source pollution is based mainly on a previous review (*20*) supplemented by some recently published field studies. Studies published in the publicly available peer-reviewed literature since 1982 reporting agriculturally derived pyrethroid concentrations were assessed for the exposure situation of surface waters. Besides the included literature, various studies, e.g. those based on monitoring programs conducted by governmental agencies, demonstrated pyrethroid concentrations in surface waters (e.g. *23*, *24*). These studies are not discussed further in this book chapter because they are not peer-reviewed and also not accesible via scientific literature databases.

During the last fifteen years, detections of pyrethroids in surface waters increased when compared to the period before 1992 (Table I). Higher frequency of detections are attributed to the replacement of organophosphorous insecticides by the more efficient synthetic pyrethroids (*25*) and to sensitive modern analytical techniques such as gas chromatography-tandem mass spectometry or liquid chromatography-tandem mass spectometry. These analytical procedures permit detection of pyrethroid concentrations in surface waters at the ng/L or ng/g level, where some authors report relevant biological effects (e.g. *26*, *27*). Weston *et al.* (*28*) estimated the *Hyallela azteca* 10-d LC_{50} values for bifenthrin, cyfluthrin, and deltamethrin in the range 3 – 6 ng/g in a low organic carbon sediment (1% organic carbon). These LC50 values are only slightly above the detection limit of 1 ng/g achieved in the same study for those pyrethroids. Therefore, Weston *et al.* (*28*) indicated the need to improve the detection limits to permit the reliable identification of pyrethroids already at environmental levels beneath ecotoxicological relevant concentrations. Regarding the high toxicity of pyrethroids for non-target organisms a further focus on the environmental concentrations in surface waters is recommended to evaluate the risk for these aquatic ecosystems.

Table I. Total no. of pyrethroid detections in agricultural water and suspended particle samples for the time periods 1980 to 1992 and 1993 to 2005 based on data published in the open literature.

Time span	*Water*	*Particles*	*∑Water + particles*
1980 – 1992	54	30	84
1993 – 2005	275	191	466

Water: n = 329; Particles: n = 221

An analysis of field studies identified runoff and spray-drift as significant transport pathways of pyrethroids to surface waters, whereas runoff during rainfall or major storm events (e.g. *29-31*) is the most important route of entry (Table II). As expected, spray-drift results more often in pyrethroid detections in the water than in the particle phase (e.g. *32*). The acute lack of information regarding the actual entry routes in field studies results in the high percentage of unattributable nonpoint-source related detections listed in Table II. Future work (e.g. additional event-related sampling programs) may help to better assess and quantify the contributions of the various potential exposure pathways to pyrethroid detections in surface waters allowing implementation of appropriate and effective landscape-level or application-related risk mitigation measures.

Table II. Routes of entry for pyrethroids into agricultural surface waters based on data published in the open literature between 1980 and 2005.

Route of entry	*Water (%)*	*Particles (%)*
Nonpoint-source	76	79.2
Aerial application	0.9	1.3
Spray-drift	7.6	4.1
Runoff	14	11.3
Irrigation (Runoff)	1.5	4.1

Water: n = 329; Particles: n = 221

After entering surface waters, pyrethroids may pose a risk to aquatic organisms depending on their particular intrinsic toxicity and their concentration in the water body. Table III lists case studies published since 1982 on the detection of different pyrethroid compounds in surface waters caused by agricultural nonpoint-source pollution to assess the field relevance of pyrethroids exposure. The reports are sorted according to the pyrethroid compound, listing first the detections in water followed by the detections in suspended particles and sediments.

A range of pyrethroid concentrations have been detected in water and particles of surface waters (Table III); maximum concentrations up to 6.2 µg/L of fenvalerate in the water phase (*31*), up to 71.3 µg/kg fenvalerate in suspended particles (*33*), and up to 302 µg/kg fenvalerate in sediments (*31*) have, for example, been detected in these three compartments. Apart from fenvalerate, the highest environmental levels of the other pyrethroid compounds listed in table III are 5 µg/L cyfluthrin (*34*), and 1.7 µg/L cypermethrin (*32*) in the water phase, 37.5 µg/kg deltamethrin in sediments (*35*), and 163 µg/kg permethrin in sediment cores, respectively (*36*). However, these results represent the maximum concentrations of detections, and values below 0.1 µg/L (e.g. *37, 38*) have also been measured in a large number of the field studies.

It is interesting to note that nearly half of the field studies listed in Table III used an event-triggered sampling program to detect pyrethroids. This is not surprising because pyrethroids originating from nonpoint-sources are present for only brief periods in small headwater environments typical for agricultural

Table III. Summary of field case studies on pyrethroid contamination in surface waters due to agricultural practice published since 1982 (modified after Schulz (*20*))

Substance	*Concentration†*	*Source*	*Sampling interval*	*Location*	*Ref.*
Cyfluthrin	0.2–5 µg/L	nonpoint	daily	Sweden	(*34*)
Cyfluthrin	0.34 µg/L	nonpoint	monthly	USA	(*37*)
Cypermethrin	0.4–1.7 µg/L	spray-drift	event	France	(*32*)
Cypermethrin	0.001-1.6 µg/L	nonpoint	monthly	USA	(*37*)
Cypermethrin	2.7 µg/kg	nonpoint	single	UK	(*35*)
Deltamethrin	0.08–2 µg/L	runoff	event	UK	(*38*)
Deltamethrin	1.4 µg/L	runoff	event	South Africa	(*39*)
Deltamethrin	0.04-0.5 µg/L	nonpoint	seasonal	Brazil	(*40*)
Deltamethrin	1.9–37.5 µg/kg	nonpoint	single	UK	(*35*)
Fenvalerate	0.2–6.2 µg/L	runoff	event	Germany	(*31*)
Fenvalerate	0.01–0.11 µg/L	runoff	biweekly	USA	(*41*)
Fenvalerate	0.11 µg/L	runoff	event	USA	(*42*)
Fenvalerate	0.02–0.9 µg/L	runoff	event	USA	(*30*)
Fenvalerate	20–70 µg/kg SP	nonpoint	seasonal	Sweden	(*43*)
Fenvalerate	33–71.3 µg/kg SP	runoff	14 d	Germany	(*33*)
Fenvalerate	35.8 µg/kg SP	runoff	event	Germany	(*44*)
Fenvalerate	302 µg/kg	runoff	event	Germany	(*31*)
Fenvalerate	10–80 µg/kg	nonpoint	seasonal	Sweden	(*43*)
Fenvalerate	0.6–3.6 µg/kg	nonpoint	single	UK	(*35*)
Fenvalerate	0.7–10.8 µg/kg	nonpoint	single	USA	(*45*)
Fenvalerate	1.0–10 µg/kg	runoff	event	Germany	(*31*)
Permethrin	0.6 µg/L	nonpoint	monthly	Sweden	(*46*)
Permethrin	0.01–0.13 µg/L	runoff	biweekly	USA	(*41*)
Permethrin	0.094 µg/L	runoff	event	USA	(*29*)
Permethrin	0.5–1.6 µg/L	runoff	event	USA	(*47*)
Permethrin	2 µg/kg SP	nonpoint	seasonal	Sweden	(*43*)
Permethrin	1–3 µg/kg	nonpoint	seasonal	Sweden	(*43*)
Permethrin	0.5–163 µg/kg	nonpoint	single	UK	(*36*)
Permethrin	18 µg/kg	nonpoint	single	UK	(*35*)

† If not mentioned otherwise, the concentrations given refer to the sum of all isomers. SP, suspended particles

landscapes and their detections is much more difficult without event-controlled sampling (*48*). Considering the high water/organic carbon (K_{OC}) partition coefficients of pyrethroid insecticides (*12*) the exposure of aquatic organisms living in the water column can be assumed to be very short (hours). Nevertheless, a number of studies exist that indicate that an exposure duration between 0.5 to 1h to pyrethroid concentrations as low as 0.001 µg/L may cause long-lasting effects on some aquatic organisms (26, 49, 50).

Finally, most of the field studies listed in Table III report pyrethroid detections in surfaces waters of the United States or Europe demonstrating a lack on information about other regions of the world, especially Africa, Asia, and South America. More research is needed to provide data for a global perspective of pyrethroid contamination of surface waters or a pre-assessment of this issue in developing countries.

Effects

Importance of biological interactions

Agricultural nonpoint-source pollution results in transient pesticide contamination of surface waters (20, *34*). Following application onto agricultural fields, chemicals with low water solubility, such as pyrethroids sorb to soil particles that may be introduced into surface waters by erosion and surface runoff during heavy rainfall conditions (*21*).

In runoff studies monitoring the pyrethroid insecticide fenvalerate (FV), values between 0.5 and 39.7 µg/L (water + sediment) were measured (*51*). River sediments contained FV concentrations between 0.6 and 3.6 µg/kg (*35*), while estuary sediments contained levels up to 100 µg/kg (*52*). Short-term peak concentrations in suspended particle samples attained approx. 900 µg/kg in edge-of-field runoff and up to 302 µg/kg in in-stream samples (*20*).

Previous single-species studies on the toxicity of FV on the caddisfly larvae *Limnephilus lunatus* Curtis reported sublethal and lethal effects between 2 µg/kg and 2000 µg/kg, respectively (*26*). Chandler (*52*) measured significant effects on the reproductive output of meiobenthic copepods at sediment FV levels of 25 µg/kg, while mortality was not detected at levels as high as 100 µg/kg. Based on comparisons with the above-cited exposure values measured in the field, only sublethal effects may be expected to result from short-term exposure to sediment-associated FV contamination.

The release of pesticides in aquatic ecosystems can result in impacts at the community level. As a result, multispecies tests have been developed to reduce uncertainties when extrapolating from the laboratory to the field (*53, 54*). To predict the effects of FV on riffle insect communities, Breneman and Pontasch

(*55*) exposed stream microcosms for a period of 30 d to continuous aqueous concentrations between 0.01 and 10 µg/L. They found significant reductions of many insect taxa at levels of 0.1 µg/L; however, the field relevance of FV exposure over such long time periods is limited. Pond mesocosms were used by Webber *et al.* (*56*) to test for ecosystem effects of esfenvalerate simulating aerial drift and runoff exposure. Following esfenvalerate application, macroinvertebrate densities were significantly lower at high-rate ponds containing 0.7 µg/L mean aqueous and 56.3 µg/kg mean bulk sediment concentration.

Multispecies studies allow for the assessment of interactions between and within species in relation to chemical exposure. This aspect has been investigated so far mainly in plankton communities (*57*). Our knowledge about FV effects in multispecies systems is based either on aqueous-phase exposure or on pond mesocosm studies. However, in most regions, streams potentially receive greater total exposure to nonpoint-source agricultural pollution than ponds and runoff represents the most important exposure scenario. Unfortunately limited information is available concerning the effects of transient peak levels of pyrethroids associated with suspended particles on stream communities.

Multispecies stream microcosms were used to test the toxicity of fenvalerate (FV) associated with suspended particles in order to simulate a typical runoff exposure scenario in a study by Schulz and Liess (*58*) used here as a case study. Stream microcosms were exposed for 1 hour in triplicate to 0.0, 13.6, 136 or 1365 µg/kg FV and effects were monitored for 93 days. Experimental design (Figure 1) allowed for detection of interspecific effects on the emergence and thus survival of the caddisfly species *Limnephilus lunatus* Curtis and of intraspecific effects on the spatial distribution of adult and juvenile *Gammarus pulex* L. (Amphipoda).

Survival of *L. lunatus* was significantly reduced in the 1365-µg/kg treatment during single-species exposures. When other species were present, survival of *L. lunatus* was significantly reduced at 136 µg/kg (Figure 2). Possible reasons for this lethal effect include alterations in food competition, predation or density due to the chemical exposure. It cannot be determined whether this interspecific effect on *L. lunatus* was caused by a specific species within the outdoor microcosm study. However, it can be deduced that *G. pulex* is at least partially responsible for this effect in view of its abundance and active behaviour. Nilsson and Otto (*59*) reported a negative effect of the presence of *G. pulex* on the growth and survival of the caddisfly *Potamophylax cingulatus* Steph. This effect was present only during conditions with food limitation as an additional stressor, indicating that competition for restricted food resources played a major role in the interference between the two species. As derived from the drift data, about 35% of the *L. lunatus* larvae left their cases following exposure in the 136- and 1365-µg/kg treatment. This case-leaving behaviour has previously been reported as a general stress response following pesticide exposure (*60, 61*). The case-leaving behaviour may have increased the predation pressure on *L. lunatus* by other species (*62*). Another potential mechanism is an increased predation rate on the larvae during their pupal stage. Wisseman and Anderson (*63*) reported pupal mortality of 20% caused by predation in various caddisfly species.

A similar increase in test system susceptibility was observed in relation to the spatial distribution of *G. pulex*. Juvenile individuals avoided areas with high numbers of adult amphipods, which may prey on the juveniles (*58*). This avoidance was significant in the control and the 13.6-μg/kg treatment, but did not occur at higher levels of exposure (Figure 3).

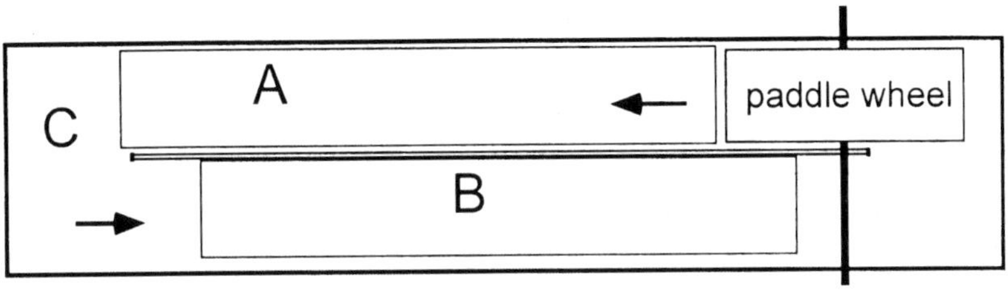

Figure 1. View from above of one stainless steel microcosm (1.2 x 0.3 m). The two boxes with 1-mm mesh on each end wall were initially provided either with Limnephilus lunatus *alone (A) or with* Limnephilus lunatus *together with adult* Gammarus pulex *and six other species (B). Juvenile* Gammarus pulex *born during the experimental period were able to leave the box and to move into the other box or into the remaining microcosm area C. Arrows denote the direction of water flow. (Reproduced with permission from reference 58. Copyright 2001 Allen Press.)*

High predation pressure by adults on the juveniles may be responsible for their spatial distibution, since it has been reported that adult male *G. pulex* feed on juveniles (*64-66*). Two sublethal effects may have caused the lack of the juveniles′ avoidance behaviour in the higher-exposure treatments. First, the juveniles, representing a sensitive life stage (*67, 68*), might have been affected at these concentrations in such a way as to reduce their potential to avoid areas with increased predation. Alternatively, the exposure may have caused a reduction of feeding activity in adult *G. pulex*, which is a well known reaction to pollutant stress in this species (*69, 70*). However, the intraspecific effects on spatial distribution of juvenile *G. pulex* were present at concentrations one order of magnitude lower than levels affecting the survival of this species (*58*).

In summary, the case study by Schulz and Liess (*58*) suggests that a field-relevant exposure design employing FV associated with suspended particles can affect aquatic macroinvertebrates. Interspecific and intraspecific interactions may influence the test results; that is, statistically significant lethal and sublethal effects are measurable at FV levels approximately an order of magnitude lower than reported effects observed in the absence of biological interactions. Biological interactions following a 1-h exposure were altered at levels of 136 µg/kg in the suspended particles added to simulate contaminated runoff, which is equivalent to approx. 10 µg/kg in the microcosm bulk sediments, if an equivalent distribution of the total amount of FV added in the total amount of sediment present is assumed. Transient peak concentrations of FV in this range have been reported to occur frequently in measured agricultural surface waters following runoff events (Table III), indicating the field-relevance of those results. It can be deduced that short-term contamination with particle-associated chemicals and biological interactions in the test system should be considered more carefully in risk assessment.

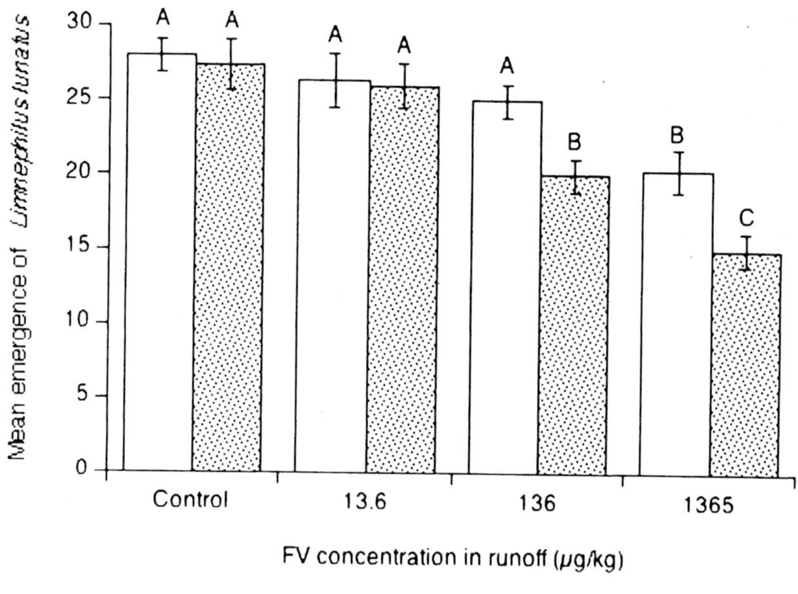

Figure 2. Mean number (±SE) of Limnephilus lunatus *emerging within 86 days after short-term runoff simulation with particle-associated fenvalerate (n = 3). White bars: single-species community, dotted bars: multispecies community. Treatments with the same letter are not significantly different (ANOVA, Scheffé's F-Test). (Reproduced with permission from reference 58. Copyright 2001 Allen Press.)*

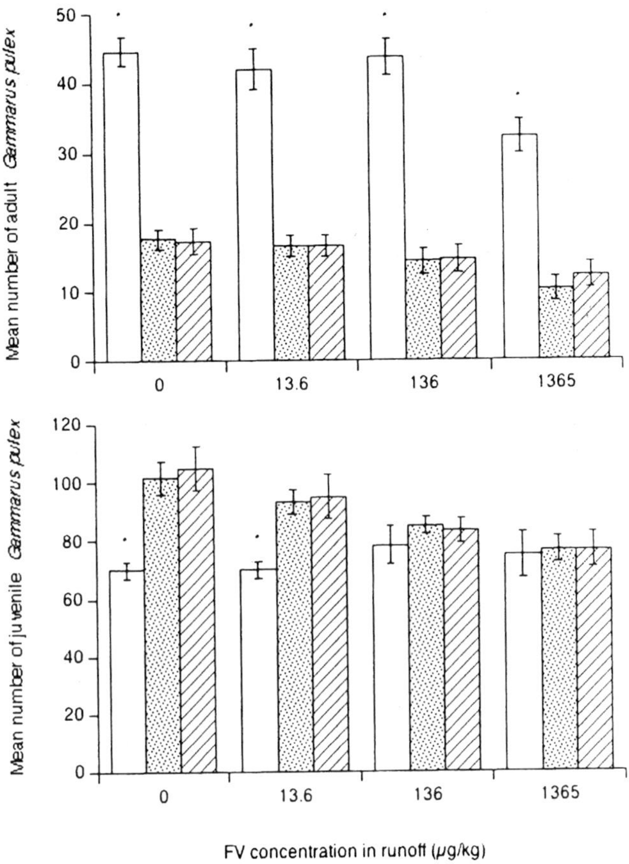

Figure 3. Mean number (±SE) of adult Gammarus pulex *(top) and juvenile* Gammarus pulex *(bottom) in different microcosm compartments at the end of the experiment 93 days after short-term runoff simulation with particle-associated fenvalerate (n = 3). White bars: box B, initially containing adult amphipods; dotted bars: box A, initially containing* Limnephilus lunatus; *hatched bars: remaining microcosm space C, initially free of animals (see also Figure 1 for further details). Asterisks indicate significant difference (ANOVA, Scheffé's F-Test) between B and A or C. (Reproduced with permission from reference 58. Copyright 2001 Allen Press.)*

Importance of abiotic interactions

Spray-drift and edge-of-field runoff are regarded as two major nonpoint-sources of pyrethroids in surface waters and typically result in short-term exposure, as the compounds are present in peak concentrations for only a few hours at most (*26*). Both may however differ considerably in their resulting exposure scenarios (*7*). Spray-drift leads to input of pesticides dissolved directly in the water phase, which is the only factor likely to influence aquatic fauna. Runoff-related input however, usually leads to increased discharge, flow velocity as well as increased levels of total suspended solids and pesticides, which may enter the surface water either as water dissolved or particle-associated chemicals (*71*). Despite reduced bioavailability through adsorption of pesticides to sediment (*15*), contaminated suspended particles have been shown to cause toxicological (*58*) as well as physical (*72*) effects on aquatic macroinvertebrates. For aquatic organisms, flow velocity is also a particularly relevant abiotic factor and has been shown to influence the toxicity of contaminants towards macroinvertebrates (*73*).

Hence, a multitude of factors can influence the aquatic environment under runoff conditions. Studies of these environmental factors have, however, usually been single-factor studies and little attention has focussed on their interactive impacts on aquatic organisms. Few studies have compared the ecotoxicological effects of typical runoff and spray-drift exposure scenarios on macroinvertebrates, and those that have, have only compared water-dissolved to particle-associated contamination (*13*), without incorporating the inherent differences in flow velocity in the experimental design. A comparative study of the sublethal effects of runoff and spray-drift related pesticide contamination that combines the various single factors is very relevant, particularly with regards to higher-tier risk assessments of pesticides.

In order to tackle the question of the importance of relevant abiotic factors, the effects of the pyrethroid insecticide, cypermethrin (CYP), increased flow speed (Flow), and increased suspended particles (Part) on drift behavior and activity of mayfly nymphs (*Baetis harrisoni*) were investigated individually and in combination in replicated (n = 5) laboratory stream microcosm experiments by Dabrowski *et al.* (*74*). The stream microcosms used in this study were similar to those described in the case study above (Figure 1; *58*), however, they were equipped with drift nets instead of boxes and contained a number of rocks as substrate (*74*). Spray-drift trials (CYP) were performed by exposing the *Baetis* nymphs to 1 μg/L cypermethrin. During runoff trials (CYP X Part) contaminated sediment containing 2000 μg/kg cypermethrin was introduced to the microcosm at a total suspended solid (TSS) concentration of 500 mg/L. Both studies were carried out as 30-minute trials under high (0.2 m/s for CYP x Flow and CYP x Part x Flow) and low flow (0.06 m/s for CYP and CYP x Part) conditions and for all cases control experiments were performed. Drift rate (drift per unit of time), drift density (drift per given volume of discharge for any treatments with increased flow) and activity were used as behavioral endpoints (*74*). Five trials were performed for each treatment.

Multi-factorial analysis of variance showed that CYP exposure significantly increased the drift (Table IV ; Figure 4), while Part and Flow trials significantly decreased the drift ($p < 0.05$). The increased drift rate in the CYP trials took place under field relevant exposure conditions, in terms of length and concentration of pesticide exposure (Table III) (*32*, *75*). This finding is in agreement with many other studies that have reported increased invertebrate drift as a response to insecticide exposure. Farmer *et al.* (*76*) found that the drift and emergence of baetid mayflies increased due to cypermethrin exposure of 0.7 g a.i./ha in experimental mesocosms. Schulz and Dabrowski (*77*) reported a significant increase in drift of *Baetis* sp. following a 0.2 µg/L exposure by the pyrethroid insecticide fenvalerate in microcosms. *B. harrisoni* is a relatively tolerant mayfly species (*78*) and is most probably relatively well adapted to the periodical return of suspended particles and hydraulic stress during heavy rainfall events in its habitats in the Western Cape of South Africa. This may explain why both elevated flow and turbidity reduced the drift. It is thus likely that the mayfly nymphs actively cope with both hydraulic and turbidity stress by remaining within the bed sediments as the activity was reduced in both setups from about 42 to 28% (*74*).

Table IV. Three-factorial analysis of variance of the effect of pesticide (CYP), suspended particles (Part) and increased flow (Flow) on *Baetis harrisoni* drift (ln(x+1) transformed)[a]

	Source	*df*	*MS*	*F*	*P*
Drift rate	CYP	1	8.519	148.874	<0.001
	Part	1	2.938	51.335	<0.001
	Flow	1	0.350	6.111	0.019
	CYP x Part	1	0.331	5.788	0.022
	CYP x Flow	1	0.037	0.650	0.426
	Part x Flow	1	0.001	0.011	0.916
	CYP x Part x Flow	1	0.000	0.000	1.000
Drift density	CYP	1	6.749	152.434	<0.001
	Part	1	2.328	52.585	<0.001
	Flow	1	3.764	85.015	<0.001
	CYP x Part	1	0.329	7.441	0.010
	CYP x Flow	1	0.264	5.964	0.020
	Part x Flow	1	0.046	1.029	0.318
	CYP x Part x Flow	1	0.000	0.000	0.994

[a] *df* = degrees of freedom; MS = mean square; *F* = likelihood ratio; *p* = probability

SOURCE: Reproduced with permission from reference 74. Copyright 2005 Allen Press.

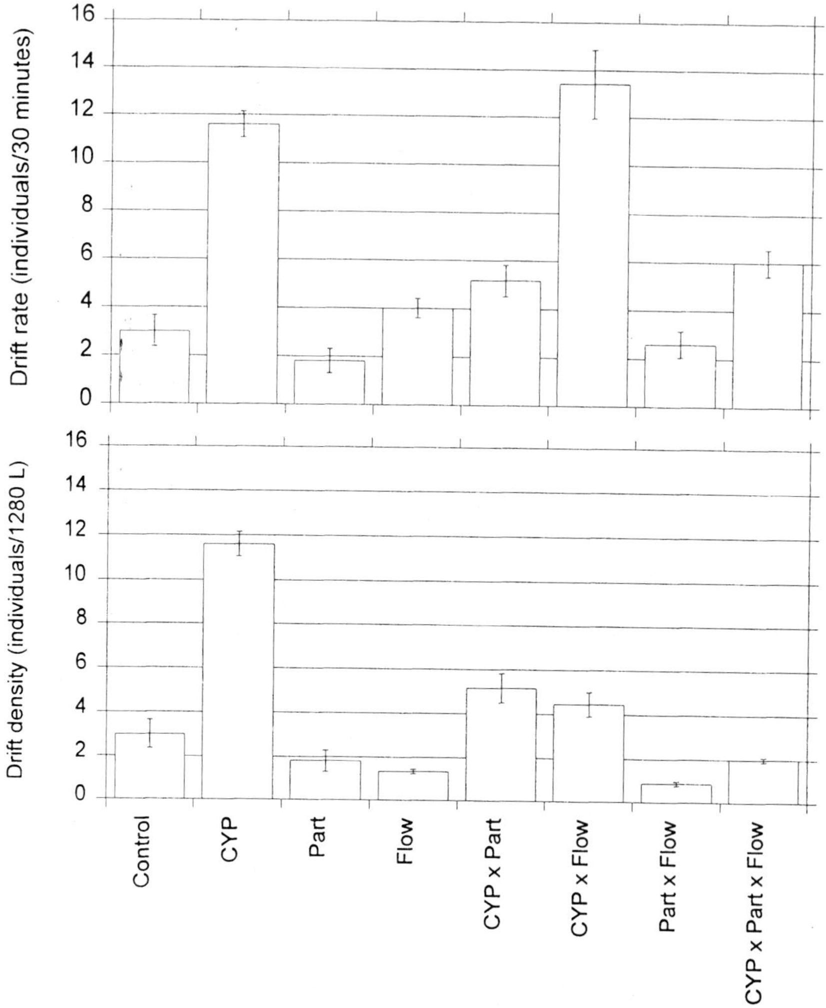

*Figure 4. Mean (± standard error, n = 5) of drifting mayflies (*Baetis harrisoni*) expressed as drift rate (*top*) and as drift density (*bottom*) in 30-min treatments either with the pyrethroid cypermethrin (CYP), suspended particles (Part), or flow increase (Flow) and their combinations. (Reproduced with permission from reference 74. Copyright 2005 Allen Press.)*

The CYP x Part and CYP x Flow trials had a significant (Table IV) antagonistic interactive effect on drift rate and density respectively with measured levels being lower than expected levels (Figure 5). For CYP x Part, this is most likely as a result of the pesticide binding strongly to sediment particles, thereby reducing its bioavailability (*79*) (*15*). Studies of the toxicity of contaminants in flow-through and static systems showed that macroinvertebrates usually react more sensitively in the flow-through systems (*80*), which is usually explained by the higher exposure at higher flow velocities. However, the case study by Dabrowski *et al.* (*74*) found an opposite result. The most likely explanation is that due to the reduction of activity at high flow rates, the exposure was also reduced. The mayfly nymphs actively positioned themselves underneath or downstream of the rock substrate available in the microcosms where the exposure due to reduced current velocities was presumably much lower than in the open water.

The case study by Dabrowski *et al.* (*74*) shows that abiotic interactions between chemicals, particles and flow rates may result in complex behavioural reactions in aquatic organisms. It indicated that mayflies reacted actively in response to flow conditions as the reduced drift was associated with a reduced activity and passively in response to pesticide exposure as an increased drift was observed, though the activity was reduced (*74*). These responses in turn can act to protect the organisms and greatly affect the extent of toxic effects of pesticides such as the tested pyrethroid cypermethrin. It may therefore be concluded, that testing under field-relevant conditions yields valuable information for a realistic description of potential adverse effects.

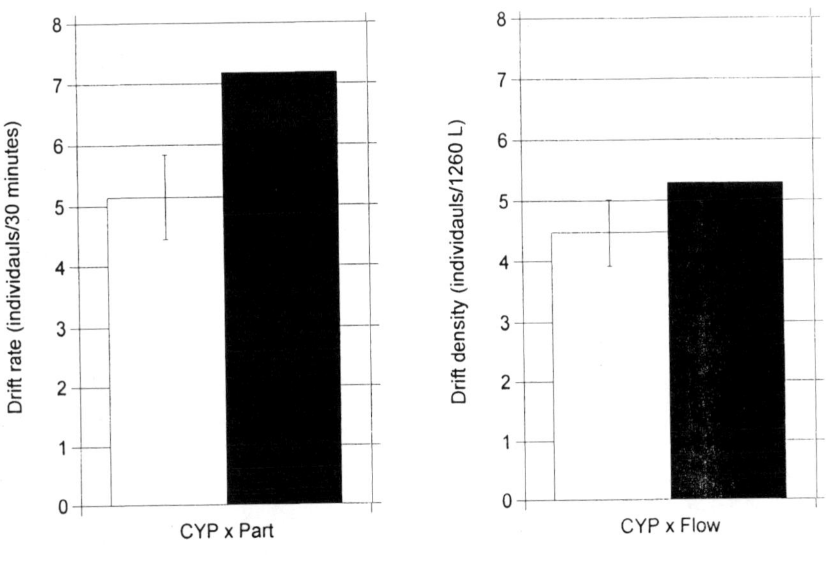

Figure 5. Mean measured (white bars ± standard error, n = *5) and expected (black bars) drift for the significant (Table IV) interaction CYP x Part and CYP x Flow. CYP = cypermethrin. (Reproduced with permission from reference 74. Copyright 2005 Allen Press.)*

Importance of complex reactions in the field

While most of the regulatory testing of agrochemicals is done either in the laboratory or in artificial experimental systems (outdoor micro- and mesocosms), the assessment of these chemicals is used for their application under field conditions and normal farming practice. It is therefore useful to consider test conditions that are as field-relevant as possible, though these types of tests usually are not repeated and are not standardized. *In situ* bioassays or active biomonitoring techniques may be valuable as they combine reproducibility with the potential to cause link and effects (*81*, *44*). In *in situ* bioassays, selected test organisms are placed in the field (*82*) so that they are exposed to the contamination that ordinarily occurs under field conditions. The results are thus considerably more relevant to the natural situation than those of laboratory experiments, especially with respect to the contamination scenario (*83*). Whereas chemical analyses in the field provide abiotic data, the results of the *in situ* bioassay are based on a toxicological response (*69*) and, accordingly, substantially more informative regarding the protection of animal and plant communities. *In situ* bioassays thus represent a link between ecotoxicological laboratory experiments and field studies. If meaningful conclusions are to be drawn, they must fulfill two requirements in particular (*44*):

"Toxicological relevance" e.g. there must be a clear relationship between the environmental stressor of interest (e.g., pesticide contamination) and the response measured in the bioassay.

"Ecological relevance" e.g. the response in the bioassay must directly or indirectly reflect responses of the same species (or the whole community) in the field.

In many studies, only the relationship between contamination and bioassay response have been examined, which suggests that *in situ* techniques may be used as a powerful tool for linking cause and effect (*84*, *30*). On the other hand, only very few studies investigated the relation between bioassay and field responses (*85*, *44*).

The following case study (*86*) is an example of a validation study to test for the toxicological and ecological relevance of an *in situ* bioassay with *Gammarus pulex* L. (Amphipoda) and *Limnephilus lunatus* Curtis (Trichoptera) deployed in an agricultural stream system (Figure 6a). Both tested species show sensitive reactions to transient insecticide input, but the effects on their population dynamics are different: while larval densities of *L. lunatus* were decreased due to elevated mortality, an active drift and avoidance behaviour enables *G. pulex* to leave contaminated sites temporarily and thus occur at high densities in contaminated streams (*75*). Details are included here only for the the amphipod results from Schulz and Liess (*86*) and focus exclusively on the question of the ecological relevance. *G. pulex* is frequently used as a test species for *in situ* bioassays (84, *87, 88*).

At each site, during the period from April 18th to July 8th, 1995, four boxes (40 x 17 x 15 cm) each containing 30 adult *G. pulex* (Amphipoda; carapace length > 6 mm) were installed in the stream (Figure 6b). The bioassay experiment began 10 days before insecticides were first applied in the study area. The boxes floated with the upper third above the water surface; the front and rear walls were made of netting (1 mm mesh) to allow water flow through (current velocity: 0.15 ± 0.04 m/s). Simultaneously to the bioassay exposure, the short-term insecticide concentrations and the stream population dynamics of both species were monitored.

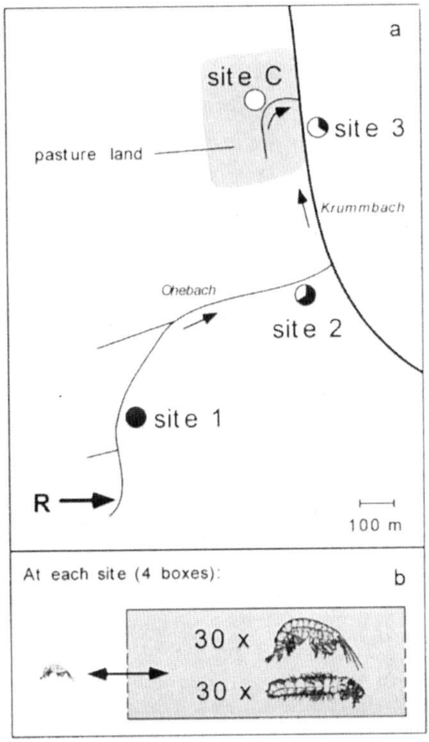

Figure 6. (a) Locations of the study area 35 km south of Braunschweig, Germany and the stations for bioassay exposure, field investigations and water quality sampling. R indicates the site at which runoff-related insecticide inputs occurred. The control site is surrounded by pasture land and served as a control with no insecticide contamination. Arrows indicate flow direction. (b) Four boxes containing 30 larvae of Limnephilus lunatus *and 30 adult* Gammarus pulex *were exposed at each site. Juvenile* G. pulex *were able to migrate into and out of the boxes. (Reproduced with permission from reference 86. Copyright 1999 Allen Press.)*

The number of surviving test organisms (adult gammarids) in the bioassay boxes was determined once a week. Juvenile gammarids, between 2.5 and 4 mm in carapace length were identifiable by their size, and their number was also determined weekly. However, it was impossible to decide whether such individuals actually immigrated or were born in the exposure boxes. In view of the rapid temporal dynamics, it was very likely that most of them were immigrants. This assumption was corroborated by the fact that the number of juveniles found in the boxes at a certain site and date was positively correlated ($R^2 = 0.59$; $p = 0.02$; $n = 9$) with the total numbers of *G. pulex* in the in-stream samples taken in parallel (*86*).

During transient insecticide inputs (duration: about 1 h; peak concentrations: 6.2 µg/L fenvalerate; 0.6 µg/L parathion-ethyl (Table V)), mortality of the exposed adult *G. pulex* in the in situ bioassay was significantly higher in the contaminated sites (e.g. site 3; Figure 7) than in the uncontaminated control tributary at site C (ANOVA, Fisher's PLSD; $p < 0.05$). Compared with other studies the measured fenvalerate concentrations were relatively high, whereas the parathion-ethyl concentrations were in the range of the values previously reported (*20*). The possible reason for these elevated levels might be that peak concentrations during runoff were measured with the event-controlled sampler. This is particularly likely for fenvalerate, which tends to be transported during short-term runoff-related input events due to its very low water solubility. Responses of the test species to insecticide inputs have previously been demonstrated in laboratory and field experiments (*75*, *89*). Microcosm studies (*55*) indicated that fenvalerate concentrations between 1 and 10 µg/L resulted in significant short-term (1 h) drift response of several macroinvertebrate species.

In the bioassays, *G. pulex* exhibited a clear decrease in the number of adult individuals when insecticides were present at the contaminated sites, while only a minor decrease occurred at the control site (Figure 7). In the field samples, however, the number of individuals of *G. pulex* present at these times decreased only slightly, but increased in general with time (*86*). Consequently, measures of the toxicity of pesticide inputs based on the responses of *G. pulex* in this bioassay may be overestimates with respect to the degree of toxicity in the field situation (*86*). The most likely reason for this is that the adult individuals in the exposure boxes cannot behave in the natural way when threatened by a toxin. As stated above, *G. pulex* is known to respond with a marked drifting behaviour, in both field and laboratory, when exposed to insecticides or other stressors (*75*, *90*, *91*). A significantly increased drift was indeed observed at site 1 during the insecticide input on May 27th. Schulz and Liess (*75*) assumed that the drift response is an active reaction and suggested that it enables the species to escape temporary stressful situations by moving to unaffected regions, with the result that *G. pulex* can maintain high population densities in contaminated waters.

Even in cases of acidification and copper contamination an elevated drift response of *G. pulex* has been observed, which was discussed as an active avoidance behaviour (*91*). The drift reaction of *G. pulex* in the field can be regarded as an adaptation to transient unfavourable conditions.

Table V. Occurrence of runoff events in the 1995 study period. Bioassays were deployed between April 18th and July 8th, 1995. First insecticide application in the catchment area was on April 28th, 1995. In the right column are the peak insecticide contaminations measured with automatic samplers placed in the stream between sites 1 and 2. At the control site, no insecticide substances could be detected at any time during the year. I_1, I_2 and I_3 are abbreviations as used in Figure 7 to indicate insecticide input events

Date	*Precipitation (mm/d)*	*Hourly peak discharge (L/s)*		*Insecticides in water samples (between sampling sites 1 and 2)*
April 19th/20th	11.5	25.1		not quantifiable
May 27th	15.6	7.3	I_1	fenvalerate: 6.2 µg/L parathion-ethyl: 0.6 µg/L
June 01st	26.3	37.9	I_2	fenvalerate: 3.3 µg/L parathion-ethyl: 0.15 µg/L
July 2nd	16.9	6.6	I_3	fenvalerate: 0.85 µg/L parathion-ethyl: 0.08 µg/L
mean[a]	1.4±2.4	5.9±4.7		not quantifiable

[a] mean values (± SD) for the week intervals during the study in which no runoff event occurred (n = 7).

SOURCE: Reproduced with permission from reference 86. Copyright 1999 Allen Press.

In summary the case study by Schulz and Liess (*86*) indicated, that during runoff events, *G. pulex* migrated from the potentially contaminated headstream section into the uncontaminated tributary, which can be regarded as a refuge and source for recolonization. This case study also indicated that significantly lower coefficients of variance in the bioassay (≤ 0.22 compared to ≥ 0.55 in the field samples) allow for a better detection of adverse effects of pesticide with this method. Hence, although the bioassay is valuable for identifying insecticide input events, supplementary field studies are recommended for a correct ecological interpretation of the results and further understanding of the complexity of reactions under field conditions.

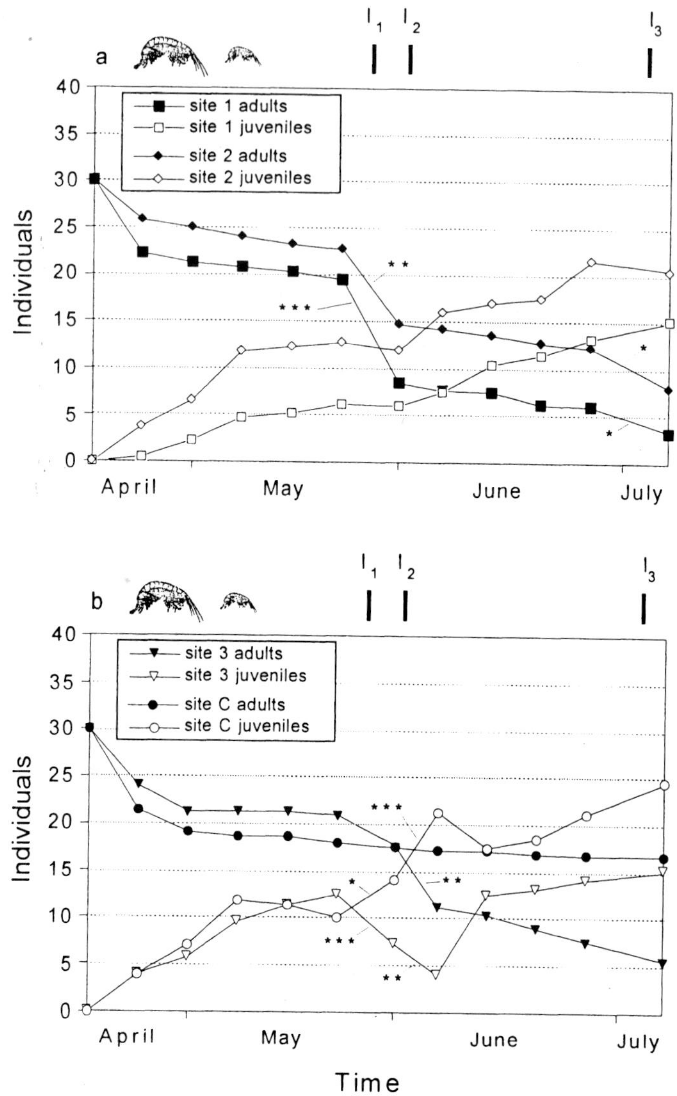

Figure 7. Mean survival rate (n = 4) of adult Gammarus pulex *and number of immigrant juvenile* G. pulex *in the in situ bioassay at the potentially contaminated stream site 3, 30 m upstream of the confluence with the uncontaminated control tributary (site C). Arrows indicate times of runoff-related insecticide inputs (see also Table V). Asterisks indicate significant (ANOVA, Fisher's PLSD; * $p < 0.05$; ** $p < 0.01$; *** $p < 0.001$) changes in the number of organisms between sampling dates. (Reproduced with permission from reference 86. Copyright 1999 Allen Press.)*

Risk Mitigation

Practical methods of controlling agricultural pollution risk were reviewed by Schulz (*20*), including both in-field (soil conservation measures, application practices, and integrated pest management) and end-of-field (buffer zones) techniques (*92*). The specific types of pesticide-related best management practices (BMPs) commonly used in the United States include reducing pesticide use, improving the timing and efficiency of application, preventing backflow of pesticides into water supplies, improved calibration of pesticide spray equipment, and IPM (*93*).

According to some authors, the suitability of buffer strips to retain mobile pesticides is questionable (*94-97*). One aspect that might restrict the effectiveness of any buffer strip is the fact that elevated rainfall intensities unavoidably lead to a relatively large proportion of water leaving the cropped land as surface runoff. This "hydrological dilemma" (*20*), means that an increase in rainfall intensity on loamy soil with a high soil moisture by a factor of three, e.g., from 10 to 30 mm, results in an increase of surface runoff by a factor of ten, from about 1 to 10 mm. This may result in unavoidable pesticide contamination of surface waters, particularly under conditions in which other mitigation measures are not applied or do not adequately produce the necessary benefit (i.e., high-quality soil areas under intensive agricultural use). In these cases, structural features of the receiving surface waters, such as vegetation coverage, may be useful in mitigating the risk of insecticide pollution.

Constructed wetlands or vegetated ditches were proposed as risk mitigation techniques. Complementing their ecological importance as ecotones between land and water (*98*) and as habitats with great diversity and heterogeneity (*99*), constructed wetlands are used extensively for water quality improvement. The concept of vegetation as a tool for contaminant mitigation (phytoremediation) is not new (*100*). Many studies have evaluated the use of wetland plants to mitigate pollutants such as road runoff, metals, dairy wastes, and even municipal wastes (*101-104*). According to Luckeydoo *et al.* (*105*), the vital role of vegetation in processing water passing through wetlands is accomplished through biomass nutrient storage and sedimentation, and by providing unique microhabitats for beneficial microorganisms. Macrophytes serve as filters by allowing contaminants to flow into plants and stems, which are then sorbed to macrophyte biofilms (*106*, *107*) .

Schulz (*20*) reviewed the published studies on the use of artificial wetlands or vegetated ditches for the mitigation of agricultural insecticides. Schulz (*20*) concluded that very few and only recent studies have dealt with wetlands or vegetated ditches as risk mitigation tools for nonpoint-source insecticide pollution and only some of them dealt with pyrethroids. However, the results obtained thus far on chemical retention and toxicity reductions are promising, and justify further investigation (*20*). A few other studies that have emphasized special aspects of pesticide fate or toxicity in wetlands (*108*, *109*) or uptake of insecticides to plants (*110-112*) corroborate the idea that aquatic macrophytes are important to insecticide risk reduction.

Bennett *et al.* (*16*) investigated drainage ditches in Mississippi as indispensable components of the agricultural production landscape. An environmental benefit of these ditches is mitigation of contaminants associated with agricultural storm water runoff. In the Mississippi Delta region there is a wide range of pesticides currently used in production farm acreage. Of these, synthetic pyrethroids are one of the main classes of insecticides used, especially in cotton and corn production. For example, bifenthrin is a fourth-generation pyrethroid insecticide. Approximately 52,000 kg active ingredient bifenthrin was applied to US corn (94%), cotton (3%), and blackberry (3%) crops in 2001 (NASS, 2003, http://www.nass.usda.gov). Lambda-cyhalothrin is another fourth-generation pyrethroid. Over 45,000 kg of lambda-cyhalothrin (as active ingredient) was applied to US cotton (54%), corn (41%) and soybean (5%) crops in 2001 (NASS, 2003, http://www.nass.usda.gov). Owing to their intrinsic toxic effects, there is regulatory concern about such compounds potentially reaching aquatic environments, especially during agricultural runoff and spray-drift events.

The purpose of the study by Bennett *et al.* (*16*) was two fold. The first objective was to evaluate the retention and partitioning (water, plant, sediment) of bifenthrin and lambda-cyhalothrin within a vegetated agricultural drainage ditch located in the Mississippi Delta, MS, USA during a simulated, worst case scenario runoff event. From these data, the relative importance of aquatic vegetation in facilitating the removal of insecticide from water was evaluated using mass balance calculations and insecticide physico-chemical properties. The second objective of the study by Bennett *et al.* (*16*) was to estimate drainage ditch lengths necessary for the effective mitigation of bifenthrin and lambda-cyhalothrin given recommended field application rates, and other rainfall and runoff variable assumptions. A controlled-release storm runoff simulation was conducted on a 650-m vegetated drainage ditch in the Mississippi Delta. Bifenthrin and lambda-cyhalothrin were released into the ditch experimentally (*16*). Samples of ditch water, sediment, and plants were collected and analyzed for pesticide concentrations following methods described by Bennet *et al.* (*113*).

Three hours after initiation of the storm runoff simulation, bifenthrin and lambda-cyhalothrin water concentrations ranged from 666 µg/L and 374 µg/L, respectively, at the inlet to 7.24 µg/L and 5.23 µg/L at 200 m downstream (Table VI). No chemical residues were detected at the 400 m sampling site. A similar trend was observed throughout the first 7 d of the study where water concentrations were elevated at the front end of the ditch (0 – 25 m) and greatly reduced by the 400 m sampling site (*16*).

Approximately 33.9% and 36.3% of the original dose of bifenthrin and lambda-cyhalothrin, respectively, remained in the water 3 h post application, and these concentrations were further reduced to 3.09% and 1.24% after 1 d. Similar results by Leistra *et al.* (*114*) demonstrated that 1.8-6.5% of the original dose of lambda-cyhalothrin applied into vegetated ditch enclosures remained after 3 d. These results indicate rapid reduction of both pesticides within the first day of application. This rapid decrease in aqueous concentrations in field conditions is of ecological relevance since pyrethroids intrinsically elicit toxic effects at extremely low concentrations in non-target aquatic organisms (*76, 58, 115*). In summary, under field conditions using vegetated ditches, bioavailability and hence exposure of aquatic organisms to pyrethroids is reduced.

Table VI. Downstream dissipation of bifenthrin and lambda-cyhalothrin concentrations (μg/L) in water at the 3-h, 12-h and 7-d sampling times (N.D. = Not Detected, below detection limits; 1.00 ng/L)

Distance	*Bifenthrin*			*Lambda-Cyhalothrin*		
(m)	*3 h*	*12 h*	*7 d*	*3 h*	*12 h*	*7 d*
0	666	10.7	0.887	375	5.29	0.250
25	235	25.9	7.76	115	11.8	2.32
50	77.2	6.33	0.178	39.1	3.44E	0.055
75	33.8	1.03	0.270	20.6	0.745	0.074
100	27.8	1.32	0.064	16.6	0.899	0.024
200	0.724	0.454	0.051	0.309	0.296	0.020
400	N.D.	0.471	N.D.	0.144	0.099	N.D.
650	N.D.	N.D.	N.D.	N.D.	N.D.	N.D.

SOURCE: Reproduced with permission from reference 16. Copyright 2005 Allen Press.

To reduce loadings and toxic effects in receiving water bodies after a runoff event, an effective vegetated drainage ditch length is required. Ditch lengths of 120 m and 280 m were estimated from the study by Bennett *et al.* (*16*) to reduce bifenthrin and lambda-cyhalothrin to 1.00% and 0.100% of initial values, respectively (Figure 8). These values are based on a worst-case scenario indicating that shorter ditch lengths may be acceptable, but depending on space limitations more conservative distances would likely be more effective. Other studies found a vegetated wetland length of 40 m to be required to reduce a worst-case runoff-related concentration of about 700 μg/L of the organophosphate insecticide methyl-parathion to levels below 0.1 μg/L (*116*).

Water, plant and sediment samples were collected throughout the study by Bennett *et al.* (*16*) to determine the relative importance of each compartment. Since it is difficult to directly compare concentrations between compartments, mass balance calculations were performed to better understand the distribution and fate of bifenthrin and lambda-cyhalothrin in this system. Using mass balance calculations it was determined that the ditch plants were the major sink and/or sorption site for the rapid dissipation of bifenthrin and lambda-cyhalothrin from the water column (Table VII). After 12 h the majority of pesticide mass (>93%) was found in the plant compartment giving evidence that this compartment was the most important and effective compartment in the mitigation of these insecticides. Other studies (*110*, *117-118*) have also shown the importance of aquatic vegetation in pesticide mitigation. It would be expected that sediments would also play an important role in this mitigation process since pyrethroids have a relatively high K_{OC} (*12*). In the study by Bennett *et al.* (*16*), however, sediments were a minor sink due to the dense plant community that limited the movement and/or partitioning of bifenthrin and lambda-cyhalothrin to the sediment compartment. Similar results were found in a microcosm study by Hand *et al.* (*110*) where aquatic plants significantly reduced the amount of lambda-cyhalothrin reaching the sediments. Moreover, in the same study, it was shown that lambda-cyhalothrin plant adsorption was virtually irreversible in turn reducing sediment partitioning.

Differences in stability between bifenthrin and lambda-cyhalothrin in the study by Bennett *et al.* (*16*) was evident from the half-lives calculated for each pesticide. Bifenthrin exhibited a half-life of 6.12 d, while the half-life for lambda-cyhalothrin was only 1.35 d. Hand *et al.* (*110*) reported similar results in a study investigating the route of metabolism of [^{14}C]lambda-cyhalothrin following adsorption to aquatic plants where lambda-cyhalothrin quickly bound to the plant surface and was readily degraded by ester cleavage. This was evident due to the rapid increase in the cyclopropane acid metabolite and lack of parent compound present in the water of their test system. Alternatively, this shorter half-life may have been attributed to alkaline hydrolysis. Studies have shown that the pH in surface waters can exceed 9 due to the photosynthesis by plants and algae (*119*). Lambda-cyhalothrin has been shown to be unstable under these basic conditions while bifenthrin has been shown to be stable (*12*).

It may be concluded, that by incorporating vegetated drainage ditches into a watershed management program, agriculture can continue to decrease potential non-point source threats to downstream aquatic receiving systems. Overall results of this case study by Bennet *et al.* (*16*) illustrate that aquatic macrophytes play an important role in the retention and distribution of pyrethroids in vegetated agricultural drainage ditches. Ditch lengths of less than 300 m were required to reduce loadings into receiving water bodies during worst-case scenario runoff events. This demonstrates the importance and effectiveness of vegetated drainage ditches as a BMP for the mitigation of pyrethroid runoff. For this tool to be optimally effective, it would need to work in parallel with other existing BMPs and farming practices (*20*) to facilitate single entry points into surrounding ditch systems. Buffer strips on the fringes of agricultural fields would also help in the funneling of runoff into these ditch entry points. In many cases where other BMPs are not available, simple vegetated drainage ditches would still be an effective tool. Another study (*17*) has already shown experimental vegetated ditches to effectively reduce the downstream-transport of lambda-cyhalothrin at concentrations of 500 μg/L. The implementation of retention ponds in agricultural watersheds was examined by Scott *et al.* (*30*) as one strategy to reduce the amount and toxicity of runoff-related insecticide pollution including fenvalerate levels up to 0.9 μg/L discharging into estuaries.

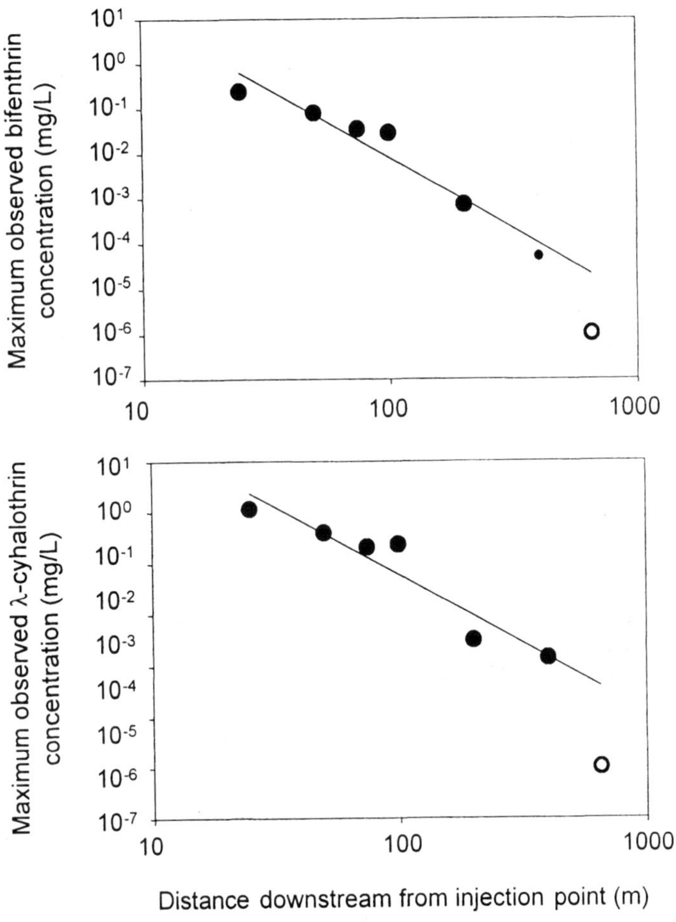

Figure 8. Least-squares regression relationships fit to log-transformed maximum observed pesticide concentration in water versus distance downstream from injection point. Concentrations at x = 650 m (open circles) are detection limits. (Reproduced with permission from reference 16. Copyright 2005 Allen Press.)

Table VII. Estimated mass of bifenthrin and lambda-cyhalothrin in the water, sediment and plant compartments relative to each sampling time (*total A.I. amended to ditch at time zero)

Bifenthrin (g)					*Lambda–Cyhalothrin (g)*				
Time	*Water*	*Plants*	*Sediment*	*Total*	*Time*	*Water*	*Plants*	*Sediment*	*Total*
0 h	-	-	-	*11.4	0 h	-	-	-	*5.70
3 h	5.78	6.29	0.039	12.1	3 h	3.10	6.13	0.062	9.29
12 h	0.718	7.22	0.033	7.97	12 h	0.353	3.76	0.011	4.13
24 h	0.191	4.03	0.011	4.24	24 h	0.106	1.59	0.037	1.70
7 d	0.134	1.93	0.063	2.13	7 d	0.041	0.067	0.018	0.126
14 d	0.045	3.00	0.053	3.10	14 d	0.007	0.209	0.013	0.229
30 d	0.004	0.199	0.002	0.206	30 d	0.001	0.034	0.042	0.078
44 d	0.001	0.049	0.001	0.051	44 d	0.006	0.091	0.050	0.148

Acknowledgements

We would like to express our gratitude to various colleagues and friends which have been involved in the work described here in different ways: James M Dabrowski, Sue K.C. Peall, Silke Bollmohr, Jenny A. Day, Adrian Reinecke, Matt T. Moore, Erin R. Bennett, Charlie M. Cooper, Sammie Smith Jr. Financial support from the Volkswagen-Foundation (I/76 177), the German Society for the Advancement of Sciences (DFG), and the USDA is greatly acknowledged. Permission to reprint Tables IV to VII and Figures 1 to 8 was kindly granted by Society of Environmental Toxicology and Chemistry, Allen Press Publishing Services.

References

1. Bowles, R. G.; Webster, J. P. G. *Crop Prot.* **1995**, *14*, 593-600.
2. Falconer, K. E. *Agric. Ecosys. Environ.* **1998**, *69*, 37-54.
3. Humenik, F. J.; Smolen, M. D.; Dressing, S. A. *Environ. Sci. Technol.* **1987**, *21*, 737-742.
4. Line, D. E.; Osmond, D. L.; Coffey, S. W.; McLaughlin, R. A.; Jennings, G. D.; Gale, J. A.; Spooner, J. *Water Environ. Research* **1997**, *69*, 844-860.
5. Loague, K.; Corwin, D. L.; Ellsworth, T. R. *Environ. Science Technol.* **1998**, *32*, 130-133.

6. Baker, J. L. In *Environmental impact of nonpoint source pollution*; Overcash, M. R., Davidson, J. M., Eds.; Ann Arbor Sci.: Ann Abor, Mich. USA, 1983; pp 275-310.
7. Groenendijk, P.; van der Kolk, J. W. H.; Travis, K. Z. In *Freshwater field tests for hazard assessment of chemicals*; Hill, I. R., Heimbach, F., Leeuwangh, P., Matthiessen, P., Eds.; Lewis Publisher: Boca Raton, 1994; pp 105-125.
8. Schiavon, M.; Perrin-Ganier, C.; Portal, J. M. *Agronomy* **1995**, *15*, 157-170.
9. Casida, J. E.; Gammon, D. W.; Glickman, A. H.; Lawrence, L. J. *Ann. Rev. Pharmacol. Toxicol.* **1983**, *23*, 413-438.
10. Coats, J. R.; Symonik, D. M.; Bradbury, S. P.; Dyer, S. D.; Timson, L. K.; Atchison, G. J. *Environ. Toxicol. Chem.* **1989**, *8*, 671-679.
11. Oros, D. R.; Werner, I. *Pyrethroid Insecticides: An analysis of use patterns, ditributions, potential toxicity and fate in the Sacramento-San Joaquin Delta and Central Valley*; SFEI Contribution 415 San Fransisco Estuary Institute: Oakland, 2005, 110 pages.
12. Laskowski, D. A. *Rev. Environ. Contamin. Toxicol.* **2002**, *174*, 49-170.
13. Chandler, G. T.; Coull, B. C.; Davis, J. C. *Mar. Environ. Res.* **1994**, *37*, 313-327.
14. Hill, I. R. In *The Pyrethroid Insecticides*; Leahey, J. P., Ed.; Taylor and Francis: London, 1985; pp 151-262.
15. Hill, R. *Pest. Sci.* **1989**, *27*, 429-465.
16. Bennett, E. R.; Moore, M. T.; Cooper, C. M.; Smith Jr., S.; Shields Jr., F. D.; Drouillard, K. G.; Schulz, R. *Environ. Toxicol. Chem.* **2005**, *24*, 2121-2127.
17. Moore, M. T.; Bennett, E. R.; Cooper, C. M.; Smith Jr., S.; Shields Jr., F. D.; Milam, C. D.; Farris, J. L. *Agric. Ecosys. Environ.* **2001**, *87*, 309-314.
18. Clark, J. R.; Goodman, L. R.; Bothwick, P. W.; Patrick Jr., J. M.; Cripe, G. M.; Moody, P. M.; Moore, J. C.; Lores, E. M. *Environ. Toxicol. Chem.* **1989**, *8*, 393-401.
19. Smith, T. M.; Stratton, G. W. *Residue Rev.* **1986**, *97*, 93-120.
20. Schulz, R. *J. Environ. Qual.* **2004**, *33*, 419-448.
21. Ghadiri, H.; Rose, C. W. *J. Environ. Qual.* **1991**, *20*, 628-634.
22. Schulz, R.; Hauschild, M.; Ebeling, M.; Nanko-Drees, J.; Wogram, J.; Liess, M. *Chemosphere* **1998**, *36*, 3071-3082.
23. Sullivan, D. J.; Richards, K. D. *U.S. Geological Survey (USGS) Fact Sheet FS-107-96* **1996**, 4 pages.
24. Bacey, J.; Starner, K.; Spurlock, F. *California Environmental Protection Agency. Sacramento, California. Report EH04-01* **2004**, 35 pages.
25. Amweg, E. L.; Weston, D. P.; Ureda, N. M. *Environ. Toxicol. Chem.* **2005**, *24*, 966-972.
26. Schulz, R.; Liess, M. *Environ. Toxicol. Chem.* **2001**, *20*, 185-190.

27. Schroer, A. F. W.; Belgers, J. D. M.; Brock, T. C. M.; Matser, A. M.; Maund, S. J.; Van den Brink, P. J. *Arch. Environ. Contamin. Toxicol.* **2004**, *46*, 324-335.
28. Weston, D. P.; You, J.; Lydy, M. J. *Environ. Sci. Technol.* **2004**, *38*, 2752-2759.
29. Bacey, J.; Spurlock, F.; Starner, K.; Feng, H.; Hsu, J.; White, J.; Tran, D. M. *Bull. Environ. Contam. Toxicol.* **2005**, *74*, 864-871.
30. Scott, G. I.; Fulton, M. H.; Moore, D. W.; Wirth, E. F.; Chandler, G. T.; Key, P. B.; Daugomah, J. W.; Strozier, E. D.; Devane, J.; Clark, J. R.; Lewis, M. A.; Finley, D. B.; Ellenberg, W.; Karnaky Jr., K. J. *Toxicol. Ind. Health* **1999**, *15*, 200-213.
31. Liess, M.; Schulz, R.; Liess, M. H.-D.; Rother, B.; Kreuzig, R. *Water Res.* **1999**, *33*, 239-247.
32. Crossland, N. O.; Shires, S. W.; Bennett, D. *Aquat. Toxicol.* **1982**, *2*, 253-270.
33. Liess, M.; Schulz, R.; Neumann, M. *Chemosphere* **1996**, *32*, 1963-1969.
34. Kreuger, J. *Sci. Total Environ.* **1998**, *216*, 227-251.
35. House, W. A.; Farr, I. S.; Orr, D. R.; OU, Z. In *Pesticides in soil and water*; Walker, A., Ed.; BCPC Monograph No. 47: Thornton Heath, 1991; pp 183-192.
36. Daniels, W. M.; House, W. A.; Rae, J. E.; Parker, A. *Sci. Total Environ.* **2000**, *253*, 81-92.
37. Cooper, C. M.; Smith Jr., S.; Moore, M. T. *Int. J. Ecol. Environ. Sci.* **2003**, *29*, 171-184.
38. Turnbull, A.; Harrison, R.; DiGuardo, A.; Mackay, D.; Calamari, D. *BCPC* **1995**, *62*, 87-92.
39. Dabrowski, J. M.; Peall, S. K. C.; Reinecke, A. J.; Liess, M.; Schulz, R. *Wat. Air Soil Pollut.* **2002**, *135*, 265-283.
40. Moraes, R.; Elfvendahl, S.; Kylin, H.; Molander, S. *Ambio* **2003**, *32*, 258-263.
41. Cooper, C. M. *Arch. Hydrobiol.* **1991**, *121*, 103-113.
42. Baughman, D. S.; Moore, D. W.; Scott, G. I. *Environ. Toxicol. Chem.* **1989**, *8*, 417-429.
43. Kreuger, J.; Peterson, M.; Lundgren, E. *Bull. Environ. Contamin.Toxicol.* **1999**, *62*, 55-62.
44. Schulz, R. In *Techniques in Aquatic Toxicology*; Ostrander, G. K., Ed.; CRC Press LLC: Boca Raton, 2005; Vol. Vol. 2, pp 427-448.
45. Cooper, C. M. *J. Freshw. Ecol.* **1991**, *6*, 237-248.
46. Kreuger, J. K.; Brink, N. In *Pesticides: Food and Environmental Implications.*; International Atomic Energy Agency: Vienna, 1988; pp 101-112.
47. Kirby-Smith, W. W.; Eisenreich, S. J.; Howe, J. T.; Luettich Jr., R. A. *Final Project Report. U.S. Environmental Protection Agency, Gulf Breeze, FL. (CR 813415)* **1992**.

48. Liess, M.; Schulz, R. In *Handbook of water analysis*; Nollet, L. M. L., Ed.; Marcel Dekker: New York, 2000; pp 1-24.
49. Lauridsen, R. B.; Friberg, N. *Environ. Toxicol.* **2005**, *20*, 513-521.
50. Heckmann, L.-H.; Friberg, N.; Ravn, H. W. *Pest Manage. Sci.* **2005**, *61*, 627-635.
51. Smith, S.; Reagan, T. E.; Flynn, J. L.; Willis, G. H. *J. Environ. Qual.* **1983**, *12*, 534-537.
52. Chandler, G. T. *Mar. Environ. Res.* **1990**, *29*, 65-76.
53. Crane, M. *Hydrobiologia* **1997**, *346*, 149-155.
54. Pontasch, K. W.; Cairns, J. *Arch. Environ. Contam. Toxicol.* **1991**, *20*, 103-112.
55. Breneman, D. H.; Pontasch, K. W. *Environ. Toxicol. Chem.* **1994**, *13*, 381-387.
56. Webber, E. C.; Deutsch, W. G.; Bayne, D. R.; Seesock, W. C. *Environ. Toxicol. Chem.* **1992**, *11*, 87-105.
57. Hanazato, T. *Environ. Pollut.* **1998**, *101*, 361-373.
58. Schulz, R.; Liess, M. *Environ. Toxicol. Chem.* **2001**, *20*, 757-762.
59. Nilsson, L. M.; Otto, C. *Hydrobiologia* **1977**, *54*, 109-112.
60. Ernst, W. R.; Jonah, P.; Doe, K.; Julien, G.; Hennigar, P. *Environ. Toxicol. Chem.* **1991**, *10*, 103-114.
61. Symons, P. E. K.; Metcalfe, J. L. *Can. J. Zool.* **1978**, *56*, 1284-1290.
62. Johansson, A. *Hydrobiologia* **1991**, *211*, 185-194.
63. Wisseman, R. W.; Anderson, N. H. In *Proceedings of the 4th International Symposium Trichoptera*; Morse, J. C., Ed.; Dr. W. Junk Publisher: The Hague, 1984; pp 455-460.
64. Dick, J. T. A.; Montgomery, I.; Elwood, R. E. *J. Anim. Ecol.* **1993**, *62*, 79-88.
65. Dick, J. T. A. *J. Zool.* **1995**, *236*, 697-706.
66. Jenio, F. *Proc. W. VA Acad. Sci.* **1979**, *51*, 67-73.
67. McCahon, C. P.; Pascoe, D. *Ecotox. Environ. Safety* **1988**, *15*, 245-252.
68. Taylor, E. J.; Maund, S. J.; Pascoe, D. *Arch. Environ. Contam. Toxicol.* **1991**, *21*, 371-376.
69. Crane, M.; Maltby, L. *Environ. Toxicol. Chem.* **1991**, *10*, 1331-1339.
70. Maltby, L. *Environ. Toxicol. Chem.* **1992**, *11*, 79-85.
71. Wauchope, R. D. *J. Environ. Qual.* **1978**, *7*, 459-472.
72. Newcombe, C. P.; MacDonald, D. D. *N. Amer. J. Manage.* **1991**, *11*, 72-82.
73. Lowell, R. B.; Culp, J. M.; Wrona, F. J. *Environ. Toxicol. Chem.* **1995**, *14*, 1209-1217.
74. Dabrowski, J. M.; Bollen, A.; Schulz, R. *Environ. Toxicol. Chem.* **2005**, *24*, 1395-1402.
75. Schulz, R.; Liess, M. *Aquat. Toxicol.* **1999**, *46*, 155-176.
76. Farmer, D.; Hill, I. R.; Maund, S. J. *Ecotoxicology* **1995**, *4*, 219-244.
77. Schulz, R.; Dabrowski, J. M. *Environ. Toxicol. Chem.* **2001**, *20*, 2537-2543.

78. Palmer, G. C.; Scherman, P. A. *Application of an artificial stream system to investigate the water quality tolerances of indigenous, South African, riverine macroinvertebrates*; Water Research Commission Report 686/1/00: Pretoria, South Africa, 2000.
79. Brittain, J. E.; Eikeland, T. J. *Hydrobiologia* **1988**, *166*, 77-93.
80. Crossland, N. O.; Mitchell, G. C.; Dorn, P. B. *Environ. Toxicol. Chem.* **1992**, *11*, 49-59.
81. Schulz, R. *Environ. Toxicol. Chem.* **2003**, *22*, 1172-1176.
82. Hopkin, S. P. In *Handbook of Ecotoxicology*; Calow, P., Ed.; Blackwell: Oxford, 1993; pp 397-427.
83. Chappie, D. J.; Burton Jr., G. A. *Environ.. Toxicol. Chem.* **1997**, *16*, 559-564.
84. Matthiessen, P.; Sheahan, D.; Harrison, R.; Kirby, M.; Rycroft, R.; Turnbull, A.; Volkner, C.; Williams, R. *Ecotox. Environ. Safety* **1995**, *30*, 111-119.
85. Jergentz, S.; Pessacq, P.; Mugni, H.; Bonetto, C.; Schulz, R. *Ecotox. Environ. Safety* **2004**, *59*, 133-141.
86. Schulz, R.; Liess, M. *Environ. Toxicol. Chem.* **1999**, *18*, 2243-2250.
87. Maltby, L.; Naylor, C.; Calow, P. *Ecotox. Environ. Safety* **1990**, *19*, 292-300.
88. Mulliss, R. M.; Revitt, D. M.; Shutes, R. B. E. *Ecotoxicology* **1996**, *5*, 209-215.
89. Schulz, R.; Liess, M. *Arch. Environ. Contam. Toxicol.* **2001**, *40*, 481-488.
90. Dermott, R. N.; Spence, H. J. *Can. J. Aquat. Sci.* **1984**, *14*, 1695-1701.
91. Taylor, E. J.; Rees, E. M.; Pascoe, D. *Aquat. Toxicol.* **1994**, *29*, 83-101.
92. Mainstone, C. P.; Schofield, K. *European Water Pollution Control* **1996**, *6*, 21-29.
93. Caruso, B. S. *Environ. Manage.* **2000**, *25*, 9-22.
94. Williams, R. D.; Nicks, A. D. *Procedings of the International Erosion Control Association Conference, Indianapolis, Indiana* **1993**, 279-287.
95. Blanco-Canqui, H.; Gantzer, C. J.; Anderson, S. H. *J. Environ. Qual.* **2006**, *35*, 1969-1974.
96. Dosskey, M. G. *Environ. Manage.* **2001**, *28*, 577-598.
97. FOCUS. *Landscape and Mitigation Factors in Aquatic Risk Assessment. Volume 2. Detailed Technical Reviews.* Report of the Focus Working Group on Landscape and Mitigation Factors in Ecological Risk Assessment, EC Document Reference SANCO/10422/2005, 2005, 434 pages.
98. Mitsch, W. J.; Gosselink, J. G. *Wetlands*; Van Nostrand Reinhold: New York, 1993; Vol. 2nd edition, 722 pages.
99. Wetzel, R. G. In *Constructed wetlands for water quality improvement*; Moshiri, G. A., Ed.; CRC Press: Boca Raton, 1993; pp 3-7.
100. Dietz, A. C.; Schnoor, J. L. *Environ. Health Perspect.* **2001**, *109*, 163-167.

101. Brix, H. *Wat. Sci. Technol.* **1994**, *30*, 209-223.
102. Cooper, C. M.; Testa, S.; Knight, S. S.; Bowen, J. J. *Assessment of a Constructed Bulrush Wetland for Treatment of Cattle Waste: 1991-1994; Final Report;* Agricultural Research Service: Oxford, Mississippi, 1995; Vol. 4, pp 1-22.
103. Scholes, L.; Shutes, R. B. E.; Revitt, D. M.; Forshaw, M.; Purchase, D. *Sci. Total Environ.* **1998**, *214*, 211-219.
104. Vymazal, J. In *Constructed Wetlands in Water Pollution Control*; Cooper, P. F., Findlater, B. C., Eds.; Pergamon Press: Oxford, 1990; pp 347-358.
105. Luckeydoo, L. M.; Fausey, N. R.; Brown, L. C.; Davis, C. B. *Agric. Ecosys. Environ.* **2002**, *88*, 89-94.
106. Headley, J. V.; Gandrass, J.; Kuballa, J.; Peru, K. M.; Gong, Y. *Environ. Sci. Technol.* **1998**, *32*, 3968-3973.
107. Kadlec, R. H.; Knight, R. L. *Treatment Wetlands*; CRC Lewis: Boca Raton, 1996, 624 pages.
108. Dieter, C. D.; Duffy, W. G.; Flake, L. D. *Environ. Toxicol. Chem.* **1996**, *15*, 308-312.
109. Spongberg, A. L.; Martin-Hayden, J. M. *Environ. Sci. Technol.* **1997**, *31*, 3161-3165.
110. Hand, L. H.; Kuet, S. F.; Lane, M. C. G.; Maund, S. J.; Warinton, J. S.; Hill, I. R. *Environ. Toxicol. Chem.* **2001**, *20*, 1740-1745.
111. Karen, D. J.; Joab, B. M.; Wallin, J. M.; Johnson, K. A. *Chemosphere* **1998**, *37*, 1579-1586.
112. Weinberger, P.; Greenhalgh, R.; Moody, R. P.; Boulton, B. *Environ. Sci. Technol.* **1982**, *16*, 470-473.
113. Bennett, E. R.; Moore, M. T.; Cooper, C. M.; Smith Jr., S. *Bull. Environ. Contamin. Toxicol.* **2000**, *64*, 825-833.
114. Leistra, M.; Zweers, A. J.; Warinton, J. S.; Crum, S. J. M.; Hand, L. H.; Beltman, W. H. J.; Maund, S. J. *Pest Manage. Sci.* **2003**, *60*, 75-84.
115. Siegfried, B. D. *Environ. Toxicol. Chem.* **1993**, *12*, 1683-1689.
116. Schulz, R.; Moore, M. T.; Bennett, E. R.; Farris, J. L.; Smith Jr., S.; Cooper, C. M. *Environ. Toxicol. Chem.* **2003**, *22*, 1262-1268.
117. Schulz, R.; Peall, S. K. C. *Environ. Sci. Technol.* **2001**, *35*, 422-426.
118. Schulz, R.; Hahn, C.; Bennett, E. R.; Dabrowski, J. M.; Thiere, G.; Peall, S. K. C. *Environ. Sci. Technol.* **2003**, *37*, 2139-2144.
119. Kersting, K.; Van den Brink, P. *Environ. Toxicol. Chem.* **1997**, *16*, 251-259.

Chapter 10

Persistence and Phase Distribution in Sediment

J. Gan[1], S. Bondarenko[1], and F. Spurlock[2]

[1]Department of Environmental Sciences, University of California, 2258 Geology Building, Riverside, CA 92521
[2]California Department of Pesticide Regulation, 1001 I Street, Sacramento, CA 95812

The pyrethroids are all strongly hydrophobic and therefore are found associated primarily with bed sediment after entry into water bodies. Therefore, their persistence and phase distribution in sediment greatly influences their fate and effects. This chapter provides an up-to-date review of data on the persistence and partitioning of pyrethroids in sediment. Information from recent studies is summarized, and half-lives ($T_{1/2}$), K_{OC} and K_{DOC} values are tabulated. Pyrethroids display differing persistence in sediment, with bifenthrin being more persistent than the other compounds. However, the bioavailable concentrations of pyrethroids decrease quickly in sediment due to the aging effect, with bioavailable $T_{1/2} \leq 2$ months, suggesting diminishing toxicity over time. K_{OC}s from earlier literature may have been underestimated due to incomplete phase separation. Recent studies using selective methods such as solid phase microextraction show that K_{OC}s and K_{DOC}s of pyrethroids are in the 10^6 range, with K_{DOC}s a few times smaller than K_{OC}. We also identify information gaps that may serve as topics for future research.

Introduction

The persistence and phase distribution (i.e., adsorption/desorption behavior) of a pesticide are two of the fundamental processes which control its fate and its effects in the environment. Researchers in earlier studies have extensively considered the degradation and persistence of various pyrethroid compounds in soil, and to a lesser degree, phase partitioning such as soil adsorption (*1-5, 7*). In comparison, only a limited number of studies have been reported for sediments. Because offsite movement such as runoff will most likely transport pyrethroid residues into the bed sediment via erosion of soil particles bearing residues and subsequent deposition, the potential ecotoxicological effects of pyrethroids are expected to depend closely on their persistence and phase distribution in sediment. In this chapter, we provide an up-to-date review of the data available for describing the persistence and degradation in sediment, the partition between the sediment and water phases as defined by K_d and K_{OC}, and the partition between the dissolved organic carbon (DOC) and water phases as defined by K_{DOC}. To provide information for comparison, selected physical-chemical properties such as aqueous solubility and K_{OW}, and soil persistence and sorption data are briefly summarized at the beginning. Information gaps that merit further research are highlighted.

Basic Properties and Behavior in Soil

Basic Properties

The values for some basic properties of pyrethroids varied drastically in literature. For instance, the water solubility values for pyrethroids cited in the Pesticide Properties Manual (*6*) are often higher than those listed in the review by Laskowski (*7*) (Table I). Given the more recent date that the Laskowski review was published and the number of sources from which the values were compiled, it is likely that the data in the Laskowski review are generally more reliable than those in many earlier references.

According to Table I, pyrethroids are essentially insoluble in water, and are strongly hydrophobic, as is apparent from their very large K_{OW} values. The extremely low solubility and strong hydrophobicity is a source of challenge to researchers due to analytical artifacts caused by the tendency for pyrethroids to sorb to surfaces of glassware, and the fact that imperfect separation between the octanol and water phases could lead to abnormally high aqueous phase concentrations and thus artificially low K_{OW}s. This likely has had an impact on

Table I. Literature solubility and K_{OW} values for pyrethroids

Compound	*Solubility (μg/L)*		K_{OW}		
	Pesticide Manual[a]	*Laskowski*[b]	*Pesticide Manual*	*Laskowski (measured)*	*Laskowski (calculated)*
Bifenthrin	100	0.014	1×10^6	6.4×10^6	7.2×10^6
Cyfluthrin	2	2.3	-	5.97×10^6	6.4×10^6
λ-cyhalothrin	5	5	10×10^6	7.0×10^6	6.1×10^6
Cypermethrin	10-200	4	4×10^6	6.54×10^6	6.1×10^6
Deltamethrin	2	0.2	2.7×10^5	4.53×10^6	6.5×10^6
Esfenvalerate	300	6	1.6×10^6	5.62×10^6	6.8×10^6
Fenpropathrin	330	10.3	1×10^6	6.0×10^6	5.7×10^6
Permethrin	200	5.5	1.26×10^6	6.9×10^6	6.9×10^6

[a] Ref. (*6*); [b] Ref. (*7*).

the data quality of some earlier measurements of the physical-chemical properties for pyrethroids, as discussed by Laskowski (*7*).

Measuring the water solubility of pyrethroids can be technically challenging because of the potential for the formation of pyrethroid suspensions during saturation of the water phase when a stirring technique is used to achieve saturation by equilibration of water with excess chemicals. Laskowski (*7*) noted that approximately half the experiments used a column saturation technique that does not produce the suspension artifact, providing water solubility values usually lower than those achieved with stirring. According to Table I, water solubility values for pyrethroids are generally of the order of 0.001-0.01 ppm, or 1 to 10 ppb, with that of bifenthrin at 0.01 ppb. Therefore, the detection of pyrethroids in surface water or porewater samples at levels higher than the specific solubility is likely a result of either an artifact of analysis or enhanced "apparent"solubility caused by the combined analysis of solubolized pyrethroids with those bound to dissolved organic carbon (DOC) and fine particles.

Difficulties similarly exist in accurately measuring the K_{OW} for pyrethroids. As noted by Laskowski (*7*), because of the high hydrophobicity of pyrethroids, chemical concentrations in the octanol phase are many orders of magnitude higher than those in the water phase, making it difficult to prevent the contamination of the water phase. This could result in apparently high water-phase concentrations that do not reflect the true partitioning behavior and thus create an artificially low K_{OW}. In Table I, both measured K_{OW} values are listed, as well as values calculated from molecular structure.,The calculated K_{OW}s are all $\geq 5.7 \times 10^6$, further demonstrating that all pyrethroids are extremely hydrophobic and tend to bind strongly to organic matter in sediment.

Persistence and Sorption in Soil

A number of different degradation studies were reported on pyrethroids using soils as media in earlier literature. When reviewing the data for the persistence of pyrethroids in soil, Laskowski (*7*) employed a rating procedure to account for the potential effects of the use of large quantities of organic solvents in soil spiking or the use of unrealistically high initial pesticide concentrations. Under aerobic conditions, the half-lives of pyrethroids ranged from a low of 3.3 d for tralomethrin to a high of 96.3 d for bifenthrin. Pyrethroids generally display half-lives under anaerobic conditions similar to those under aerobic conditions. Therefore, in aerobic or anaerobic soils, pyrethroids have short to moderate persistence, and variations between the different pyrethroids suggest that bifenthrin is relatively more persistent than the other pyrethroids.

Measuring the sorption of pyrethroids to soil is prone to several complications that likely have contributed to the relatively low K_{OC}s reported in earlier studies (Table II). As previously mentioned, pyrethroids tend to sorb to surfaces of glass or plastic centrifuge tubes and other containers. In addition, because K_d is the ratio of chemical concentration in soil (C_s) over that in water (C_w), incomplete phase separation may lead to an exaggerated C_w and consequently an artificially low K_d or K_{OC}. One cause for incomplete separation, as observed in Lee et al. (*8*), includes the inability of centrifugation to exclude all fine particles and DOC from the aqueous phase. Because pyrethroids are preferentially sorbed on fine particles and DOC over large particles, and it is generally these smaller particles that remain suspended, a very small quantity of fine particles and DOC remaining in the supernatant after centrifugation could increase C_w by many times. Laskowski (*7*) used a rating procedure to account for the variation in the quality of sorption data. The K_{OC}s listed in Table II are from the analysis of 392 adsorption (K_d) values. Data for esfenvalerate are absent from the table because sorption experiments were not available for this chemical at the time of the publication. The K_{OC} values in Table II indicate that all the pyrethroids are sorbed exceptionally strongly to soil, with the exception of fenpropathrin, which has a K_{OC} lower than the rest. Laskowski (*7*) further noted that, contrary to most findings, the expression of sorption as K_{OC} in Table II had little or no impact on reducing the variability of sorption from one soil to another. This suggests that qualitative differences in soil OC may have greatly impacted the K_{OC}, and a single K_{OC} may not apply across different soil types for the same pyrethroid. It must be noted that although K_{OC}s in the Laskowski review were generally higher than the earlier values, these data were obtained using conventional batch equilibration methods. As discussed later, some degree of continuing underestimation in those "higher quality" batch-determined K_{OC}s is likely.

Table II. Literature K_{OC} values for pyrethroid adsorption in soil

Chemical	*PAN Pesticides Database*[a]	*USDA-NRCS (WIN-PST)*[b]	*Laskowski*[c]
Bifenthrin	6,314	2.4×10^5	2.4×10^5
Cyfluthrin	8,930	1.0×10^5	1.2×10^5
λ-cyhalothrin	2,341	1.8×10^5	3.3×10^5
Cypermethrin	82	4.2×10^3	3.1×10^5
Deltamethrin	6,291	1.9×10^5	7.0×10^5
Esfenvalerate	-	5.3×10^3	-
Fenpropathrin	-	5.0×10^3	0.4×10^5
Permethrin	2.3×105	1.0×10^5	2.8×10^5

[a] Ref. (*9*); [b] Ref (*10*); [c] Ref. (*7*).

Degradation and Persistence in Sediment

Knowledge of pyrethroid degradation and persistence in sediments is limited compared to soils. Earlier studies using mesocosms often stopped after measuring the dissipation of pyrethroids from the water column without further following their degradation and persistence in sediment. However, several recent studies have examined the degradation and persistence of pyrethroids in sediment in more depth.

In a published study, we incubated field-contaminated sediments at room temperature and followed pesticide dissipation using exhaustive solvent extraction to measure the total sediment concentration. The sediments were collected from three different locations along a runoff drainage ditch at a commercial nursery in southern California (*11*). Due to continuous onsite use, the sediments contained elevated levels of bifenthrin and permethrin. The sediment samples were incubated under either flooded aerobic or flooded anaerobic conditions. Pesticide dissipation over time was fitted to a first-order decay model to estimate the first-order rate constant k (d^{-1}) and half-life ($T_{1/2}$) (Table III). Under aerobic conditions, noticeable differences in persistence were observed between the different pesticides and all dissipation rates were slower than in soil. Bifenthrin exhibited similar persistence in the different sediments, with $T_{1/2}$ ranging 428-483 d, or 12-16 months. Degradation of permethrin isomers under the same conditions was markedly faster than for bifenthrin, with a $T_{1/2}$ of 98-142 d (3-4.7 months) for *cis*-permethrin and 60-312 d (2-10 months) for *trans*-permethrin. Therefore, under aerobic conditions, while permethrin showed intermediate persistence in the sediments at 20 °C, bifenthrin was much more persistent.

The degradation of bifenthrin was slightly enhanced under anaerobic conditions when compared to the aerobic treatments, with the $T_{1/2}$ ranging 251-498 d (8-16 months) in the same sediments. Degradation of *cis*-permethrin was inhibited under anaerobic conditions when compared to the aerobic treatments, with the $T_{1/2}$ extended from 98-142 d (or 3-4.7 months) to 209-380 d (or 7-13 months) under anaerobic conditions. Therefore, the oxidation state of sediment may affect the persistence of pyrethroids in sediment. Overall, the selected pyrethroids exhibited intermediate to long persistence in sediment, and bifenthrin was apparently more persistent than permethrin (*11*).

Table III. First-order rate constant k (d^{-1}) and half-life $T_{1/2}$ (d) for degradation of bifenthrin and permethrin isomers in sediments under aerobic conditions

Sediment[a]	Bifenthrin		cis-Permethrin		trans-permethrin	
	k	$T_{1/2}$	*k*	$T_{1/2}$	*k*	$T_{1/2}$
			Aerobic			
104 M	0.0016	428	0.0049	142	0.0022	312
166 M	0.0016	436	0.0051	137	0.0031	223
210 M	0.0014	483	0.0071	98	0.0116	60
			Anaerobic			
104 M	0.0014	498	0.0033	209	0.0025	276
166 M	0.0028	251	0.0028	245	0.0043	160
210 M	0.0025	280	0.0018	380	0.0040	175

[a] Sampled at different distances along a drainage channel.

In a more recent study (unpublished data), we spiked four pyrethroids (bifenthrin, fenpropathrin, cyfluthrin and lambda-cyhalothrin) into two sediments and incubated the spiked sediments at room temperature under flooded aerobic conditions. One sediment was collected from San Diego Creek (SDC) in southern California and contained OC at 1.4%. The other sediment was sampled from a pond in Black Mountains (BM) in Paso Robles (central California) and contained OC at 5.0%. Pesticide dissipation in the sediment over time was fitted to a first-order decay model to estimate k and $T_{1/2}$. The selected pyrethroids showed differential degradation rates under the same conditions, with bifenthrin being the most persistent in both sediments, and cyfluthrin being relatively the least persistent (Table IV). The $T_{1/2}$ values of bifenthrin ranged from 11 to 21 months, while with the exception of cyfluthrin in SDC sediment ($T_{1/2}$ = 1 month), those of the other pyrethroids were mostly 3 to 5 months (Table IV). The persistence of bifenthrin observed in this study was in close agreement with that found in the previous degradation study using field-aged sediments. Therefore, although more pyrethroid compounds need to be included in future studies, bifenthrin appears to be one of the most persistent members of the pyrethroid family, which may partly contribute to its more frequent detections than the other pyrethroids in stream sediments. However, other pyrethroids have moderate and sometimes long persistence, which, along with other factors such as use patterns, may explain their presence in sediment.

Table IV. First-order rate constants (d^{-1}) and half lives (d) for dissipation of total chemical concentration (k, $t_{1/2}$) and rapidly desorbing concentration (k', $T_{1/2}'$) of pyrethroid compounds in sediments under aerobic conditions

Compound	*SDC*		*BM*	
	k	$T_{1/2}$	k	$T_{1/2}$
Bifenthrin	1.10×10^{-3}	629	2.07×10^{-3}	335
Cyfluthrin	1.76×10^{-2}	30	5.76×10^{-3}	120
Fenpropathrin	7.76×10^{-3}	89	4.57×10^{-3}	152
Cyhalothrin	7.29×10^{-3}	95	4.42×10^{-3}	157
	k'	$T_{1/2}'$	k'	$T_{1/2}'$
Bifenthrin	1.21×10^{-2}	56	1.10×10^{-2}	63
Cyfluthrin	2.88×10^{-2}	24	1.38×10^{-2}	50
Fenpropathrin	1.90×10^{-2}	36	1.26×10^{-2}	55
Cyhalothrin	1.85×10^{-2}	37	1.25×10^{-2}	55

To understand the role of microbial degradation in pyrethroid degradation, we isolated a large number of bacteria strains capable of degrading bifenthrin and permethrin from field-contaminated sediments (*12*). In solution media, the selected bacteria strains were able to effectively degrade both bifenthrin and permethrin, with the $T_{1/2}$ ranging from 1.3 to 5.5 d for bifenthrin, and from 1.5 to 3.3 d for permethrin isomers. However, we further observed that in the presence of sediment, the ability of the same bacteria for degrading bifenthrin or permethrin greatly decreased, and the inhibition was attributed to the strong adsorption of these compounds to the sediment phase. Therefore, even though microbial degraders may be ubiquitous in sediment, the persistence of pyrethroids in sediment can be prolonged due to their strong affinity for the solid phase and consequentially reduced bioavailability.

Information Gaps

From the current state of knowledge on the degradation and persistence of pyrethroids in sediment, the following gaps exist. First, knowledge on pyrethroid persistence in sediment is incomplete and readily comparable half-life values are not available for all pyrethroids. In addition, different studies used different sediments or incubation conditions, which makes the comparison of parameters such as $T_{1/2}$ difficult. Therefore, it is necessary to evaluate the persistence of most or all pyrethroids under the same conditions while using the same sediments. This may be achieved by spiking low levels of mixtures of all pyrethroids into the same sediment. Likewise the relationship between % C in sediments and pyrethroid biodegradation should be investigated, since bioavailability to microbial degradation may be similar to bioavailability to other sediment organisms such as benthic invertebrates. Information from such a study will provide valuable information on the relative persistence of the different pyrethroid compounds, which will help to predict which pyrethroids will likely appear and accumulate in sediment.

The second noticeable gap is the lack of knowledge on the degradation and persistence of pyrethroids in urban compartments where the application of pyrethroid products first occurs. For instance, although the general perception is that application of pesticides around houses or on lawns contribute to pesticide runoff in urban areas, as pesticides from such uses may deposit onto impervious concrete surfaces and be subject to runoff, there is little data to validate this assumption. A study is needed to understand the dissipation and persistence of pyrethroids on concrete surfaces as a function of seasonality, application methods and formulations. Such knowledge will not only improve our understanding of how pyrethroids and other pesticides move from residential areas to urban streams, but may also reveal options that can be useful for reducing runoff-facilitated transport.

Another important topic is the need to distinguish the persistence of pyrethroids as the total chemical concentration from its persistence as the bioavailable concentration. As will be discussed in more detail in the following section, the strong hydrophobicity of pyrethroids accentuates the importance of bioavailability. That is, the persistence of pyrethroids in sediment is more appropriately expressed in terms of bioavailable concentration, and which will change as the sediment-bound materials age. A few methods are available for measuring bioavailable concentrations of hydrophobic compounds in sediments, including sequential extractions with Tenax (*13*), and use of solid phase microextraction (SPME) (*14*). In a recent study (unpublished data), we used a sequential Tenax extraction procedure to measure the rapid desorption concentration (C_{rapid}) of four pyrethroids (bifenthrin, fenpropathrin, cyfluthrin and lambda-cyhalothrin) in two sediments, with C_{rapid} representing the bioavailable concentration for hydrophobic compounds, as described by Hulscher et al. (*15*). We observed that C_{rapid} decreased more rapidly than the total chemical concentration as seen in the above degradation studies. Table IV shows that if the half-lives ($T_{1/2}$') were calculated for the decline in $C_{rapid,}$ they were shortened to ≤2 months for all of the pyrethroids considered, including bifenthrin. It may be argued that concentrations such as C_{rapid} are much more meaningful for predicting sediment toxicity and should be used instead of the total chemical concentration when considering the potential for effects. Therefore, even though bifenthrin is more persistent than the other pyrethroids when measured as the total chemical concentration, the potential bioavailability or the apparent sediment toxicity would diminish rather quickly. Assuming $T_{1/2}$' as 2 months, the bioavailable concentration in pore water arising from sediment-borne pyrethroids would decrease by 97% in 10 months, although conventional analysis would suggest limited dissipation had taken place. Further studies on this topic should include other pyrethroids as well as other methods to measure the bioavailable concentration. Predicted changes of sediment toxicity over time, or the aging effect, should be corroborated with toxicity assays.

Phase Distribution in Sediment

Sorption on Sediment (K_d and K_{OC})

Underestimation and artifacts

The sorption coefficient K_d is a fundamental parameter for predicting the offsite movement potential of a given contaminant, and the actual pore water exposure concentration that benthic organisms might experience. K_d is usually measured using the so-called batch equilibration method, from which K_{OC} is then derived. However, as shown in recent studies, artifacts may occur in such measurements due to incomplete phase separation for strongly hydrophobic compounds such as pyrethroids. This effect has likely contributed to the artificially low and scattered K_{OC} values reported for pyrethroids in earlier literature (Table II).

Underestimation of K_d occurs because centrifugation does not completely remove the fine particles and DOC from the aqueous phase, and that the fraction of pyrethroids associated with the unseparated fine particles and DOC is counted as part of the aqueous phase concentration C_w when the supernatant is extracted with a solvent. The artificially enhanced C_w leads to underestimated K_d. The following equations can be used to illustrate the cause for this artifact. The apparent aqueous-phase concentration C_w as measured by the conventional liquid-liquid extraction (LLE) method consists of the freely dissolved concentration C_{free} and the DOC-complexed fraction:

$$C_w = C_{free} + C_{DOC}[DOC] \quad (1)$$

where C_{DOC} is the DOC-adsorbed concentration, and $[DOC]$ is the DOC content of the aqueous phase (i.e., supernatant) upon phase separation. Therefore, the measured K_d may be written in relation to $[DOC]$ as:

$$K_d = \frac{C_S}{C_w + C_{DOC}[DOC]} \tag{2}$$

where C_S is the sediment sorbed phase concentration. From eq. 2, K_d can then be expressed in relation to the "true" $K_{d\text{-true}}$ which is equal to C_S/C_w:

$$K_d = \frac{C_S / C_w}{1+(C_{DOC}/C_w)[DOC]} = \frac{K_{d-true}}{1+K_{DOC}[DOC]} \tag{3}$$

Thus, the measured K_d may deviate from $K_{d\text{-true}}$ by a factor of $K_{DOC}[DOC]$. From eq. 3, operational conditions such as the solid-to-solution ratio and centrifugation speed and time, can all influence $[DOC]$ in the supernatant. It was likely that different solid-to-solution ratios and centrifugation conditions were used in earlier studies, which had resulted in highly variable and generally underestimated K_ds and K_{OC}s for pyrethroids (Table II). K_{OC} values for soils cited in Laskowski (*7*) are much larger than those from earlier sources. These values were measured using a very low solid-to-solution ratio (1 g to 100 ml) (personal communications with Pyrethroid Working Group members), which to some degree could have lessened the artifact by decreasing $[DOC]$ in the aqueous phase, resulting in improved measurements.

We recently carried out studies to specifically demonstrate the underestimations caused by the conventional batch method and to generate more accurate K_{OC}s for pyrethroids. In Lee et al. (*8*), K_ds and K_{OC}s of bifenthrin and permethrin were determined using two different methods to derive C_w. The overall procedure was the same as the conventional batch method, but the supernatant was analyzed concurrently by LLE and solid-phase microextraction (SPME). SPME is a relatively new sampling technique in which the analyte is enriched onto a polymer fiber via diffusion and analyzed by GC. Studies show that SPME selectively detects C_{free} in various aqueous solutions (*16,17*). K_ds and K_{OC}s were derived using spiked sediments or field contaminated sediments. After centrifugation, various levels of DOC (4.1-16.2 mg/L) were present in the supernatant, suggesting incomplete phase separation between water and the sorbent phases. When the supernatant was analyzed by LLE, the $K_{d\text{-LLE}}$ values for the same compound were consistently smaller than the $K_{d\text{-SPME}}$ values as given by SPME analysis. From the spiked sediments, $K_{d\text{-LLE}}$ was smaller than $K_{d\text{-SPME}}$ by 2-5 times for the same pesticide-sediment combination (Table V).

Measurements from the field-contaminated sediments showed even greater deviations, with the $K_{d\text{-}LLE}$ being smaller by 4-24 times (Table VI). Results from Lee et al. (*8*) clearly showed that underestimation of K_d (and K_{OC}) by the conventional method was due to sorption to DOC that was not excluded from the aqueous phase by centrifugation. The degree of underestimation was dependent on the source and amount of DOC and may be generally significant for compounds with $K_{DOC} > 10^4$.

Table V. K_d values of bifenthrin and permethrin measured by liquid-liquid extraction (LLE) or solid-phase microextraction (SPME)

Sediment	*Bonita Creek*		*San Diego Creek*		*Field sediment*	
	K_d	r^2	K_d	r^2	K_d	r^2
			Bifenthrin			
LLE	3.4×10^4	0.88	1.2×10^4	0.87	3.6×10^3	0.77
SPME	6.5×10^4	0.78	4.3×10^4	0.90	5.8×10^3	0.89
			cis-Permethrin			
LLE	1.4×10^4	0.78	1.3×10^4	0.85	2.6×10^3	0.90
SPME	7.5×10^4	0.83	5.1×10^4	0.86	7.1×10^3	0.88
			trans-Permethrin			
LLE	1.5×10^4	0.89	1.4×10^4	0.85	1.7×10^3	0.81
SPME	3.7×10^4	0.78	4.4×10^4	0.88	3.2×10^3	0.90

In another study (unpublished data), we systematically evaluated the effect of solid-to-solution ratio and centrifugation speeds on K_d measurement for pyrethroids. Four compounds (bifenthrin, cypermethrin, esfenvalerate and cyfluthrin) and two sediments were used. Again, the procedure followed the conventional batch equilibration approach, except that the aqueous phase was simultaneously analyzed by LLE and SPME. Four solid-to-solution ratios (mass to volume) were considered for each treatment, including 1:4, 1:10, 1:50, and 1:100, while the centrifugation speed was kept constant at 10,000 rpm. Similarly, four centrifugation speeds (1500, 300, 7000 and 10000 rpm) were considered for each treatment, while the solid-to-solution ratio was kept at 1:4.

From Table VII, as the solid-to-solution ratio decreased (i.e., from 1:4 to 1:100), the $K_{d\text{-}LLE}$ consistently increased, while $K_{d\text{-}SPME}$ remained relatively constant. The difference between $K_{d\text{-}LLE}$ and $K_{d\text{-}SPME}$ became increasingly smaller as the solid-to-solution ratio increased. At 1:100, $K_{d\text{-}LLE}$ and $K_{d\text{-}SPME}$ for the same pesticide-sediment combination differed only by a few fold, suggesting that when a very small solid-to-solution ratio is coupled with a high centrifugation speed, underestimation by $K_{d\text{-}LLE}$ becomes less significant. As the

Table VI. K_d values of bifenthrin and permethrin in nursery runoff sediments measured by liquid-liquid etxraction (LLE) or solid-phase microextraction (SPME)

	0 M	*104 M*	*140 M*	*240 M*
		Bifenthrin		
LLE	$(4.8\pm0.6)\times10^2$	$(3.9\pm0.8)\times10^3$	$(1.2\pm0.2)\times10^4$	$(2.1\pm0.1)\times10^4$
SPME	$(4.9\pm0.2)\times10^3$	$(3.6\pm0.3)\times10^4$	$(1.4\pm0.1)\times10^5$	$(1.2\pm0.2)\times10^5$
		cis-Permethrin		
LLE	$(4.1\pm0.7)\times10^2$	$(1.8\pm0.2)\times10^3$	$(5.3\pm0.5)\times10^3$	$(8.4\pm0.8)\times10^3$
SPME	$(8.9\pm1.2)\times10^3$	$(6.5\pm0.6)\times10^3$	$(3.5\pm0.4)\times10^4$	$(6.3\pm1.4)\times10^4$
		trans-Permethrin		
LLE	$(5.2\pm1.0)\times10^2$	$(7.3\pm1.9)\times10^2$	$(1.9\pm0.5)\times10^3$	$(3.3\pm0.9)\times10^3$
SPME	$(9.8\pm1.4)\times10^3$	$(6.6\pm0.7)\times10^3$	$(2.8\pm0.4)\times10^4$	$(4.6\pm1.4)\times10^4$

solid-to-solution ratio was decreased, [*DOC*] in the centrifugation supernatant decreased (Table VII), suggesting clearly that the cause for the improved K_d measurement was due to an increased removal of DOC from the aqueous phase. In Figure 1, $K_{d\text{-}LLE}$ values were further plotted against the solid-to-solution ratio. The dependence of $K_{d\text{-}LLE}$s on the solid-to-solution ratio was linear for most treatments.

The effect of centrifugation speed on K_d measurements is shown in Table VIII for bifenthrin. As centrifugation speed was increased from 1500 to 10000 rpm, $K_{d\text{-}LLE}$ increased by several fold. However, even at 10,000 rpm, there were still large differences between $K_{d\text{-}LLE}$ and $K_{d\text{-}SPME}$ for the same sediment. This was due to the use of a high solid-to-solution ratio (1:4). Again, when the $K_{d\text{-}LLE}$s were plotted against the centrifugation speed, there was a clear linear relationship (Figure 2). Therefore, to obtain the improved K_d measurements by the conventional approach, it is critical to employ a high centrifugation speed along with a small solid-to-solution ratio. However, even under optimum conditions, significant differences are still expected between $K_{d\text{-}LLE}$ and $K_{d\text{-}SPME}$. Therefore, it is likely that the K_{OC} values given in Laskowski's review are still underestimates.

Compilation of K_{OC} data

Over the last few years, we have used SPME in the course of several different experiments to measure K_ds and K_{OC}s of various pyrethroids. As discussed above, SPME-measured K_{OC} values are likely improvements over literature-cited values, including those in Laskowski (7). Table IX shows a compilation of K_{OC}s for the various pyrethroids measured using SPME and

Table VII. Effect of solid-to-solution ratio on K_d of bifenthrin measured by LLE ($K_{d\text{-}LLE}$) and SPME ($K_{d\text{-}SPME}$) with a constant centrifugation speed of 10,000 rpm

Solid-to-solution ratio	*DOC (ppm)*	$K_{d\text{-}LLE}$ ($\times 10^3$)	$K_{d\text{-}SPME}$ ($\times 10^3$)	*Ratio*
		Freshwater sediment		
1:100	40.9±4.3	1.54±0.09	3.11±0.17	0.495
1:50	34.1±1.3	1.23±0.11	3.17±0.30	0.397
1:10	71.3±1.4	0.64±0.16	3.70±1.11	0.173
1:4	68.5±3.4	0.36±0.03	3.93±0.05	0.092
		Saltwater sediment		
1:100	27.1±1.0	4.83±0.43	17.44±1.34	0.277
1:50	31.3±0.4	4.12±0.34	13.74±1.42	0.299
1:10	62.3±0.8	2.07±0.16	19.60±2.29	0.106
1:4	122.2±3.8	0.77±0.03	26.52±2.99	0.029

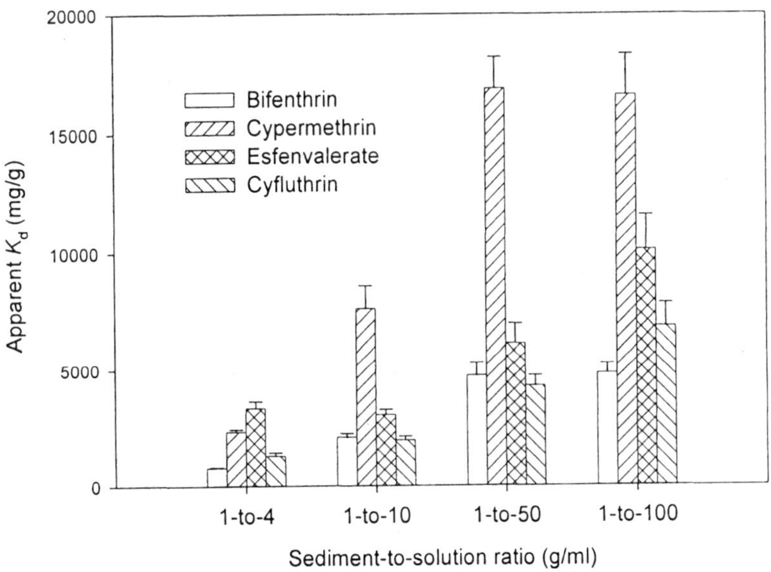

Figure 1. Effect of solid-to-solution ratio (mass to volume) on derived K_ds by LLE (centrifugation speed was at 10,000 rpm)

Table VIII. Effect of centrifugation speed on K_d of bifenthrin measured by LLE ($K_{d\text{-}LLE}$) and SPME ($K_{d\text{-}SPME}$) with a constant soil:solution ratio of 1:4

Centrifugation speed (rpm)	*DOC (ppm)*	$K_{d\text{-}LLE}$ ($\times 10^3$)	$K_{d\text{-}SPME}$ ($\times 10^3$)	*Ratio*
		Freshwater sediment		
1500	12.58±0.19	0.11±0.1	1.17±0.05	0.095
4000	11.76±0.21	0.16±0.1	1.40±0.03	0.114
7000	11.83±0.37	0.25±0.01	2.14±0.12	0.117
10000	11.49±0.16	0.49±0.05	2.52±0.71	0.194
		Saltwater sediment		
1500	78.41±3.81	0.42±0.03	10.08±1.16	0.042
4000	70.71±1.75	1.05±0.08	13.77±0.28	0.076
7000	74.27±1.74	1.41±0.16	18.91±0.83	0.075
10000	69.83±1.51	1.92±0.06	20.63±1.71	0.093

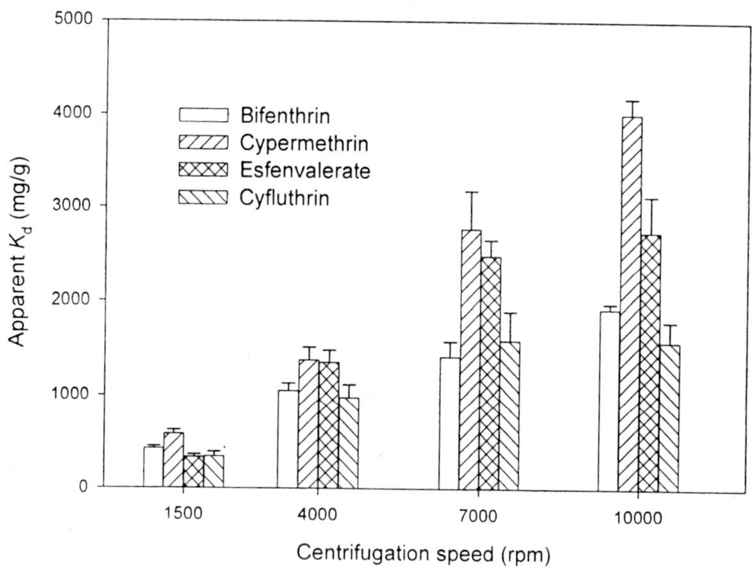

Figure 2. Effect of centrifugation speed on derived K_ds by LLE (solid-to-solution ratio was 1:4)

exclusively with sediments. K_{OC} values from Laskowski (7), which were derived from soils, are also included for comparison. It is clear that for a few compounds, K_{OC}s from Laskowski (7) and SPME measurements are similar, but for the other pyrethroids, K_{OC}s given by SPME are substantially higher.

Table IX. Comparison of K_{OC}s for pyrethroids from Laskowski (7) and recent studies using SPME

Compound	*Mean SPME*	*n*[a]	*Corrected SPME*[b]	*Laskowski*	*ratio*[c]
Bifenthrin	5.4×10^5	12	2.2×10^6	2.4×10^5	9.2
Cyfluthrin	7.5×10^5	4	3.0×10^6	1.2×10^5	25
Cypermethrin	4.1×10^5	4	1.6×10^6	3.1×10^5	5.2
Deltamethrin	3.7×10^5	2	1.5×10^6	7.0×10^5	2.1
Esfenvalerate	10.9×10^5	4	4.4×10^6	-	-
Fenpropathrin	1.2×10^5	2	0.5×10^6	0.4×10^5	12
Λ-Cyhalothrin	1.7×10^5	2	0.7×10^6	3.3×10^5	2.1
c-Permethrin	4.6×10^5	10	1.8×10^6	2.8×10^5	6.4
t-Permethrin	4.1×10^5	10	1.6×10^6	2.8×10^5	5.7

[a] *n*, number of sediments or treatments used for K_{OC} computation; [b] The measured values were correct by a factor of four to account for systematic error in SPME analysis; [c] Ratio of the corrected K_{DOC} over Laskowski K_{DOC} for the same compound.

However, the K_{OC}s derived from SPME measurements are likely still underestimated due to a systematic error in the SPME sampling procedure used. The underestimation occurs because in the process of SPME sampling, very fine suspended sediment particles in the aqueous phase may become attached to the SPME fiber and introduced into the GC inlet, resulting in an increased response (i.e., C_w) and a decreased K_d. A method to alleviate this bias, as demonstrated for measuring PAHs in sediment porewater by Hawthorne et al. (*18*), is to add a flocculent such as alum to eliminate the fine particles before SPME sampling. We evaluated this method in one of the K_d experiments (unpublished data). As shown in Table X, after addition of alum to remove the fine particles, the SPME-detected concentrations significantly decreased. The decrease ranged from 3.2 to 7.3 (mean = 4.8)-fold for San Diego Creek sediment (SDC), and from 2.5 to 6.9 (mean = 4.0) for the marine sediment. Therefore, the "true" K_{OC}s are likely higher than those given by SPME by 4-5 times (Table IX). After correcting the SPME-derived K_{OC}s by a factor of 4, K_{OC}s for pyrethroids generally fall in the low 10^6 range, with fenpropathrin displaying a smaller K_{OC} than the other pyrethroids (Table IX). The corrected K_{OC}s are 2-25 fold higher than the values cited in Laskowski's review. It must be noted that different numbers of sediments or treatments were used to calculate the average K_{OC} values in Table IX. Therefore, more confidence should be given to bifenthrin and permethrin, for which K_{OC}s were found to be 2-3 × 10^6 (Table IX).

Table X. Aqueous concentrations detected by SPME without and with pretreatment of alum to remove fine particles

	San Diego Creek sediment		*Marine sediment*	
Compound	*SPME*	*Alum-SPME*	*SPME*	*Alum-SPME*
Bifenthrin	0.06±0.02	0.019±0.008	0.013±0.004	0.005±0.0017
Fenpropathrin	0.23±0.02	0.050±0.002	0.062±0.007	0.009±0.0014
Cyhalothrin	0.08±0.01	0.014±0.004	0.013±0.002	0.003±0.0006
c-Permethrin	0.09±0.01	0.021±0.004	0.016±0.006	0.006±0.0015
t-Permethrin	0.16±0.02	0.022±0.004	0.027±0.006	0.005±0.0014
Cyfluthrin	0.10±0.01	0.025±0.005	0.017±0.001	0.005±0.0010
Cypermethrin	0.11±0.02	0.025±0.003	0.024±0.008	0.005±0.0009
Esfenvalerate	0.08±0.04	0.018±0.005	0.010±0.003	0.004±0.0013
Deltamethrin	0.08±0.02	0.014±0.003	0.010±0.003	0.003±0.0003

Sorption to Dissolved Organic Carbon (K_{DOC})

Measurement methods

Dissolved organic carbon (DOC) is ubiquitous in natural surface waters and present at elevated levels in sediment porewater. The partition of pyrethroids between water and DOC phases is another important parameter that is critical for both the transport and effects of pyrethroids. For instance, the transport of pyrethroids in surface streams is likely facilitated by DOC, because DOC, by definition, is miscible with water and may move over a long distance. This may act as the most important mechanism for long-range offsite movement of pyrethroids in streams and rivers. Additionally, the K_{DOC} is expected to play an important role in the phase partitioning and thus bioavailability of pyrethroids in sediment. A sediment may be considered to consist of a particulate solid phase (bulk sediment), porewater DOC phase and water phase. According to the equilibrium partition theory (EqP), sediment toxicity exposure is related to the freely dissolved concentration in sediment porewater, which corresponds directly to the OC-normalized sediment concentration (*19*). However, as demonstrated above, because K_{OC} may vary greatly across sediment types due to different sediment OC properties, the OC-normalization approach carries with it inherent uncertainties. Recognizing this limitation, the recent EPA methods for deriving equilibrium sediment guidelines suggest to measure C_{free} or use K_{DOC} to calculate C_{free} from the total porewater concentration C_{pw} (*20*).

The measurement of K_{DOC} has been a topic of many studies for hydrophobic compounds. Because the DOC and water phases are physically inseparable, different chemical methods have been used to detect C_{free}. These methods include dialysis membrane (*21*), reversed phase SPE (*22*), fluorescence quenching (*23*), and SPME (*24*). Because a SPME fiber selectively detects C_{free} in an aqueous medium, K_{DOC} may be derived from simultaneous determination of C_{free} by SPME and C_w by LLE, as shown in the following relationship:

$$K_{DOC} = \frac{(C_w - C_{free})/[DOC]}{C_{free}} \tag{4}$$

The above approach may be applied to various water samples, such as surface water samples, sediment porewater samples and the supernatant samples from a batch equilibrium experiment to estimate K_{DOC}.

Compilation of K_{DOC} data

In our recent studies involving the use of SPME to evaluate the bioavailability of pyrethroids, a number of K_{DOC} data sets have been generated for the various pyrethroids. Table XI is a compilation of these K_{DOC}s from published (*8*, *25-29*) and unpublished studies. It is evident that compared to the K_{OC} values listed in Table IX, the K_{DOC} is generally smaller by several fold for the same compound, ranging from the high 10^5 to the low 10^6 (Table XI). However, the number of water samples or treatments used for deriving these K_{DOC}s varied greatly. Therefore, more confidence should be given to K_{DOC} values for bifenthrin, cyfluthrin and permethrin (Table XI) that had the greatest number of replicates of measurement, which all showed K_{DOC}s close to 1.0×10^6.

Information Gaps

The use of SPME helped identify the principal reasons for the underestimated K_{OC}s and K_{DOC}s for pyrethroids. A review of existing K_{OC} and K_{DOC} data, including those derived from recent SPME determinations, suggests at least three deficiencies that merit further investigation. First, in the use of SPME for measuring K_{OC}s and K_{DOC}s, only a limited number of sediments and

Table XI. K_{DOC} values for pyrethroids estimated from surface water and sediment porewater samples

Compound	*Mean*	*n*[a]	*Corrected*[b]
Bifenthrin	4.5×10^5	15	1.8×10^6
Cyfluthrin	2.9×10^5	16	1.2×10^6
Cypermethrin	2.6×10^5	2	1.0×10^6
Deltamethrin	5.3×10^5	2	2.1×10^6
Esfenvalerate	3.4×10^5	2	1.4×10^6
Fenpropathrin	1.8×10^5	2	0.7×10^6
λ-Cyhalothrin	5.1×10^5	2	2.0×10^6
c-Permethrin	2.0×10^5	21	0.8×10^6
t-Permethrin	2.1×10^5	21	0.8×10^6

[a] *n*, number of sediments or treatments used for K_{OC} computation; [b] The measured values were corrected by a factor of four to account for systematic error in SPME analysis.

treatments have been considered for most of the pyrethroids (Table IX and Table XI). Because sediment OC properties can vary greatly from sediment to sediment, no single value can be considered "typical," and because it is impossible to test more than a limited number of sediment types, it is important to derive useful and functional mean and representative ranges for K_{OC}s and K_{DOC}s using a spectrum of sediments. Secondly, as mentioned above, the current SPME method needs to be coupled with a pre-treatment procedure, such as the use of alum to remove fine particles, to further improve the accuracy for measuring K_{OC} and K_{DOC}. In addition, the strong hydrophobicity and relatively long persistence suggest that aging is an important factor to consider for pyrethroids under field conditions. The dependence of K_{OC} and K_{DOC} on residue contact time should be evaluated through extended laboratory incubation experiments, and further verified with field-contaminated sediment samples.

References

1. Kaufman, D.D.; Russell, B.A.; Helling, C.S.; Kayser, A.J. *J. Agric. Food Chem.* **1981**, *29*, 239-245.
2. Hill, B.D.; Inaba, D.J. *J. Agric. Food Chem.* **1991**, *39*, 2282-2284.
3. Sakata, S.; Mikami, N.; Yamada, H. *J. Pestic. Sci.* **1992**, *17*, 181-189.
4. Skidmore, M.W.; Kirkpatrick, D.; Shaw, D. *Pestic. Sci.* **1994**, *42*, 101-107.
5. Baskaran, S.; Kookana, R.S.; Naidu, R. *Pestic. Sci.* **1999**, *55*, 1222-1228.
6. *The Pesticide Manual*; Worthing, C.R.; Hance, R.J., Eds.; 9th edition; The British Crop Protection Council: Surrey, U.K., 1991.

7. Laskowski, D.A. Rev. *Environ. Contam. Toxicol.* **2002**, *174*, 49-170.
8. Lee, S.J.; Gan, J.; Liu, W.P.; Anderson, M.A. *Environ. Sci. Technol.* **2003**, *37*, 5597-5602.
9. *PAN Pesticide Database*, URL http://www.pesticideinfo.org/Index.html
10. *WIN-PST Pesticide Properties Database*, URL http://www.wcc.nrcs.usda.gov/pestmgt/winpst.html#pst%20ppd
11. Gan, J.; Lee, S.J.; Liu, W.P.; Haver, D.; Kabashima, J.N. *J. Environ. Qual.* **2005**, *34*, 836-841.
12. Lee, S.J.; Gan, J.; Kim, J.S.; Kabashima, J.N.; Crowley, D. *Environ. Toxicol. Chem.* **2004**, *23*, 1-6.
13. Cornelissen, G.; Van Noort, P.C.M.; Govers, H.A.J. *Environ. Toxicol. Chem.* **1997**, *16*, 1351-1357.
14. Mayer, P.; Vaes, W. H. J.; Hermens, J. L. M. *Anal. Chem.* **2000**, *72*, 459-464.
15. ten Hulscher, T.E.M.; Postma, J.; den Besten, P.J.; Stroomberg, G.J.; Belfroid, A.; Wegener, J.W.; Faber, J.H.; van der Pol, J.J.C.; Hendriks, A.J.; van Noort, P.C.M. *Environ. Toxicol. Chem.* **2003**, *22*, 2258-2265.
16. Pawliszyn, J. *Solid-Phase Microextraction: Theory and Practice;* Wiley: New York, 1997.
17. Vaes, W. H. J.; Mayer, P.; Oomen, A. G.; Hermens, J. L. M.; Tolls, J. *Anal. Chem.* **2000**, *72*, 639-641.
18. Hawthorne, S.B.; Grabanski, C.B.; Miller, D.J.; Kreitinger, J.P. *Environ. Sci. Technol.* **2005**, *39*, 2795-2803.
19. Di Toro, D.M.; Zarba, C.S.; Hansen, D.J.; Berry, W.J.; Cowan, C.E.; Pavlou, S.P.; Allen, H.E.; Thomas, N.A.; Paquin, P.R. *Environ. Toxicol. Chem.* **1991**, *10*, 1541-1583.
20. U.S. Environmental Protection Agency. *Methods for the Derivation of Site-Specific Equilibrium Partitioning Sediment Guidelines (ESGs) for the Protection of Benthic Organisms: Nonionic Organics* [Draft]; EPA-822-R-02-042, U.S. EPA, Washington, DC, **2002**.
21. Akkanen, J.; Kukkonen, J. V. K. *Environ. Toxicol. Chem.* **2003**, *22*, 518-524.
22. Landrum, P. F.; Nihart, S. R.; Eadie, B. J.; Gardner, W. S. *Environ. Sci. Technol.* **1984**, *18*, 187-192.
23. Backhus, D. A.; Gschwend, P. M. *Environ. Sci. Technol.* **1990**, *24*, 1214-1223.
24. Ramos, E.U.; Meijer, S.N.; Vaes, W.H.J.; Verhaar, H.J.M.; Hermens, J.L.M. *Environ. Sci. Technol.* **1998**, *32*, 3430-3435.
25. Liu, W.P.; Gan, J.; Lee, S.J.; Kabashima, J.N. *Environ. Toxicol. Chem.* **2004**, *23*, 7-11.
26. Yang, W.C.; Spurlock, F.C.; Liu, W.P.; Gan, J. *J. Agric. Food Chem.* **2006**, *54*, 3967-3972.

27. Yang, W.C.; Hunter, W.; Gan, J.; Spurlock, F.C. *Environ. Toxicol. Chem.* **2006**, *25*, 1585-1591.
28. Yang, W.C.; Spurlock, F.C.; Liu, W.P.; Gan, J. *Environ. Toxicol. Chem.* **2006**, *25*, 1913-1919.
29. Bondarenko, S.; Putt, A.; Poletika, N.; Gan, J. *Environ. Toxicol. Chem.* **2006**, *25*, 3148-3154.

Chapter 11

Predicted Runoff Loads of Permethrin to the Sacramento River and Its Tributaries

S. Dasgupta[1], J. M. Cheplick[1], D. L. Denton[2], J. J. Troyan[3], and W. M. Williams[1]

[1]Waterborne Environmental Inc., 897-B Harrison Street SE, Leesburg, VA 20175
[2]U.S. Environmental Protection Agency, Region IX, Sacramento, CA 95814
[3]Sacramento Regional County Sanitation District, 10545 Armstrong Avenue, Suite 101, Mather, CA 95655

A probabilistic modeling assessment was conducted to identify potential sources of permethrin loadings to the Sacramento River and its tributaries from runoff and furrow irrigation drainage. A Geographical Information System (GIS) was used to construct approximately 6,956 model simulations representing unique combinations of soil, land use, and permethrin use within the study area. Simulations were conducted using the U.S. Environmental Protection Agency's Pesticide Root Zone Model (PRZM). Information about permethrin use was obtained from the California Department of Pesticide Regulation's (CDPR's) Pesticide Use Reporting (PUR) database. Simulations were conducted for 30-years of historical weather to evaluate runoff loadings under a range of potential low, moderate, and high rainfall events. Mass loadings are presented in terms of temporal probability of occurrence. Areas predicted to have high loading may be candidates for detailed analysis, monitoring, or mitigation.

Introduction

Permethrin is a synthetic pyrethroid that has been identified as one of 22 "high relative risk pesticides" by California's Central Valley Regional Water Quality Control Board (*1*). To quantify the potential movement of permethrin to aquatic habitats, a probabilistic modeling study was conducted to estimate potential loadings of permethrin to the Sacramento River and its tributaries in terms of spatial and temporal probability of occurrence.

Materials and Methods

Pesticide losses for this study were calculated as edge-of-field loadings from runoff and erosion induced by both rainfall and furrow irrigation drainage. No attempt was made to model the transport or conveyance in creeks, streams, and rivers. The Pesticide Root Zone Model (PRZM) was selected for this study based on its ability to simulate the interaction factors relevant in the fate and transport of permethrin within the agricultural landscape and based on the preference for its use by the U.S. Environmental Protection Agency (USEPA) (*2*). PRZM is a dynamic, compartmental model developed by USEPA for use in simulating water and chemical movement in unsaturated soil systems within and below the plant root zone (3). The hydrologic component of PRZM simulates the physical processes of rainfall, runoff, infiltration, erosion, and evapo-transpiration. The chemical transport component of PRZM calculates pesticide uptake by plants, surface runoff, erosion, decay, vertical movement, foliar loss, dispersion and retardation. PRZM includes the ability to simulate pesticide metabolites and irrigation.

For this study, 6,956 individual PRZM simulations were conducted. Simulations were defined by the intersection of land areas designating different combinations of soil, land use, weather, chemical use, irrigation, and application dates within the Sacramento River watershed study area (Figure 1).

Chemical Applications

Permethrin use records were obtained from the Pesticide Use Reporting (PUR) database (*4*), accessed from the California Pesticide Information Portal (CalPIP). The PUR database contains detailed information about chemical applications (application dates, application amounts, application method, and others) at the section (1 square mile) resolution.

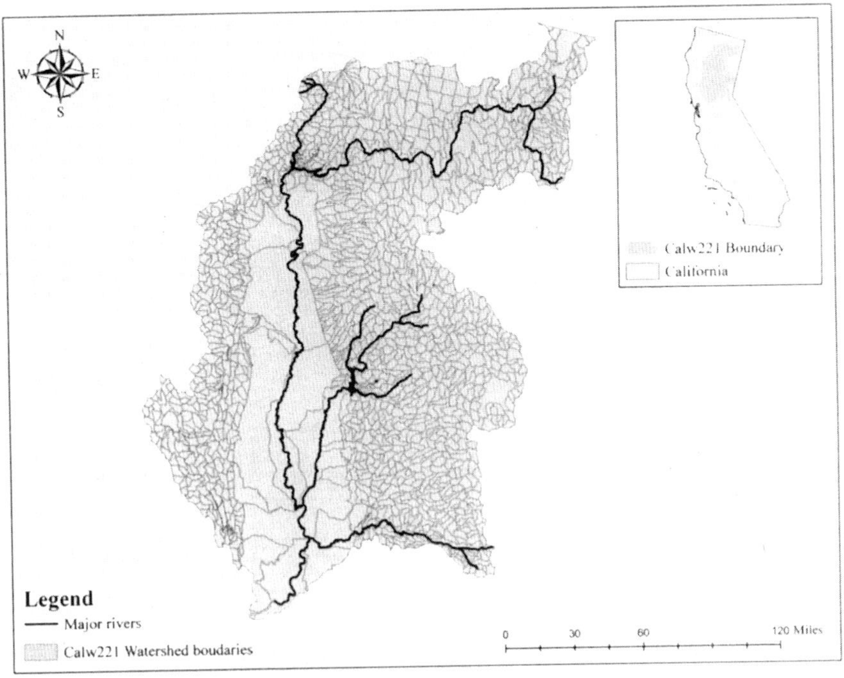

Figure 1. Sacramento River Watershed with subbasin delineation. (Obtained from the California Interagency watershed map (Calwater 221) (5). (See Page 1 of color inserts.)

Chemical Environmental Fate Properties

Environmental fate properties for permethrin (Table I) were obtained from the U.S. Department of Agriculture, Agricutural Research Servie (USDA-ARS) Pesticide Property database (*6*). For the PRZM simulations, the combined soil photolysis and aerobic soil half-life was used in the model. Foliar degradation was assumed to occur at the rate given by soil photolysis.

Soil data and Land Use

Soil parameters were identified from the State Soil Geographic (STATSGO) database (*7*). The STATSGO data set is a digital general soil association map developed by the National Cooperative Soil Survey and distributed by the USDA's Natural Resources Conservation Service (formerly Soil Conservation Service). The STATSGO soil regions within the study area are illustrated in

Table I. Environmental fate properties for permethrin

Property	*Value*
CAS Number	52645-53-1
Empirical formula	$C_{21}H_{20}Cl_2O_3$
Molecular weight, g/mole	391.3
Vapor pressure, mm Hg	2.10E-8
Aqueous solubility, ppm	0.006
Henry's Law Constant, atm-m^3/mole	1.93E-10
Soil Koc, mL/g	39300
Soil photolysis half-life, days	33.00
Aerobic soil half-life, days	30.00
Combined soil photolysis and aerobic soil half-life, days	15.71
Anaerobic soil half-life, days	108.00
Hydrolysis at pH 7	Stable

Figure 2. There were up to 18 unique soil types within a single STATSGO soil polygon. The land uses at the section level were intersected with the soil polygons within a GIS framework to identify soil types. Since, it was not possible to identify the exact spatial location of a soil type within a single PUR section, all soil types associated with STATSGO polygon that intersected a PUR section were used for modeling. Results were scaled in proportion to the percentage a specific soil existed in a STATSGO polygon to reflect the relative probability that a given soil maybe associated with a PUR section. The land use data for the study area was obtained from the Pesticide Use Reporting (PUR) database (*4*), accessed from the California Pesticide Information Portal (CalPIP). Land uses were grouped into seven categories, namely corn, fruit, grain, grass, nut, vegetable, and vineyard.

Crop Parameters

Cropping dates for emergence, maturation, and harvest and other crop parameters for interception storage, maximum coverage, active root depth, aerial coverage, maximum canopy height, and others were derived from USEPA Standard Tier 2 scenarios (*2*).

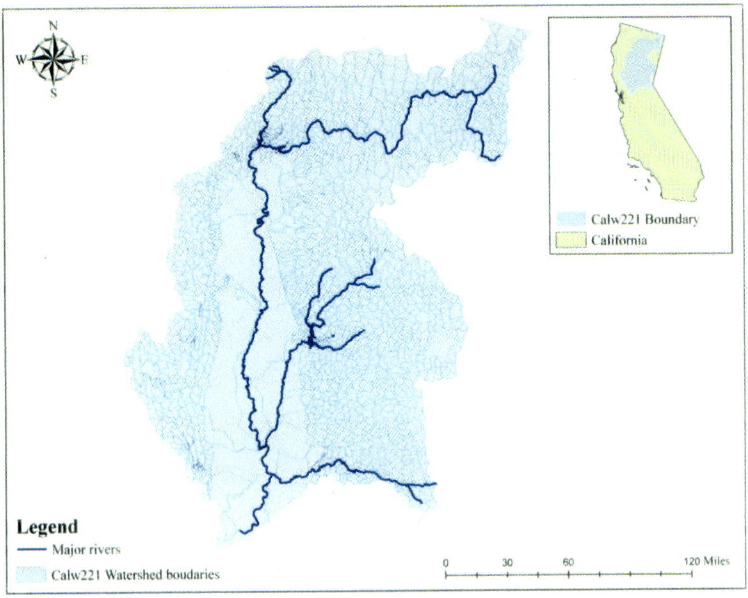

Figure 11.1. Sacramento River Watershed with subbasin delineation. (Obtained from the California Interagency watershed map (Calwater 221) (5).

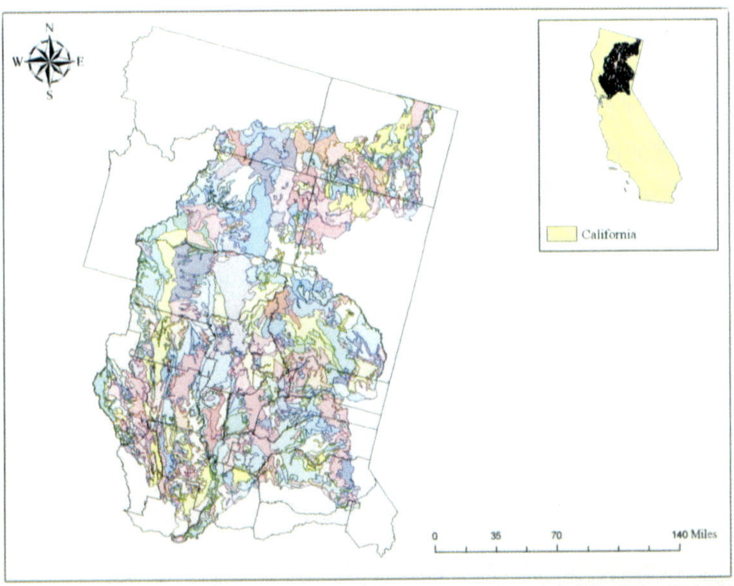

Figure 11.2. STATSGO Soil polygons within the Sacramento River Watershed.

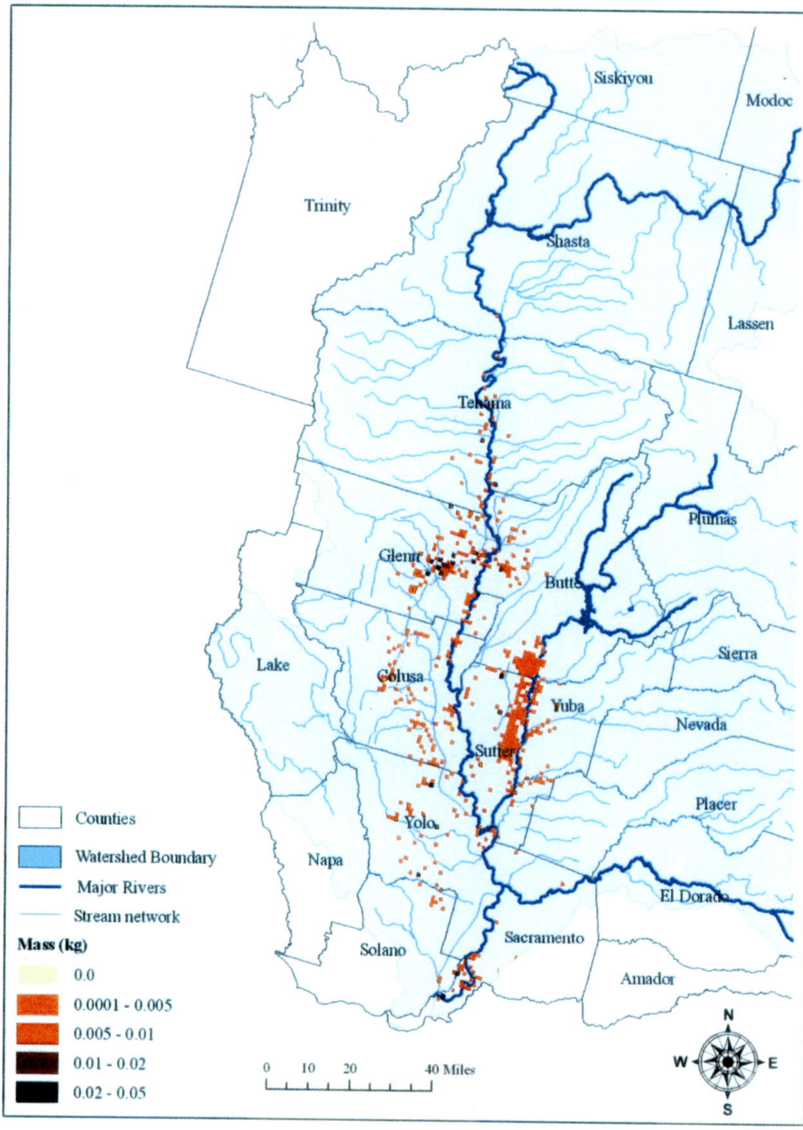

Figure 11.3. PUR cell-based 50th percentile mass loadings for permethrin (kg)

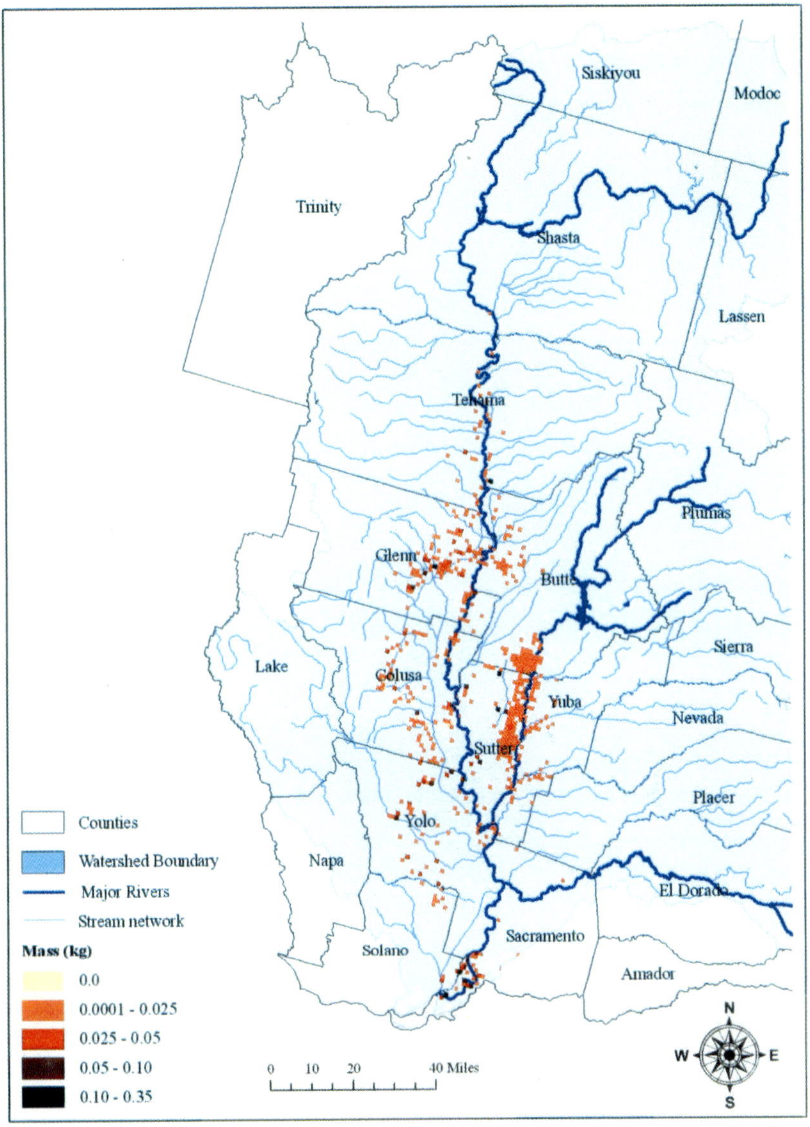

Figure 11.4. PUR cell-based 90th percentile mass loadings for permethrin (kg)

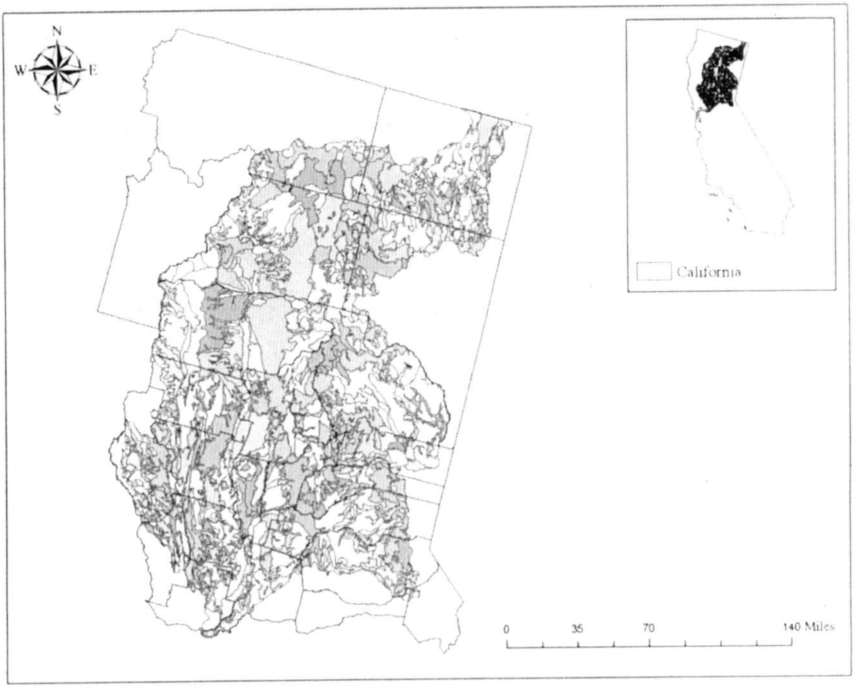

Figure 2. STATSGO Soil polygons within the Sacramento River Watershed. (See page 1 of color inserts.)

Weather Data

Simulations were conducted for 30-years of historical weather (1961-1990) to evaluate runoff loadings under low, moderate, and high rainfall events. Five weather stations (Sacramento, CA; San Francisco, CA; Reno, NV, Medford, OR, and Santa Maria, CA) were used to account for weather variability in the study area. The weather data was obtained from USEPA's Center for Exposure Assessment Modeling (CEAM) in PRZM-ready format (*8*). CEAM developed these weather files by associating National Oceanic and Atmospheric Administration (NOAA) primary weather stations to Major Land Resource Areas (MLRAs). MLRAs are a classification system developed by the USDA to represent areas of similar climate, geomorphology, and natural resources. PUR sections were assigned a specific weather station based on the MLRA and STATSGO polygon in which it resides. The majority of simulations were associated with the Sacramento weather station (W23232).

Irrigation

Corn and vegetables are generally irrigated using furrow irrigation within the Sacramento watershed. Other crops (fruit, grain, vineyard, grass, and nut) are irrigated by other methods, including drip irrigation and micro-sprinkler irrigation systems. Corn generally requires the application of 3-3.5 acre-feet of water which is applied over 5-9 irrigation events throughout the season (*9*). Tomato production was used as the prototype crop for assessing irrigation for vegetables. Across California, an approximation for furrow irrigated fields would be 2.5-3 acre-feet of water to be applied throughout the season with 7-14 events (*10*). It was suggested by local expert (*11*) that runoff from furrow irrigated fields in the Sacramento River watershed can range between 10 to 30 percent of total water applied for irrigation. An approach was developed to simulate tailwater releases and chemical losses in irrigation water.

Design of Furrow Irrigation system

Key water balance guidelines that were incorporated into the study are highlighted below:

- Amount of water that infiltrates the system should match overall water requirements for both corn and tomatoes.
- A total volume of 2.7 acre-feet of water should be applied for both crops over 9 irrigation events which constitutes 0.3 acre-feet of water for each application event.
- The amount of water generated as runoff by the system should be between 10 to 30 percent of total applied water.
- The irrigation system should yield an overall efficiency between 60 to 75 percent.

The latter assumption was based on the information from Schwab et al., (*12*), that a gently sloping, well leveled and uniformly graded field usually has a furrow irrigation efficiency of 60 to 75 percent. Efficiency (as related to the water balance of the system) refers to the fraction of water that actually infiltrates into the system to the total amount of water that is applied to the system.

Calibration of PRZM to Simulate Furrow Irrigation

Irrigation within the PRZM model is activated when the average root zone soil moisture falls below a threshold value f_c defined by the user as a fraction of the available water capacity (PCDEPL). The soil moisture deficit is given by:

$$D = (\theta_{fc} - \theta_z) * Z_r \quad (1)$$

in which D is soil moisture deficit (cm), θ_z is the average root-zone soil moisture content (cm^3 cm^{-3}), θ_{fc} is the average root-zone soil moisture content at field capacity (cm^3 cm^{-3}), and Z_r is the root-zone depth (cm). The amount of soil moisture deficit (D) is added per unit area to the system as irrigated water by the PRZM model.

Several input parameters in the PRZM model were calibrated to achieve the water balance guidelines for the furrow irrigation system:

- Pan evaporation factor (PFAC)
- Soil evaporation moisture loss (ANETD)
- Universal soil loss cover management factor (USLELS)
- SCS Runoff curve number (CN)
- Fraction of available water capacity (PCDEPL)
- Irrigation application rate (RATEAP)
- Leaching factor as a fraction of irrigation water depth (FLEACH)

Since most furrows are set up as parallel strips between row crops, it was decided that a ratio of 7:3 be used (for ground applications) to divide the chemical/pesticide application mass between crops and furrows. In other words, 70 percent of the chemical/pesticide mass was assumed to be applied over crops and the remaining 30 percent was assumed to be applied over furrows within the particular area. For aerial applications however, a higher ratio of 1:1 was used meaning the total mass of the chemical/pesticide was equally distributed (50 percent for each) between crops and furrows. The rationale for this being that when a chemical is applied aerially, there is a more uniform distribution of the chemical between the crops and furrows as compared to when only the crops are targeted using a ground application technique.

Two sets of PRZM runs were conducted for 'corn' and 'vegetable'. The first run included either 70 or 50 percent (depending on ground or aerial application method) of the applied mass within the crop area. The second run included either the remaining 30 or 50 percent of the applied mass within the furrow.

Results and Discussion

Calibration results

Calibrated input parameters are compared to original input parameters in Table II. Calibrations were conducted for 30-years of historical weather (1961-

1990) to evaluate runoff loadings under a range of soil moisture conditions using four different weather stations: Sacramento, CA (W23232), San Francisco, CA (W23234), Reno, NV (W23185), and Santa Maria, CA (W23273). The 30-year average values were used for comparing variations in results for the four weather stations (Table III). The first column lists various irrigation and water budget parameters that were used for comparisons. The second column contains values that are based on the furrow irrigation design and are consistent with observed practices in the Sacramento River watershed. The subsequent columns illustrate variations in parameters when different weather station data are used for the PRZM simulations. The model predicts irrigation frequency (number of irrigation events), and components of the water balance (amount of irrigation, amount of runoff) with reasonable accuracy when compared to design results. The predicted amount of infiltration is greater due to the inclusion of daily precipitation events in the model which was excluded from the initial water balance guidelines.

Table II. Calibrated PRZM parameters for furrow irrigation

PRZM Parameter	*Variable name*	*Original value*	*Calibrated value*
Pan evaporation factor (dimensionless)	PFAC	0.7	0.5
Soil evaporation moisture loss (cm)	ANETD	17	25
Universal soil loss equation cover management factor (dimensionless)	USLEC	0.915	0.400
SCS Runoff curve number (dimensionless)	CN	84	60
Fraction of available water capacity (dimensionless)	PCDEPL	0.55	0.15
Irrigation application rate (cm/hr)	RATEAP	0.00	0.44
Leaching factor as a fraction of irrigation water depth (dimensionless)	FLEACH	0.10	0.00

Predicted Permethrin Loadings

Runoff and eroded losses were converted into combined annual loads for each simulation and then aggregated to the county scale for tabular reporting (Table IV) and to the township scale (36 square miles) for mapping (Figure 3 and Figure 4). In Table IV, applied mass is based on 2003 application data. Pesticide use is listed based on amount of aerial application (kg), amount of ground application (kg), percentage aerial application, and total application (kg). Predicted loads are listed based on amounts (kg) and percentages.

For those land uses that were subjected to furrow irrigation (corn and vegetable) the losses of both portions (70 percent for crop and 30 percent for furrow) were added together to generate total annual loads. The Weibull plotting position (*13*) was used to calculate the 50th and 90th percentile annual load for each aggregation level. The 50th and 90th percentiles express pesticide loadings into a temporal probability context (i.e., frequency of occurrence). For example annual loads at the 90th percentile values are estimated to occur on average once in a 10-year period. The 50th percentile values have a recurrence interval of 2-years. Results indicate that highest loads occur around the tributaries and streams of the major rivers within the study area as opposed to the smaller water bodies and segments. Moreover, predicted loads are concentrated within nine counties, namely Butte, Colusa, Glenn, Sacramento, Solano, Sutter, Tehama, Yolo, and Yuba.

Table III. Calibration results for different weather stations

Irrigation parameters and water budget for corn	*Design*	*Predicted (PRZM) parameters for different weather stations*			
		W23232	W23234	W23273	W23185
# of irrigation events per season	5 - 9	8.97	8.97	7.37	8.67
Amount of irrigation (cm)	95.76	94.69	67.23	77.79	91.52
Amount of infiltration (cm)	82.26	121.16	103.55	95.16	93.33
Amount of runoff (cm)	12.69	16.25	12.40	12.59	14.48
Runoff as % of total water applied (irrigation only)	17.21	17.21	18.66	16.21	15.83
Evapotranspiration (cm)		59.36	48.50	51.20	54.52
Sediment eroded (tonnes/ha)		2.14	1.55	1.69	1.99
Irrigation parameters and water budget for tomato	*Design*	*Predicted (PRZM) parameters for different weather stations*			
		W23232	W23234	W23273	W23185
# of irrigation events per season	5 - 9	7.87	5.90	6.87	8.13
Amount of irrigation (cm)	95.76	83.07	62.30	72.51	85.89
Amount of infiltration (cm)	82.26	112.17	99.87	91.40	89.96
Amount of runoff (cm)	12.69	14.43	11.63	11.69	13.51
Runoff as % of total water applied (irrigation only)	17.21	17.50	18.92	16.21	15.80
Evapotranspiration (cm)		57.87	48.40	51.43	55.12
Sediment eroded (tonnes/ha)		1.05	0.83	0.84	1.01

Table IV. Permethrin applications and predicted loadings

County	*Applied Mass (kg)*		*Mass Loadings (kg)*		*Mass Loadings (% of applied mass)*	
	Total	*% Aerial*	*50th*	*90th*	*50th*	*90th*
Butte	725.6	11.7	0.03	0.33	<0.01	0.05
Colusa	886.5	62.4	0.09	0.31	0.01	0.04
Glenn	784.1	46.2	0.40	1.36	0.05	0.17
Sacramento	348.7	0.5	0.18	1.08	0.05	0.31
Shasta	0.02	0.0	0.00	0.00	<0.01	0.04
Solano	185.5	2.5	0.03	0.10	0.01	0.05
Sutter	1152.7	8.6	0.22	0.68	0.02	0.06
Tehama	239.3	6.5	0.09	0.29	0.04	0.12
Yolo	445.6	32.7	0.155	0.91	0.04	0.20
Yuba	616.0	0.1	0.034	0.40	<0.01	0.06

Uncertainty

Models are mathematical representations of complex physical, chemical, and biological processes. Some level of uncertainty is inherent in any modeling study because of simplifications required in representing the system as a prototype, limitations in data used to configure the model, and in the predictive capabilities of the models themselves. Areas of greatest uncertainty from a model setup standpoint relate to accurate knowledge and characterization of the field systems summarized by the PUR database and the fate and transport of permethrin under local conditions.

It was not possible to identify the exact spatial location of a soil type within a single PUR section. Therefore, all soils associated with a STATSGO polygon that intersected a PUR section were used for modeling. Results were scaled in proportion to the percentage a specific soil existed in a STATSGO polygon. The soil actually underlying the application field may not represent this distribution.

In interpreting results, consideration should be given to mitigation factors that exist in the watershed that were not specifically represented. Simulations reflect edge-of-field loadings and do not represent reductions caused by natural configurations in the landscape or anthropogenic management practices (e.g., buffers). Again, the influence of these factors can be addressed in a refined assessment. These factors can also be incorporated in the model to predict load reductions that may occur if mitigation procedures were to be implemented.

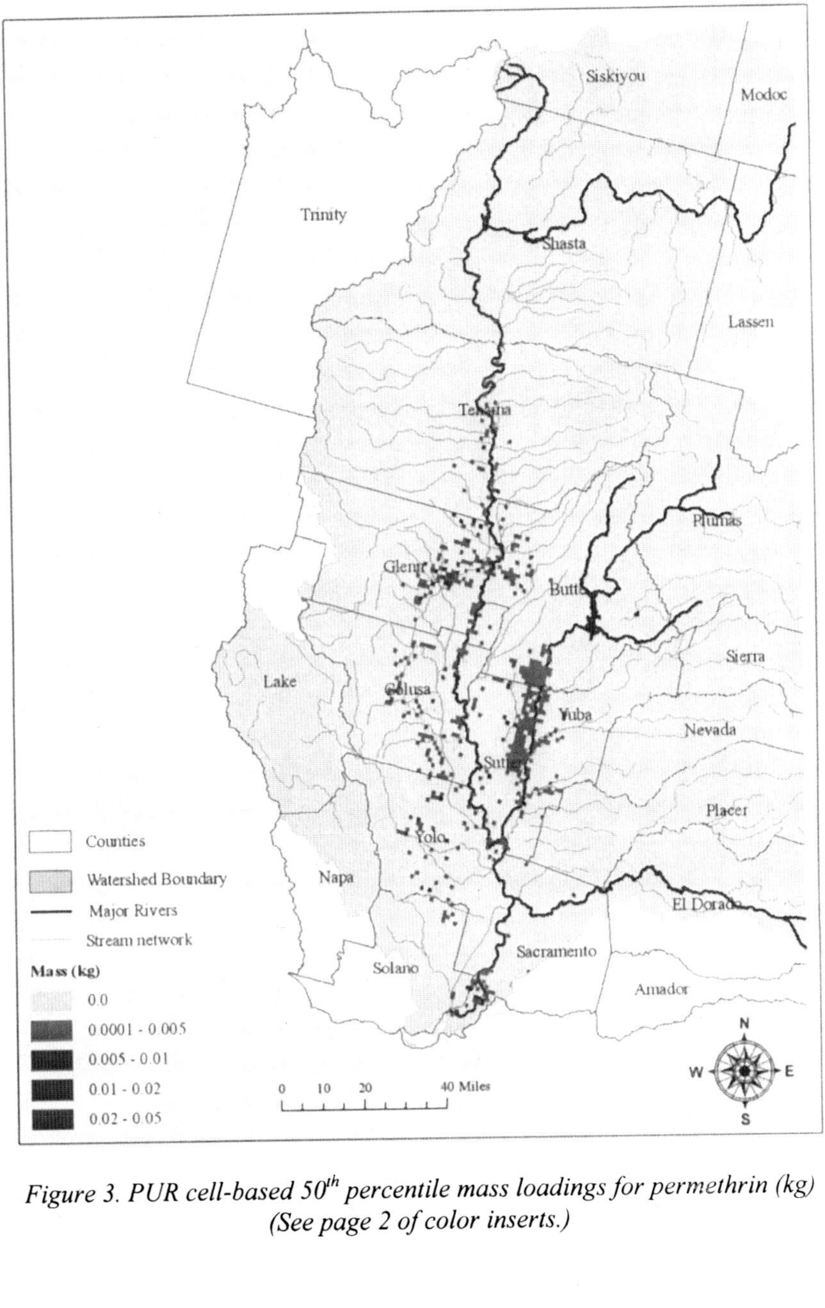

Figure 3. PUR cell-based 50th percentile mass loadings for permethrin (kg) (See page 2 of color inserts.)

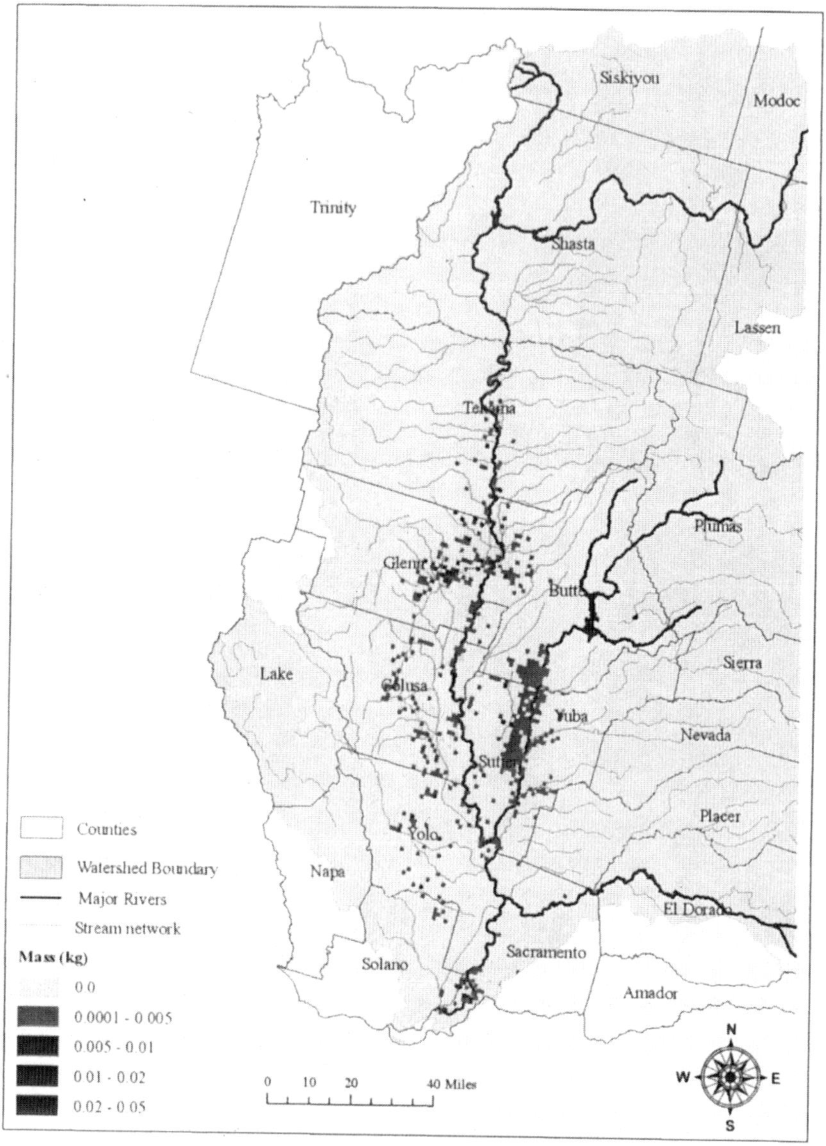

Figure 4. PUR cell-based 90th percentile mass loadings for permethrin (kg) (See page 3 of color inserts.)

Chemical use varies from year to year from crop rotations and pest pressures. This study addressed spatial probability of occurrence based on reported pesticide applications from 2003 and temporal probability based on historical weather variability over a 30-year period (1961 through 1990). Additional application years could be included in this type of assessment.

Calibration/validation of actual pesticide runoff was not conducted as part of this study. However, PRZM is widely used and has had site-specific validation (*14,15*)

These areas of uncertainty are disclosed to assist in the interpretation of the results. This study should not be expected to predict accurate pesticide losses from individual fields for the reasons discussed above. Rather, the study is best used to identify areas likely to contribute to pesticide loadings to aquatic systems. These areas are likely to be priority areas for future monitoring, mitigation, or more in-depth analysis.

Conclusions and Recommendations

Pesticide loadings to the Sacramento River and its tributaries were estimated in terms of spatial and temporal probability of occurrence. Loadings were predicted for permethrin using USEPA's Pesticide Root Zone Model, PRZM, Version-3.12.2. Simulations reflect the 2003 year usage of permethrin within the study area and a 30-year historical rainfall period.

Results indicated that predicted (50th and 90th percentile) mass loadings were concentrated within nine counties (Butte, Colusa, Glenn, Sacramento, Solano, Sutter, Tehama, Yolo, and Yuba) within the Sacramento Valley and these regions were distributed around the tributaries and streams of the major rivers within this region. The daily predictions from each simulation have been retained in the event that other end points are desired in the future (e.g., as monthly or seasonal loads, by crop, or aggregated to other geographical areas of analysis).

Results indicate counties and watersheds that may be generating the highest loadings of chemicals in streams and rivers in the Sacramento River watershed. These areas may be candidates for more detailed analysis, monitoring, or mitigation. Future research and additional studies focused on monitoring and mitigation efforts within the watershed could include:

- Applying the approach to the other chemicals identified by the Central Valley Regional Water Quality Control Board as high relative risk pesticides.
- Use of the model to evaluate the relative benefits of mitigation measures and best management practices (BMPs).

- Field-scale monitoring studies to characterize the fate and transport of specific chemicals under localized conditions. Such studies would monitor relevant parameters including application efficiency and drift, residue levels in the field at measurement intervals sufficient for characterizing pesticide dissipation rates. Hydrologic measurement should include precipitation, pan evaporation, and soil moisture. Studies at this detail would also provide data sets for model validation.
- Spatial characterization of agricultural fields and their proximity to nearby ditches, streams, and other potential receiving water bodies from remotely sensed imagery. This could be used to improve estimates of drift loads.

Acknowledgements

This study was conducted under the Sacramento River Toxic Pollutant Control Program (SRTPCP) under a grant funded by the U.S. Environmental Protection Agency (USEPA) Region 9 to the Sacramento Regional County Sanitation District (SRCSD). The authors would like to acknowledge the following individuals for providing information and peer review during the course of the study: Stephanie Fong, Central Valley Regional Water Quality Control Board (CVRWQCB) and John Oram, Ph.D., San Francisco Estuary Institute (SFEI).

References

1. Central Valley Regional Water Quality Control Board (CVRWQCB). Relative Risk Evaluation for Pesticides Used in the Sacramento Valley, 2006.
2. U.S. Environmental Protection Agency. Pesticide Root Zone Model Field and Orchard Crop Scenario Metadata. Pesticides: Science and Policy, 2004. URL http://www.epa.gov/oppefed1/models/water/metadata.htm.
3. Suárez, L.A. PRZM_3, A model for predicting pesticide and nitrogen fate in the crop root and unsaturated soil zones: users manual for release 3.12.2, National Exposure Research Laboratory, Office of Research and Development, U.S. Environmental Protection Agency, Athens, Georgia, 2005.
4. California Department of Pesticide Regulation. California Pesticide Information Portal Database. URL http://calpip.cdpr.ca.gov/cfdocs/calpip/prod/main.cfm (accessed September 2005).
5. California Spatial Information Library. California Interagency Watershed Map of 1999 (CalWater 2.2.1). URL http://gis.ca.gov/meta.epl?oid=22175 (accessed September 2005).

6. U.S. Department of Agriculture. The ARS Pesticide Properties Database. Agricultural Research Service. URL http://www.ars.usda.gov/Services/docs.htm?docid=14199 (accessed September 2005).
7. U.S. Department of Agriculture. State Soil Geographic (STATSGO) Database. National Cartography and GIS Center. URL http://www.essc.psu.edu/soil_info/index.cgi?soil_data&statsgo (accessed September 2005).
8. U.S. Environmental Protection Agency. Center for Exposure Assessment Modeling (CEAM). Meteorological Data. July 2002. URL http://www.epa.gov/ceampubl/tools/metdata/index.htm (accessed September 2005).
9. University of California Cooperative Extension, 2002. Trial results of field corn production for Yolo/Solano/Sacramento counties. 2002. URL ucce.ucdavis.edu/files/filelibrary/2557/10282.pdf
10. Le Strange, M.; Schrader, W.L.; Hartz, T.K. 2000. Fresh-market tomato production in California. Publication 8017. University of California Division of Agriculture and Natural Resources. 2002. URL http://anrcatalog.ucdavis/edu/pdf/8017.pdf
11. Hansen, B. Extension Irrigation Specialist. University of California, Davis. Department of Land, Air, & Water Resources. (Phone conversation on November 13, 2006).
12. Schwab G.O.; Fangmeier, D.D.; Elliot, W.J.; Frevert, R.K. *Soil and Water Conservation Engineering*. Fourth edition. John Wiley & Sons, Inc.: USA, 1993.
13. Haan, C.T. *Statistical Methods in Hydrology*. Iowa State University Press: Ames, IA, 1977.
14. *FIFRA Environmental Model Validation Task Force Final Report*. Jones, R.L; Russell, M.H., Eds; American Crop Protection Association: Washington, DC, 2001 URL http://femvtf.com/femvtf/Files/ FEMVTFbody.pdf (accessed June 2007).
15. Snyder N.J.; Williams, W.M. Exposure assessment model for diazinon sources in the Sacramento River Basin's Main Drainage Canal. Sacramento River Watershed Program, 2004. URL www.sacriver.org/ subcommittees/ag/documents (accessed June 2007).

Chapter 12

Chiral Selectivity in the Environmental Fate of Pyrethroids

Sujie Qin[1], Weiping Liu[2], and Jay Gan[1]

[1]Department of Environmental Sciences, University of California, 2258 Geology Building, Riverside, CA 92521
[2]Research Center of Environmental Sciences, Zhejiang University of Technology, Hangzhou 310032, People's Republic of China

Pyrethroids are a family of chiral pesticides with a large number of stereoisomers. Following the discovery of significant differences in non-target toxicities between pyrethroid enantiomers, enantioselectivity during degradation and analysis have been investigated in recent studies to better understand their fate and ecotoxicological effects. Enantioselective degradation occurs frequently for pyrethroids in various field matrices and under laboratory conditions. Enantiomerization, which may contribute to the enantio-selectivity in the environmental fate and effects of pyrethroids, is common in pyrethroids with the α-cyano carbon when exposed to alcohols and other conditions. Results suggest that the fate and effects of individual stereoisomers, instead of the racemic compound, should be considered for pyrethroids to better predict their environmental impacts. In addition, rigorous analytical methods and pesticide handling procedures should be developed to prevent the artifacts caused by enantiomerization.

Introduction

Chirality arises from the asymmetric positions in a molecular structure or derives from substituted rings due to steric hindrance as in the case of α-HCH and chlordane. For instance, when four different groups are attached to a tetrahedral atom (usually carbon, but also phosphorus, and other tetrahedral substituted atoms) in a molecule, the compound becomes chiral and contains a pair of enantiomers. Chirality is a very popular phenomenon in nature. It exists in almost all biological systems and is considered to play a very important role in various fields of chemistry. The significance of chirality has long been recognized in relation to the relative biological activity of the individual enantiomers of natural compounds and synthetic drugs (*1,2*). Enantiomers are chiral pairs of molecules that are nonsuperimposable mirror images of each other. Since all biological systems constitute chiral environments and most enzymatic pathways are stereoselective, enantioselectivity occurs when chiral compounds are introduced into a biological system, from the basic building blocks of life such as amino acids, carbohydrates and lipids, to the layout of the human body (*3*). For example, it is well known that only L-amino acids are of nutritional value for animals. While L-glutamate is used as a food flavor enhancer, the D-isomer does not have any such property (*1*).

Many anthropogenic chemicals of environmental concern, including some phenoxy acid herbicides, organophosphorus insecticides, polychlorinated biphenyls and some DDT derivatives, are chiral. These compounds are typically synthesized and applied to the environment as racemates (i.e., 50:50 mixtures of enantiomers when there is one chiral position) (*1*). Chiral compounds account for approximately 25% of all pesticides used commercially and for 26% of the total value of the world pesticides market (*4*). However, among these compounds, those sold in single isomer form contribute only 7% of the market value (*4*). Upon entering the environment, enantiomers of the same compound show identical physical and chemical properties, which makes them identical in abiotic environmental processes. However, individual stereoisomers may interact differently with enzymes and other naturally occurring chiral molecules, which leads to enantioselectivity in microbial transformation and environmental persistence (*5,6*).

Pyrethroids are widely used for controlling insects in crop production and around households. With the restriction of the usage of organophosphate insecticides, the use of pyrethroids is expected to increase further. Although pyrethroids are highly hydrophobic which makes them immobile in soil, they may still find their way into an aquatic system via runoff or soil erosion (*7,8*). This makes them a significant environmental concern because most pyrethroids possess high acute toxicity to aquatic organisms, often at a concentration less than 0.5 ppb (*9-11*).

All pyrethroids contain two or three chiral centers, making them a family of pesticides with the largest number of enantiomers. Chirality of pyrethroids may arise from the acid moiety, the alcohol moiety, or both (*12,13*). In the development history of pyrethroids, significant enantioselectivity has been widely observed in insecticidal activity for the enantiomers from the same compound (*12-15*) and recently, studies show that enantioselectivity also exists in their aquatic toxicity (*16,17*). These findings prompted the interest to further understand the enantioselectivity of pyrethroids in biologically-mediated environmental processes, especially degradation.

Enantioselectivity in Pyrethroid Degradation

Enantioselective biodegradation has been studied for a number of chiral pesticides under both field and laboratory conditions. However, most of the test compounds in those studies were legacy chlorinated insecticides. Studies show that for these compounds, one enantiomer is often preferentially degraded over the other enantiomer, providing evidence for the enantiomer-specific environmental fate of chiral contaminants (*18-22*). For instance, in the laboratory study of metalaxyl degradation in soil, the fungicidally active *R*-enantiomer degraded more rapidly than the inactive S-enantiomer, resulting in residues enriched with *S*-metalaxyl when the racemic compound was incubated (*22*). The occurrence of enantioselectivity in the degradation of pyrethroids was systematically investigated in a few recent studies, using widely used pyrethroids such as bifenthrin, permethrin, cypermethrin and cyfluthrin.

Enantioselective Degradation in Field Samples

Enantioselective degradation can be evaluated by several means: comparing the concentration of individual enantiomers, comparing the change of stereoisomer profiles from the original values, or studying the changes in the enantiomer ratio. In one study (*23*), a set of runoff samples were taken from a runoff discharge channel at a nursery site located in Orange County, CA. Concentrations of individual enantiomers were determined for *cis*-bifenthrin and *cis*-permethrin by solid-phase microextraction (SPME) coupled with chiral selective analysis using gas chrompagraphy (GC) (Table I). The concentration of the *1R-cis* enantiomer was consistently higher than that of the *1S-cis* enantiomer for both *cis*-bifenthrin and *cis*-permethrin, although the commercial formulations were racemic mixtures.

The enantiomer ratio (*ER*) is defined as the concentration of one enantiomer of a chiral compound divided by the concentration of the other enantiomer (*24*). *ER* is often used to evaluate the enantioselective behavior of chiral compounds. In another study (*17*) in which aged field sediment samples were analyzed, *ER* was used to show whether enantioselective degradation occurred, resulting in a

Table I. Enantiomer compositions of *cis*-bifenthrin and *cis*-permethrin in runoff water samples (µg L^{-1})

Water samples	*Compounds*			
	SS-bifenthrin	*RR-bifenthrin*	*SS-permethrin*	*RR-permethrin*
1	0.35±0.04	1.02±0.05	ND	0.17±0.05
2	0.08±0.04	0.12±0.05	0.42±0.04	0.72±0.5
3	0.05±0.04	0.11±0.05	ND	0.11±0.04
4	0.07±0.05	0.15±0.04	ND	ND
5	ND	0.10±0.04	ND	ND
6	ND	0.07±0.05	ND	ND

Ref. (*23*).

significant deviation from the original *ER* value for *cis*-bifenthrin (1.02) and *cis*-permethrin (0.99). The dried sediment samples were taken from sediment that had accumulated from surface runoff over a 4-year period. The results summarized in Table II show that *ER* increased with depth of the aged sediment samples. In the surface layer, the average *ER* value for *cis*-bifenthrin was about 1.0, but it increased to 1.11 in the 15-30-cm layer and further to 1.32 in the 30-45-cm layer, suggesting a faster degradation of *1S-cis*-enantiomer. A similar but stronger trend was also observed for *cis*-permethrin (Table II). The results in Table II demonstrate that the rate of the enantioselective degradation are not always the same, but may be influenced by soil and environmental conditions. In this study, differences in moisture content (relatively low in the surface layer) and oxidation state with depth (more aerobic in the surface layer) were thought to have affected the enantioselectivity of the microbial transformation.

Even the direction of enantioselectivity in degradation may be different for the same pyrethroid due to environmental conditions. The analysis of sediment samples from the Newport Bay-San Diego Creek Watershed in CA (*25*) showed that for *cis*-bifenthrin, permethrin and cyfluthrin, when enantioselective degradation was observed, not only the rate but also the direction of enantioselectivity was not always the same and appeared to depend on the sampling location and environmental conditions. The variation in the direction and rate of enantioselectivity for pyrethroid enantiomers was in agreement with previous observations made for other chiral compounds (*18*) and is reasonable because there may be variations in the microbial populations as a result of the influence by environmental factors including plant cover, soil types, soil pH, and the soil oxidation state, among other things.

Table II. Concentrations of cis-bifenthrin and cis-permethrin enantiomers and enantiomer found in a sediment pond next to a nursery in California (μg kg^{-1})

Sample	*(Z)-cis-Bifenthrin*			*cis-Permethrin*		
	RR	*SS*	*ER*	*RR*	*SS*	*ER*
A(0-15cm)	69.7±1.3	68.1±1.2	1.02±0.02	26.7±1.9	25.3±1.6	1.05±0.02
A(15-30cm)	93.3±1.3	84.0±1.1	1.11±0.03	45.6±1.6	35.0±1.7	1.30±0.02
A(30-45cm)	50.3±10.6	38.3±1.5	1.32±0.03	15.8±1.5	10.3±1.4	1.54±0.03
B(0-15cm)	152.9±2.3	149.6±2.4	1.02±0.03	43.3±1.2	41.5±1.4	1.02±0.03
B(15-30cm)	177.2±1.5	159.9±2.5	1.11±0.03	65.5±1.2	48.8±1.1	1.33±0.02
C(0-15cm)	143.2±2.4	139.9±2.4	1.32±0.02	50.6±1.3	48.2±1.2	1.05±0.03
C(15-30cm)	189.9±2.5	170.4±3.1	1.11±0.03	80.7±1.5	60.3±1.6	1.32±0.03
Formulation	-	-	1.02±0.03	-	-	0.99±0.02

Ref. (*17*).

Enantioselective Degradation under Laboratory Controlled Conditions

The environmental fate of *cis*-bifenthrin, permethrin and cypermethrin was assessed through laboratory incubation experiments with soil slurries and individual microbial strains (*26, 27*). For *cis*-bifenthrin and permethrin, in addition to the preferential degradation of *trans*-permethrin over *cis*-permethrin, the authors found that the *1S-cis* enantiomer was preferentially degraded over *1R-cis* enantiomers for both *cis*-bifenthrin and *cis*-permethrin. Enantioselectivity was more pronounced for *cis*-permethrin than for *cis*-bifenthrin, and also varied by microbial strains (Table III). A similar observation was also obtained for cypermethrin. For the same microbial strain degrader, the *trans* diastereomers were consistently degraded preferentially over the *cis* diastereomers and one stereoisomer, *1R-cis-αS* had a relatively longer half-life than the other *cis* stereoisomers. The degradation of cypermethrin and fenvalerate isomers by isolated soil bacteria was also examined by Sakata et al. (*28*). Bacteria strains were isolated by a dilution plate method from two Japanese soils. A large proportion of the isolated soil bacteria showed stereospecificity in the main degradation route, that is, ester hydrolysis of cypermethrin and fenvalerate isomers. The *trans* and *αS* isomers of cypermethrin were degraded faster than the *cis* and *αR* isomers. For fenvalerate, more rapid degradation was also observed for *αS* isomers. The results from this study also showed that there was a difference between the *1R* and *1S* isomers of cypermethrin and between *2R* and *2S* isomers of fenvalerate. In this study, the cell-free extracts were further prepared from four bacteria strains and fractionated by gel filtration or ion-exchange column chromatography. The enzyme assay showed that the presence of several enzyme fractions capable of preferentially degrading the *αS* isomers, whereas the two enzyme fractions degraded the *trans* isomers faster than the *cis* isomers. These results demonstrate that enantioselective degradation of pyrethroid insecticides in soils is primarily due to soil microorganisms through various esterase enzymes that can hydrolyze pyrethroid isomers with a high stereospecificity.

Table III. First-order rate constants for *1R-cis-* (k_{RR}) and *1S-cis-* (k_{SS}) enantiomers of *cis*-bifenthrin and *cis*-permethrin and enantiomer ratio doubling time ($T_{ER=2}$) for biodegradation by isolated bacteria[a]

	Bacteria	k_{RR} *(1/h)*	K_{SS} *(1/h)*	$T_{ER=2}$ *(h)*
cis-BF	*BF-6**	0.0135	0.0172	187
	BF-24	0.0032	0.0037	1380
	*BF-28**	0.0146	0.0171	277
cis-PM	*PM-1**	0.0200	0.0408	33
	*PM-2**	0.0118	0.0338	31
	*PM-5**	0.0121	0.0315	36

[a] An asterisk denotes a significant difference between k_{RR} and k_{SS} at $p < 0.05$. Ref. (*26*).

Enantioselectivity in pyrethroid degradation was also evaluated in whole soils or sediments under controlled laboratory conditions. In a published study (*27*), cypermethrin was spiked in a sediment (from a nursery site in Irvine, CA) with no native cypermethrin residue. The samples were incubated at room temperature under aerobic conditions and changes in isomer composition were investigated. The results revealed that the *cis* diastereomers were considerably more persistent than the *trans* diastereomers and the half life ($T_{1/2}$) for *1R-cis-αS* (74.5 d) was the longest, whereas the half life of *1R-trans-αS* was the shortest (33.3 d).

In another published study (*25*), the degradation of the individual enantiomers of selected pyrethroids was determined in a soil and a sediment. Dissipation of the selected individual enantiomers of *cis*-bifenthrin, permethrin, and cypermethrin was measured at ambient temperature under either sterilized, aerobic, or anaerobic conditions. The degradation rate of both enantiomers of permethrin and cypermethrin was always similar under sterilized conditions, and no significant difference was found between the estimated $T_{1/2}$ values, suggesting that the enantioselective degradation, when observed, was caused by microbial transformations. However, under aerobic or anaerobic conditions, enantioselective degradation occurred to some extent for all these compounds.

Even though *R-cis*-BF degraded slightly faster in aerobic conditions, no significant difference was detected through the statistic analysis. Under anaerobic conditions, *S-cis*-BF was degraded faster than *R-cis*-BF in both soil and sediment, and the difference was significant at $\alpha = 0.05$. For enantiomers from *cis*-permethrin, the $T_{1/2}$ values were similar between the enantiomers and no significant difference was found for either aerobic or anaerobic treatments. However, *1R-cis-αS* CP was degraded more rapidly than *1S-cis-αR* in both the soil and sediment under either aerobic or anaerobic conditions, and the difference was significant at $\alpha = 0.05$ (Table IV).

Table IV. Half-life values ($T_{1/2}$, d) for the degradation of enantiomers of permethrin and cypermethrin in soil and sediment[a]

Soil/Sediments	*Permethrin*		*Cypermethrin*	
	1R-cis-	*1S-cis-*	*1S-cis-αR*	*1R-cis-αS*
			Sterilized	
Arlington Soil	141±9	139±11	116±4	120±8
San Diego Creek Sediment	139±10	139±11	120±2	120±11
			Aerobic	
Arlington Soil	124±11	102±12	71±7 *	63±7 *
San Diego Creek Sediment	124±9	126±11	85±13 *	53±6 *
			Anaerobic	
Arlington Soil	114±7	102±12	120±14 *	71±14 *
San Diego Creek Sediment	99±7	122±11	154±10 *	136±13 *

[a] An asterisk indicatessignificant difference between enantiomers at $\alpha = 0.05$. Ref. (*25*).

Another similar study (*29*) examined the degradation rates of pyrethroid isomers in two Japanese upland soils under aerobic conditions. Four isomers of permethrin and fenvalerate, eight isomers of cypermethrin, and the seven isomers of deltamethrin were included in the study. The samples were incubated in the dark at room temperature. It was observed that in the cases of cypermethrin, deltamethrin and permethrin, the *trans* isomers degraded more rapidly than the corresponding *cis* isomers. For the *trans* isomers of cypermethrin and deltamethrin, *1S* isomers degraded faster than the corresponding *1S* isomers. Meanwhile, for their *cis* isomers, there was no consistent tendency in the degradation rates of the *1R* and *1S* isomers. The degradation rate of cypermethrin and deltamethrin isomers decreased in order of (*trans, αS*) > (*trans, αR*) > (*cis, αS*) > (*cis, αR*). For fenvalerate, the *2R* isomers degraded faster than the corresponding *2S* isomers. Another study on fenvalerate also showed enantioselective degradation in soil slurries under simulated laboratory conditions (*30*).

Enantioselectivity in Degradation Pathways of Pyrethroids

Many earlier studies have been reported on the degradation pathways of pyrethroids in soils or in mammals. However, almost all of these studies treated racemic pyrethroids as single compounds, failing to acknowledge the potential enantioselectivity. Kaneko et al. (*31*) was the first study in the literature that evaluated the difference between the isomers of pyrethroids in their degradation pathways, but only at the diastereomer level. In this study, the difference between the degradation of *cis* and *trans* permethrin in two soils was investigated. The results showed that 4'-OH-permethrin was detected in a larger amount from *cis* permethrin than from *trans* permethrin. Another common degradation product of pyrethroids, 3-phenoxybenzoic alcohol (PBalc), was formed in a larger amount with *trans* permethrin than *cis* permethrin. A similar study was carried out for fenvalerate, cypermethrin and deltamethrin (*29*). In soils treated with the *trans* isomers of cypermethrin and deltamethrin or the αS epimer of fenvalerate, the ester cleavage products, 3-phenoxybenzoic acid (PBacid) and $^{14}CO_2$, were detected in larger amounts than the corresponding *cis* isomers and αR epimer, respectively, whereas the desphenyl derivative, the 4'-OH derivative and the soil bound ^{14}C, were found in larger amounts with *cis* isomers of cypermethrin and deltamthrin or αR epimer of fenvalerate.

We recently completed a study on permethrin to understand enantio-selectivity in its degradation pathways (*32*). Individual ^{14}C-permethrin enantiomers, with ^{14}C labeled in the acid or alcohol moiety, were separated on a chiral HPLC and spiked into soil and sediment. Significantly more ($p = 0.01$) bound residues were formed with the *S*-enantiomer than the *R*-enantiomer, while the loss of the *R*-enantiomer to mineralization was greater than that for *S*-enantiomer ($p < 0.05$). This difference was consistent for both *cis* and *trans* permethrin. In soils spiked with ^{14}C-carbonyl labeled permethrin, cyclopropanic acid (Cl_2CA) was detected in all the samples, ranging from 1.4 to 13.9% of the applied activity. For *cis*-permethrin, Cl_2CA was formed in significantly ($p < 0.05$) larger fractions for the *S*- than for the *R*-enantiomer. However, between the two enantiomers from *trans*-permethrin, no significant difference was found in their relative fractions. In the soils spiked with ^{14}C-alcohol labeled permethrin, PBalc and PBacid were detected in most cases at small percentages, usually < 5% of the applied activity, suggesting a relatively rapid transformation of PBalc to PBacid and further to more simple derivatives. The data from the 14 d and 56 d treatments also showed that with *cis*-permethrin enantiomers, the amount of all three hydrolysis products increased over incubation time, whereas it decreased or remained unchanged for *trans*-permethrin, indicating that the hydrolysis products of *cis*-permethrin were more persistent than those of *trans*-permethrin and thus may retain biological activity for a longer time in the environment. These results together suggest that for pyrethroids, enantioselectivity may be reflected not only in the environmental dissipation of the parent enantiomers, but also in the kinetics of formation of intermediate transformation products contributing to the overall enantioselective mineralization.

Findings from the above studies show that selective degradation occurred not only between the *cis* and *trans* diastereoisomers of pyrethroids, but also between the *R* and *S* enantiomers from the same diastereoisomer. The direction and rate of the enantioselective degradation may be due to selectivity in both biologically-mediated hydrolysis and subsequent transformation pathways. Different enantiomers may undergo hydrolysis at different rates when incubated under different conditions. In previous studies, some metabolites of pyrethroids are known to have enhanced toxicity to certain non-target organisms when compared to the parent compound. For instance, in a study by Tyler et al. (*33*), pyrethroid metabolites of environmental degradation were found to have the potential to interact with steroid hormone receptors. Some metabolites of permethrin displayed both estrogenic and antiandrogenic activity with potencies more than 100-fold greater than the parent compound. Therefore, the enantioselectivity in the degradation pathways of pyrethroids merit further investigation to better understand the associated sub-lethal toxicological risks.

Enantiomerization of Pyethroids in Abiotic Processes

Isomer conversion, or enantiomerization, is another important process that must be considered for chiral compounds. This may occur in both biological and abiotic processes, potentially influencing the activity and contributing to the side-effects of chiral pesticides in the environment (*34, 35*). For instance, Muller and Buser (*34*) observed that when enantiopure mecoprop and dichlorprop enantiomers were incubated, significant enantiomerization occurred with the formation of the *R*-enantiomers from the *S*-enantiomers, and vice versa. Since enantiomerization may proceed in both directions with equilibrium constants favoring one enantiomer, when racemic compounds are used and analyzed, it would be difficult to detect the role of enantiomerization in enantioselective behaviors. Enantiomerization may be identified only when the individual enantiomers are incubated and analyzed.

Isomerization is a very common phenomenon for pyrethroids. Many factors can lead to the isomerization of pyrethroids, including light, heat, and polar solvents. However, so far there is no direct observation of enantiomerization occurring in biological processes and most isomerization of pyrethroids has been evaluated in abiotic processes. Reviewing previous data on isomerization suggests that most earlier studies were conducted at the diastereomer level, not at the enantiomer level, apparently due to the lack of chiral separation methods at that time.

Isomerization Induced by Light

Natural pyrethrins have excellent insecticidal properties, but are very unstable in air and light. Synthetic pyrethroids generally have an enhanced stability in light. However, most of them still undergo various photochemical reactions. In a previous study (*36*), decamethrin (deltamethrin) in various solvents or in the solid phase was irradiated with sunlight or with UV light (λ = 290-320 nm). Isomerization between *cis* and *trans* isomers resulted from photolysis of decamethrin in various solvents (methanol, ethanol, 2-propanol, acetonitrile-water at 3:2 (v/v), cyclohexane, hexane, and acetone) irradiated with UV and was the major reaction of sunlight irradiation in the solid phase on glass or silica gel. An additional process for diluted solutions in methanol exposed to sunlight involved racemization at the α carbon position. Similar results were observed for permethrin (*37*) when *cis* or *trans* permethrin in solution or in the solid phase was irradiated with UV light ($\lambda > 290$ nm) or sunlight. In addition to photodecomposition, the permethrin isomers underwent an extensive isomerization of the cyclopropane ring in hexane and methanol, with a faster rate in hexane. At equilibrium, the more thermodynamically stable *trans* isomer constituted 65-70% of the isomer mixture. More rapid isomerization of permethrin was found in water and a water-acetone (49:1, v/v) mixture. The isomerization of the cyclopropane ring was the predominant reaction, reaching equilibrium in less than four hours in water and at > 1 h in aqueous acetone. However, the exposure of the permethrin isomers on soil to sunlight for 48 days resulted in relatively little isomerization at the cyclopropane ring compared to that in the solution.

In another study (*38*), the isomerization of deltamethrin in different treatments was evaluated on the enantiomer level. Deltamethrin (*1R-cis-αS*), as a thin film on glass, or in hexane solution, was irradiated outdoors in the bright summer sunshine for 5 days. The treatment resulted in the formation of *1S-cis-αS, 1R-trans-αS* and *1S-trans-αS* enantiomers. No isomerization occurred in hexane in the dark, suggesting that the isomerization of deltamethrin in hexane under outdoor conditions was caused by the sunlight. *1R-cis-αS* deltamethrin was also kept in a Hamilton Harbour water with sunlight irradiation for 5 days. *1S-cis-αS, 1R-tranis-αS* and *1S-trans-αS* enantiomers were observed with or without the addition of the microbial inhibitor mercuric chloride. Similar observations were also made in distilled water and on potato leaves when *1R-cis-αS* deltamethrin was exposed to sunlight.

Isomerization Induced by Heat

Audino et al. (*39*) studied solid phase *cis*-permethrin's stability to heat in relation to inorganic salts. Samples were heated at 210 °C in an oven in the dark and showed that, in the absence of potassium chlorate (the salt present in smoke-generating formulations of these pyrethroids) *cis*-permethrin was not isomerized, whereas when this salt was present, thermal enantiomerization occurred by converting *cis*-permethrin to *trans*-permethrin. Other salts of the type KXO_3 or N_aXO_3, where X is a halogen or nitrogen, also led to a considerable thermal enantiomerization in this direction. Different enantiomerization rates were observed for different inorganic salts. For instance, in the presence of $KClO_3$, after 30 min at 210 °C, around 47% *cis*-permethrin was enantiomerized to *trans*-permethrin. The enantiomerization rate was ~23% for KNO_3 and ~11% for $KBrO_3$ (*39*). Similar results were also obtained for deltamethrin and *β*-cyfluthrin when they were heated to 210 °C in the presence of potassium chlorate.

In another study (*40*), the thermal behavior of pyrethroids in smoke-generating formulations was investigated. A basic smoke-generating mixture is composed by potassium chlorate: dextrin: kaolin (16:10:74 by weight) with the addition of a solid pyrethroid (15 g kg^{-1}) and various amounts of foaming agents. When the formulations were combusted, *cis-trans* isomerization occurred, and the addition of foaming agents such as cyanoguanidine (CNG) or azodicarbonamide (ADC) partially inhibited the isomerization process. Results from this study on permethrin also showed that the percentage of transformation depended on the initial isomer ratio and as a general trend, the isomerization processes led to *cis-trans* equilibrium. *β*-Cypermethrin, an isomeric mixture enriched in one diastereomer, produced another diastereomer upon combustion in the same study (*40*).

The enantiomerization of pyrethroids induced by heat is of great interest because heat is widely present in GC analysis for pyrethroids and may cause artificially biased results. In one published study (*41*), isomer conversion during GC analysis was evaluated at the enantiomer level for *cis*-bifenthrin, permethrin, cypermethrin and cyfluthrin. Individual enantiomers of these compounds were injected and resolved on a chiral GC column with different inlet temperatures (160 to 260 °C). The results showed that no enantiomerization occurred to *cis*-bifenthrin and permethrin enantiomers, suggesting that the chirality on the cyclopropyl ring was not affected by the inlet heat. However, for cypermethrin and cyfluthrin, an isomer conversion occurred at the *α*C position, leading to the change of a stereoisomer to an epimer belonging to a different diastereomer, rather than to the corresponding enantiomer in the same diastereomer. In addition, the degree of enantiomerization was reduced when the GC inlet temperature decreased (*41*).

Isomerization in Water, Solvents or Their Mixtures

In earlier studies, the isomerization of pyrethroids was widely observed in water, some pure solvents, or their mixtures, but the studies were limited to the diastereomer level. For instance, the isomerization of cyfluthrin, which has four enantiomer pairs (i.e., diastereomers), was investigated in different solvents *(42)*. In aprotic solvents such as hexane, acetonitrile and dichloromethane at room temperature and in the absence of light, the cyfluthrin diastereomers were found to be stable. However, when cyfluthrin diastereomers were incubated in methanol or methanol-water mixtures, a fairly rapid isomer conversion was observed. In the same study, an experiment was carried out to investigate the temperature dependence of the isomerization. When pure diastereomers of cyfluthrin were incubated for 20 h at 2 °C, they were stable in both hexane and methanol. However, in methanol at 10 °C, traces of another diastereomer could be detected. At 22 °C, a 1:1 mixture of two diastereomers was formed after just 20 h. Similar observations were made in a study for deltamethrin isomerization in alcohols at the diastereomer level (*43*). Deltamethrin reacted to some of the aliphatic alcohols and to a lesser degree, with acetone and acetonitrile in the dark whereas no effect was observed with hexane, diethyl ether, ethyl acetate and a few other solvents.

In some published studies, the isomerization of some pyrethroids was also observed in water even without sunlight. Maguire (*38*) incubated *1R-cis-αS* deltamethrin in a natural water sample in the dark. Isomerization was observed and the *1S-cis-αS* enantiomer was produced, with or without the microbial inhibitor (mercuric chloride). The fact that the isomerization was found in both sterile and non-sterile solutions was evidence that the water matrix may lead to the isomerization of some pyrethroids. In a more recent study (*41*), the stability of several pyrethroid enantiomers in water was investigated at room temperature. No isomer conversion occurred for *cis*-bifenthrin and permethrin, whereas as the incubation time increased, another enantiomer was gradually formed for cypermethrin and cyfluthrin enantiomers. The enantiomerization consistently occurred at the *α* carbon position and led to change of a stereoisomer to an epimer.

We recently carried out studies to specifically demonstrate the enantiomerization of pyrethroids as induced by different organic solvents (*n*-hexane, methylene chloride, isopropanol, acetone and methanol) and water-solvent (methanol, acetone and isopropanol) mixtures at the enantiomer level (*44*). The individual enantiomers of *cis*-permethrin were stable in all treatments. However, for cypermethrin enantiomers, rapid enantiomerization was observed in isopropanol and methanol (Table V), but not in *n*-hexane, acetone, or methylene chloride. After 4 d at room temperature, 18-39% conversions occurred for the different cypermethrin stereoisomers in isopropanol and methanol, and the enantiomerization invariably took place at the α carbon position and bond rotation did not occur at the C1 or C3 position on the cyclopropyl ring.

Table V. Kinetic constant k_1 (d^{-1}) and standard error for interconversion of cypermethrin enantiomers in selected organic solvents at room temperature

Enantiomer	*Isopropanol*	*Methanol*
1R-cis-αR	0.120±0.007	0.217±0.017
1S-cis-αS	0.096±0.034	0.258±0.066
1R-trans-αR	0.185±0.010	0.279±0.023
1S-trans-αS	0.145±0.022	0.340±0.008

Ref. (*44*).

In this study, rapid enantiomerization was observed for all cypermethrin enantiomers in all solvent-water (1:1) mixtures, reaching equilibrium in one day, except for *1R-cis-αR*-CP and *1S-cis-αS*-CP in the acetone-water mixture (1:1, v/v), which took around 4 d to reach an apparent equilibrium. At equilibrium, about 45% of the starting enantiomer was converted to the corresponding epimer. From the results of this study, it is evident that the presence of water as a co-solvent substantially enhanced the conversion of cypermethrin enantiomers for all selected solvents when compared to that in pure solvents. The effect of the solvent-water ratio was further studied by evaluating the enantiomerization of cypermethrin in methanol-water mixtures of different ratios (Table VI). Compared to the rate of enantiomerization of cypermethrin in pure methanol, enantiomerization was significantly faster in methanol-water mixtures at the 9:1 or 1:1 ratio, but slower at the 1:9 ratio, suggesting that the extent of enantiomerization was greatly influenced by water as a co-solvent.

Based on the above findings on the enantiomerization of pyrethroids, caution should be applied during chemical analysis to reduce analytical artifacts, which may lead to biased results. When the samples are analyzed, caution should be used to avoid the use of inappropriate solvents, exposure to light or heat, and also to account for abiotic isomerization when interpreting enantiomer data from natural waters. In addition, to obtain accurate information on enantiomer compositions in environmental samples, it is important to evaluate the possibility of enantiomerization during sample analysis and handling.

Enantiomerization also has another important implication for chiral compounds. Enantiomer-pure or enantiomer-enriched products are viewed as "green chemistry" options over the conventional use of racemic products. However, if extensive enantiomerization occurs before, during or after application, it would be pointless to use the enantiopure products because they could racemize prior to deployment of the desired biological activity. Moreover, if enantiomerization occurs, it would be incorrect to estimate the toxicity or other side effects only based on the initial composition of the products used. Therefore, a detailed knowledge of the environmental behavior of the biologically active chiral compounds must include a good understanding of

Table VI. Extent of enantiomerization of cypermethrin after incubation in methanol-water mixtures at different ratios (solvent-to-water, v/v) (% converted, with 50% as complete conversion)

Starting enantiomer		10:0	9:1	1:1	1:9
1R-cis-αR	1 d	16.3±1.0	39.9±0.1	37.6±0.7	8.4±0.1
	2 d	27.1±1.8	42.6±0.4	44.6±0.3	9.6±0.8
	4 d	38.3±0.5	43.1±0.1	44.8±0.0	20.0±0.3
1S-cis-αS	1 d	12.5±0.1	38.1±0.6	37.7±0.0	6.7±0.9
	2 d	28.0±2.1	39.7±0.1	41.1±0.1	9.1±0.6
	4 d	32.6±0.8	40.7±0.6	41.8±0.3	16.1±1.0
1R-trans-αR	1 d	15.9±1.4	38.9±0.0	40.4±0.1	8.8±0.5
	2 d	27.1±0.3	39.7±0.7	40.5±0.1	17.7±0.3
	4 d	34.5±0.6	40.5±0.8	40.4±0.1	23.2±1.0
1S-trans-αS	1 d	21.4±0.3	39.4±0.2	40.5±1.0	9.7±0.1
	2 d	28.6±1.7	38.6±0.6	40.7±0.1	14.7±0.6
	4 d	34.9±0.6	40.3±0.3	42.0±0.0	24.0±0.2

Ref. (*44*).

enantiomerization and its magnitudes relative to other processes such as enantioselective degradation.

Conclusion

Pyrethroid insecticides have been widely used for the last several decades. Although a great deal of information is available on their fate and transformation in the environment and organisms, enantioselectivity has not been adequately considered in terms of their environmental behavior and risk. Based on the above information, enantiomers from the same pyrethroid compound may behave differently in the environment and it is important to use the information of individual stereoisomers instead of any information about the total chemical for better predicting ecotoxicity derived from pyrethroid residues in the environment. Another important topic, enantiomerization, which may also contribute to the enantioselectivity of pyrethroids, should also be carefully considered.

Safe use of pesticides is instrumental to the sustainability of U.S. agriculture. Chiral compounds currently make up about 25% of all agrochemicals in the market, and this ratio is likely to increase as more natural product-like compounds with complex structures are being developed and introduced into use (*4*). Along with this trend, better knowledge of chiral pesticides at the enantiomeric level is urgently needed and merit further investigation.

References

1. Kurihara, N.; Miyamoto, J.; Paulson, G.; Zeeh, B.; Skidmore, M.; Hollingworth, R.; Kuiper, H. *Pure & Appl. Chem.* **1997**, *69*, 2007-2025.
2. Ladik, J. J.; Szekeres, Z. *J. Molecular Modeling* **2006**, *12*, 462-467.
3. McConathy, J.;, Owens, M. *J. Clinical Psychiatry* **2003**, *5*, 70-73.
4. Williams, A. *Pestic. Sci.* **1996**, *46*, 3-9.
5. Wong, C. *Anal. Bioanal. Chem.* **2006**, *386*, 544-558.
6. Garrison, A. W. *Environ. Sci. Technol.* **2006**, *40*, 16-23.
7. Weston, D. P.; You, J.; Lydy, M. J. *Environ. Sci. Technol.* **2004**, *38*, 2752-2759.
8. Gan, J.; Lee, S. J.; Liu, W. P.; Haver, D. L.; Kabashima, J. N. *J. Environ. Qual.* **2005**, *36*, 834-841.
9. Coats, J. R.; Symonik, D. M.; Bradbury, S. P.; Dyer, D. D.; Timson, L. K.; Atchison, G. J. *Environ. Toxicol. Chem.* **1989**, *8*, 671-679.
10. Hill, I. R. *Pestic. Sci.* **1989**, *27*, 429-465.
11. Mian, L. S.; Mulla, M. S. *J. Agric. Entomol.* **1992**, *9*, 73-98.
12. *Chirality in Agrochemicals;* Kurihara, N.; Miyamoto, J. Eds; Wiley & Sons: Chichester, England, 1998.
13. *Chiral pollutants: Distribution, toxicity and analysis by chromatography and capillary electrophoresis;* Ali, I.; Aboul-Enein, H. Y. Eds; Wiley & Sons: Chichester, England, 2004.
14. Elliot, M. E.; Farnham, A. W.; Janes, N. F.; Soderland, D. M. *Pestic. Sci.* **1978**, *9*, 112-116.
15. *Synthetic pyrethroid insecticides: Structures and properties;* Naumann, K. Eds; Springer-Verlag Berlin Heidelberg: New York, 1990.
16. Leicht, W.; Fuchs, R.; Londershausen, M. *Pestic. Sci.* **1996**, *48*, 325-332.
17. Liu, W. P.; Gan, J. Y.; Schlenk, D.; Jury, W. A. *PNAS* **2005**, *102*, 701-706.
18. Buser, H.; Muller, M. *Environ. Sci. Technol.* **1995**, *29*, 664-672.
19. Buser, H.; Muller, M. *Environ. Sci. Technol.* **1995**, *29*, 2031-2037.
20. Buser, H.; Muller, M. *Environ. Sci. Technol.* **1998**, *32*, 626-633.
21. Buerge, I.; Poiger, T.; Muller, M.; Buser, H. *Environ. Sci. Technol.* **2003**, *37*, 2668-2674.
22. Buser, H.; Muller, M.;Poiger, T.; Balmer, M. *Environ. Sci. Technol.* **2002**, *36*, 221-226.
23. Liu, W.; Gan J. *J. Agric. Food Chem.* **2004**, *52*, 736-741.
24. Vetter, W.; Schurig, V. *J. Chromatogr.* **1997**, *774*, 143-275.

25. Qin, S.; Budd, R.; Bondarenko, S.; Liu, W.; Gan, J. *J. Agric. Food Chem.* **2006**, *54*, 5040 -5045.
26. Liu, W.; Gan, J.; Lee, S.; Werner, I. *Environ. Toxicol. Chem.* **2005**, *24*, 1861-1866.
27. Liu, W.; Gan, J.; Lee, S.; Werner, I. *J. Agric. Food Chem.* **2004**, *52*, 6233-6238.
28. Sakata, S.; Mikami, N.; Yamada, H. *J. Pesticide Sci.* **1992**, *17*, 181-189.
29. Sakata, S.; Mikami, N.; Yamada, H. *J. Pesticide Sci.* **1992**, *17*, 169-180.
30. Faraoni, M.; Messina, A.; Polcaro, C. M.; Aturki, Z.; Sinibaldi, M.; *J. Liquid Chromatogr Rel Technol.* **2004**, *27*, 995-1012.
31. Kaneko, H.; Ohkawa, H.; Miyamoto, J. *J. Pesticide Sci.* **1978**, *3*, 43-51.
32. Qin, S.; Gan, J. *J. Agric. Food Chem.* **2006**, *54*, 9145-9151.
33. Tyler, C. R.; Beresford, N.; Woning, M. V.; Sumpter, J. P.; Thorpe, K. *Environ. Toxic. Chem.* **2000**, *4*, 801-809.
34. Muller, M.; Buser, H. *Environ. Sci. Technol.* **1997**, *31*, 1953-1959.
35. Jarman, J. L.; Jones, W. J.; Howell, L. A.; Garrison, A. W. *J. Agric. Food Chem.* **2005**, *37*, 2668-2674.
36. Ruzo, L. O.; Holmstead, R. L.; Casida, J. E. *J. Agric. Food Chem.* **1977**, *25*, 1385-1394.
37. Holmstead, R. L.; Casida, J. E.; Ruzo, L. O.; Fullermer, D. G. *J. Agric Food Chem.* **1978**, *26*, 590-595.
38. Maguire, R. J. *J. Agric Food Chem.* **1990**, *38*, 1613-1617.
39. Audino, P.; Licastro, S. A.; Zerba, E. *Pestic. Manag. Sci.* **2001**, *58*, 183-189.
40. Audino, P.; Licastro, S. A.; Zerba, E. *Pestic. Sci.* **1999**, *55*, 1187-1193.
41. Liu, W.; Qin, S.; Gan, J. *J. Agric Food Chem.* **2005**, *53*, 3814-3820.
42. Leicht, W.; Fuchs, R.; Londershausen, M. *Pestic. Sci.* **1996**, *48*, 325-332.
43. Perschke, H.; Hussain, M. *J. Agric. Food Chem.* **1992**, *40*, 686-690.
44. Qin, S.; Gan, J. *J. Agric. Food Chem.* (in press)

Aquatic Toxicology

Chapter 13

Ecological Risk Characterization for the Synthetic Pyrethroids

Mah T. Shamim, Michael D. Hoffmann, Jose Melendez, and Mohammed A. Ruhman

U.S. Environmental Protection Agency, Office of Pesticide Programs, Environmental Fate and Effects Division, 1200 Pennsylvania Avenue, NW, Washington, DC 20460

In its reevaluation of pesticides under the Federal Insecticide, Fungicide, and Rodenticide Act (FIFRA), the Environmental Protection Agency (EPA) examines all major routes of exposure. As individual pyrethroids have been reevaluated, a distinct pattern of similarities and differences have emerged among them. The synthetic pyrethroids are characterized not only by similar environmental fate and transport properties, but also by their common toxicity endpoints. This chapter includes a discussion of these trends and provides innovative modeling approaches to estimate the exposure of pyrethroids to aquatic organisms in wastewater and freshwater from use of these pesticides in agricultural and urban environments. Also included is a discussion of potential risks to non-target aquatic and terrestrial organisms.

NOTE: This document has been subjected to review by the Environmental Fate and Effects Division in the Office of Pesticide Programs and approved for publication. Approval does not signify that the contents reflect the views of the Agency, nor does mention of trade names or commercial products constitute endorsement or recommendation for use.

Introduction

Under the Federal Insecticide, Fungicide, and Rotendicide Act (FIFRA), EPA evaluates the use of a pesticide before it can be sold or distributed in commerce. EPA will register or approve a new pesticide if scientific data show that, when used according to label directions, it will not cause unreasonable effects on human health and the environment. Under the 1988 Amendments to FIFRA, EPA is also required to review and re-register older pesticides to ensure that they meet current health, safety, and environmental standards. The Environmental Fate and Effects Division (EFED) in EPA's Office of Pesticide Programs (OPP) is responsible for conducting ecological risk assessments, endangered species risk assessments and drinking water exposure assessments (to be used in the dietary human health risk assessments) in support of the registration and reregistration of pesticides. The registered pesticides involved in these two programs cover a wide range of compounds with many belonging to specific classes of compounds.

One of the major classes of pesticides that have been introduced over the past three decades for a variety of insecticidal uses is the synthetic pyrethroids. The pyrethroids share similar modes of action and are considered neurotoxins that act on the axons in both the peripheral and central nervous system. The primary biological effects of pyrethroids on insects and vertebrates are caused by an inhibition of the correct firing of neurotransmitters that deliver a signal from one cell to another via voltage gated calcium ($Ca^{++)}$ and sodium ($Na^{+)}$ channels) (*1*). Two different types of pyrethroids are recognized: Type I and Type II. This classification is based on differences in basic structure (presence or absence of a cyano group in the alpha position) and the symptoms of acute poisoning in rodents. In general, the symptoms of synthetic pyrethroids poisoning follow the typical pattern of nerve poisoning: (a) excitation, (b) convulsions or tremor, (c) paralysis, and (d) death. Marked differences exist in the duration of action on sodium channel gate, particularly between Type I and Type II pyrethroids. Pyrethroids with an alpha-cyano group, such as cyfluthrin, esfenvalerate, and cypermethrin, produce a more prolonged transient increase in sodium permeability of the nerve membrane during excitation than do Type I pyrethroids such as bifenthrin.

This chapter describes the approaches and methodologies for estimating exposure and characterizing ecological risk from the use of synthetic pyrethroids in both agricultural and non-agricultural environments. It also includes a discussion of the relative toxicity and fate of several synthetic pyrethroids as well as resulting risks from their use in specific settings.

Regulatory History

Registration

Since the early to late 1970s, nearly 22 synthetic pyrethroids, including enriched isomers, have been registered and used for controlling various insects in agriculture and non-agriculture settings. In 1971, 1975, and 1977 resmethrin, sumithrin, and permethrin were registered, respectively. In 1990, six synthetic pyrethroids were conditionally registered on cotton as substitutes for organophosphate, carbamate and organochlorine pesticides: bifenthrin, cyfluthrin, cypermethrin, esfenvalerate, lambda-cyhalothrin and tralomethrin. More pyrethroids were added later, including zeta-cypermethrin, deltamethrin, fenpropathrin, and gamma-cyhalothrin.

Most of the synthetic pyrethroids were classified as restricted use pesticides because of their high toxicity to fish and other aquatic organisms. In order to support new registration actions and maintain existing ones, the Environmental Protection Agency (EPA) issued data call-in (DCI) notices in 1985 for chemical specific aquatic toxicological data. In 1990, the Pyrethroid Working Group (PWG) was formed to address data needs for the conditional registration of these chemicals. The PWG included representatives from EPA, the National Cotton Council, and the major manufacturers of synthetic pyrethroids. In this process, manufacturers of pyrethroids were initially requested to submit an interim risk reduction proposal to mitigate aquatic risk. The proposal was to include labeling changes to reduce aquatic exposure such as buffer zones; educational programs that targeted growers, applicators, and consultants and publicized label changes; and field studies to monitor the effectiveness of buffer zones.

In February, 1999, EPA presented a preliminary assessment entitled "Sediment Toxicity and Fate of Synthetic Pyrethroids" to the FIFRA Scientific Advisory Panel (SAP) (*2*). The purpose of the meeting was to seek the SAP's advice concerning the adequacy of the studies conducted by the PWG to characterize the partitioning, bioavailability, and toxicity of synthetic pyrethroids. The SAP also considered issues associated with assessment of risks from exposure of non-target aquatic organisms to synthetic pyrethroids in sediments. The PWG has developed additional data to address the conditional registration of the synthetic pyrethroids. This includes new mobility studies for nine synthetic pyrethroids, using the solid phase micro-extraction (SPME) technique. It also includes acute and chronic sediment toxicity studies for four synthetic pyrethroids. The Agency is reviewing the submitted fate and toxicity data and will use these data to address the uncertainties in the risk assessments.

Evaluation of New and Existing Uses of Pyrethroids

Since the conditional registration of synthetic pyrethroids on cotton, more than 50 new agricultural and non-agricultural uses have been submitted and evaluated by the Agency for registration. Additionally, under the re-registration program, the Agency completed the re-evaluation of all existing food uses of these chemicals by August 2006 as mandated by Congress. More specifically the Agency completed Reregistration Eligibility Decisions (REDs) for cypermethrin, permethrin, and resmethrin and ecological risk assessments for the proposed new uses of bifenthrin, cyfluthrin, lambda-cyhalothrin, cypermethrin, zeta-cypermethrin, and prallethrin. An outline of EFED's approach to develop ecological risk assessments in support of the registration or re-registration of pesticides can be found in the document entitled, "Overview of the Ecological Risk Assessment Process in the Office of Pesticide Programs, Environmental Protection Agency (*3*)."

Use Characterization

Use characterization is known to include analysis of use and/or usage. Although use is examined in all regulatory actions, usage can only be examined for pesticides that have been in use for some time. Therefore, use involves examination of suggested label use for new products/crops (that is a potential use), while usage involves examination of current/past use for already registered products, namely current/historical use. Use characterization is an important first step in estimating exposure to pesticides.

Use analysis relies on information associated with each use pattern and formulation (e.g., liquids, granular, and ULV), application dates and rates, application procedures (including equipment and specific parameters such as efficiency, droplet size and release height), timing and number of applications, and re-treatment intervals. In most cases, labels contain a range of values, and the baseline ecological risk assessment depends on the conservative values. Therefore, at this level, exposure is estimated based on labeled maximum application rates, maximum number of applications/year, minimum re-treatment intervals, and only label stated mitigation measure(s) (e.g., drift/runoff buffers, boom height, and wind speed). It is noted, however, that when parameters are not specifically stated on the label, reasonable conservative estimates are used (referred to as default values). Based on the labeled uses, an application window can usually be inferred, and an application date that gives the most conservative exposure is chosen for the assessment.

Usage involves analysis of historical use data and is executed for regulatory actions that involve previously registered chemicals (e.g., for new uses and/or reregistration). Although the main exposure estimates are based on active labeled use, other possible exposures are also characterized based on the analysis of current (or typical) application rates, procedures, and timing. Furthermore, effects of some possible mitigation measures on estimated exposure are also considered.

In the U.S., use patterns for pyrethroids can be divided into agricultural and non-agricultural or urban. The highest agricultural uses are typically reported on field crops, while urban uses are typically reported for structural and public health pest control. Although no national data are available for pyrethroids urban use, California data suggest that the total pounds used in agriculture constitute only 25% of the total pyrethroid use. Therefore, information and data needed to quantify urban use of pyrethroids exposure are important.

Agricultural Use and Usage Patterns

Pyrethroids are used as insecticides, miticides, and/or acaricides effective against a broad range of pests. Most pyrethroids are labeled for use on agricultural crops. The list of pyrethroids used in agriculture includes bifenthrin, permethrin, cypermethrin, cyfluthrin, fenpropathrin, lambda-cyhalothrin, esfenvalerate, tralomethrin, and deltamethrin. These chemicals are formulated in common crop use formulations such as emulsifiable concentrates (EC), granules (G), water-dispersible granules (WDG), capsule suspensions, wettable powders (WP), water soluble bags (WSB), and as ultra low volume sprays (ULV). Pyrethroids are applied through most agrochemical application procedures, which include aerial, ground, airblast, and chemigation. Table I is a summary of important labeled application parameters for various uses in agriculture.

Table I. Pyrethroid use in agriculture: a summary of labeled-application.

Chemical	Application Parameters (Maximums with Rates in lb a.i./A)			Minimum Intervals (day)
	Single Rate	Number	Total Rate/Season [a]	
Bifenthrin	0.10	5	0.50	3
Permethrin	0.20	8	1.60	4
Cypermethrin	0.10	6	0.60	3
Cyfluthrin	0.05	10	0.50	3
Fenpropathrin	0.40	2	0.80	7

[a] In most labels the total/year is not given and for some crops such as lettuce in California, exposure is estimated assuming more than one crop a year, resulting in significantly higher exposure.

The list of labeled agricultural use patterns is extensive and includes two main field crops (corn and cotton); other field crops (soybean, alfalfa, sunflower, peanuts, cereals); vegetables (sweet corn, fruiting vegetables, head/stem and leafy vegetables, tuberous/corm vegetables, bulb vegetables, and cucurbits); fruit and vine trees (grapes, nut trees, peaches, and berries); and ornamental and turf. New uses have also been added to the list or are in the process of being reviewed for registration.

Pyrethroid usage on agricultural crops is useful in understanding the extent of use from available national and state data. Potential usage can be easily inferred from yearly crop acreage statistics; however, real usage can only be obtained from historical use data. For pyrethroids, national usage data are available for 1997 (*4*) and for specific years, crop groups, and program states (*5*). Additionally, statewide data are available for California from 1991 to 2005 (*6*). The use distributions of various pyrethroid chemicals and crops, which are summarized[1] in Figure 1a and 1b, respectively, are based on the amount of pyrethroid active ingredients used and on the 1997 national data. Data indicate that permethrin and tefluthrin are the main pyrethroids used in agriculture, followed by cyfluthrin, esfenvalerate, bifenthrin, and cypermethrin (Figure 1a). The use is apparently concentrated on field crops such as corn and cotton, followed by vegetables and fruits (Figure 1b).

In 1997, approximately 26 million acres were treated with about 2.4 million lbs of active ingredients, resulting in an average pyrethroid rate of nearly 0.09 lbs a.i./Acre. The total annual minimum and maximum rates, which are listed in Table II, were obtained from the U.S. Department of Agriculture National Agricultural Statistics Service (USDA-NASS) (*5*) and were compared to labeled seasonal rates for selected pyrethroids. With the exception of cyfluthrin, data suggest that the chemicals were applied at lower rates than those permitted by the label. Although this may be the case, the accuracy of these annual "survey rates" would depend largely on the accuracy of the surveys. Survey rates may be used, with some uncertainty, to obtain other possible exposure estimates.

The extent of pyrethroid usage in 1997 is presented in Figure 2. The map shows the national 1997 usage areas for eight pyrethroids (total acres treated with permethrin, tefluthrin, cyfluthrin, esfenvalerate/ fenvalerate, bifenthrin, cypermethrin, tralomethrin, and fenpropathrin). As expected, high usage areas are in the corn and cotton belts in addition to areas where vegetables, fruits, and tree nuts are grown (California, Florida, Georgia, and North and South Carolina).

Also, USDA-NASS has an agricultural chemical use database, which was examined for pyrethroid usage. Statistics present in this database were based on surveys conducted on farm use for targeted crops in chosen years (2002-2006) and states (program states) (Figure 3) (*7*). In comparison to 1997, it appears that more pyrethroid active ingredients (lbs a.i.) were used in 2002-2006 on vegetables and cotton and less on corn, with no change in fruit trees.

Table II. Maximum annual "survey rates" compared to maximum labeled rates [a]

Chemical	Range of Total Rate in lb a.i./Acre		Calculated Annual Rate As Percent of Seasonal Label Rate
	Seasonal (labeled)	Annual (calculated)	
Bifenthrin	0.04-0.50	0.01-0.20	40
Permethrin	0.60-1.60	0.03-1.37	87
Cypermethrin	0.40-0.60	0.03-0.26	43
Cyfluthrin	0.04-0.50	0.01-0.66	132
Fenpropathrin	0.60-0.80	0.08-0.48	60

[a] Labels for most pyrethroids include buffer zones that vary in type/width with application type.

Additionally, the distribution of usage between various pyrethroids was affected with marked reductions in permethrin and tefluthrin balanced by increases in lambda-cyhalothrin and cypermethrin.

Trends in use patterns for pyrethroids in California were obtained for 1991 to 2005 from the PANNA database and plotted in Figure 4. The data suggest that acreage treated with pyrethroids increased from nearly 2 million acres in 1991 to nearly 3.5 million acres in 1995. This increase was followed by a general decrease to approximately 2.3 million acres in the year 2001. Finally a slight increase occurred from the 2.3 million acres in 2001 to nearly 2.9 million acres in the year 2005. In contrast to the erratic changes in total pyrethroids treated acreage, there was a steady increase in total pounds applied through two periods: the first from 1991 to 1998 (increase from 165,000 to 283,000 lbs) and the second from 1999 to 2005 (increase from 205,000 to 224,000 lbs). It is also noted that pyrethroid use rates for California were near the low side of the reported national rate, which could be a reflection of differences in the dominant California crop use patterns compared to those nationally.

Non-agricultural Use and Usage Patterns

The list for non-agricultural use of pyrethroids includes many categories such as structural; household indoor/outdoor; landscaping; public health; animal husbandry; food processing; and commercial, storage, recreational, and uncultivated non-agricultural areas. Table III summarizes data on important pyrethroid chemicals with non-agricultural use patterns along with information on end use products and use purpose (s). In general, non-agricultural pesticides are applied by individuals for many indoor/outdoor household uses; by municipalities or public health authorities for vector control pesticides; and by

(a) By chemical (based on lbs a.i. used in the treatment)

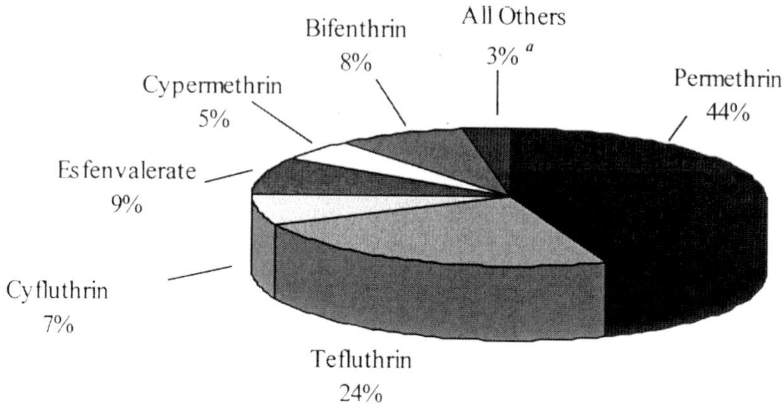

[a] All Others inlude: Tralomethrin + deltamethrin (3%), and fenpropathrin (<1%).

(b) By crop (based on lbs a.i. used in the treatment)

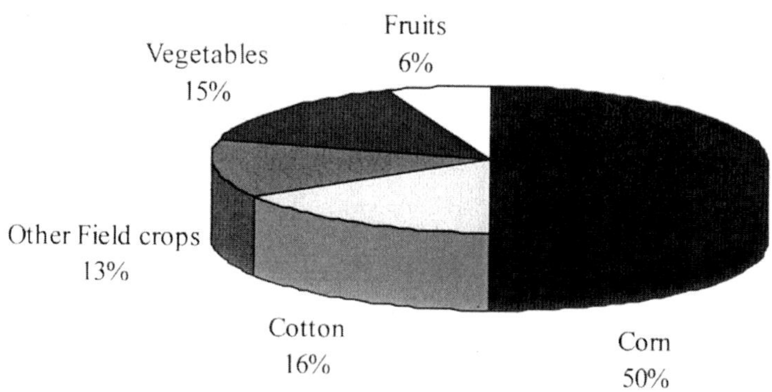

Figure 1. Pyrethroid usage based on 1997 NCFAP data: (a) by chemical and (b) by crop.

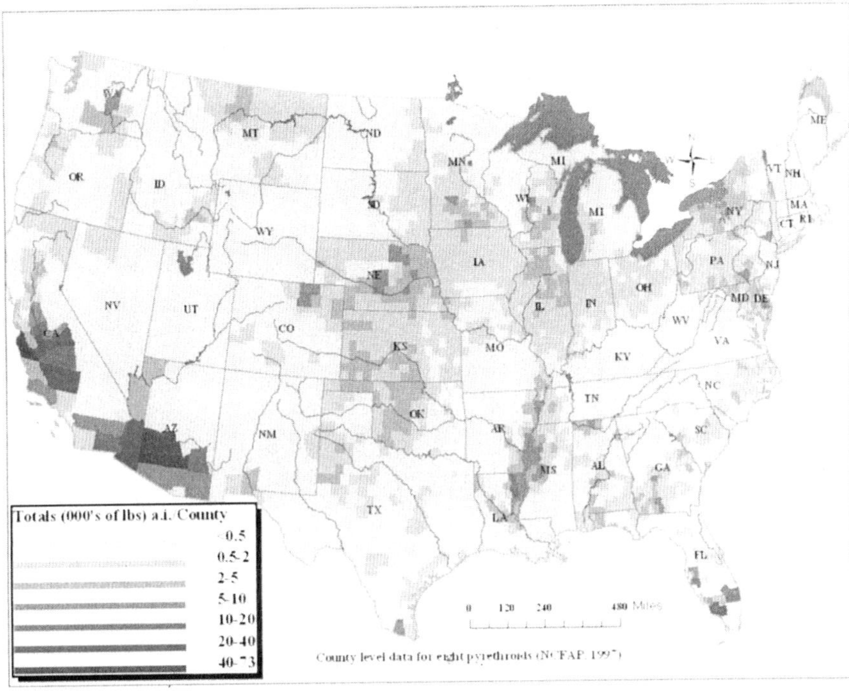

Figure 2. Pyrethroid usage areas in the USA in 1997.

certified professional pesticide applicators to control pests including termites and structural pests in homes, commercial and industrial premises, and many other public areas.

There are a large number of pyrethroid end use products in the consumer market with varied formulations and label parameters such as rate, application procedure, and frequency of application. One of the important labeled uses of pyrethroids is mosquito control. Label parameters for resmethrin, permethrin, and prallethrin are similar in their single application rates (0.007 to 0.008 lb a.i. /A). Other important parameters such as number of applications and application intervals were not specified in the labels possibly because such parameters are dependent on local conditions.

Typical rates were reported to equal half of the maximum label rate (0.0035 lb a.i. /A) with the maximum number of applications ranging from 25 to 50/season applied twice a week (i.e., 4-day intervals). Exposure assessment for the unique ULV application requires more information than that which is included on the label such as application timing, boom height, desired spray characteristics (droplet size), and width of the spray buffer. In this respect, it is

(a) By chemical (based on lbs a.i. used in the treatment)

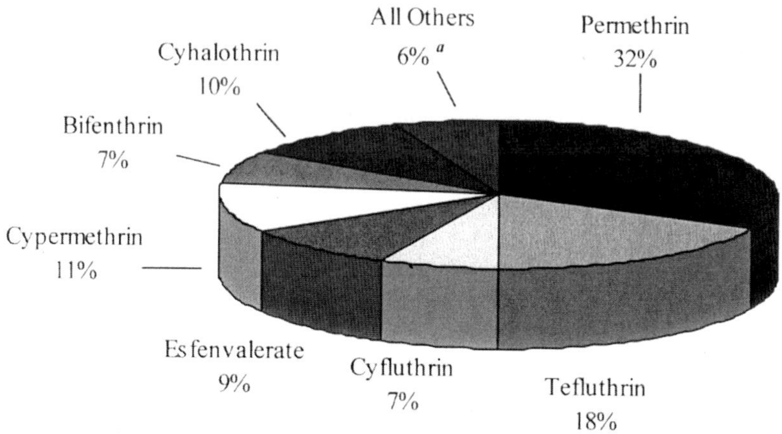

[a] All Others include: Fenpropathrin (3%), and Tralomethrin + Deltamethrin (3%).

(b) By crop (based on lbs a.i. used in the treatment)

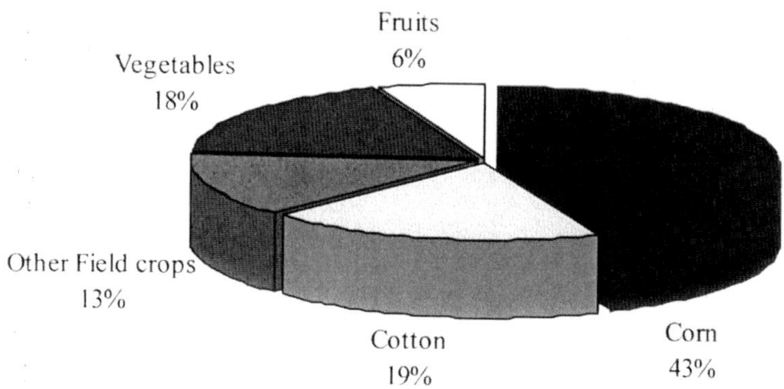

Figure 3. Pyrethroid usage based on 2002-2006 USDA/NASS data (average data extrapolated to 100% crop presentation): (a) by chemical and (b) by crop.

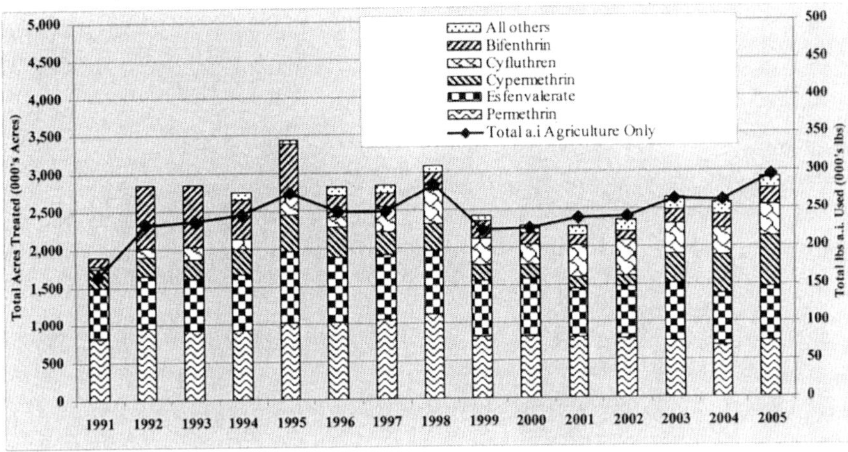

Notes: X-axis = Usage year; others are: Deltamethrin, Fenpropathrin, Resmethrin, Tetramethrin, and Tralomethrin

Figure 4. Pyrethroid usage in California by chemical in acres treated.

expected that exposure from this type of application will differ from year to year and from one region to another. For example, the typical application period lasts nearly 3 months in northern states, while it lasts up to 10 months in the south where more applications are needed.

Information regarding pyrethroid usage on non-agricultural sites is useful in understanding the extent of this use from available national and state data. However, national scale usage data is not available for all non-agricultural use sites because of the difficulties in obtaining such data for major pesticide products such as those purchased by consumers for use indoors or outdoors and by public health authorities and professional pest control applicators for vector and structural pest control applications. Examples of available data are sporadic and include data from a pilot project conducted from 1997 to 2002 on national sales trends for pounds of indoor-use permethrin products and usage in animal husbandry. The indoor-use data suggest that the active ingredient of permethrin products sold for indoor use was relatively level (around 20,000 lbs air.), while there was a three-fold overall increase for outdoor-use (from 2,700 to 10,000 lbs a.i.) (*8*). The animal husbandry usage data on cattle, swine, sheep, and associated facilities suggest that pyrethroids use in animal health constitutes 12% of the total amount of pesticides used in this sector and is equal to 13% of the amount of pyrethroids used in agriculture nationally (*9*).

In contrast to the non-availability of national comprehensive yearly pesticide usage statistics, California has published such statistics yearly since 1991. These

Table III. Non-Agricultural pyrethroid use: chemicals, formulations and use purposes.

Chemical [a]	Formulations/Products	Purposes
Permethrin Cypermethrin Cyfluthrin Lambda-cyhalothrin Bifenthrin	Water or oil based liquids; aerosols (flying and crawling insects); emulsifiable concentrates; foggers; granules; special sprays (cracks/crevices, bedding sprays, wasp and hornet killer); pet sprays and shampoos; mixtures with fertilizers (e.g., lawn & garden summer fertilizer plus insect killer); perimeter sprays; termite treatment; lawn & garden insect killer (granular), micro encapsulated, dust (e.g., livestock dust), wettable powders; Pre-treated clothing and garment	Pest control in many indoor/outdoor and other settings such as structural (home, hotels, hospitals, restaurants, commercial, industrial, institutional and storage, food handling and processing facilities); transportation systems; right of way; commodity fumigation, Home, garden, and landscaping; personal and Pets hygiene; and animal husbandry. Pests include: ants, roaches, flies, mosquitoes, wasps, hornets, spiders, fire ants, termites, flea, tick, lice, outdoor home and garden insects, and other outdoor/domestic animal pests such as flies, lice, grubs, and others
Resmethrin Permethrin Prallethrin	ULV (ultra low volume spray), Fogging concentrate	Disease vector control (mosquito abatement) for outdoor residential and recreational area such as parks, woodlands, swamps, marshes, overgrown areas and golf courses, mosquito abatement districts, municipalities, urban, suburban and rural areas

[a] Listed are top chemicals, however, other chemicals are also used, including fenpropathrin, deltamethrin, tefluthrin, phenothrin, esfenvalerate, allethrin, tralomethrin, tetramethrin, and imiprothrin.

yearly statistics are important in identifying trends in use patterns for pyrethroids in California. In addition, these data can be used, with limitations to obtain national inferences for similar use patterns, elsewhere after taking into consideration varied environmental factors that influence pest pressure. Data obtained from the database and usage items listed in Table IV were considered as non-agricultural use patterns. It is noted that household usage items were not included in the database. Data for the most recent year suggest that the major non-agricultural use pattern is structural pest control (97%), followed by landscaping (2.3%) and public health (0.4%). All other uses constitute only <0.5%. The data also suggest that the most important chemicals used in non-agricultural settings are permethrin and cypermethrin, with bifenthrin and cyfluthrin use as a distant second.

Finally, yearly California statewide usage data are summarized in Figure 5. The graph shows that the total amount of pyrethroid actives used in non-agricultural treatments increased steadily from 0.087 million lbs a.i in 1991 to nearly 0.699 million lbs a.i in 2005. The two major pyrethroids used were permethrin and cypermethrin, and the increased total pyrethroid usage was probably related to the apparent increase in their use from 1991 to 2005. Esfenvalerate use appears to cease after 1997, replaced most likely by cyfluthrin until 2000, and then replaced by cyfluthrin and bifenthrin until 2005.

Comparison between agriculture and non-agriculture usage (1991-2005) reveals a trend of increase use for both agricultural and non-agricultural usage though the increase was variable and much less pronounced in the years prior to 2000 (total and non-agricultural use lines are parallel in Figure 5). Data for 2005 suggest that agricultural use was only 25% of the total pyrethroid use.

Exposure Characterization

EPA bases its baseline risk assessment on the deterministic risk quotient (RQ) method in which a point estimate of exposure is divided by a point estimate of effects or toxicity. In this equation, the exposure concentration, which is represented by the estimated environmental concentration "EEC", is the predicted concentration of a pesticide within an environmental compartment based on estimates of quantities released, discharge patterns, and fate and transport of the pesticide as well as the nature of the specific receiving ecosystem.

Synthetic pyrethroids are relatively insoluble in water, are unlikely to volatilize or to hydrolyze at environmentally relevant pH's, do not photo-degrade on soil (except for cyfluthrin), and are not expected to leach. The major routes of dissipation for these compounds are aqueous photolysis for the first generation pyrethroids and aerobic/anaerobic soil metabolism for majority of the first and second generation pyrethroids. Because of their low mobility,

Table IV. Pyrethroids: Non-agricultural usage items and important chemicals for each use.

Usage Item	Chemicals totaling >90% [a]	Pyrethroid Usage	
		lbs a.i. used	% total
Structural Pest Control	Permethrin, Cypermethrin, Bifenthrin, Cyfluthrin	679,685	97.02%
Landscape	Permethrin, Bifenthrin, Cypermethrin	15,921	2.27%
Public Health Pest Control	Permethrin, Resmethrin, Phenothrin	2,708	0.39%
Animal Husbandry	Cyfluthrin	1,169	0.17%
Regulatory	Bifenthrin, Cyfluthrin, Resmethrin	493	0.07%
Right of Way	Bifenthrin, Cypermethrin	440	0.06%
Vertebrate Pest Control	Permethrin, Cypermethrin	79	0.01%
Buildings	Cypermethrin	54	0.01%
Unknown	Cypermethrin	10	0.00%
Food Processing	Resmethrin, Cyfluthrin, Bifenthrin	9	0.00%
Other Fumigation	Cyfluthrin, Resmethrin	4	0.00%
Commodity Research	Cypermethrin, Resmethrin	3	0.00%
Household Sites	Cyfluthrin	1	0.00%
Commercial Area	Bifenthrin	1	0.00%
Commodity Fumigation	Cyfluthrin, Deltamethrin	1	0.00%
County Sales	Bifenthrin, Cyfluthrin	0	0.00%
Uncultivated Non-Agricultural Area	Bifenthrin	0	0.00%
Storage Areas	Cyfluthrin-B	0	0.00%
Recreational Area	Deltamethrin	0	0.00%
	Total: All chemicals	700,578	100 %

[a] Chemicals listed for each usage item constitute >90% of the total usage.

pyrethroids are not expected to move to sub-surfaces and leach to ground water. They may reach aquatic environments via erosion during rain events. Their high K_{OC}'s indicate that they will partition with the soil, be transported offsite, and reach adjacent bodies of water carried on the solid phase. The chemicals may also reach aquatic systems by spray drift. As the chemicals reach these systems, they are expected to remain adsorbed to suspended solids or organic matter in

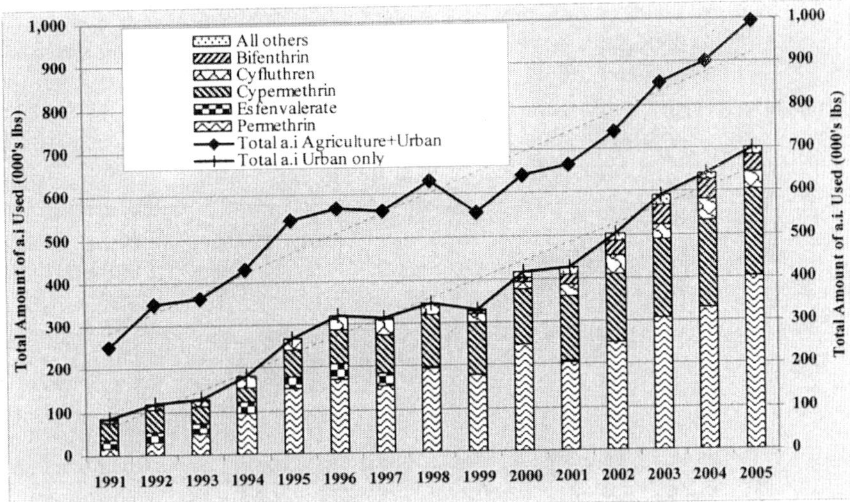

Notes: X-axis = Usage year; others are: Deltamethrin, Fenpropathrin, Resmethrin, Tetramethrin, and Tralomethrin

Figure 5. Pyrethroids: Agricultural usage in California by chemical (lbs a.i used).

the water column, or partition into the sediment phase. Since it was observed that for seven of the twelve synthetic pyrethroids examined, the anaerobic metabolism is slower than the aerobic metabolism, and there is a potential for repeated applications, these chemicals may remain in the sediments and the benthic layer for prolonged periods of time. The sediments, which are often anaerobic, may become repositories of the chemicals.

Pyrethroid pesticides are insecticides or insecticides/acaricides that represent a challenge in selecting and quantifying possible exposure scenarios. Chemicals in this class are associated with varied and unique fate and transport properties, a multitude of use patterns, formulations, and application types in varied environmental settings.

The following section gives a brief description of the methods that are employed, by the Environmental Fate and Effects Division (EFED), to estimate exposure to aquatic and terrestrial compartments of the environment.

Fate and Transport Characterization

A large number of variables affect the fate and transport of synthetic pyrethroids in the environment. Some of these variables are chemical dependent such as solubility, vapor pressure, sorption strength, hydrolytic/photolytic characteristics, and susceptibility to biodegradation. Other types of variables are associated with application practices such as timing and place of application (region of the U.S.) and agricultural practices such as cropping and irrigation methods. A summary of the environmental fate properties of the synthetic pyrethroids can be found in Table V below.

Except for prallethrin, the synthetic pyrethroids have low solubility. (Prallethrin is 8.03 ppm, while all others range from 0.014 ppb for bifenthrin through 2.32 ppb for cyfluthrin to 84 ppb for tralomethrin.) The chemicals also have moderate to low vapor pressure. (The moderate range is from 1.8×10^{-7} mm Hg for bifenthrin and fenpropathrin to 6.0×10^{-5} mm Hg for tefluthrin, while the low range is 1.8×10^{-11} mm Hg for tralomethrin.) In general, Henry's Law Constants indicate a moderate to low potential for these chemicals to volatilize from moist surfaces. The moderate range is from 1.80×10^{-4} Atm-m^3 mol^{-1} for bifenthrin, fenpropathrin, and tefluthrin to 3.7×10^{-6} Atm-m^3 mol^{-1} for cyfluthrin, while the low range is 1.9×10^{-10} Atm-m^3 mol^{-1} for tralomethrin. Synthetic pyrethroids have a high tendency to bind to organic matter in the sediments and in the particulate suspended in the water, resulting in a substantial reduction of volatilization even for those chemicals with moderate Henry's Law Constants. All synthetic pyrethroids are highly non-polar as indicated by their high log K_{OW}. The Log octanol/water coefficients (K_{OW}) are 4.5-5.0 for prallethrin and deltamethrin, 5.1-6.0 for resmethrin, cyfluthrin, fenvalerate, tralomethrin, and fenpropathrin and 6.1-7.0 for permethrin, bifenthrin, cypermethrin, lambda-cyhalothrin, and tefluthrin.

The fate of pyrethroids in various compartments of the environment can be deduced from their laboratory abiotic and biotic reactions. All synthetic pyrethroids are relatively stable to abiotic hydrolysis, except in alkaline environments. In alkaline media, resmethrin, permethrin, bifenthrin, fenvalerate, and tralomethrin show the highest stability, while fenpropathrin, lambda-cyhalothrin, and tefluthrin show moderate reactions (half-life "$t_{½}$" range=13->30 days at pH 9). Others such as prallethrin, cyfluthrin, cypermethrin, and deltamethrin are highly susceptible to alkaline hydrolysis ($t_{½}$ range = 2-5 days at pH 9). This shows that most pyrethroids are highly stable in water bodies under normal environmental conditions (pH 7). In contrast to abiotic hydrolysis, first generation pyrethroids such as prallethrin and resmethrin show high reactivity towards photolytic reaction ($t_{½}$ = hours to 0.5 day). Others such as cyfluthrin and fenvalerate show moderate reactivity ($t_{½}$ = 4.5 and 6 days). However, all the other pyrethroids are more resistant to photolysis. In comparison to aqueous photolysis, photodegradation on soil does not appear to be a major dissipation route for synthetic pyrethroids ($t_{½}$ = 26 days to stable). A noticeable exception is cyfluthrin, which is reactive to photodegradation on soil, with a half-life of 5 days. This indicates that with a few exceptions (prallethrin, resmethrin, cyfluthrin and fenvalerate) most pyrethroids are very stable under the sun light in both terrestrial and aquatic environments.

Aerobic soil metabolism is an important dissipation mechanism for many chemicals, but it is a major route of dissipation for only one pyrethroid, prallethrin ($t_{½}$ = 3 days). For other pyrethroids, resistance to aerobic soil biotransformation varies widely and can be categorized into relatively low persistence ($t_{½}$ = <30 days for deltamethrin and permethrin), moderate persistence ($t_{½}$ = 30-60 days for tefluthrin and tralomethrin), and high persistence ($t_{½}$ = 61-198 days for cypermethrin, fenvalerate, cyfluthrin, bifenthrin, fenpropathrin, and resmethrin). Pyrethroids are also susceptible to anaerobic biotransformation in soil and aquatic systems. It has been found that in certain instances, the rate of biotransformation in anaerobic conditions is similar to the rate in aerobic condition (e.g., cypermethrin, deltamethrin, fenpropathrin, and tefluthrin). Most of the time, however, the anaerobic metabolism is slower than aerobic metabolism (e.g., prallethrin, permethrin, resmethrin, bifenthrin, fenvalerate, lambda-cyhalothrin, and tralomethrin). Finally, the rate of anaerobic metabolism is faster for only one pyrethroid: cyfluthrin ($t_{½}$ = 30 days for anaerobic soil metabolism *vs.* 84 days for the aerobic soil metabolism). This suggests that most pyrethroids show moderate to high stability in soil but are more persistent in anaerobic environments such as sediments or aquatic systems with low oxygen content.

For the synthetic pyrethroids, the FIFRA Scientific Advisory Panel has determined that the K_{OC} (organic carbon adsorption coefficient) is a better predictor of the mobility of these chemicals than the K_d (adsorption coefficient) (*2*). Based on the K_{OC}, all the synthetic pyrethroids can be characterized as having very low mobility. Prallethrin and resmethrin, which are first generation synthetic pyrethroids, are slightly mobile according to the FAO mobility classification (respective average K_{OC}'s of 1,616 and 2,533 mg L^{-1}) (*10*). On the other hand, all of the other pyrethroids can be characterized as hardly mobile to immobile. Given the low solubility and mobility of bifenthrin, it is impossible to draw a Freundlich curve; therefore, values are reported for single test solutions. The same problem occurred with cyfluthrin. At this time, the Pyrethroids Working Group (PWG) has submitted new mobility studies for nine synthetic pyrethroids, using the solid phase micro-extraction (SPME) technique. This method should yield better estimates of K_d's and K_{OC}'s for these types of chemicals. The Agency is evaluating the quality and usefulness of these data.

The terrestrial field dissipation study is designed to reflect all routes of dissipation (degradation and transport) for a chemical: volatilization, hydrolysis, photolysis, biodegradation, plant interception/uptake, adsorption, and leaching. Depending on the degree of importance of these routes, the fate of the chemical is decided. Although terrestrial field dissipation studies were conducted on small plots and under controlled conditions, there was high variability in the results for certain synthetic pyrethroids (*e.g.*, bifenthrin 35-345 days, n=11; fenpropathrin 8-144 days, n=5). This variability within one chemical may have been caused by different environmental conditions such as soil aerobicity, pH, temperature, light intensity, soil humidity, rainfall, run-off/erosion intensity, and the execution protocol among others.

Table V. Environmental Fate Properties of the Synthetic Pyrethroids. [a]

Property	Prallethrin	Resmethrin	Permethrin	Bifenthrin	Cyfluthrin	Cypermethrin	Deltamethrin	Fenpropathrin	Fenvalerate	Lambda-Cyhalothrin	Tefluthrin	Tralomethrin
Solubility (ppm)	8.03	3.79E-2	5.5E-3	1.4E-5	2.32E-3	4.0E-3	2.00E-4	1.03E-2	6.0E-3	5E-3	2E-2	8.40E-2
Vapor pressure (mmHg)	9.99E-8	<3E-9	1.5E-8	1.8E-7	1.5E-8	1.3E-9	9.32E-11	5.48E-6	1.50E-9	1.5E-9	6.0E-5	1.8E-11
Henry's Law Const. atm-m^3/mol	4.92E-9	1.3E-7	1.4E-6	7.2E-3	3.7E-6	2.5E-7	3.1E-7	1.80E-4	1.4E-7	2.4E-7	1.6E-3	1.9E-10
log K_{OW}	4.49	5.43	6.1.	6.5	5.97	6.60	4.53	6.0	5.62	7.0	6.5	5.05
Hydrolysis $t_{1/2}$ in days at pH 5	Stable	89	Stable	Stable	Stable	Stable	Stable	Stable	Stable	Stable	Stable	Stable
Hydrolysis $t_{1/2}$ in days at pH 7	Stable	168	Stable	Stable	185	Stable	Stable	Stable	Stable	Stable	Stable	Stable
Hydrolysis $t_{1/2}$ in days at pH 9	4.9	127	238	Stable	1.9	2.1	2.5	16	Stable	13	>30	Stable
Photolysis in Water $t_{1/2}$ in days	0.6	1.5E-2	80	Stable	4.5	36.2	Stable	Stable	6	25	>31	<0.5
Photodegradation on Soil $t_{1/2}$ in days	25.9	NA	106	126	5.3	172	71	Stable	Stable	13-16% degraded after 35 d.	>31	Stable
Aerobic Soil Metabolism $t_{1/2}$ in days	3	197.5	23	146	84.2	60.3	19	152	83	48	<31	45 for total res. 3.2 for delta.

Anaerobic Soil Metabolism $t_{1/2}$ in days	NA	NA	204	Stable	Approx. 30	57.2	34	186	274	NA	>30	155
Aerobic Aquatic Metabolism $t_{1/2}$ in days	17	36.5	41	NA	NA	9.5	72.0	NA	NA	34	NA	NA
Anaerobic Aquatic Metabolism $t_{1/2}$ in days	34	682	144	NA	NA	15.2	NA	NA	NA	128	NA	NA
K_{OC}	2533 (1361-3769)	1616 (510-3180)	71,950 (28,200-194,000)	237,000 (131,000-302,000)	123,900 (73,480-180,700)	141,700 (20,800-328,000)	2,913,000 (705,000-3,140,000)	34400 (22100-51100)	441,500 (estimated value)	333,000 (110,000-724,000)	152,000 (68,000-425,000)	483,000 (44,000-877,000)
K_d	24.27 (9.22-51.51)	5.5 (0.9-8.7)	381 (344-446)	3570 (992-5430)	1370 (1116-1793)	1030 (657-1900)	2968 (1840-5830)	359 (184-662)	N/A	4350 (1970-7610)	1140 (493-2300)	3330 (197-8780)
BCF Whole Fish	1160X	2700X	510-610X	6090X	854X	448X	698X	830X	503-3340	448X	2051X	490X
Terrestrial Field Dissipation	waived	waived	30 (17-43)	156 (35-345)	<<32 (<<30-<<32)	7.7 (3.2-12.2)	111 (60.8-209)	76 (8-144)	14	39 (14-63)	124 (28->215)	2.0 parent (1.5-2.4); 15.4 total residues (12.4-19.8)

[a] In instances where multiple values were available for various fate parameters, the average was taken and is reported here. Values reported in parenthesis are ranges.

Within the synthetic pyrethroids family, the terrestrial field dissipation half-lives lie within a wide range of values from 1.5 days for tralomethrin to 345 days for bifenthrin. However, in general, it is observed that among all of the routes of dissipation mentioned above, the aerobic soil metabolism appears to be the one that drives observed trends in terrestrial field dissipation half-lives. For example, bifenthrin, which is one of the most persistent synthetic pyrethroids in the laboratory (aerobic soil metabolism 146 days, n=6), was the most persistent chemical in the field (terrestrial field dissipation 156 days, n=11). This result though is expected because bifenthrin has no other major routes of dissipation (stable to hydrolysis, stable to photolysis in water, very little photodegradation on soil, stable to anaerobic metabolism and low mobility). On the other hand, aerobic soil metabolism is not the only predictor of the persistence of the synthetic pyrethroids in the field. Three synthetic pyrethroids presented lower persistence in the field than was predicted by aerobic soil metabolism. The three chemicals were cyfluthrin (terrestrial field dissipation half-lives <32 days), cypermethrin (average terrestrial field dissipation half-life = 7.7 days), and esfenvalerate (terrestrial field dissipation = 14 days).

In general, degradation of the synthetic pyrethroids involves the breakdown of the ester linkage of the pyrethroid structure. Usually, resultant degradates are not degradates of concern with the exception of tralomethrin, whose major degradate is deltamethrin, another synthetic pyrethroid.

As indicated above, all the synthetic pyrethroids have log K_{OW}'s in the range of 4.5 to 7.0, indicating high potential for bioaccumulation. In fish bio-accumulation studies, all synthetic pyrethroids were found to bioaccumulate with BCF's that range from below 1000X (permethrin, cyfluthrin, cypermethrin, deltamethrin, fenpropathrin, lambda-cyhalothrin, and tralomethrin with BCF's in the range of 448 to 854X) to as high as 6,090X (prallethrin, resmethrin, bifenthrin, fenvalerate, and tefluthrin with BCF's in the range of 1,160 to 6,090X). The highest bio-concentration factor is for bifenthrin with relatively slow depuration (43-53% depurated after 42 days). In general, for other synthetic pyrethroids, the rates of depuration vary from chemical to chemical. It is relatively rapid for prallethrin, cyfluthrin, and fenpropathrin, but relatively slow for resmethrin, bifenthrin, and tefluthrin.

Measures of Exposure in Aquatic Systems

Pyrethroids are applied in both agricultural and urban settings. In agricultural settings, applications are directed towards target crop(s). During application, a varied amount (relatively small amount) of the applied pesticide may be carried by drift to adjacent non-target areas, including aquatic systems that may result in exposure. The majority of the pesticide that reaches targeted crops will be subjected to the dissipation process that includes both degradation (hydrolysis, photolysis, and aerobic/anaerobic soil metabolism) and transport (wash-off, runoff, and erosion). Aquatic systems may be receiving parent pesticide and/or transformation product(s) as they are transported from the treated field by runoff water in soluble form (runoff) and/or solid-adsorbed form (erosion). Pesticide and transformation products reaching aquatic systems result in exposure that is expected to change with time as a result of dissipation processes in the aquatic system.

In urban areas, use patterns of pyrethroids may be broadly characterized as either indoor/outdoor home uses or public health vector control uses. Indirect exposure to aquatic systems may be limited for most indoor home use but is expected to be more significant from outdoor home/public health use as pyrethroids may be carried into aquatic systems by drift, run-off, and/or erosion. Public health use labels for pyrethroids instruct avoidance of direct application to water bodies (streams, ponds, and lakes). However, it may not be possible to avoid spraying over such water bodies, and in some cases, overspray is permitted (over swamps and tidal marshes).

Aquatic exposure may be measured directly by monitoring or estimated by modeling. Monitoring programs may be conducted at a national, state, or regional scale to measure levels of pesticides in the water body adjacent to application site shortly after application, or may be un-targeted to detect levels of pesticides in the water body not necessarily near the application site geographically or temporally. Several targeted or un-targeted monitoring programs have analyzed water samples for pyrethroids. Data points may be found in a variety of databases: USGS-NAWQA (*11*), EPA-STORET (*12*), California Department of Pesticide Regulation surface water database (CA-DPR), and may also be obtained through targeted monitoring programs conducted by the pesticide registrant to fulfill EPA requirements for registration and re-registration (*13*).

Use of monitoring, as a measure of exposure, is important but will not be discussed here as it is usually used for higher tier assessments. Rather, exposure modeling will be discussed here, as it is the major procedure used in estimating aquatic exposure for baseline risk assessments. Even in baseline risk assessments, however, available monitoring data are generally evaluated and may be used in characterizing uncertainties associated with modeling.

Measures of aquatic exposure in agricultural and urban settings are included here for the following use patterns: agricultural crops, public health vector control, and home use. Under these topics, modeling procedures resulting in measurement of aquatic EECs for both water and sediment is discussed.

Agricultural Application: Insecticides/Acaricides

Most pyrethroids are used as agricultural insecticides (e.g., cyfluthrin, tralomethrin/deltamethrin, esfenvalerate/fenvalerate, tefluthrin, and tralomethrin) and insecticides/acaricides (e.g., bifenthrin, lambda-cyhalothrin, cypermethrin, fenpropathrin, and permethrin). In this case, aquatic exposure concentrations are estimated based on the Agency's aquatic Tier 2 linked PRZM and EXAMS models with its graphical user interface (pe4v01.pl) (*14*). In this protocol, PRZM (Pesticide Root Zone Model; version 3.122 released May 2005) simulates processes in a 10-hectare treated field and EXAMS (Exposure Analysis Modeling System; version 2.98.04.06 released April 2005) uses flux of runoff/erosion/chemical loadings to predict resultant EECs.

Modeled EECs are 1 in 10-year high-end values that represent the concentrations that might be found in the various compartments of the one-hectare farm pond (L x W x D= 100 x 100 x 20 meter). The thickness of the active sediment in the pond is assumed to be 0.05 meter. In these simulations, site-specific flows are achieved through using labeled application parameters (maximum application rate/number and minimum application intervals), crop scenarios (defines specific soil, site and crop parameters), and 30-year weather data for modeled locations. In standard simulations, application efficiency and spray drift are currently equal to default values of 99-95% for application efficiency (99% for ground application and 95% for aerial application) and 1-5% for spray drift (1% for ground application and 5% for aerial application). Drift is calculated for each application as a fraction of a one-hectare application rate with no consideration to the presence of a buffer zone. For most pesticides, standard simulations are usually adequate and resultant water column EECs are reasonable indications of surface water exposure. In the case of pyrethroids, it may prove necessary to execute extra analyses of the data to estimate concentrations in pore water and benthic sediment. Also, it may be necessary to execute additional PRZM/EXAMS simulations in order to obtain data necessary to analyze sensitivity of modeled EECs to factors such as drift. These analyses are described below under the topic: Special analyses/simulations.

Public health vector mosquito control: adulticides

Some pyrethroids (e.g., resmethrin, prallethrin and permethrin) are used as adulticides and are typically applied via spray methods with considerably small droplets (ultra low volume or ULV). ULV application methods result in the formation of mists necessary to prevent immediate deposition of the pesticide in order to better ensure the pesticide comes into direct contact with the insects in flight (more efficacious). Pyrethroids used as adulticides can be applied by ground methods using either backpack or truck mounted equipment or by using special aircraft. Measures of exposure from aerial applications will be discussed in this chapter because they appear to result in higher ecological exposure than ground application.

Aquatic EECs are estimated using Tier 2 PRZM/EXAMS modeling with a representative turf scenario as described earlier. Turf scenarios are chosen to represent applications on golf courses, parks, campsites, and athletic fields. According to the label, these locations represent the sites where adulticides are typically used. In this modeling exercise, the default values of application efficiency (95%) and spray drift (5%), currently used for applications to agricultural crops, are not considered appropriate for aerial applications of a mosquito adulticide. Instead, levels of efficiency/drift are estimated using the AGricultural DISPersal model (AGDISP v. 8.15) (*15*). This model estimates the fraction of pesticide deposited on the treated area (to represent the application efficiency parameter for PRZM/EXAMS) and that deposited downwind to a predefined area, such as an adjacent body of water. The latter is estimated for a 208.7 ft wide water body located downwind of the treated area (to represent the level of spray drift for PRZM/EXAMS).

Fractions of pesticide deposited on treated/downwind areas (PRZM/EXAMS application efficiency/spray drift) are estimated using the AGDISP deposition assessment "toolbox." The AGDISP simulation is executed using AGDISP/USDA default parameters (such as those associated with aircraft, spray equipment, swath width, and wind direction); available chemical specific label data (wind speed, stability, required spray material characteristics, and spray volume); parameters representing local conditions (such as canopy presence, ambient temperature and relative humidity); and finally the 208.7 ft width of the downwind deposition. In this exercise, some parameters, such as boom height and droplet size, can be changed to investigate possible effects on exposure. Following the determination of various application efficiencies and drift, the PRZM/EXAMS simulations are executed to arrive at water and sediment exposure EECs.

Special analyses/simulations

Sediment exposure is important for pyrethroids, as they are lipophilic compounds with low water solubility that have a propensity to readily partition into the sediment. When the pyrethroids reach a body of water by spray drift or runoff, they are expected to remain adsorbed to suspended solids or organic matter in the water column or partition in the sediment phase, as indicated by their generally high K_{OC} values. In addition, the sediment can act as a reservoir for a number of these compounds, as they are expected to persist and accumulate in the benthic layer over prolonged periods of time. Therefore, sediment-bound pyrethroids could present a toxicity risk for benthic aquatic life and aquatic ecosystems in general.

Given the aquatic toxicity of most pyrethroids, it is important to run additional PRZM/EXAMS simulations beyond standard ones. These simulations should be designed to identify the relative contribution of drift and erosion to water/sediment EECs in addition to analyzing effects of buffers and variation in pond depth on modeled water/sediment EECs. Such analyses provide invaluable information that can be used to reduce exposure uncertainties and/or to suggest mitigation measures. The following is a brief description of the special analyses and simulations that are often used for measuring exposure to pyrethroids:

- ***Pore water/benthic sediment EECs:*** A new graphical interface is being developed to create output files for 1 in 10 years pore water/sediment

EECs similar to those currently produced for the water column. In the meantime, one of the output files of EXAMS contains these data in a table designated as "Table 20." This file can be used to obtain pore water and sediment EECs for any and all of the PRZM/EXAMS simulations. As will be shown later, risk quotients can be calculated to determine possible risks.

- ***Spray drift and buffer strips:*** Peak EECs are most affected by spray drift although the relative importance of this contribution is largely dependent on the use pattern (i.e., the crop scenario). The objective of this exercise is to obtain an idea of the contribution of spray drift and runoff to peak EECs. For this purpose, special theoretical runs are performed for selected scenarios with drift set to zero. The peak EECs are then compared to the peak EECs generated with the standard simulation, and the contribution of each component, spray drift and runoff, can be estimated. Scenarios with the highest percent of EEC attributable to spray drift are further analyzed with reduced drift assumptions, especially if these reductions can be achieved by using buffers. Effects of buffer widths on reducing drift can be analyzed using AgDRIFT and suggestions on possible mitigation measures may result from such analyses (*16*).

Indoor Home Application: Household Products Containing Pyrethroid Insecticides

Residue from household consumer products could potentially be disposed into domestic wastewater and end up in aquatic systems. To assess this route of exposure, EFED utilized the Office of Pollution Prevention and Toxics (OPPT) consumer exposure model, Exposure and Fate Assessment Screening Tool (E-FAST) (*17*). The Down-the-Drain module of E-FAST is specifically designed to address all sources of a chemical that could potentially be disposed into domestic wastewater from a "down-the-drain" application. This model provides screening level estimates of chemical residues in surface water that may result from household uses and the disposal of these consumer products into wastewater. The model uses input parameters that include annual production volume of the pesticide and takes into account the fraction of the chemical removed during wastewater treatment. The assumptions of the model state that in a given year, the entire production volume of the chemical in question is parceled out on a daily per capita basis to the U.S. population and converted to a daily mass release per capita (e.g., gm/person/day). In other words, the daily per capita release of the chemical to a wastewater treatment facility in grams/person/day can be expressed as[a]:

---

[a] Publicly available data on the average of the total volume of the active ingredient (in kilograms) of products aimed for uses that could potentially reach household drains in the US.

$$\frac{\textbf{Production volume (Kg/year)} \times 1000\text{g/Kg} \times 1\text{ year/365 days}}{\textbf{US population (persons)}}$$

This mass is diluted into the average daily volume of wastewater released per person per day to arrive at an estimated concentration of target chemical in wastewater prior to entering a treatment facility. The target chemical concentration in untreated wastewater is then reduced by the fraction removed during the wastewater treatment process before release into a river or stream. This fraction can be obtained by utilizing the EPIWIN program (*18*). A Stream Dilution Factor is the volume of the receiving stream flow divided by the volume of the wastewater released from the Publicly Owned Treatment Works (POTW). The resulting values are used for ecological effects: acute/chronic estimated environmental concentrations (EECs).

Measures of Exposure in Terrestrial Systems

Terrestrial birds and mammals may be exposed to pyrethroids shortly after their application through oral and/or dietary exposure to vegetative plant material or insects when foraging in the treated fields for nesting material or food. The EFED terrestrial exposure model T-REX is used to estimate exposures and potential risks to avian and mammalian species. Input values for avian and mammalian toxicity as well as chemical application and foliar dissipation half-life data are required to run the model, which provides estimates of environmental concentrations and risk quotients (RQs). Specifically, the model provides estimates of concentrations (upper-bound and mean) of chemical residues on the surface of different types of foliage and insects that may be dietary sources of exposure to avian, mammalian, reptilian, or terrestrial-phase amphibian receptors. Baseline assessments use upper-bound predicted residues as the measure of exposure.

The initial surface residue concentration (ppm) is estimated by multiplying the application rate (pounds active ingredient per acre) by a value specific to each food item. These values (termed the Hoerger-Kenaga estimates), along with a more detailed discussion of the methodology implemented by T-REX ,can be found elsewhere (*19*). EECs are determined at any time following application and are based on first order kinetics using the foliar half-life as the rate constant. For multiple applications, the EEC is determined by adding the mass on the surface immediately following the application to the mass of the chemical still present on the surfaces on the day of the next application. The chemical specific foliar half-life determined from foliar dissipation guideline studies is used as input to the model. In the absence of such data, a conservative default foliar half-life of 35 days based on open literature studies is used (*20*).

Uncertainties in the terrestrial EECs are usually associated with lack of data on dissipation from foliar surfaces. Characterization of such uncertainties is important especially if there is a reason to suspect the applicability of the 35-day half-life. In this case, it may prove useful to bracket exposure with a lower estimate that uses a reasonable shorter half-life value.

Effects Characterization

In baseline ecological risk assessments conducted by the Agency, the effects characterization encompasses the types and magnitude of effects a pesticide can produce under varying exposure levels, and is based on an effects profile that describes and interprets the available toxicity information for various plants and animals. Although the available information may include open-literature studies, incidents information, and effects monitoring data, the majority of data typically available for quantitative incorporation in a baseline risk assessment are registrant-submitted ecological effects studies, which are performed on a limited number of organisms in the following broad groupings:

- Birds (mallard duck and bobwhite quail) used as surrogates for terrestrial-phase amphibians and reptiles,
- Mammals (laboratory rat),
- Freshwater fish (bluegill sunfish, rainbow trout, and fathead minnow) used as surrogates for aquatic-phase amphibians,
- Freshwater invertebrates (*Daphnia magna*),
- Estuarine/marine fish (sheepshead minnow),
- Estuarine/marine invertebrates (*Crassostrea virginica* and *Mysidopsis bahia*).

As can be seen, only two surrogate species for birds are used to represent all bird species and serve as a surrogate for terrestrial-phase amphibians and reptiles. Three species of freshwater fish are used to represent all freshwater fish species and serve as a surrogate for aquatic-phase amphibians. One estuarine/marine fish species is used to represent all estuarine/marine fish. The surrogate species for terrestrial invertebrates is the honey bee (*Apis mellifera*). For freshwater invertebrates the surrogate species is usually the waterflea (*Daphnia magna*), and for estuarine/marine invertebrates the surrogate species are mysid shrimp (*Mysidopsis bahia*) and eastern oyster (*Crassostrea virginica*). These four species are used to represent all invertebrate species.

Although aquatic and terrestrial plant toxicity data are also often required for the registration and re-registration of pesticides, there are currently very limited data available for all of the synthetic pyrethroids. However, based on their neural toxic mode of action, the dearth of studies demonstrating adverse effects of any pyrethroid to plants, and the fact that no incident reports have reliably linked any compound in the synthetic pyrethroid class of compounds to phytotoxic effects, it appears that the relative likelihood of risk to aquatic and terrestrial plants is low compared to that posed to aquatic and terrestrial animals. Therefore, similar to all of the ecological risk assessments conducted by the Agency for synthetic pyrethroids, this document focuses on animal toxicity and risk.

An acute and a chronic endpoint are selected from the most sensitive species tested within each of the very broad taxonomic groups listed above, and the endpoints are used to estimate the toxicity of a pesticide to that group. Any additional toxicity data that may be identified from alternate sources (e.g., searches conducted using the ECOTOX database (*21*)) may be available and considered once they have been reviewed according to the Agency's study protocols and risk assessment guidance to determine their quality, reliability, and utility in the risk assessment (*22*, *23*, *24*, *25*). All toxicity data used in the Agency's ecological risk assessments must meet the data quality classification of "supplemental" or "acceptable."

Over the course of evaluation of available toxicity data, the risk assessment team may encounter other effects data that provide: (1) additional information on existing toxicity endpoints commonly used in the screening risk assessment, (2) insight on endpoints not routinely considered for risk estimation, and/or (3) effects data on specific additional taxonomic groups. Professional judgment is used and documented by the risk assessment team to determine whether available data on other toxicological endpoints are included in the risk assessment. This evaluation may include (a) reference to data quality objectives for specific types of studies, (b) the degree to which adequate documentation is available to evaluate the technical merit of the data, and (c) whether the data are applicable to the assessment endpoints established for the risk assessment. To decide if the data are applicable to assessment endpoints, the risk assessment team uses professional judgment and available lines of evidence to determine if the toxicological endpoints can be linked to assessment endpoints in a reasonable and plausible manner (*3*).

Regardless of the extent of data beyond that required under FIFRA, a set of the most sensitive endpoints identified from all acceptable studies will be selected for use in the risk assessment. The Electronic Code of Federal Regulations (e-*CFR*): Title 40 Protection of Environment, Parts §158.490 (*26*), §158.540 (*27*), and §158.590 (*28*) specify the suite of studies that the Agency may request to determine the risks of a pesticide to wildlife, aquatic organisms, and plants. Although the data requirements may vary on a chemical-by-chemical basis, depending on the expected usage scenarios and exposure regimes, the toxicity endpoints presented in Table VI are typically reported and used to quantitatively estimate levels of risk for the synthetic pyrethroids in baseline risk assessments.

Table VI. Typical toxicity endpoints that are used to quantitatively estimate levels of risk for the synthetic pyrethroids in screening-level risk assessments.

Acute Toxicity to:	**Toxicity Endpoint**
Birds	LD_{50} (single oral dose) and LC_{50} (subacute dietary).
Mammals	LD_{50} from single oral dose test.
Aquatic Animals	EC_{50} or LC_{50} for freshwater fish and invertebrates and estuarine/marine fish and invertebrates acute toxicity tests.
Chronic Toxicity to:	**Toxicity Endpoint**
Birds	NOEC for 21-week avian reproduction test.
Mammals	NOEC for two-generation reproduction test.
Aquatic Animals	NOEC for freshwater fish and invertebrates and estuarine/marine fish and invertebrate's early life-stage or full life-cycle tests.

Although the above toxicity endpoints are routinely used to estimate risk, they do not represent a limitation on the types of toxicity endpoints that may be considered in the risk assessment. However, given the varying breadth of studies available for each of the synthetic pyrethroids, these are the taxonomic groups and toxicity data that will be considered for the purposes of this chapter since they are most commonly available to the Agency. Unless otherwise stated, all data presented are based on studies with the technical grade active ingredient that have been formally reviewed by Agency scientists and that meet the data quality classification of "supplemental" or "acceptable."

Terrestrial Animal toxicity

Categories of acute toxicity ranging from "practically nontoxic" to "very highly toxic" have been established for non-target insects (based on LD_{50} values for honey bees), avian species (based on and LD_{50} and LC_{50} values), mammals (based on LD_{50} values), and aquatic organisms (based on LC_{50} and EC_{50} values) (*29*). These categories are presented in Table VII.

Table VII. Categories of Acute Toxicity to Various Taxonomic Groups

Toxicity to Non-Target Beneficial Insects	
LD_{50} (mg/kg-bw)	**Toxicity Category**
< 2	Highly toxic
2 – 11	Moderately toxic
> 11	Practically non-toxic
Dietary Toxicity to Avian Species	
LC_{50} (ppm)	**Toxicity Category**
< 50	Very highly toxic
50 – 500	Highly toxic
501 – 1000	Moderately toxic
1001 – 5000	Slightly toxic
> 5000	Practically non-toxic
Acute Oral Toxicity to Avian and Mammalian Species	
LD_{50} (mg/kg-bw)	**Toxicity Category**
< 10	Very highly toxic
10 – 50	Highly toxic
51 – 500	Moderately toxic
501 – 2000	Slightly toxic
> 2000	Practically non-toxic
Toxicity to Aquatic Animals	
LC_{50} or EC_{50} (ppm)	**Toxicity Category**
< 0.1	Very highly toxic
0.1 – 1	Highly toxic
>1 – 10	Moderately toxic
>10 – 100	Slightly toxic
> 100	Practically non-toxic

The members of the synthetic pyrethroid class of compounds can be generally characterized as having relatively high acute toxicity to terrestrial invertebrates, and low to moderate acute toxicity to terrestrial vertebrates (Table VIII). For instance, based on the compounds for which data are available (Table VIII), acute contact toxicity data for honey bees suggest that the synthetic pyrethroids can be classified as highly toxic to non-target beneficial insects, with LD_{50} values ranging from 0.0015 (deltamethrin) to 0.129 (tralomethrin).

On the other hand, according to the most sensitive endpoints from studies available to the Agency, acute dietary data demonstrate that synthetic pyrethroids range from practically non-toxic (LC_{50} >10,000 ppm for permethrin) to slightly toxic (LC_{50} = 1280 ppm for bifenthrin) to avian species (Table VIII). Although acute oral data show that synthetic pyrethroids range from practically non-toxic

(LD_{50} = >9869 mg/kg-bw for permethrin) to moderately toxic (LD_{50} = 75 mg/kg-bw for resmethrin) to avian species, only two of the twelve pyrethroids considered in this chapter were moderately toxic (resmethrin and esfenvalerate; LD_{50} = 75 and 381 mg/kg-bw, respectively). Of the remaining ten, three were slightly toxic (prallethrin, fenpropathrin, and bifenthrin; LD_{50} = >1000, 1089, and 1800 mg/kg-bw, respectively), and the rest were classified as practically non-toxic.

Although the acute toxicity classification of the synthetic pyrethroids to mammals ranges from highly toxic (LD_{50} = 16.2 mg/kg-bw for cyfluthrin) to practically non-toxic (LD_{50} = 8900 mg/kg-bw for permethrin), only three of the twelve pyrethroids classified in this chapter were considered as highly toxic to mammals (cyfluthrin, tefluthrin, and fenpropathrin; LD_{50} = 16.2, 21.8, and 48.5 mg/kg-bw, respectively; Table VIII). Of the remaining nine, seven were classified as moderately toxic (bifenthrin, lambda-cyhalothrin, deltamethrin, tralomethrin, esfenvalerate, cypermethrin, and prallethrin; LD_{50} = 53.8, 56, 66.7, 84.9, 87.2, 247, and 460 mg/kg-bw, respectively), and the remaining two were classified as practically non-toxic (resmethrin and permethrin; LD_{50} = 4639 and 8900 mg/kg-bw, respectively).

In terms of reproductive toxicity, the range in the magnitude of effects exhibited by avian species is bound at the lower end by the NOAEC and LOAEC values of 12 ppm and 60 ppm, respectively, for resmethrin, and at the upper end by the NOAEC and LOAEC values of 500 ppm and >500 ppm, respectively, for permethrin. Some of the observed effects at the reported LOAEC levels for all pyrethroids considered here (Table VIII) include reductions in eggshell thickness, percent of normal hatchlings from live 3-week embryos, number of eggs laid, embryo viability and hatchability, and 14-day survivor weight, and increased number of cracked eggs and early embryonic deaths. However, it should be made clear that although reproductive toxicity studies with avian species indicate that there is potential for reproductive effects from chronic exposure to this class of compounds, there is considerable uncertainty surrounding the characterization of effects because of the limitations of the data set available to the Agency. Of the twelve synthetic pyrethroids considered here, half were unable to elicit toxic effects up to the highest concentrations tested (bifenthrin, lambda-cyhalothrin, cypermethrin, deltamethrin, permethrin, and tefluthrin); however, all of these studies were conducted with test concentrations ≤ 500 ppm, and four of these were conducted with test concentrations ≤ 100 ppm (bifenthrin, lambda-cyhalothrin, cypermethrin, and tefluthrin).

The NOAEC and LOAEC values for mammalian reproductive toxicity range from <15 (lowest concentration tested) and 15 ppm for tralomethrin, respectively, to 1000 and 3000 ppm for permethrin, respectively. Some of the observed effects at the listed LOAECs (Table VIII) include decreased mean body weight and body weight gain of parents and offspring; decreased food consumption, litter size, mating index, and reproductive viability index;

increased stillbirths, mortality and clinical signs of neurotoxicity in parents and offspring; incidence of skin lesions and subcutaneous hemorrhage; increased liver weights and microscopic findings in liver, kidney, thyroid and pituitary in parents. However, it should be noted that two of the twelve synthetic pyrethroids considered in this assessment were not evaluated at test levels low enough to achieve a NOAEC, and that one of these compounds was the one with the most sensitive NOAEC (tralomethrin); therefore, the lower bound on the range for reproductive toxicity of synthetic pyrethroids to mammals presented here is not a conservative one.

Aquatic Toxicity

Water column

The most sensitive pyrethroid aquatic toxicity data available to the Agency are presented in Table IX for selected taxonomic groups. Based on these data, the entire group of synthetic pyrethroids can be classified as very highly toxic to all aquatic animals. However, the toxicity data for invertebrates suggests that they are often at least an order of magnitude more sensitive than fish (Figure 6). More specifically, acute LC_{50} values range from 0.06 ppb (tefluthrin) to 12 ppb (prallethrin) for freshwater fish, and 0.13 ppb (tefluthrin) to 26 ppb (prallethrin) for estuarine/marine fish. Data for invertebrates demonstrate acute LC_{50} ranges of 0.0036 ppb (cypermethrin) to 6.2 ppb (prallethrin) for freshwater invertebrates, and 0.0022 ppb (cyfluthrin) to 3.9 ppb (prallethrin) for estuarine/marine invertebrates.

In terms of chronic toxicity, estimated NOAEC and LOAEC levels for freshwater fish range from 0.004 and 0.008 ppb for tefluthrin, respectively, to 3 and >3 (no observed effects) for prallethrin, respectively. Some effects observed at the listed LOAECs for freshwater fish include reduced reproduction, egg production, number of fry, hatchability, growth (adult and larval), and survival (adult and larval).

Estimated NOAEC and LOAEC values for freshwater invertebrates range from 0.0013 and 0.0029 ppb for bifenthrin, respectively, to 0.65 and 1.3 for prallethrin, respectively. Some effects observed at the listed LOAECs for freshwater invertebrates include reduced survival, growth (length), and reproduction (number of young per female per reproductive day and total offspring). However, it should be noted that chronic data are lacking for cypermethrin and resmethrin. Because the most sensitive acute toxicity value for freshwater invertebrates is based on data for cypermethrin (LC_{50} = 0.0036 ppb), it appears reasonable to conclude that the lower bound for the range of chronic toxicity exhibited by the synthetic pyrethroids may be underestimated by the range estimated in this chapter based on actual data.

Table VIII. Most sensitive terrestrial toxicity data available to the Agency for the synthetic pyrethroids.

Chemical	Birds			Mammals		Insects
	LC_{50} [a]	LD_{50} [b]	NOAEC/ LOAEC [a]	LD_{50} [b]	NOAEC/ LOAEC [a] (NOAEL/ LOAEL [b])	LD_{50} [c]
Bifenthrin	1280	1800	75/>75	53.8	60/100 (3/5)	0.015
Cyfluthrin	>5000	>2000	250/>250	16.2	50/150 (2.5/7.5)	0.037
Lambda-cyhalothrin	3948	>3950	30/>30	56	30/100 [d] (1.5/5)	0.038
Cypermethrin	>2634	>2000	50/>50	247	100/500 (5.0/25)	0.023
Deltamethrin	>5620	>2250	450/>450	66.7	80/320 (5/21)	0.0015
Esfenvalerate	4894	381	<25/25	87.2	<75/75 (<3.75/3.75)	0.017
Fenpropathrin	9026	1089	25/125	48.5	40/120 (3/8.9)	N/A
Prallethrin	>5620	>1000	120/360	460	600/3000 (31/156)	0.028
Permethrin	>10000	>9869	500/>500	8900	1000/3000 (50/150)	0.024
Resmethrin	>5000	75	12/60	4639	500/1000 (34.8/70.8)	0.063
Tralomethrin	4214	2510	<100/100	84.9	<15/15 (<0.75/0.75)	0.129
Tefluthrin	2317	4190	25/>25	21.8	15/50 (0.75/2.5)	N/A

[a] Units are ppm.
[b] Units are mg/kg-body weight.
[c] LD_{50} values for non-target insects are based on contact toxicity studies and the units are µg/bee.
[d] A mammalian reproduction study is not available for lambda-cyhalothrin. This data point was based on a study with lambda-cyhalothrin.

Table IX. Most sensitive aquatic toxicity data available to the Agency for the synthetic pyrethroids [a].

Chemical	Freshwater Fish		Freshwater Invertebrates		Estuarine/ Marine Fish		Estuarine/ Marine Invertebrates	
	LC_{50}	NOAEC/ LOAEC	EC_{50}	NOAEC/ LOAEC	LC_{50}	NOAEC/ LOAEC	EC_{50}	NOAEC/ LOAEC
Bifenthrin	0.15	N/A	1.6	0.0013/ 0.0029	17.5	N/A	0.004	N/A
Cyfluthrin	0.068	0.01/ 0.0177	0.14	0.02/ 0.041	4.05	0.025/ 0.084	0.0022	0.000/ 0.0004
Lambda-cyhalothrin	0.21	0.031/ 0.062	0.007	0.002/ 0.0035	0.807	0.25/0.38	0.005	0.000/ 0.0005
Cypermethrin	0.39	0.14/0.33	0.0036	N/A	0.95	N/A	0.00475	0.000781/ 0.00197
Deltamethrin	0.91	0.017/ 0.035	3.5	0.0041/ 0.0089	0.58	N/A	0.0037	N/A
Esfenvalerate	0.18	0.09/0.21	0.9	0.052/ 0.079	N/A	N/A	N/A	N/A
Fenpropathrin	2.20	0.06/0.14	0.53	0.22/0.35	3.1	N/A	0.021	0.012/ 0.024
Prallethrin	12	3/>3	6.2	0.65/1.3	26	N/A	3.9	N/A
Permethrin	0.79	0.30/0.41	0.039	0.039/ 0.084	2.2	<10 /10	0.018	0.011/ 0.024
Resmethrin	0.28	0.32/0.59	3.11	N/A	11	1.90/4.05	1.34	N/A
Tralomethrin	1.60	0.046/ 0.098	0.038	0.0044/ 0.009	2.5	N/A	0.84	N/A
Tefluthrin	0.06	0.004/ 0.008	0.07	0.0079/ 0.024	0.13	N/A	0.053	N/A

[a] All data are in ppb and are based on toxicity testing with the technical grade active ingredient (TGAI).

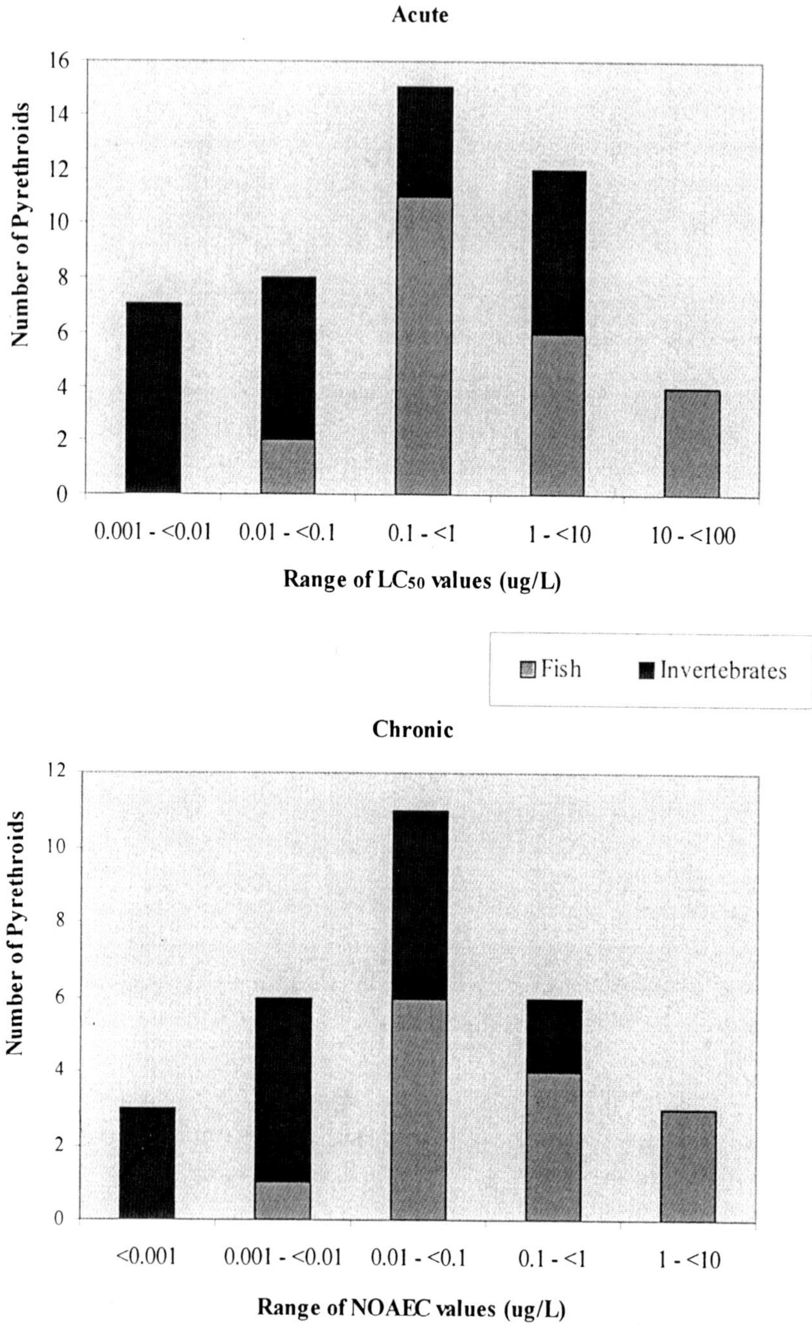

Figure 6. Summary of the number of pyrethroids for which the most sensitive acute and chronic toxicity endpoints fall within the specified ranges.

For estuarine/marine fish, estimated NOAEC and LOAEC levels range from 0.025 and 0.084 ppb for cyfluthrin, respectively, to <10 and 10 for permethrin, respectively. Effects associated with the listed LOAECs for estuarine/marine fish include decreased weight of survivors and reduced larval and juvenile survival. Because there were effects observed in the lowest concentration tested in the study on which the permethrin NOAEC is based, the upper-bound of the chronic toxicity range of synthetic pyrethroids is based on the non-definitive endpoint for permethrin. For this reason and because the most sensitive acute toxicity value for permethrin (LC_{50} = 2.2 ppb) is less than one-fourth of its reported NOAEC, this upper-bound estimate of chronic toxicity for permethrin may not be reliable. Therefore, it would probably be more appropriate to place an upper-bound on the estimates of chronic toxicity of the synthetic pyrethroids to estuarine/marine fish using the NOAEC and LOAEC of 1.90 and 4.05 ppb for resmethrin, respectively. In addition, it should be noted that only four of the twelve pyrethroids considered in this chapter currently have chronic toxicity estimates for estuarine/marine fish. More importantly, there are currently no chronic toxicity data for prallethrin and tefluthrin. Because the range of acute toxicity values for estuarine/marine fish is bound by data for these two chemicals, it seems reasonable to conclude that the range of chronic toxicity exhibited by the synthetic pyrethroids would likely expand if chronic data for these compounds were included.

Estimated NOAEC and LOAEC values for estuarine/marine invertebrates range from 0.0002 and 0.0004 ppb for cyfluthrin, respectively, to 0.012 and 0.024 ppb for fenpropathrin, respectively. Some effects observed at the listed LOAECs for estuarine/marine invertebrates include reduced survival, growth, number of young per female per reproductive day, and number of young per treatment. However, it should be noted that only five of the twelve compounds considered in this chapter currently have chronic toxicity estimates for estuarine/marine invertebrates; and there is significant uncertainty surrounding the range of chronic toxicity exhibited by the synthetic pyrethroids to this taxonomic group.

Sediment

As mentioned previously, sediment-bound pyrethroids could present a serious toxicity risk for benthic aquatic life and aquatic ecosystems in general. However, evaluating the risk to aquatic life from this route of exposure for this class of compounds is generally problematic because of the lack of adequate sediment toxicity and exposure data. In fact, only benthic invertebrate toxicity data are available only for cypermethrin; the EC_{50} for this chemical was estimated to be 3.6 ppb sediment (0.00257 ppb pore water). Therefore, in order to assess the potential for pesticide risk to aquatic benthic systems, EFED has adopted an approach based on the equilibrium partitioning (EqP) theory that is used by the Agency's Office of Water (OW). This extrapolation method is useful for estimating potential sediment exposure values, as well as sediment toxicity values that can be used in a baseline risk assessment.

The EqP theory is based on the hydrophobicity and concentrations of the chemical normalized to organic carbon (OC) in sediment and holds that a nonionic chemical in sediment partitions between sediment organic carbon, pore water, and benthic organisms (*30*). At equilibrium, if the concentration in any phase is known, then the concentration in the other phases can be predicted through the organic/ carbon soil partition coefficient (K_{oc}). This key component (i.e., the K_{oc}), is constant for every chemical and represents the ratio of the chemical concentration in water to the concentration in organic carbon. The document, "Technical Basis for the Derivation of Equilibrium Partitioning Sediment Benchmarks (ESBs) for the Protection of Benthic Organisms: Nonionic Organics," demonstrates that biological responses of benthic organisms to nonionic organic chemicals in sediments are different when the sediment concentrations are expressed on a dry weight basis, but similar when expressed on a μg chemical/g organic carbon basis ($\mu g/g_o$) (*31*). Similar responses were also observed across sediments when pore water concentrations were used to normalize biological availability. The Technical Basis Document further demonstrates that if the toxic effect concentration in water is known (e.g., LC_{50}), the effect concentration in sediment on a $\mu g/g_{oc}$ basis can be predicted by multiplying the effect concentration in water by the chemical K_{oc}:

$$\mathbf{(LC_{50}\ \mu g/L\ x\ K_{oc}\ L/kg_{oc}\ x\ 1\ kg_{oc}/1000g_{oc} = LC_{50}\ \mu g/g_{oc})}$$

In order to assess possible toxic pesticide exposure to aquatic organisms from sediments, EFED uses the PRZM/ EXAMS model to generate EECs from sediment and pore water based on the principles of the equilibrium partitioning theory. By relying on sediment and/or pore water output values, EFED has two ways for calculating RQ values for sediments by using pore water exposure values and bulk sediment values.

The calculations that rely on pore water can be calculated by dividing the PRZM/ EXAMS output value for pore water concentrations by the most sensitive measures of effect produced in bioassays with the TGAI of the synthetic pyrethroid of interest (e.g., LC_{50}). However, for all pyrethroids except cypermethrin, sediment toxicity data were unavailable for use in the most current risk assessments. Therefore, EFED assumed that benthic organisms are no more sensitive to toxic compounds than those organisms living in the water column. Subsequently, RQ values for pore water exposure were calculated based on the measures of effect from standard water column studies in the absence of sediment toxicity data (e.g., LC_{50}, EC_{50}):

$$\mathbf{EEC\ pore\ water\ \mu g/L\ /\ LC_{50}\ \mu g/L}$$

If sediment effects data are available (LC_{50} ug/kg_{OC}), RQs can be produced by using the PRZM/ EXAMS sediment output value for sediment.

$$\mathbf{EEC\ sediment\ \mu g/\mu g_{oc}\ /\ LC_{50}\ \mu g/kg_{oc}}$$

In the case of cypermethrin, toxicity data were available in terms of bulk sediment, but not in terms of pore water concentrations. Subsequently, the equivalent concentrations in pore water were calculated based on the equilibrium partitioning theory and the definition of the K_{oc}:

$$\mathbf{K_{oc}\ (L/kg_{OC}) = Sediment\ Concentration\ in\ OC\ (mg/kg_{OC}) \div Pore\ Water\ Concentration\ (mg/L)}$$

The Agency recognizes that actual sediment toxicity data are needed to fully characterize the effects to these animals, and that there is uncertainty regarding the amount of sorbed pesticide that may contribute to the toxicity of organisms. In addition, it is recognized that the model-generated sediment/pore water values may be an overestimation or underestimation of actual exposure values noted in the laboratory or field, and that inherent variations in biological uptake as result of differences in feeding pathways, organism life stage, and population-level effects versus individual responses could affect the toxicity of sorbed pesticide to benthic organisms. Agency efforts in conjunction with the Pyrethroid Working Group (PWG) have helped to begin addressing data gaps associated with the analysis of the potential ecological risk of synthetic pyrethroid insecticides to aquatic benthic organisms. Following the recommendations of a Science Advisory Panel (*2*) and based on comments from the Agency and Oak Ridge National Laboratories (ORNL), sediment testing was performed on four representative registered synthetic pyrethroids (cypermethrin, esfenvalerate, bifenthrin and cyfluthrin). These sediment toxicity data have been submitted by the PWG and will be reviewed and incorporated into the Agency's risk assessments during the Registration Review Process.

Enhanced Toxicity in the Formulation

The Agency routinely conducts baseline risk assessments for animals on an active-ingredient basis, and subsequently, the majority of toxicity data received by the Agency is for active ingredients alone. However, the Agency regulations have provisions for the request of additional data on formulated products. Specifically, 40 *CFR* 158.75 (*32*) allows the Agency to request additional data if routinely required data are inadequate to evaluate the potential of a pesticide product to cause adverse effects on the environment, and 40 *CFR* 158.202 (*33*) indicates that acute aquatic animal toxicity testing of formulations may be required if any of the following conditions are met:

- Active ingredient LC_{50}/EC_{50} values are equal to or less than the maximum expected environmental concentration or the estimated environmental concentration in aquatic systems when the product is used as directed;
- The end-use product is applied directly to water when used as directed; or

- An ingredient in the end-use product is expected to enhance the toxicity of the active ingredient or is toxic itself to aquatic organisms.

For some of the members of the synthetic pyrethroid class of compounds, all of these conditions are met. The first requirement is met for most, if not all of the synthetic pyrethroids, as indicated by aquatic exposure modeling for maximum use scenarios for each pesticide. The second requirement is met for a number of the pyrethroids that are used as mosquito adulticides (e.g., resmethrin, permethrin, prallethrin) and whose labels permit the application of theses pesticides over shallow bodies of water such as swamps and tidal marshes. Lastly, there is significant reason to believe that the toxicity of some synthetic pyrethroid formulations is greater than the toxicity of the technical grade active ingredients alone due to the presence of potentially synergistic compounds (e.g., PBO) in combination with synthetic pyrethroids in the formulations.

A synergist is a compound that when added to a pesticide product increases the potency of the active ingredient by effectively binding or suppressing an organism's detoxification system. This binding of critical mixed function oxydase (MFO) enzymes and the reduction in the detoxification system, enhances the pesticidal properties of the other active ingredient by allowing it to reach its critical target sites more efficiently. Many pyrethroid products, including repellants and pediculicides (lice killers), foggers and garden sprays, and adulticides used for mosquito control, contain synergists. The agricultural and mosquito adulticide uses of these formulations are the most likely to involve potential toxic exposure to aquatic organisms.

For example, resmethrin has generally effective "knockdown" potential, but its toxicity is usually enhanced in formulations such as Scourge®, in combination with a synergist such as piperonyl butoxide (PBO). Piperonyl butoxide enhances the pesticidal properties of the other active ingredient by competitive inhibition of detoxifying enzymes (cytochrome P450). Resmethrin is the only pyrethroid for which the Agency has acceptable toxicity data available for both formulations containing synergists and the TGAI, and for this chemical all aquatic taxonomic groups generally have a similar level of sensitivity.

Based on these analyses, the Agency has identified the need to further explore and characterize these effects for other formulations containing members of the synthetic pyrethroid class of compounds that are expected to exhibit similar enhanced toxicity.

Risk Characterization

The objective of the baseline ecological risk assessment is to identify, quantify, and/or characterize the risk from pesticide application and its subsequent release into the environment. The Agency has completed baseline risk assessments for a number of synthetic pyrethroids. These include Reregistration Eligibility Decisions (REDs) for permethrin (*34*), cypermethrin (*35*), and resmethrin (*36*) and new use assessments for bifenthrin, cyfluthrin, lambda-cyhalothrin, cypermethrin, zeta-cypermethrin, and prallethrin. To evaluate the potential risk to non-target organisms from the use of pyrethroids in these assessments, quantitative risk quotients (RQ) for aquatic invertebrates, fish, reptiles, amphibians, birds, and mammals are calculated from the ratio of estimated environmental concentrations (EECs) to acute ancd chronic toxicity values (e.g. LC_{50}, EC_{50}, LD_{50}, NOAEC, NOAEL). The risk quotients are then compared to the Agency's Levels of Concern (LOC). These LOCs are the Agency's interpretative policy used to identify potential risk to non-target organisms and the consequent need to consider regulatory action. The acute and chronic LOCs for all animals above which the Agency has concerns for acute risk to non-listed species are 0.5 and 1.0, respectively. The acute listed species LOCs for aquatic (i.e., aquatic invertebrates, fish, and aquatic phase amphibians) and terrestrial (i.e., birds, mammals, reptiles, and terrestrial phase amphibians) animals above which the Agency has concerns for acute risk to listed species are 0.05 and 0.1, respectively. For non-target beneficial insects, the Agency does not currently assess quantitative risk; instead, results of laboratory studies are used for a qualitative evaluation of risk.

These assessments show a risk concern for aquatic species (e.g., fish, invertebrates, aquatic-phase amphibians) exposed to pyrethroids that reach surface water via drift and erosion/runoff accompanied by rain events. In addition, because pyrethroids have an affinity to bind to particulate and organic matter, there is a potential for acute and chronic risk to aquatic invertebrates that reside in the sediment. For avian species, the Agency has concluded that pyrethroids do not appear to pose an acute risk because of their relatively low toxicity to these species, however, there is a potential for chronic risk to birds, especially those that consume large amounts of grass. There is also a potential for acute and chronic risk to mammals, which feed on short grass, tall grass, broadleaf plants, and large insects since both acute and chronic RQs exceed the level of concern for most pyrethroids. The Agency is also concerned with the potential for risk to non-target insects, including honeybees and other insect pollinators, as well as several beneficial insects such as predatory wasps. Risk to aquatic and terrestrial plants were not assessed because data are not available, and the Agency does not expect the pyrethroid mode of action to be a phytotoxic concern. The risk extends to agricultural uses and mosquito abatement, as well as other non-agricultural uses. Certain urban uses such as pyrethroid-containing drugs (both prescribed and over-the-counter), pre-treated clothing, and pet products were also evaluated by the Agency by modeling the expected residues that could occur in an aquatic system from domestic uses.

Risk to Aquatic Organisms

In the Agency's ecological risk assessments, it is generally assumed that the major points of exposure for aquatic receptors are through direct contact with the water column, sediment, and interstitial (pore) water contaminated with spray drift and erosion/runoff from treated areas. Therefore, evaluation of risk to aquatic organisms is generally approached in the Agency's synthetic pyrethroid ecological risk assessments by dividing the aquatic system into two general compartments, the water column and the benthos.

The first compartment, the water column, is defined as the aquatic area between the surface and the benthos and represents an area where organisms are free swimming. However, it is generally understood that although water column-dwelling organisms may spend the majority of their time at the surface or in mid-water, they are not restricted to this compartment alone and may still feed and/or breed on the benthos. Direct pesticide contact (e.g., gill lamella, ingestion, and integument) to these organisms is assumed to result from the pesticide in the surface water. The toxicity assessment endpoints are the acute LC_{50} and chronic NOAEC, which are generated through standard water column toxicity tests.

The second major aquatic compartment to be considered is the benthos, which is composed of sediments and an area six inches above the sediment (epibenthos). The benthos is composed of a diversity of aquatic invertebrates (e.g., insect larvae, crustaceans, mollusks) and species of fish (e.g., catfish, loachs), as well as certain critical life stages of organisms that reside in the water column. The benthos is also the initial breeding strata and nursery area for several species of fish, especially commercial species such as salmonides. The benthos can also be a source of food items for several species of fish that are actively feeding on the organisms in the sediment and/or capturing organisms that are emerging from this area. Therefore, exposure to sediment and pore water contaminated with pyrethroids can result in a direct impact to aquatic life through respiration, ingestion, dermal contact, as well as indirect impact through alterations of the food chain.

In developing its pyrethroid ecological risk assessment, the Agency has focused on use patterns that represent maximum as well as typical application rates. The uses include agricultural crops, mosquito abatement, and certain other non-agricultural uses that involve exposure to wastewater. Although modeling scenarios for non-agricultural uses are not as well developed as for agricultural uses, the Agency assumes that exposure from these uses can present risk to aquatic organisms because of the potential for drift or erosion/runoff to adjacent aquatic areas. In order to evaluate the potential for risk from agricultural uses of pyrethroids, the aquatic model, PRZM/EXAMS, was used to generate the estimated environmental concentrations (EECs) in the water column by assessing drift and erosion/runoff exposure potential. The EECs were divided by acute and chronic effects endpoints ((e.g. LC_{50}, EC_{50}, LD_{50}, NOAEC, NOAEL) to develop Risk Quotients (RQs) for a wide variety of crop scenarios.

The results of these analyses show that acute and chronic RQs exceed the levels of concern (LOCs) for freshwater fish and invertebrates and estuarine/marine fish and invertebrates for most of the pyrethroids. For freshwater fish, the RQs exceed the level of concern for acute risk for all pyrethroids evaluated (RQs range from 419 to 1.9). The only exception is prallethrin which is an adulticide use. Also, chronic RQs for freshwater fish exceed the level of concern for all pyrethroids with the exception of cypermethrin (RQs range from 408 to 1.7). For estuarine/marine fish, the risks are not as pronounced as for freshwater fish. However, among all the pyrethroids evaluated, only bifenthrin did not exceed the level of concern for acute risk to estuarine/marine fish (RQs range from 118 to 0.7). The other two exceptions are prallethrin and resmethrin which are adulticide uses. Similarly chronic RQs for estuarine/marine fish are less pronounced than freshwater fish. Only lambda-cyhalothrin, cyfluthrin, and permethrin exceed the level of concern for chronic risk to estuarine/marine fish (RQs range from 15 to 1.2) (Figures 7 and 8).

Similar analyses for invertebrates show that freshwater invertebrates exceed the level of concern for acute risk for all pyrethroids with the exception of prallethrin, an adulticide use (RQs range from 558 to 2.8). Also chronic risk to freshwater invertebrates exceeds the level of concern for all pyrethroids evaluated (RQs range from 13,000 to 33.9). A comparison between the acute and chronic LOCs for freshwater invertebrates shows that chronic RQs could be several orders of magnitude higher than the acute RQs for a number of pyrethroids. For example for bifenthrin, the chronic RQ is 4000 times higher than the acute RQ. However in case of lambda-cyhalothrin, the chronic RQ is only 5 times higher than the acute RQ. For estuarine/marine invertebrates, the RQs exceed the level of concern for acute risk for all pyrethroids with the exception of fenpropathrin and prallethrin (an adulticide use) (RQs range from 5300 to 3.4). Chronic risk to estuarine/marine invertebrates exceeds the level of concern for all pyrethroids evaluated (RQs range from 17,959 to 120). Again the chronic RQs for estuarine/marine invertebrates are higher than acute RQs. For example the chronic RQ for lambda-cyhalothrin is 3.4 times higher than acute RQ. Two exceptions are permethrin and cypermethrin where the chronic RQs are two times lower than the acute RQs (Figures 7 and 8).

In general, both acute and chronic RQs for freshwater and estuarine/marine invertebrates appear to be several orders of magnitude greater than the RQs for freshwater and estuarine/marine fish. This holds true for cyfluthrin, lambda-cyhalothrin, cypermethrin, and permethrin where acute and chronic RQs for both freshwater and esturaine/marine invertebrates are one to three orders of magnitude greater than the freshwater and estuarine/marine fish. One exception is esfenvalerate where acute RQs for freshwater and estuarine/marine fish (RQs are 204 and 118, respectively) are higher than acute RQs for freshwater and estuarine/marine invertebrates (RQs are 41 and 44, respectively). The analysis also shows that the acute and chronic RQs for estuarine/marine invertebrates are

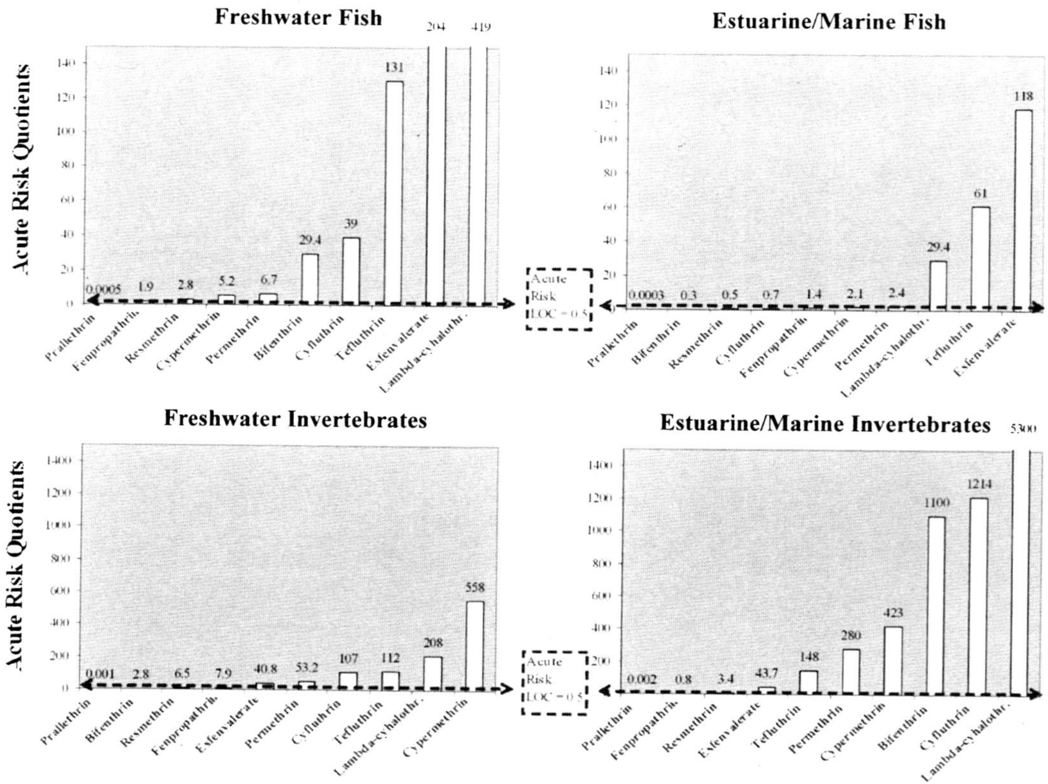

Figure 7. Maximum acute aquatic risk quotients reported in all of the EPA's ecological risk assessments for various synthetic pyrethroid pesticides.

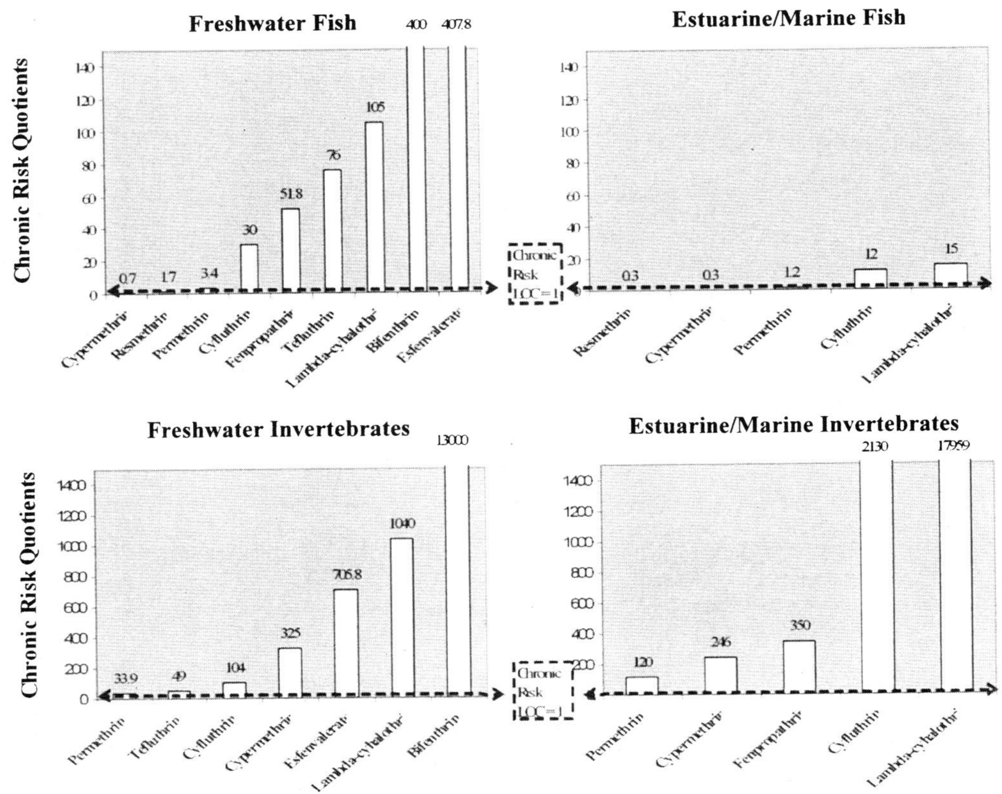

Figure 8. Maximum chronic aquatic risk quotients reported in all of the the EPA's ecological risk assessments for various synthetic pyrethroid pesticides.

one to two orders of magnitude greater than the acute and chronic RQs for freshwater invertebrates. This is true for bifenthrin, cyfluthrin, lambda-cyhalothrin, and permethrin. One exception is cypermethrin, where the acute and chronic RQs for freshwater invertebrates are slightly higher than the acute and chronic RQs for estuarine/marine invertebrates (Figures 7 and 8).

The evaluation of toxic risk to the benthos compartment was approached by assuming equilibrium partitioning of pyrethroids between the sediment and interstitial water. The estimated environmental concentrations (EECs) for pore water were model-generated. Since sediment toxicity data were not available for most of the pyrethroids, the Agency relied on the most sensitive water column effects values (i.e., LC_{50}, EC_{50}, and NOAEC). The RQs for assessing potential risk to sediment reflect pore water EEC values divided by the most sensitive water column toxicity values. The Agency assumed that there is no difference in sensitivity between water column organisms and comparable benthic organisms regarding toxicity to pyrethroids. These assumptions are supported by EPA's Office of Water which used a similar approach and assumption in developing environmental sediment guidelines for the protection of benthic organisms (*31*). Although this approach focused on pore water, the Agency understands that there can be various routes of benthic exposure. Several species feed and are in direct contact with the sediment proper, while others may have more contact with pore water. Some are buried in the sediment, while others spend a lot of time at the surface of the sediment. Based on this approach, both acute and chronic risk for most of the pyrethroids exceeds the level of concern for freshwater and estuarine/marine invertebrates. The Agency plans to re-calculate the RQs after sediment toxicity data becomes available for these pyrethroids.

Cypermethrin is the only pesticide that had available sediment toxicity data. Based on these data, risk estimates for benthic organisms show that cypermethrin poses acute and chronic risks to benthic organisms. This estimate is based on acute and chronic RQs that were calculated using both sediment and pore water EECs. All acute RQs exceeded an LOC of 0.5 (sediment RQs 7-48; pore water RQs 2-12) for all modeled crops. All chronic RQs exceeded an LOC of 1 (sediment RQs 35-244; pore water RQs 9-60) for all modeled crops. The crops, cotton, pecans, and lettuce were modeled using six PRZM/EXAMS scenarios. The potential for cypermethrin to pose acute and chronic risk to sediment dwelling organisms is further supported by results of field studies. Results of these studies show variable immediate and longer-term effects on freshwater invertebrates and benthic organisms exposed to cypermethrin, ranging from no observable effects to catastrophic drift and profound decreases in abundance and diversity.

Buffer Zone Analysis

The Agency has performed a preliminary buffer zone/spray drift analysis in order to evaluate the extent to which the buffer zone imposed by the registrants might mitigate the level of spray drift reaching bodies of water for selected chemicals and scenarios under various conditions. For permethrin, the North Dakota (ND) corn scenario was run with standard input for ground and aerial applications and various buffer zones with three levels of drift to evaluate their effects in reducing potential risk to aquatic systems. The level of spray drift simulated a high-end drift scenario (high boom height, high wind speed, and small droplet size), a low-end drift scenario (low boom height, low wind speed, large droplet size) and a typical drift scenario, to bracket the spectrum of possibilities. The Agency calculated the acute risk quotients (RQs) for the most sensitive species for permethrin for all the buffer zones at all levels of spray drift and found that the RQ values for all the specific drift scenarios yielded acute risk to estuarine/ marine invertebrates. The low-end drift scenario yielded lower levels of drift and consequently lower EECs and RQs. With a large buffer zone (100-300 ft), the RQs were of a similar order of magnitude as those produced by ground applications. A typical application with a buffer zone of 150 ft (as proposed on the label for aerial applications) yielded an acute RQ 69% smaller than the typical application with no buffer zone. Further increasing the buffer zone to 200 ft yielded an acute RQ that is 74% smaller than the typical application with no buffer zone. It was determined that for this particular ND crop scenario for permethrin, the level of drift was an important component of the overall peak EECs.

For cypermethrin, the North Carolina (NC) cotton scenario was selected and a similar analysis was performed. As in the previous case, all scenarios showed that the RQs for cypermethrin for estuarine/marine invertebrates exceeded all the LOCs. However, the low-end drift scenario with no buffer zone yielded an RQ of a similar order of magnitude as the one produced by ground application. One major finding of this set of analyses is that the NC cotton crop scenario did not have a major drift component (9.6% for NC cotton for cypermethrin vs. 70.9% for the ND corn for permethrin). Another scenario with a small drift component is Minnesota (MN) potatoes, while California (CA) lettuce has a high drift component. The other components to the peak EEC are erosion and runoff. Given the physical-chemical characteristics of the synthetic pyrethroids, a minimal contribution from runoff is expected. For the NC cotton scenario, a buffer zone of 150 ft reduces the acute RQ by about 14.1% for the typical drift scenario, compared to the equivalent conditions with no buffer zone. Note that the percent reduction is not as dramatic as that for permethrin.

This buffer zone analysis was conducted for drift buffers only. The current version of the PRZM/EXAMS model cannot evaluate runoff buffers or the reduction in EECs from the mandatory vegetative filter strips included in the labels of most synthetic pyrethroids. The scope of the analysis is limited to two chemicals and two crop scenarios. It constitutes an overview, with these scenarios considered representative of all the areas where the synthetic pyrethroids are applied aerially. With this analysis, it was found that the crop scenarios have certain limitations: one has a large component of drift while the other has a small component of drift. Ideally, various synthetic pyrethroids need to be explored with various crop scenarios, and the data analyzed to fully evaluate the impact of drift buffers.

Down-the-Drain Assessment

In order to address the exposure of permethrin and resmethrin to domestic wastewater and its potential release to Publicly Owned Treatment Works (POTWs), the Agency relied on the Office of Pollution Prevention and Toxics' (OPPT) consumer exposure model, Exposure and Fate Assessment Screening Tool (E-FAST). The "down-the-drain" module of E-FAST is especially designed to address all sources of a chemical that could potentially be disposed to domestic wastewater from a "down-the-drain" application. This model provides screening-level estimates of chemical residues in surface water that may result from household uses and the disposal of consumer products into wastewater. This model does not include degradation or partitioning; however, it includes dilution in the pipes and the receiving waters. For permethrin, the EECs were derived using three real levels of removal in the treatment plant, while for resmethrin, results were obtained assuming a conservative level of removal modeled by EPI Suite and assuming no removal.

For permethrin, the exposure to urban environment (wastewater) results from domestic uses; these include drugs (both prescribed and over-the-counter), pretreated clothing, pet products, and products for the treatment of clothes. The RQs generated through these combined uses show that permethrin residues in surface waters are a potential acute risk to estuarine/ marine invertebrates (RQ 2.33). Chronic risk to both freshwater and estuarine/marine fish appears to be limited; however, this scenario also triggers concern for risk to listed species. Resmethrin also has a number of products for urban use that could potentially reach domestic wastewater. The results obtained by the Agency, using "down-the-drain" module suggest that resmethrin residues released to aquatic systems should not cause acute or chronic risk to fish, or acute risk to aquatic invertebrates. Chronic risk to invertebrates could not be assessed because toxicity data were not available. Assuming that the degree of removal is zero, there are no exceedances to any of the levels of concern.

Mosquito Abatement

Mosquito adulticides are applied as mists (very small droplets) with the intent that the pesticide will linger in the air as a fog and that the eventual deposition onto the target area will be relatively slow. However, the level of drift can be substantial and has the potential for eventual contact with adjacent bodies of water. In order to estimate risk, the Agency calculated the level of drift for such circumstances using the AGDISP model, and the EECs for the bodies of water were calculated using the Florida (FL) or Pennsylvania (PA) turf scenarios. In the case of resmethrin, which has formulation data, the toxicity end points used for these scenarios were based on the formulated product. The Agency also evaluated such variables as boom height, buffer zone, application rate, and droplet size. When the boom height and the application rate are fixed, it is observed that a variable buffer zone (0-150 ft) does not significantly alter the acute exposure to aquatic areas. For permethrin, a boom height of 75 ft and the maximum application rate result in acute and chronic risk to freshwater (RQs are 7.6 and 4.8 respectively) and estuarine/marine invertebrates (RQs are 15.7 and 16.9 respectively). For resmethrin, these application parameters result in acute risk for freshwater and estuarine/ marine invertebrates (RQs are 6.5 and 3.4 respectively). However, use of resmethrin results in acute and chronic risk for only freshwater fish (RQs are 2.8 and 1.7 respectively) under the 75-ft boom height assumption. There are no acute and chronic LOC exceedances for estuarine/marine fish. No toxicity data regarding chronic effects to aquatic invertebrates is available for resmethrin. For prallethrin, a boom height of 50 ft and the maximum application rate result in no exceedance of the LOCs for all aquatic animals.

Risk to Terrestrial Organisms:

Terrestrial exposure tends to localize near the application site, extended somewhat by spray drift downwind. In some cases, long-range aerial transport can extend this range for a volatile compound. However, in the case of pyrethroids, volatility is not an issue of concern. Therefore, to assess risks to terrestrial organisms (e.g., birds and mammals), the Agency focused on the potential for exposure to pyrethroid residues on food items. An evaluation of potential risks was conducted using the T-REX model, which provides estimates of concentrations of chemical residues on different types of food items that may be sources of dietary exposure to avian, mammalian, reptilian, or terrestrial-phase amphibian receptors (*19*). The exposure of most pyrethroids to forage material does not appear to present acute risk to nonendangered birds. However, acute RQs for mammals exceed the level of concern for bifenthrin, cyfluthrin, lambda-cyhalothrin, and esfenvalerate (RQs range from 26 to 1.3). The use of all pyrethroids on agricultural crops presents chronic risk to mammals (RQs range from 165 to 2.2). In contrast, the use of only a few pyrethroids results in chronic risk to birds. These include bifenthrin, lambda-cyhalothrin, esfenvalerate and fenpropathrin (RQs range from 19 to 1.9). Non-agricultural uses of resmethrin and prallethrin (adulticide uses) do not present acute risk to birds or mammals and chronic risk to mammals. The only exceedance of LOC from adulticide uses is from resmethrin resulting in chronic risk to birds (RO 2.3) (Figure 9).

Although, the Agency does not derive RQ values for non-target insects, risks can be assessed quantitatively. Pyrethroid toxicity data show that most are

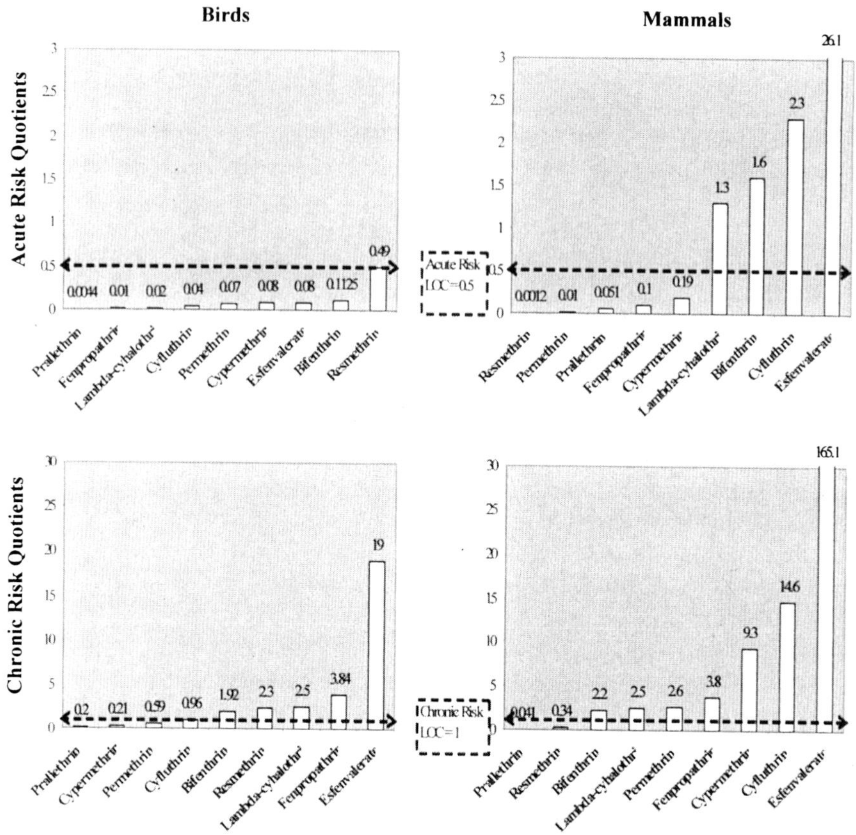

Figure 9. Maximum acute and chronic terrestrial risk quotients reported in all of the the EPA's ecological risk assessments for various synthetic pyrethroid pesticides.

very highly toxic to honeybees on both a contact and oral basis. In addition, certain pyrethroids have also been shown to be highly toxic to earthworms. Based on these results, acute risks to non-target insects and terrestrial invertebrates are anticipated for the uses of pyrethroids.

Conclusions

The Agency has identified acute and chronic risk concerns to aquatic organisms from the use of pyrethroids on a wide variety of agricultural crops. This concern includes risk to federally listed threatened and endangered and non-listed fish, aquatic crustaceans, snails, clams, invertebrates, and amphibians.

Although these compounds bind readily to particulate and organic carbon in the water column, possibly limiting bioavailability in the water column after 24 to 48 hours, the bound residues settle onto the benthos, increasing their concentrations in the sediment. Therefore this media can serve as a repository of pesticide residues that can result in a direct toxic risk concern for benthic and epibenthic aquatic organisms (e.g., early life-stage of many invertebrates and fish, as well as crabs and shrimp). Based on the persistence and toxicity of pyrethroids, water soluble and sediment bound pyrethroids present a potential for direct acute and chronic toxic risk to aquatic life in the water column and in the benthos (e.g., invertebrates). In addition, there is a potential for indirect sublethal risk through food chain alterations that can affect freshwater and estuarine/marine fish and invertebrates. Development of sublethal effects, exhibited in the studies submitted to the Agency includes erratic swimming, partial/complete loss of equilibrium, immobility, lethargy, and darkened pigmentation. While the Agency does not have studies that specifically link these sublethal effects to survival and reproduction, there is a potential that fish and aquatic invertebrates could become more vulnerable to other pressures. Sublethal effects and subsequent indirect effects, if any, are anticipated to occur at concentrations lower than those triggering acute risks to aquatic animals.

Relative to aquatic species, exposure of terrestrial organisms is expected to result in much lower acute and chronic risk to birds. This is not true for mammals, though, where exposure to most pyrethroids results in exceedance for both acute and chronic LOCs. The Agency is also concerned with the potential for risk to terrestrial invertebrates such as non-target insects, including honeybees and other insect pollinators, as well as earthworms. Risk to aquatic and terrestrial plants have not been assessed because of the lack of plant data. Also, the Agency does not expect the pyrethroid mode of action to be a phytotoxic concern.

The Agency plans to re-evaluate all synthetic pyrethroids as part of Registration Review which requires re-evaluation of all registered pesticides in the next fifteen years. The synthetic pyrethroids are scheduled for registration review in the next four to five years. The Agency's revised risk assessment will include integration of additional data submitted by the registrants (especially sediment toxicity data) as well as public literature data that become available after the last risk assessments are completed. This review will specifically focus on assessing the risk to sediment dwelling organism from pyrethroid exposure to contaminated sediment and will also include endangered species risk assessments.

Acknowledgements

The authors would like to convey their special thanks to Dr. Tom Bailey, Mr. George LaRocca, Mr. Michael Knuth, Ms. Arthur-Jean Williams and Ms.

Karen McCormack for their peer review support and valuable comments for improving this document.

References

1. Matsumura, F. *Toxicology of Insecticides*, 2nd ed.; Plenum Press: New York, NY, 1985; pp 147.
2. Federal Insecticide, Fungicide, and Rodenticide Act (FIFRA), Scientific Advisory Panel (SAP) Meeting, February, 1999. http://www.epa.gov/oscpmont/sap/meetings/1999/february/final.pdf
3. U.S. Environmental Protection Agency. 2004. Overview of the Ecological Risk Assessment Process in the Office of Pesticide Programs. Office of Prevention, Pesticides, and Toxic Substances. Office of Pesticide Programs. Washington, D.C. January 23, 2004. http://www.epa.gov/espp/consultation/ecorisk-overview.pdf
4. The National Center for Food and Agricultural Policy (NCFAP), National Pesticide Use Database, 1997 national summary of pesticide use in crop production by active ingredient and crop http://www.ncfap.org/database/national/default.asp
5. U.S. Department of Agriculture National Agricultural Statistics Service (USDA-NASS), Agricultural Chemical Use Database (by state) http://www.pestmanagement.info/nass/act_dsp_statcs2_state.cfmm
6. The Pesticide Action Network North America (PANNA), PAN Pesticides Database: Data were obtained based on chemical/use search of the California Pesticide Use Report (PUR). http://www.pesticideinfo.org/Search_Chemicals.jsp#ChemSearch http://www.pesticideinfo.org/Search_Use.jsp#SearchCAUse
7. U.S. Department of Agriculture National Agricultural Statistics Service (USDA-NASS), 2004-07. Agricultural chemical usage: vegetable crops (2005), fruit crops (2006), and field crops (2007).
8. Bekarian, N. *et. al.* 2006. Use of point-of-sale data to track usage patterns of residential Environmental Health; Global Access Science Source 5:15.
9. U.S. Department of Agriculture National Agricultural Statistics Service (USDA-NASS), 2000-06. Agricultural chemical usage: Sheep, Cattle and cattle facilities (2000), swine (2005), and Dairy cattle (2006).
10. Food and Agriculture Organization of the United Nations. 2000. FAO PESTICIDE DISPOSAL SERIES 8. Assessing Soil Contamination: A Reference Manual. Appendix 2. Parameters of pesticides that influence processes in the soil. Editorial Group, FAO Information Division: Rome. http://www.fao.org/DOCREP/003/X2570E/X2570E00.htm
11. U.S. Geological Survey National Water Quality Assessment Data Warehouse http://infotrek.er.usgs.gov/traverse/f?p=NAWQA:HOME:2044713762346848

12. EPA-STORET Database
http://www.epa.gov/storet/dbtop.html
13. CA-DPR Database
http://www.cdpr.ca.gov/docs/sw/surfcont.htm
14. http://www.epa.gov/oppefed1/models/water/
15. Teske, M.E., Thistle, H.W., and G.C. Ice. 2003. Technical Advances In Modeling Aerially Applied Sprays. Transactions of the ASAE. 46: 985-996.
16. AgDRIFT®: A computer program provided by the Agency's OPP, as a product of the Cooperative Research and Development Agreement (CRADA), between the EPA's Office of Research and Development, and the Spray Drift Task Force (SDTF, a coalition of pesticide registrants formed to develop a comprehensive database of off-target drift information in support of pesticide registration requirements).
17. U.S. Environmental Protection Agency. 1999. Exposure and Fate Assessment Screening Tool (E-FAST) Beta Version Documentation Manual. Prepared for USEPA Office of Pollution Prevention and Toxics Exposure Assessment Branch; Prepared by Versar, Inc. Springfield, VA. This manual can be found in the following website:
http://www.epa.gov/opptintr/exposure/docs/efast.htm
18. EPIWIN: A part of EPISUITE A computer program provided by the Agency's and developed by the Agency's OPPT and Syracuse Research Corporation.
http://www.epa.gov/opptintr/exposure/pubs/episuite.htm
19. U.S. Environmental Protection Agency, Office of Pesticide Programs, Environmental Fate and Effects Division. 2006. Terrestrial Residue Exposure Model (T-REX).
http://www.epa.gov/oppefed1/models/terrestrial/
20. Willis, G.H., and L.L. McDowell. 1987. Pesticide Persistence on Foliage in Reviews of Environmental Contamination and Toxicology. 100:23-73.
21. U.S. Environmental Protection Agency. 2006. ECOTOX User Guide: ECOTOXicology Database System. Version 4.0.
Available: http:/www.epa.gov/ecotox/
22. U.S. Environmental Protection Agency. 2007. Electronic Code of Federal Regulations (e-*CFR*): Title 40 Protection of Environment, Part §158.108 Relationship of Pesticide Assessment Guidelines to data requirements. Updated July 16, 2007.
http://ecfr.gpoaccess.gov/cgi/t/text/text-idx?sid=1c63592f490baa0976d9c45c34094110&c=ecfr&tpl=/ecfrbrowse/Title40/40tab_02.tpl
23. U.S. Environmental Protection Agency. 2007. Electronic Code of Federal Regulations (e-*CFR*): Title 40 Protection of Environment, Part §158.70 Acceptable protocols. Updated July 16, 2007.
http://ecfr.gpoaccess.gov/cgi/t/text/textidx?sid=1c63592f490baa0976d9c45c34094110&c=ecfr&tpl=/ecfrbrowse/Title40/40tab_02.tpl

24. U.S. Environmental Protection Agency. 2007. Electronic Code of Federal Regulations (e-*CFR*): Title 40 Protection of Environment, Part §158.80 Acceptability of data. Updated July 16, 2007. http://ecfr.gpoaccess.gov/cgi/t/text/textidx?sid=1c63592f490baa0976d9c45c34094110&c=ecfr&tpl=/ecfrbrowse/Title40/40tab_02.tpl
25. U.S. Environmental Protection Agency, Office of Prevention, Pesticides and Toxic Substances (OPPTS). OPPTS Harmonized Test Guidelines, Series 850 Ecological Effects Test Guidelines. http://www.epa.gov/opptsfrs/home/guidelin.htm
26. U.S. Environmental Protection Agency. 2007. Electronic Code of Federal Regulations (e-*CFR*): Title 40 Protection of Environment, Part §158.490 Wildlife and aquatic organisms data requirements. Updated July 16, 2007. http://ecfr.gpoaccess.gov/cgi/t/text/textidx?sid=1c63592f490baa0976d9c45c34094110&c=ecfr&tpl=/ecfrbrowse/Title40/40tab_02.tpl
27. U.S. Environmental Protection Agency. 2007. Electronic Code of Federal Regulations (e-*CFR*): Title 40 Protection of Environment, Part §158.540 Plant protection data requirements. Updated July 16, 2007. http://ecfr.gpoaccess.gov/cgi/t/text/textidx?sid=1c63592f490baa0976d9c45c34094110&c=ecfr&tpl=/ecfrbrowse/Title40/40tab_02.tpl
28. U.S. Environmental Protection Agency. 2007. Electronic Code of Federal Regulations (e-*CFR*): Title 40 Protection of Environment, Part §158.590 Nontarget insect data requirements. Updated July 16, 2007. http://ecfr.gpoaccess.gov/cgi/t/text/textidx?sid=1c63592f490baa0976d9c45c34094110&c=ecfr&tpl=/ecfrbrowse/Title40/40tab_02.tpl
29. Brooks, H.L., Gates D.E. 1972. Insecticides. Cooperative Extension Service, Kansas State University, Manhatten, Kansas. 36 pgs.
30. Di Toro DM, Zarba CS, Hansen DJ, Berry WJ, Swartz RC, Cowan CE, Pavlou SP, Allen HE Thomas NA, Paquin PR. 1991. Technical basis for establishing sediment quality criteria for nonionic organic chemicals using equilibrium partitioning. Environ Toxicol Chem 10:1541-1583.
31. U.S. Environmental Protection Agency. 2003. Technical basis for the derivation of equilibrium partitioning sediment benchmarks (ESBs) for the protection of benthic organisms: Nonionic organics. EPA-600-R-02-014. Office of Research and Development, Washington, DC. (draft)
32. U.S. Environmental Protection Agency. 2007. Electronic Code of Federal Regulations (e-*CFR*): Title 40 Protection of Environment, Part §158.75 Requirements for additional data. Updated July 16, 2007. http://ecfr.gpoaccess.gov/cgi/t/text/textidx?sid=1c63592f490baa0976d9c45c34094110&c=ecfr&tpl=/ecfrbrowse/Title40/40tab_02.tpl
33. U.S. Environmental Protection Agency. 2007. Electronic Code of Federal Regulations (e-*CFR*): Title 40 Protection of Environment, Part §158.202 Purposes of the registration data requirements. Updated July 16, 2007.

http://ecfr.gpoaccess.gov/cgi/t/text/textidx?sid=1c63592f490baa0976d9c45c34094110&c=ecfr&tpl=/ecfrbrowse/Title40/40tab_02.tpl

34. U. S. Environmental Protection Agency, Office of Pesticide Programs. April 2006. Reregistration Eligibility Decision (RED) for Permethrin. http://www.epa.gov/REDs/permethrin_red.pdf
35. U. S. Environmental Protection Agency, Office of Pesticide Programs. June 14, 2006. Reregistration Eligibility Decision (RED) for Cypermethrin. http://www.epa.gov/REDs/cypermethrin_red.pdf
36. U. S. Environmental Protection Agency, Office of Pesticide Programs. June 2006. Reregistration Eligibility Decision (RED) for Resmethrin. http://www.epa.gov/REDs/resmethrin_red.pdf

Chapter 14

Effects of Pyrethroid Insecticides on Aquatic Organisms

I. Werner[1] and K. Moran[2]

[1]Aquatic Toxicology Program, School of Veterinary Medicine, Department of Anatomy, Physiology and Cell Biology, University of California at Davis, Davis, CA 95616 (email: iwerner@ucdavis.edu)
[2]TDC Environmental, LLC, 4020 Bayview Avenue, San Mateo, CA 94403 (email: kmoran@tdcenvironmental.com)

Introduction

Most aquatic invertebrates and fish are highly susceptible to synthetic pyrethroid insecticides (*1, 2*). All pyrethroids are potent neurotoxicants that interfere with nerve cell function by interacting with voltage-dependent sodium channels as well as other ion channels, resulting in repetitive firing of neurons and eventually causing paralysis (*3, 4*). Exposed organisms may exhibit symptoms of hyperexcitation, tremors, convulsions, followed by lethargy and paralysis. Pyrethroids occur mostly as mixtures of stereoisomeric forms, and the toxicity of individual isomers can vary (*5*). There are two groups of pyrethroids with distinctive poisoning symptoms, type I and type II. Type II pyrethroids are distinguished from type I pyrethroids by an alpha-cyano group in their structure. While type I pyrethroids (e.g. permethrin, cismethrin) exert their neurotoxicity primarily through interference with sodium channel function in the central nervous system, type II pyrethroids (e.g. deltamethrin, esfenvalerate, cypermethrin, bifenthrin) can affect additional ion-channel targets such as chloride and calcium channels (*6*). Pyrethroids also modulate the release of acetylcholinesterase in the brain's hippocampus region (*7*), and can inhibit ATPases (*8*). In addition, these compounds can disrupt hormone-related functions (*9, 10*). In mammals, pyrethroids decrease progesterone and estradiol production (*11*), eliciting estrogenic effects in females and anti-androgenic effects in males (*12, 13*). Breakdown products of pyrethroids have been shown to be more potent endocrine disruptors than their parent compounds (*13, 14*).

Furthermore, pyrethroids have been shown to inhibit cell cycle progress (*15*), cause cell stress (*16*), and have immunosuppressive effects (*17, 18*). Additional long-term effects may be caused by damage to respiratory surfaces, and interference with renal ion regulation (*3*).

Acute Toxicity

Acute toxicity is defined as a significant reduction in survival of the exposed organisms within a relatively short time (minutes to days), and is expressed as the species-specific median lethal concentration (LC50). For pyrethroid insecticides, most known 96-h LC50s for fish, aquatic insects and crustaceans are well below 1 µg/L (Table I), whereas molluscs are relatively insensitive to these chemicals and can bioaccumulate them (*2*). Crustaceans such as amphipods are among the most sensitive taxa. Little is known about oligochaetes, but available data indicate that this group is much less sensitive than crustaceans or insects (*19, 20*). In a hazard assessment performed by the California Department of Fish and Game (*21*), water quality criteria for cypermethrin and permethrin were derived according to US EPA guidelines (*22*). The proposed final acute values and criterion maximum concentrations were 0.003 and 0.002 µg/L, respectively, for cypermethrin, and 0.059/0.002 µg/L (freshwater/saltwater) and 0.03/0.001 µg/L (freshwater/saltwater), respectively, for permethrin.

Sublethal Toxicity

Sublethal toxic effects can occur at exposure levels far below the concentrations that cause lethality (Table II), and can have severe consequences for the fitness, reproductive success and survival of aquatic organisms, ultimately leading to population-level effects (*23*). Sublethal biological responses include altered behavior, reduced growth, immune system effects, reproductive/endocrine effects, histopathological effects as well as biochemical responses. However, direct links of these responses to higher-level effects are often difficult to establish. Nevertheless, sublethal toxic effects can have far-reaching consequences in the aquatic environment, where organisms are often simultaneously exposed to many different stressors (*24*). Effects of sublethal environmental stress can be evaluated at several levels of biological organization, from molecular processes to growth and reproduction, that may impact overall population size and community interactions. Some physiological endpoints commonly tested include hematological and immunological parameters (e.g., hematocrit, plasma cortisol concentrations), assessments of liver and gill structure and function (e.g., liver somatic index, mixed function

Table I. Summary of aquatic toxicity data for selected pyrethroids (lowest values).

Species	Lambda-Cyhalothrin		Bifenthrin		Cyfluthrin		Cypermethrin		Deltamethrin		Esfenvalerate		Permethrin	
	Test	µg/L	Test	µg/L	Test	µg/L	Test	µg/L	Test	µg/L	Test	µg/L	Test	µg/L
Invertebrates														
Waterflea, *Ceriodaphnia dubia*			48-h LC50	0.07	48-h LC50	0.14					96-h LC50[4]	0.3	48-h LC50	0.55
Waterflea, *Daphnia magna*	48-h EC50[7] 21-d NOEC[7]	0.36 0.002	48-h LC50 48-h EC50	0.32 1.6	48-h LC50 48-h EC50	0.17 0.025	24-h LC50 48-h LC50[3] 48-h EC50	0.53 0.13 1.00	24-h LC50 48-h LC50 96-h LC50 96-h EC50	0.11 0.037 0.01 0.003	48-h LC50[1] 48-h LC50 48-h EC50	0.24 0.27 0.15	48-h LC50[1] 72-h LC50 96-h LC50 96-h EC50	0.075 6.8 0.3 0.039
Waterflea, *Daphnia pulex*													3-h LC50 48-h LC50 72-h LC50	9,200 2.75 0.08
Copepod, *Cyclops* sp.	48-h EC50[7]	0.3												
Mayfly, *Hexagenia bilineata*													96-h LC50[3]	0.1
Mayfly, *Procloeon* sp.			48-h LC50[9]	0.084									48-h LC50[9]	0.090
Mayfly, *Cloeon dipterum*	48-h EC50[7]	0.038					72-h EC50[3] 96-h LC50[3]	0.006 0.023 0.03						
Isopod, *Asellus aquaticus*	48-h EC50[7]	0.026					72-h LC50[3]	0.008						
Midge, *Chironimus dilutus*			96-h LC50[9]	26.15									96-h LC50[9]	10.45
Midge, *Chironimus riparius*	48-h EC50[7]	2.4					48-h LC50[3]	0.007						

Midge, *Chironimus plumosus*													48-h EC50[3]	0.56
Grass shrimp, *Palaemonetes pugio*							96-h LC50[3]	0.016						
Oligochaeta							48-h LC50[11]	>100						
Hyalella azteca	48-h EC50[7]	0.0023	96-h LC50[9]	0.009			48-h LC50[3]	0.005			42-D LOEC 96-h LC50[6]	0.05 0.008	96-h LC50[9]	0.021
Gammarus pulex	48-h EC50[7] 0.5-h LC50[8]	0.014 5.69												
Gammarus daiberi											96-h LC50[6]	0.033		
Gammarus pseudolimnaeus													96-h LC50[3]	0.17
Crayfish, *Orconectes immunis*													96-h LC50[10]	0.08
Mysid shrimp (B) *Americamysis bahia*			96-h LC50	0.004	96-h LC50	0.00242	96-h LC50	0.005	96-h LC50	0.0017	96-h LC50[1]	0.038	96-h LC50[3]	0.02
Pink shrimp (S, juv.), *Penaeus duorarum*							96-h LC50[3]	0.036					96-h LC50[3]	0.22
Stone crab (S), *Menippe mercenaria*													96-h EC50[3]	0.018
Fiddler crab (S), *Uca pugilator*													96-h LC50[3]	2.39
Penaeus sp. (S)							96-h LC50	0.036					96-h LC50	0.17
Oyster, *Crassostrea virginica* (S, B)			48-h EC50 (embryo)	285	96-h EC50	2.69	96-h EC50	370	96-h EC50	8.2			48-h EC50 96-h EC50	1000 40.7
Oyster, *Crassostrea gigas* (S, B)	48-h EC50[2] (larvae)	590.00					48-h LC50	2,270					48-h EC50	1,050

Sources: All unmarked values from *(89)*; [1] *(104)*; [2] *(105)*; [3] *(21)*; [4] *(94)*; [5] *(27)*; [6] Werner I., unpublished data; [7] *(106)*; [8] *(41)*; [9] *(107)*; [10] *(108)*; [11] *(19)*
(S) saltwater species

(B) brackish water species

Continued on next page.

Table I. (Continued). Summary of aquatic toxicity data for selected pyrethroids (lowest values).

Species	Lambda-Cyhalothrin		Bifenthrin		Cyfluthrin		Cypermethrin		Deltamethrin		Esfenvalerate		Permethrin	
	Test	μg/L	Test	μg/L	Test	μg/L	Test	μg/L	Test	μg/L	Test	μg/L	Test	μg/L
Vertebrates														
Fathead minnow *Pimephales promelas*	96-h LC50[7]	0.70	96-h LC50[1]	0.26	96-h LC50[1]	2.49					24-h LC50 48-h LC50 96-h LC50	0.24 0.24 0.22	24-h LC50 96-h LC50[1]	5.4 2
Rainbow trout *Oncorhynchus mykiss*	96-h LC50[2] 96-h LC50[2]	0.54 0.24	96-h LC50	0.15	48-h LC50 96-h LC50	0.57 0.3	12-h LC50 24-h LC50 48-h LC50 96-h LC50	2.5 5 5 0.39	24-h LC50 48-h LC50 96-h LC50	0.7 0.5 0.25	96-h LC50[1] 96-h LC50	0.26 0.07	24-h LC50 48-h LC50 96-h LC50	4.3 6 0.62
Carp, *Cyprinus carpio*	96-h LC50[7]	0.50					96-h LC50[3]	0.9						
Mosquitofish, *Gambusia affinis*	24-h LC50[2] 24-h LC50[2]	0.18 0.08												
Atlantic salmon *Salmo salar*													96-h LC50[3]	17
Chinook salmon, *Onchorynchus tshawytscha*											96-h LC50[5]	0.1-1.0		
Coho salmon, *O. kisutch*													96-h LC50[3]	3.2
Brook trout, *Salvenlinus fontinalis*													24-h LC50 96-h LC50[3]	4 3.2
Sacramento splittail, *Pogonichthys macrolepidotus*											96-h LC50[4]	0.50		

Sheepshead minnow (S) *Cyprinodon variegatus*	28-d NOEC[7]	0.25	96-h LC50[3]	17.8	96-h LC50	4.05	96-h LC50	0.73	96-h LC50	0.36	96-h LC50[1]	430	96-h LC50	7.8
Atlantic silverside, (S) *Menidia menidia*													96-h LC50[3]	2.2
Inland silverside, (S) *Menidia beryllina*													96-h LC50	27.5
Bluegill (S), *Lepomis macrochirus*	96-h LC50 96-h LC50	0.42 0.21	144-h LC50[3]	0.35	96-h LC50	0.87	96-h LC50[3]	1.78	96-h LC50	0.36	96-h LC50[1]	0.26	24-h LC50 96-h LC50[3]	6.6 2.5
Plants														
Selenastrum capricornutum	96-h EC50[7]	>1000												
Skeletonema costatum														

Source: All unmarked values from *(89)*; [1] *(104)*; [2] *(105)*; [3] *(21)*; [4] *(94)*; [5] *(27)*; [6] Werner I., unpublished data; [7] *(106)*
(S) saltwater species
(B) brackish water species

Table II. Reported sublethal effects of several several pyrethroids on aquatic species.

Pyrethroid	Species	Life-Stage/Test Duration	Effect	Effect Concentration (μg/L)	Source
Lambda-Cyhalothrin	*Gammarus pulex*	Adult/ 30 min	EC10 (Pair formation) EC50 (Pair formation)	0.04 0.20	(*41*) "
Cypermethrin	*Daphnia magna*	Adult/6 h	LOEC (Decrease in feeding efficiency and swimming ability)	0.1	(*40*)
	Mysid shrimp, *Americamysis bahia*	28 d	LOEC (fecundity) NOEC (fecundity) LOEC (growth)	0.0028 0.0015 0.00078	(*39*) " "
	Fathead minnow, *Pimephales promelas*	Larvae/30 d	LOEC (growth) NOEC (growth)	0.33 0.15	(*21*) "
	Rainbow trout, *O. mykiss*	-	LOEC (behavior)	0. 68	(*39*)
	Bluegill sunfish, *Lepomis macrochirus*	-	LOEC (behavior)	<2.2	(*39*)
	Atlantic salmon, *Salmo salar*	Gamets/5 d Adult/5 d	LOEC (fertilization success) Impaired olfactory function	0.1 <0.004	(*34*) "
	Korean rockfish, *Sebastes schlegeli*	52 g/8 wk	Changes in blood parameters	0.041	(*33*)
Esfenvalerate	*Daphnia carinata*	Adult	Reduced fecundity	0.05	(*47*)
	Midge, *Chironomus tentans*	Larvae/14-16 d	EC10 Mobility EC50 Mobility	0.078 0.21	(*103*) "
	Fathead minnow, *Pimephales promelas*	Larvae/96 h	Reduction in hepatic glycogen NOEC Swimming performance	0.20 0.13	(*36*) "
	Fathead minnow, *Pimephales promelas*	Larvae/4 h	Swimming performance	0.7	(*43*)
	Bluegill, *Lepomis macrochirus*	Juvenile/90 d Young-of-the-Year Adult Embryos/ Larvae	LOEC (behavior) NOEC (behavior) Growth Delayed spawning Reduced larval survival	0.025 0.010 0.08 1.0 1.0	(*109*) " (*45*) " "
	Medaka, *Oryzias latipes*	Adult/7 d	Stress protein (hsp) increase	21 μg/g (diet)	(*30*)
	Chinook salmon, *Oncorhynchus tshawytscha*	Juvenile/96 h	Alteration of immune response Stress protein (hsp) increase	0.08 0.01	(*18*) (*28*)
Permethrin	Daphnid	Adult	LOEC (fecundity)	<0.01	(*46*)
	Sheepshead minnow, *Cyprinodon variegatus*	28 d	LOEC (growth) NOEC (growth)	22 10	(*110*) "

oxidases enzyme induction), energetics (e.g., RNA/DNA ratios, swimming performance, feeding and growth rates), and behavioral and nervous system function (e.g., temperature tolerance, swimming performance, altered predator-prey interactions).

Biochemical and Physiological Effects

The use of biochemical and physiological biomarkers is widespread in aquatic toxicology, partly because their induction is more sensitive to stress than traditional indices such as growth inhibition (*25, 26*). Some of these sublethal stress responses divert an organism's energy away from normal metabolic functions and can result in "higher-level" effects such as growth inhibition or reduced reproductive success.

Induction of heat-shock proteins (hsp) occurred in liver of juvenile Chinook salmon (*Oncorhynchus tshawytscha*) following exposure to sublethal concentrations of esfenvalerate (*27, 28, 29*). Werner et al. (*30*) measured elevated levels of hsp in medaka (*Oryzias latipes*) after feeding on a diet containing 21 µg/g esfenvalerate. Hsp indicate the occurrence of significant protein damage in cells and tissues, and increased expression of these proteins has been linked to abnormal development in larval sturgeon (*31*), as well as an increase in energy expenditure in juvenile steelhead trout (*32*).

An eight-week exposure of Korean rockfish (*Sebastes schlegeli*; mean fish wt: 52 g) to cypermethrin had significant effects on a number of blood parameters (*33*). Red blood cell count, hemoglobin and hematocrit were significantly reduced after exposure to 0.041 µg/L cypermethrin. The activity of several enzymes and serum osmolality were also altered. Reduced levels of serum total protein, albumin, cholesterol, lysozyme activity and significantly higher serum concentrations of glucose, bilirubin and malondialdehyde were attributed to an increased demand for energy by fish under stress. Moore and Waring (*34*) demonstrated that the pyrethroid cypermethrin impaired olfactory function in Atlantic salmon after a 5-day exposure to <0.004 µg/L. Fish (*Heteropneustes fossilis*) chronically exposed to 1.44 µg/L cypermethrin exhibited decreased blood plasma calcium levels, and degeneration of branchial cells (*35*).

Tissue and Organ Damage

Histopathological lesions in the liver were observed in the Sacramento splittail (*Pogonichthys macrolepidotus*, *29*) shortly (1 wk) after 96-h exposure to sublethal concentrations of organophosphate and pyrethroid insecticides. Fish recovered from these lesions, but showed high (delayed) mortality rates, grew slower and showed signs of cellular stress even after a 3 month recovery period.

A significant reduction in liver glycogen levels of fathead minnow (*Pimephales promelas, 36*) was observed after 96-h exposure to 0.20 µg/L esfenvalerate. Likewise, Haya and Waiwood (*37*) found a depletion of glycogen stores in liver and muscle for starving juvenile Atlantic salmon exposed to fenvalerate. The loss of glycogen (a secondary stress response) should be regarded as a nonspecific response signifying stress and has been linked to changes in cortisol during exposure to various stressors (*38*).

Swimming Performance and Behavior

Abnormal behaviors produced by contaminants include changes in preference or avoidance, activity level, feeding, performance, learning, predation, competition, reproduction and species-specific social interaction such as aggression. Such changes can have significant consequences for fitness, survival and reproductive success of an individual. For example, many neurotoxic compounds cause abnormal swimming behavior or compromise swimming ability in fish and other aquatic animals (*39, 40, 41*). In the field, such changes can directly translate into increased vulnerability to predation or decreased food intake.

Because pyrethroids are potent neurotoxins, behavioral endpoints may be among the most sensitive and ecologically relevant measurable parameters to assess their sublethal toxicity. Little and Finger (42) describe swimming behavior of fish exposed to a variety of contaminants ranging from pesticides (e.g., DDT, carbaryl, methyl parathion) to metals (e.g., zinc, copper, cadmium), and found that changes in swimming behavior were detected at exposures as low as 0.7 to 5% of the chemical's LC50 values.

Sublethal effects of acute cypermethrin exposure on swimming behavior were assessed in studies in rainbow trout and bluegill sunfish (39) The sublethal signs of toxicity included rapid and erratic swimming, partial/complete loss of equilibrium, jaw spasms, gulping respiration, lethargy, and darkened pigmentation. For the two studies, the acute NOEC (no observed effect concentration) values for swimming behavior were only slightly lower than the LC50 value; in rainbow trout, the acute NOEC and LC50 values were 0. 68 µg/L and 0.8 µg/L, respectively, and in bluegill sunfish, the acute NOEC and LC50 values were <2.2 µg/L and 2.2 µg/L, respectively. This indicates that toxic effects occur and progress rapidly once a certain pyrethroid concentration is exceeded. However, mortality may be delayed when exposure times are very short, on the order of several hours. For example, Floyd et al. (43) report significant effects on swimming ability of fathead minor larvae after 4-h exposures to 0.7 µg/L esfenvalerate, while no mortality occurred during this time at exposure concentrations up to 20 µg/L esfenvalerate. When delayed survival was measured after a 4-h exposure plus a 20-h recovery period in control water, the LC50 was 2.04 µg/L esfenvalerate.

In waterflea, the sublethal signs of pyrethroid toxicity include immobilization and decreased movement in response to stimulation. Acute NOEC values for the sublethal effects of cypermethrin range from 0.085 µg/L to 0.14 µg/L. Christensen et al. (40) showed that environmentally relevant, brief (6 h) exposures to 0.1 µg/L cypermethrin decreased feeding efficiency and swimming ability of Daphnia magna. Animals recovered after 3 days in clean water. A 30-min pulse exposure of Gammarus pulex to lambda-cyhalothrin (41) significantly impaired pair formation (pre-copula), with EC10 (30 min) and EC50 (30 min) values of 0.04 and 0.2 µg/L. Significant mortality was observed at 0.3 µg/L, with an LC50 (30 min) of 5.69 µg/L. Sublethal effects (lethargy, erratic swimming behavior, loss of equilibrium, and surfacing) of cypermethrin in estuarine/marine invertebrates were also reported in two studies of mysid shrimp (39): Acute NOEC values for sublethal effects range from 1.7 to 2.3 ng/L and are approximately 2 to 3-fold lower than the corresponding LC50 values of 5.5 and 5.9 ng/L, respectively.

Reproductive Toxicity and Endocrine Disruption

Pyrethroids were shown to have steroid receptor-binding activity *in vitro* (*14*). Their effects on the endocrine system are not uniform. While Fenpropathrin and permethrin act as weak estrogen agonists, allethrin and cypermethrin have antiestrogenic as well as antiandrogenic activity. Cyfluthrin and fenvalerate showed very weak antiestrogenic activity, but several metabolites and products of environmental degradation of permethrin and cypermethrin had up to more than 100-fold greater potencies than the parent compound (*13, 14, 44*). In mammals, pyrethroids affect sperm concentration, motility and morphology (*10*).

In fish, Moore and Waring (35) demonstrated that the pyrethroid cypermethrin reduced the fertilization success in Atlantic salmon after a 5-day exposure to concentrations of 0.1 µg/L. In a study on bluegill sunfish, Tanner and Knuth (45) found delayed spawning and reduced larval survival after two applications of 1 µg/L esfenvalerate.

Day (46) showed that concentrations of <0.01 µg/L permethrin and other pyrethroids reduced reproduction and rates of filtration of food by daphnids. A concentration of 0.05 µg/L esfenvalerate also led to a significant decrease in reproductive success (number of neonates) of Daphnia carinata (47). Reynaldi and Liess (48) demonstrated that fenvalerate delayed the age at first reproduction in Daphnia magna, and reduced fecundity at a LOEC (lowest observed effect concentration) of 0.1 µg/L (complete mortality occurred at 1 µg/L). Population growth rate was inhibited at 0.6 µg/L (24 h), and recovery occurred after 21 d. Results of chronic toxicity studies in mysid shrimp show that exposure to cypermethrin had adverse effects on reproductive parameters: For a decrease in the number of young, a chronic NOEC value of 1.5 ng/L was reported in two studies (39).

Growth

Growth integrates a suite of biochemical and physiological effects into one endpoint that can often be associated with individual fitness. Results of chronic toxicity studies in mysid shrimp show that exposure to technical grade cypermethrin had adverse effects on growth parameters. For decreased growth and length, the chronic NOEC value reported was 0.78 ng/L. In a mesocosm study on bluegill sunfish, Tanner and Knuth (*45*) found that young-of-the-year growth was reduced by 57, 62 and 86% after two applications of 0.08, 0.2 and 1 μg/L esfenvalerate, respectively. Floyd et al. (*43*) showed that feeding and growth was significantly reduced in fathead minnow larvae exposed for 4 h to 0.7 μg/L esfenvalerate.

Immune System Effects

The immune response of fish and invertebrates plays a key role in the control of aquatic diseases, fitness and reproductive success. Pesticides are among those contaminants identified to cause immunosuppressive effects on fish (*49, 50*), but few studies have established the correlation between pyrethroids and disease resistance. Zelikoff et al. (*51*) found reduced disease resistance in fish exposed to the pyrethroid permethrin. Clifford et al. (*18*) demonstrated that the susceptibility of juvenile Chinook salmon to Infectious Hematopoietic Necrosis Virus (IHNV) was dramatically increased in fish exposed to 0.08 μg/L esfenvalerate. Eder et al. (*27*) found that exposure to 0.08 ppb esfenvalerate for 96 h altered the transcription of immune-system messenger molecules (cytokines) in juvenile Chinook salmon (*Oncorhynchus tshawytscha*). Cytokines regulate the innate and adaptive immune systems and are produced in response to infection or an inflammatory insult. Activation of interleukin-6, a key inflammatory cytokine, by cyfluthrin was also reported in human astrocytes (*52*).

Population Level Effects

Pyrethroids are generally of very low water solubility and high lipophilicity, and therefore are rapidly adsorbed to particulate material and other surfaces. Adsorption occurs on the order of hours in sediment-laden solutions under ideal laboratory mixing conditions (*53*) or in systems like farm ponds that contain relatively large amounts of organic matter (*54*); however, in typical streams, where less ideal mixing conditions exist, adsorption may occur over a period of days rather than hours (*55*). In the adsorbed state their bioavailability to aquatic organisms is reduced (*56, 57*). Therefore, for water column exposures field experiments of short duration or pulse exposure experiments are believed to be more environmentally realistic than LC50 data. Below we summarize the results of such field and pulse studies.

Field Studies

Studies on the effects of cypermethrin on fish in streams and ponds, where pyrethroid application rates ranged from 0.011 lb a.i./A (*58*) to 0.0623 lb a.i./A (*59, 60*), found no acute toxicity (expressed as mortality) on fish populations, but sublethal effects including loss of equilibrium, lethargy, and muscle tetany were reported following a single application of 0.011 lb a.i./A. Sublethal pathological changes in fish were observed for 26 days following the application and were attributed to direct exposure to cypermethrin as well as to dietary exposure from ingestion of dead and dying invertebrates.

In field studies assessing the effects of cypermethrin on aquatic invertebrates and benthic populations, results show that exposure to cypermethrin at application rates to water surfaces ranging from 0.00025 lb a.i./A (61) to 0.125 lb a.i./A (39) caused significant decreases is abundance and diversity of aquatic invertebrate populations. Effects include catastrophic drift within 0-90 minutes after application of cypermethrin (59, 62, 63), and decreased abundance and diversity of macroinvertebrates over several weeks to several months (61, 62, 64, 65, 66). Plecoptera and ephemeroptera comprised 89-92% of the invertebrate drift immediately after spraying (58). Soon after treatment, concentrations of cypermethrin associated with the surface layer of the water column and emergent vegetation were much greater than those associated with deeper water and benthic sediment. Downward dispersion of cypermethrin was relatively limited. Only 8-16% of cypermethrin applied to the water surface was subsequently found in the water column (59).

Field studies on the effects of esfenvalerate also demonstrated detrimental effects on aquatic systems (2 ha pond) by reduction or elimination of many crustaceans, chironomids, juvenile bluegills and larval cyprinids at exposure levels of 1 μg/L (45, 67). Esfenvalerate exposures of 1 and 5 μg/L resulted in drastic reductions or elimination of most crustaceans, chironomids, juvenile bluegills (Lepomis macrochirus), and larval cyprinids. Abundance of some copepod and insect genera declined at esfenvalerate concentrations of 0.08 to 0.2 μg/L, and these effects were apparent up to 53 d. Some invertebrate communities were able to recover by day 25 in enclosures containing concentrations of less than or equal to 0.2 μg/L esfenvalerate (67).

Roessink et al. (68) compared the fate and effects of the pyrethroid insecticide lambda-cyhalothrin in mesotrophic (macrophyte-dominated) and eutrophic (phytoplankton-dominated) ditch microcosms (0.5 m3). Lambda-cyhalothrin was applied three times at one-week intervals at concentrations of 10, 25, 50, 100, and 250 ng/L. The highest concentration was selected based on a 5% drift emission from a field application of 0.015 kg/ha of lambda-cyhalothrin (as "Karate" formulation) into a ditch with a depth of 0.3 m. The rate of dissipation of lambda-cyhalothrin in the water column of the two types of test systems was similar. After 24 h, 30% of the amount applied remained in the water phase. Initial, direct effects were observed primarily on arthropod taxa. Threshold levels for transient direct toxic effects were similar (10 ng/L) between the two mesotropohic and eutrophic test systems. At treatment levels of 25 ng/L and higher, apparent population and community responses occurred. At treatments of 100 and 250 ng/L, the rate of recovery of the macroinvertebrate community was lower in the macrophyte-dominated systems, primarily because of a prolonged decline of the amphipod Gammarus pulex. This species occurred at high densities only in the macrophyte-dominated enclosures. Indirect effects (e.g., increase of rotifers and microcrustaceans) were more pronounced in the plankton-dominated test systems, particularly at treatment levels of 25 ng/L and higher.

Hill et al. (69) reviewed approximately 75 freshwater field studies with pyrethroid insecticides. The studies were carried out in natural/farm ponds, streams or rivers (bifenthrin, cyfluthrin, cypermethrin, deltamethrin, fenvalerate and permethrin), rice paddies (cypermethrin, lambda-cyhalothrin and permethrin), ponds for farming fish and crayfish (fenvalerate and permethrin), lake limnocorral enclosures (fenvalerate and permethrin), pond littoral enclosures (cypermethrin, esfenvalerate and permethrin) and outdoor pond microcosms or mesocosms (bifenthrin, cyfluthrin, cypermethrin, deltamethrin, esfenvalerate, lambda-cyhalothrin, permethrin and tralomethrin). The authors concluded that the spectrum of acute biological effects of these products in bodies of water, at application rates equivalent to a single "drift-entry" of 1-5% of the USA labeled maximum use-rate (applied as multiple treatments), is limited to the zooplankton and macroinvertebrate crustaceans and to some of the aquatic insects.

Van Wijngaarden et al. (70) reviewed 18 microcosm and mesocosm studies on eight pyrethroids (cyfluthrin, cypermethrin, deltamethrin, esfenvalerate, fenvalerate, lambda-cyhalothrin, permethrin and tralomethrin). The exposures included single and multiple applications; all except one were performed in stagnant systems. The authors concluded that recovery of sensitive endpoints usually occurs within 2 months of the last application when peak pyrethroid concentrations remain lower than (0.1 x EC50) of the most sensitive standard test species. Amphipoda and Hydacarina were the taxa most sensitive to pyrethroid insecticides, followed by Trichoptera, Copepoda, Ephemeroptera and Hemiptera (Table III).

Table III. Reported negative effects on various taxonomic groups as a result of repeated application of pyrethroids in aquatic microcosms and mesocosms.

	TU_{mso} 0.001-0.01	0.01-0.1	0.1-1	1-10
Amphipoda	-	100% (1)	100% (11)	100% (7)
Isopoda	-	-	80% 5)	100% (2)
Copepoda	0% (1)	60% (5)	56% (16)	73% (11)
Cladocera	0% (1)	0% (2)	50% 10)	86% (7)
Ostracoda	0% (1)	0% (1)	50% (2)	-
Trichoptera	0% (1)	67% (3)	86% (7)	83% (6)
Ephemeroptera	0% (1)	50% (6)	82% (17)	85% (13)
Diptera	0% (1)	33% (6)	82% (17)	100% (13)
Hemiptera	0% (1)	50% (2)	67% (6)	100% (2)
Odonata	0% (1)	33% (3)	36% (11)	50% (10)
Coleoptera	0% (1)	0% (2)	64% (11)	60% (10)
Hydracarina	0% (1)	100% (1)	100% (1)	-
Fish	0% (1)	0% (5)	33% (6)	83% (6)
Rotifera	0% (1)	0% (3)	0% (13)	0% (11)
Mollusca	0% (1)	0% (3)	0% (12)	0% (10)
Annelida	0% (1)	0% (2)	0% (11)	0% (6)
Turbellaria	0% (1)	0% (1)	0% (7)	0% (3)
Plants	0% (1)	0% (5)	0% (13)	8% (12)

The effects are arranged according to toxic units and expressed as a percentage (%) of the cases (n) in which a reduction in numbers or biomass of one or more taxa within a taxonomic group was reported. Table is reproduced from Van Wijngaarden et al. (2005) with kind permission from Springer Science and Business Media. TU_{mso}=toxic units= pyrethroid concentration divided by the EC50 of the most sensitive standard test species *(Daphnia magna, Pimephales promelas, or Onchorynchus mykiss)*.

Organism-Specific Factors Affecting Pyrethroid Toxicity

Critical Life Stages

Gender and reproductive stage will notably influence the effects of substances that interact with the endocrine system, such as synthetic pyrethroids and their breakdown products. An organism's trophic level will determine its susceptibility to predation after being negatively affected by contaminants. Behavioral characteristics (e.g. complex reproductive strategies) can modify the effects of toxic chemicals on the individual. However, information on life-stage or gender-specific susceptibility to pyrethroids is scarce. The available data suggests that toxicity is dose-related and that, in general, smaller organisms and earlier life-stages are more sensitive than larger and adult organisms. For example, <24-h old Daphnia magna (Cladocera) were about 10 times more sensitive to cypermethrin than 6-d old adult cladocerans (19). Calanoid copepod nauplii (Acartia tonsa) were 28 times more sensitive to cypermethrin than adults, with 96-h LC50s of 0.005 µg/L and 0.142 µg/L (measured concentrations) for nauplii and adults, respectively (71). In this study, gender differences were also observed: During the first 24 h of exposure, male adult copepods were about twice as sensitive as females.

Fish embryos appear to be less sensitive to pyrethroids than larvae. A study on the toxicity of lambda-cyhalothrin to Chinook salmon (Onchorhynchus tsahwytscha) showed no detectable effects on mortality, hatching success, or larval survival when embryos were exposed to nominal concentrations ranging from 0.3-5.0 µg/L during development. The estimated 96-h LC50 for Chinook salmon fry, on the other hand, was 0.15 µg/L (72); thus, Chinook salmon fry were at least 33 times more sensitive to lambda-cyhalothrin than embryos.

The 48-h LC50 of deltamethrin for carp (*Cyprinus carpio*) embryos was 0.21 µg/L, while the respective LC50 for carp larvae was 0.074 µg/L (*73*). Similarly, topsmelt (*Atherinops affinis*) embryos survived 30-d exposure to 3.2 µg/L fenvalerate, while 0.82 µg/L fenvalerate caused complete mortality of exposed topsmelt fry (*74*). Later-stage (stage 34) medaka embryos were the most sensitive embryonal stage to cypermethrin, probably due to partial degradation of the chorion at this time in development (*75*).

Nutritional Status

Low nutritional status may result in increased susceptibility of organisms to pyrethroids. Barry et al. (*47*) showed that esfenvalerate toxicity to *Daphnia carinata* increased significantly with decreasing food concentration. Fenvalerate decreased survival and growth of *Daphnia magna* in the week following a 24-h pulse exposure at 1.0 µg/L (*76*). Age at first reproduction increased, with adverse effects on fecundity. Low food conditions exacerbated the effects of fenvalerate exposure on juvenile survival and growth during the first week, and reduced the significant effect concentration from 0.6 μg/L (high food availability) to 0.3 μg/L. No mortality occurred during the 24-h fenvalerate exposure, but complete mortality was observed at 3.2 μg/L after a 6-d recovery period in control water.

Environmental Conditions and Pyrethroid Toxicity Relationship

Temperature

Water temperature is perhaps the most important factor affecting biochemical and physiological processes of individual organisms. It affects contaminant transformation and excretion rates. Temperature is inversely related to pyrethroid toxicity (*77*). This negative temperature dependence of pyrethroid action has in the past been ascribed to the slow metabolism of pyrethroids at low temperature. Recent studies showed that this effect is mostly due to the increased sodium current flow through (i.e., increased sensitivity of) nerve cell membranes at low temperature (*78*).

In natural aquatic systems, surface water temperature is often lower than standard laboratory toxicity testing temperatures. For example, the standard temperature for aquatic toxicity testing of sediment-dwelling invertebrates is 23°C (*79*). This is well above temperatures in creeks that can serve as habitat for salmonids and other cold water fish species, for which preferred creek average temperatures are commonly below 20°C (e.g., 14-17 °C for coho salmon, *80*).

Suspended Sediment

In their dissolved state, pyrethroids are readily bioavailable to aquatic organisms. In the adsorbed state their bioavailability to aquatic organisms is reduced. Yang et al. (*56*) showed that the presence of suspended sediment (200 mg/L) reduced toxicity of pyrethroids to *Ceriodaphnia dubia* by a factor of 2.5-13. However, the degradation of pyrethroids bound to sediment particles is considerably slower than in soil. For example, the half-life of bifenthrin is reported to be 8-17 months (20°C) in sediments (*81*), and 42-96 days in soil (*82*).

Dabrowski et al. (*83*) conducted artificial stream microcosm trials by exposing mayfly nymphs (*Baetis harrisoni*) to 1 ppb of cypermethrin. Results demonstrated that exposure to cypermethrin increased mayfly drift significantly under either high turbidity (suspended particles) or high flow conditions, but drift was reduced in the presence of both increased flow and suspended particles.

Organic Matter

Yang et al. (*57*) showed that dissolved organic matter (DOM) at 10 mg/L reduced permethrin toxicity to *Ceriodaphnia dubia* as well as bioaccumulation by *Daphnia magna* by approximately a factor of 2.

Exposure Conditions

The exposure regime (concentration, duration and frequency) is an important factor affecting toxicity. Multiple brief exposures within a given time period to a specific contaminant concentration may not have the same toxic effect as one continuous exposure over the same time period. High magnitude exposures of short duration may be enough to cause population level impacts, while low magnitude, long duration exposures may have no impact at all.

Forbes and Cold (*84*) found that even very brief (1-h) exposures to environmentally realistic concentrations of esfenvalerate during early larval life-stages of the midge *Chironomus riparius* can have measurable population level effects on larval survival and development rates. For surviving organisms, no lasting effects on fecundity or egg viability were observed. Brief (30 min) pulse exposures to lambda-cyhalothrin (nominal conc. 0.05-10 μg/L; *85*) in an in-stream mesocosm study demonstrated that macroinvertebrate drift increased significantly after each exposure. *Gammarus pulex*, Ephemeroptera and Simuliidae were predominantly affected. Structural change in the community was found at 5 and 10 μg/L, and recovery occurred within approximately two weeks.

Joint Interactions with Other Chemicals and Stressors

Pre-exposure or simultaneous exposure to other contaminants, disease or stressful environmental conditions such as salinity and temperature may considerably alter the physiological condition and therefore susceptibility of the organism, as well as modify the toxicity of a given contaminant. Organisms in the environment often experience many stressors simultaneously, including those of a physical, biological, and chemical nature (24). Chemical analysis of surface water conducted by the U.S. Geological Survey under the National Water Quality Assessment Program indicates that pesticide mixtures are contaminating surface waters. More than 50% of all stream samples tested contained five or more pesticides (86). In addition, many other contaminants such as heavy metals, PAHs and PCBs are often present in aquatic environments. When large numbers of chemicals are included in the mixture experiments, an additive response is typically found (24). It is therefore evident that mixtures must be considered to be the most common exposure scenario when evaluating the ecological effects of contaminants.

PBO

The synergist piperonyl butoxide (PBO) is commonly added to pyrethroid and pyrethrin formulations to enhance the toxic effects of the active ingredient. PBO functions by inhibiting a group of enzymes (mixed-function oxidases), which are involved in pyrethroid detoxification. PBO can enhance the toxicity of pyrethroids by 10-150 times (*87*). Recently, 3-4-fold enhancement of pyrethroid toxicity to amphipods has been reported (*88*). The 96-h LC50 of PBO for rainbow trout is 2.4 ppb (*89*). PBO in concentrations less than 1 ppm can reduce fish egg hatchability and growth of juvenile fish. Weston et al. (*90*) demonstrated that PBO concentrations in urban creeks after watershed-wide treatment with a pyrethrins/PBO mixture were high enough to enhance toxicity of pyrethroids already existing in creek sediments from general urban pesticide use, effectively "reactivating" pyrethroids already present in the environment. In a study on juvenile (90 d old) striped bass (*Morone saxitalis*), Rebach (*91*) determined 24-h and 96-h LC50s of 32.9 and 16.4 ppb for a 1:1 mixture of PBO and permethrin. No LC50 information for this species is available for permethrin alone.

Pesticide Formulations

Inert ingredients of various pesticide formulations, such as emulsifiers, solvents and surfactants may influence the environmental fate, mobility and the toxicity of pyrethroids. Overall, water-insoluble pesticides applied in emulsion formulations have higher storm- and irrigation runoff potential than water-soluble pesticides (*92, 93*). In a study on stormwater runoff from a stonefruit orchard treated with esfenvalerate in formulation (Asana), runoff from the first storm after application was highly toxic to fathead minnow and rainbow trout larvae, and toxicity to invertebrates was still present in runoff from the third storm after application (*94, 95*). In addition to increasing the risk of exposure, inert ingredients may be biologically active (*96*). For example, a household formulation of bifenthrin reduced the viability of rodent nerve cell cultures, wheres bifenthrin alone did not (*97*). Commercial formulations of bifenthrin (Talstar, Kiros EV) were more toxic to human cell cultures than bifenthrin alone (*98*). In a comparative study on the toxicity of two commercial formulations of permethrin on brook trout (*Salvelinus fontinalis*), Permanone 31-66™, a permethrin formulation containing 31.28% w/w permethrin and 66% w/w PBO, was almost three times more toxic than Permanone Technical Insecticide™, which is >92% w/w permethrin (*99*).

Pyrethroid-Other Insecticides

According to the published literature the toxicity of many pesticide combinations is at least additive. In some cases pesticide mixtures, particularly those involving insecticides, have been shown to be synergistic, with reported increases in toxicity of up to 100-fold (*100*). However, these effects are species, time and dose dependent and are therefore difficult to predict routinely. For pyrethroid – organophosphate (OP) mixtures, greater than additive toxicity is to be expected given that P450-activated OPs will inhibit esterases, thus decreasing an organism's ability to detoxify pyrethroids. OPs are increasingly used in combination with pyrethroids because they can synergistically increase the effects of pyrethroids, expecially where pest populations have developed resistance (Perry et al., 2006). Denton et al. (*101*) demonstrated that exposure to the pyrethroid esfenvalerate and the OP diazinon resulted in greater than additive toxicity in fathead minnow larvae. Similarly, mixtures of esfenvalerate and the OP chlorpyrifos resulted in greater than additive toxicity in fathead minnow (*102*). Synergistic toxic effects have also been observed between pyrethroids and carbamates. Permethrin and the carbamate propoxur elicited greater than additive toxicity in the mosquito *Culex quinquefasciatus* (*103*).

Pyrethroid-Infectious Agents

Clifford et al. (*18*) showed that susceptibility of juvenile Chinook salmon to Infectious Hematopoietic Necrosis Virus (IHNV) was significantly increased when 6-week old fish were exposed to a sublethal concentration of esfenvalerate (0.08 ppb). Of juveniles exposed to both esfenvalerate and to IHNV, 83% experienced highly significant ($p<0.001$) mortality ranging from 20% to 90% at 3 days post-viral exposure. This early mortality was not seen in any other treatment group. In addition, fish exposed to both esfenvalerate and IHNV died 2.4 to 7.7 days sooner than fish exposed to IHNV alone. Results from this study show that accepted levels of pollutants may not cause acute toxicity in fish, but may be acting synergistically with pathogens to compromise survivorship of fish populations through immunologic or physiologic disruption.

Summary

Aquatic organisms, in particular insects, crustaceans and fish, are highly sensitive to pyrethroid insecticides. Acute toxicity to fish and aquatic invertebrates is generally observed at concentrations below 1 μg/L, and sublethal effects have been reported at low ng/L concentrations. Although it is difficult to model sublethal responses to toxicants and predict ecotoxicological impact or risk, measures of sublethal effects are likely to be as important, or more important, than the measures of acute or chronic lethal effects to accurately assess the consequences of contaminant exposure. The primary mechanism of toxic action is often not the only toxic effect a chemical can exert on target and non-target species. For example, neurotoxic pesticides may impair the immune system or exhibit hormonal effects, or can alter behavior with negative effects on predator avoidance or reproductive success. Many of these chemical side effects are poorly understood or unknown.

Although toxic concentrations of pyrethroids in sediments of surface waters have been reported, there is presently limited information on their temporal and spatial distribution, as well as concentrations of pyrethroids in the water column. One of the major limitations for obtaining data on the sources and quantities of pyrethroids in the environment, in particular in water samples, is the sensitivity of the existing analytical chemistry techniques. Because pyrethroids are toxic at extremely low concentrations (low to mid parts per trillion range) monitoring data that are based on insufficiently low detection limits are of little use. In fact, such data can convey a false sense of safety with regard to the potential toxic effects of pyrethroid contamination on aquatic ecosystems, especially if multiple pyrethroids are present simultaneously. Due to their relatively short half-lives and hydrophobic nature, pyrethroid concentrations in larger water bodies are expected to be generally ephemeral, especially in the water column. Higher toxicity and reduced degradation rates at low temperatures may render pyrethroids a greater risk to aquatic life during the winter period, which—along with winter rains and associated stormwater runoff —has the potential to make winter applications of pyrethroids more important environmentally than summer applications.

Critical data gaps on pyrethroids exist. To provide much needed information the following questions should be answered in future work: How do sublethal toxic effects affect the ecological fitness of organisms and populations? - Do pyrethroids interact with other stressors or chemical contaminants to induce toxicity? – What are the effects of inert ingredients in pyrethroid formulations on toxicity and environmental fate and transport?

References

1. Smith, T.M.; Stratton, G.W. *Residue Reviews* 1986, 97:93-120.
2. Clark, J.R.; Goodman, L.R.; Borthwick, P.W.; Patrick, J.M.Jr.; Cripe, G.M.; Moody, P.M.; Moore, J.C.; Lores; E.M. *Environ. Toxicol. Chem.* 1989, 8:393-401.
3. Bradbury, S.P.; Coats, J.R. *Environ. Toxicol. Chem.* 1989, 8:373-380.
4. Shafer, T.J.; Meyer, D.A. *Toxicol. Appl. Pharmacol.* 2004, 196 (2): 303-318.

5. Liu W.P., Gan J.Y., Lee S., Werner I. *Environ. Toxicol. Chem.* 2005, 24(8):1861-1866.
6. Burr, S.A.; Ray, D.E. *Toxicol. Sci.* 2004, 77 (2): 341-346.
7. Hossain, M.M.; Suzuki, T.; Sato, I.,; Takewaki, T.; Suzuki, K.; Kobayashi, H. *Toxicol. Appl. Pharmacol.* 2004, 197 (3): 319-320 729.
8. Litchfield, M.H. In *The Pyrethroid Insecticides;* Leahay, J.P.; Ed.; Taylor & Francis, Philadelphia, PA, USA, 1985; pp 99–150.
9. Go, V.; Garey, J.; Wolff, M.S.; Pogo, B.G.T. *Environ. Health Perspect.* 1999, 107:173-177.
10. Perry, M.J.; Venners, S.A.; Barr, D.B.; Xu, S. 2006. *Reproduct. Toxicol.* 2007, 23(1):113-118.
11. Chen, J.F; Chen, H.Y.; Liu, R.; He, J.; Song, L.; Bian, Q.; Xu, L.C.; Zhou J.W.; Xiao, H.; Dai, G.D.; Wang, X.R. *Biomed. Environ. Sci.* 2005, 18(2):108-116.
12. Kim, S.S.; Lee, R.D.; Lim, K.J.; Kwack, S.J.; Rhee, G.S.; Seok, J.H.; Lee, G.S.; An, B.S.; Jeung, E.B.; Park, K.L. *J. Reprod. Develop.* 2005, 51(2):201-210.
13. Sun, H.; Xu, X.-L.; Xu, L.-C.; Song, L.; Hong, X.; Chen, J.-F.; Cui, L.-B.; Wang, X.-R. *Chemosphere* 2007, 66 (3): 474-479.
14. Tyler, C.R.; Beresford, N.; Van Der Woning, M.; Sumpter, J.P.; Thorpe, K. *Environ. Toxicol. Chem.* 2000, 19(4):801-809.
15. Agarwal, D.K.; Chauhan, L.K.S.; Gupta, S.K.; Sundararaman, V. *Mutation Research* 1994, 311:133-138.
16. Kale, M.; Rathore, N.; John, S.; Bhatnagar, D. 1999. *Toxicol. Lett.* 1999, 105:197-205.
17. Madsen, C.; Claesson, M.H.; Ropke, C. *Toxicol.* 1996, 107:219-227.
18. Clifford, M.A.; Eder, K.J.; Werner, I.; Hedrick, R.P. *Environ. Toxicol. Chem.* 2005, 24:1766-1772.
19. Sherratt, T.N.; Roberts, G.; Willliams, P.; Whitfield, M.; Biggs, J.; Shillabeer, N.; Maund, S.J. *Environ. Toxicol. Chem.* 1999, 18(11):2512-2518.
20. Lozano, S.J.; O'Halloran, S.L.; Sargent, K.W.; Brazner, J.C. *Environ. Toxicol. Chem.* 1992, 11(1):35-47.
21. California Department of Fish and Game (CDFG), Hazard assessment of the synthetic pyrethroid insecticides bifenthrin, cypermethrin, esfenvalerate, and permethrin to aquatic organisms in the Sacramento-San Joaquin River System. Administrative Report 00-06, California Department of Fish and Game, Office of Spill Prevention and Response, Sacramento, California, USA, 2000.
22. U.S. Environmental Protection Agency (USEPA), Guidelines for deriving numerical national water quality criteria for the protection of aquatic organisms and their uses. Office of Research and Development, Washington, D.C., USA, 1985.

23. Carson, R. *Silent Spring*; Houghton Mifflin Publishers, Boston, MS, USA, 1962; 368 pp.
24. Lydy, M.; Belden, J.; Wheelock, C.; Hammock, B.; Denton, D. *Ecol. Soc.* 2004, 9(6): 1.
25. Feder, M.E.; Hofmann, G.E. In *Annual Review of Physiology;* Hoffman, J.F., Ed.; Annual Reviews Inc.: Palo Alto, CA, USA, 1999; Vol. 61, pp 243-282.
26. *Biomarkers: Biochemical, Physiological and Histological Markers of Anthropogenic Stress*; Huggett, R.J.; Kimerle, R.A.; Mehrle, Jr. P.M.; Bergman, H.L.; Eds.; Lewis Publishers, Boca Raton, FL, USA, 1992; pp. 346.
27. Eder, K.J.; Leutenegger, C.M.; Wilson, B.W.; Werner, I. *Mar. Environ. Res.* 2004, 58:809-813.
28. Eder, K.J.; Koehler, H.-R.; Werner, I. *Environ. Toxicol. Chem.* 26(6):1233-1242.
29. Teh, S.J.; Deng, D.F.; Werner, I.; Teh, F.C.; Hung, S.S.O. 2004. *Mar. Environ. Res.* 2004, 59:203-216.
30. Werner, I.; Geist, J.; Okihiro, M.; Rosenkranz, P.; Hinton D.E. *Environ. Res.* 2002, 54:609-614.
31. Werner, I.; Linares-Casenave, J.; Van Eenennaam, J.P.; Doroshov, S.I. *Environ. Biol. Fish* 2006, DOI 10.1007/s10641-006-9070-z.
32. Viant, M.R.; Werner, I.; Rosenblum, E.S.; Gantner, A.S.; Tjeerdema, R.S.; Johnson, M.L. *Fish Physiol. Biochem.* 2004, 29:159-171.
33. Jee, J.H.; Masroor, F.; Kang, J.C. *Aquaculture Res.* 2005, 36 (9), 898-905
34. Moore, A.; Waring, C.P. *Aquat. Toxicol.* 2001, 52:1-12.
35. Mishra, D.; Srivastav, S.K.; Srivastav, A.K. *Ecotox. Environ. Safety* 2005, 60(2):193-197.
36. Denton, D.L. Ph.D. thesis. University of California, Davis, CA, USA, 2001.
37. Haya, K.; Waiwood, B.A. In *Aquatic Toxicology;* Nriagu, J.O.; Ed; Wiley, New York, NY, USA, 1983; pp 307-333.
38. Wedemeyer, G.A.; Barton, B.A.; McLeay, D.J. In *Methods for Fish Biology ;* Schreck C.B. ; Moyle, P.B, ; Eds. ; American Fisheries Society, Bethesda, MD, USA, 1990 ; pp 451-489.
39. U.S. Environmental Protection Agency (USEPA), 2005. EFED Risk Assessment for the Reregistration Eligibility Decision (RED) on Cypermethrin. (values reflect corrections in June 9 2006 "Addendum to the EFED RED Chapter for Cypermethrin").
40. Christensen, B.T.; Lauridsen, T.L.; Ravn, H.W.; Bayley, M. *Aquat. Toxicol.* 2005, 73(2):210-220.
41. Heckmann, L.H.; Friberg, N.; Ravn, H.W. *Pest Managem. Sci.* 2005, 61(7):627.
42. Little, E.E.; Finger, S.E. *Environ. Toxicol. Chem.* 1990, 9 (1): 13-19.
43. Floyd, E.Y.; Geist, J.P.; Werner, I. *Environ. Toxicol. Chem.* In review.
44. McCarthy, A.R.; Thomson, B.M.; Shaw, I.C.; Abell, A.D. *J. Environ. Monit.* 2005, 8:197-202.

45. Tanner, D.K.; Knuth, M.L. *Arch. Environ. Contam. Toxicol.* 1996, 31:244-251.
46. Day, K. E. 1989. *Environ. Toxicol. Chem.* 1989, 8, 411-416
47. Barry, M.J.; Logan, D.C.; Ahokas, J.T.; Holdway, D.A. *Ecotoxicol. Environ. Safety* 1995, 32( )3, 273-279.
48. Reynaldi, S.; Liess, M. *Environ. Toxicol. Chem.* 2005, 24(5):1160-1164.
49. Banerjee, B.D. *Toxicol. Lett.* 1999, 107:21-31.
50. Austin, B. *J. Appl. Microbiol.* 1999, 85:234S-242S.
51. Zelikoff, J.T. 1998. *Toxicol.* 1998, 129:63-71.
52. Mense, S.M.; Sengupta, A.; Lan, C.; Zhou, M.; Bentsman, G.; Volsky, D.J.; Whyatt, R.M.; Perera, F.P.; Zhang, L. *Toxicol. Sci.* 2006, 93(1):125-135.
53. Maund, S.J.; Hamer, M.J.; Lane, M.C.G.; Farrelly, E.; Rapley, J.H.; Goggin, U.M.; Gentle, W.E. *Environ. Toxicol. Chem.* 2002, 21 (1): 9-15.
54. *The Pyrethroid Insecticides*; Leahy, J.P.; Ed.; Taylor & Francis, Philadelphia, PA, USA, 1985.
55. Capel, P.D.; Larson, S.J.; Winterstein, T.A. *Hydrological Processes* 2001, 15: 1251-1269.
56. Yang, W.; Spurlock, F.; Liu, W.; Gan, J. *Environ. Toxicol. Chem.* 2006, 25(7):1913-1919.
57. Yang, W.; Spurlock, F.; Liu, W.; Gan, J. 2006b. *J. Agric. Food Chem.* 2006, 54(11):3967-3972.
58. Davies, P. E.; Cook, L. S. J. *Aquat. Toxicol.* 1993, 27(3-4), 201-24.
59. Crossland, N. O. *Aquat. Toxicol.* 1982, 2(4), 205-22.
60. Crossland, N.O.; Shires S.W.; Bennett D. *Aquat. Toxicol.* 1982, 2:253-270.
61. Mulla, M.S.; Navvab-Gojrati, H.A.; Darwazeh, H.A. *Mosquito News* 1978, 38(1):90-96.
62. Farmer, D.; Hill, I.R.; Maund, S.J. *Ecotoxicology* 1995, 4(4):219-244.
63. Mohsen, Z.H.; Mulla, M.S. *Environ. Entomol.* 1982, 11(2):390-398.
64. Kedwards, T.J.; Maund, S.J.; Chapman, P.F. *Environ. Toxicol. Chem.* 1999, 18(2):149-157.
65. Kedwards, T.J.; Maund, S.J.; Chapman, P.F. *Environ. Toxicol. Chem.* 1999, 18(2):158-166.
66. Mulla, M.S.; Darwazeh, H.A.; Ede L. *Mosqito News* 1982, 42(4):583-590.
67. Lozano, S.J.; O'Halloran, S.L.; Sargent, K.W.; Brazner, J.C. *Environ. Toxicol. Chem.* 1992, 11(1):35-47.
68. Roessink, I. G.; Arts, H.P.; Belgers, J. D. M.; Bransen, F.; Maund, S. J.; Brock, T. C. M. *Environ. Toxicol. Chem.* 2005, 24(7):1684–1696.
69. Hill, I.R.; Shaw, J.L.; Maund, S.J. In *Freshwater Field Tests Hazard Assess. Chem., [Proc. Eur. Workshop Freshwater Field Tests];* Hill I.R.; Ed.; Lewis Publisher, Boca Raton, FL, USA, 1994; pp 249-71.
70. Van Wijngaarden, R.P.A.; Brock, T.C.M.; Van den Brink, P.J. *Ecotoxicology* 2005, 14(3):355-380.
71. Medina, M.; Barata, T.; Telfer, C.; Baird, D.J. 2002. *Arch. Environ. Contam. Toxicol.* 2002, 42:17-22.

72. Phillips, J.P. Master's thesis, University of California, Davis, CA, USA, 2005.
73. Koprucu, K.; Aydin, R. *Pestic. Biochem. Physiol.* 2004, 80(1):47-53.
74. Goodman, L.R.; Hemmer, M.J.; Middaugh, D.P.; Moore, J.C. *Environ. Toxicol. Chem.* 1992, 11:409-414.
75. Gonzalez-Doncel, M.; Fernandez-Torija, C.; Hinton, D.E.; Tarazona, J.V. *Arch. Environ. Contam. Toxicol.* 2005, 48(1):87-98.
76. Pieters, B. J.; Paschke, A.; Reynaldi, S.; Kraak, M.H.S.; Admiraal, W.; Liess M. *Environ. Toxicol. Chem.* 2005, 24(9):2254–2259.
77. Motomura, H.; Naharashi, T. *J. Membrane Biol.* 2000, 177:23-39.
78. Narahashi, T.; Ginsburg, K.S.; Nagata, K.; Song, J.H.; Tatebayashi, H. *Neurotoxicol.* 1998, 19(4-5):581-90.
79. U.S. Environmental Protection Agency (USEPA), Methods for Measuring the Toxicity and Bioaccumulation of Sediment-Associated Contaminants with Freshwater Invertebrates. EPA 600/R-99/064. Second Edition, 2000.
80. Sullivan, K.; Martin, D.J.; Cardwell, R.D.; Toll, J.E.; Duke, S. 2000. An analysis of the effects of temperature on salmonids of the Pacific Northwest with implications for selecting temperature criteria. Sustainable Ecosystems Institute, Portland, Oregon, USA; URL http://www.sei.org (June 2007).
81. Gan, J.; Lee, S.; Liu, W.; Haver, D.; Kabashima, J. *J. Environ. Qual.* 2005, 34:836-841.
82. Laskowski, D.A. *Rev. Environ. Contam. Toxicol.*, 2002, 174:49-170.
83. Dabrowski, J.M.; Bollen, A.; Schulz, R. *Environ. Toxicol. Chem.* 2005, 24(6):1395–1402.
84. Forbes, V.E.; Cold, A. *Environ. Toxicol. Chem.* 2005, 24(1):78-86.
85. Heckmann, L.H.; Friberg, N. *Environ. Toxicol. Chem.* 2005, 24(3):582-590.
86. U.S. Geological Survey (USGS). Pesticides in surface and ground water of the United States: summary of the results of the National Water Quality Assessment Program (NAWQA). U.S. Geological Survey, Washington, D.C., USA, 1998.
87. Wheelock, C.E.; Miller, J.L.; Miller, M.J.; Gee, S.J.; Shan, G.; Hammock, B.D.; *Environ. Toxicol. Chem.* 2004, 23(11):2699–2708.
88. Amweg, E.L.; Weston, D.P.; Johnson, C.S.; You, J.; Lydy, M.J. *Environ. Toxicol. Chem.* 2006, 25(7):1817-1825.
89. U.S. Environmental Protection Agency (USEPA), Ecotox (Acquire) database: http://www.epa.gov/ecotox/, accessed October 2005.
90. Weston, D.P.; Amweg, E.L.; Makebri, A.; Ogle, R.S.; Lydy, M .J. *Environ. Sci. Technol.* 2006. 40(18): 5817-5822.
91. Rebach, S. *Bull. Environ. Contam. Toxicol.* 1999, 62:448-454
92. Wauchope, R.D. *J. Environ. Qual.* 1978, 7(4): 459-472.
93. Moran, K. Insecticide Market Trends and Potential Water Quality Implications. Report for the San Francisco Estuary Project. TDC Environmental, San Mateo, CA, USA;
URL http://tdcenvironmental.com/Pesticides.html, 2003.

94. Werner, I.; Deanovic, L.A.; Hinton, D. E.; Henderson, J. D.; Oliveira, G.H.; Wilson, B.W.; Osterli, P.; Krueger, W.; Wallender, W. W.; Oliver, M. N.; Zalom, F. G. *Bull. Environ. Contam. Toxicol.* 2002, 68(1):29-36.11.
95. Werner, I.; Zalom, F.G.; Oliver, M.N.; Deanovic, L.A.; Kimball, T.; Henderson, J.D.; Wilson, B.W.; Krueger, W.; Wallender, W.W. *Environ. Toxicol. Chem.* 2004, 23(11):2719-2726.
96. Cox, C.; Surgan, M. *Environ. Health Perspect.* 2006, 114(12):1803-1806.
97. Tran, V.; Hoffman, N.; Mofunanay, A.; Pryor, S.C.; Ojugbele, O.; McLaughlin, A. 2006. *Med. Sci. Monitor* 2006,12:BR57-62.
98. Skandrani, D.; Gaubin, Y.; Vincent, C.; Beau, B.; Murat, J.C.; Soleilhavoup, J.-P.; Croute, F. Relationship between toxicity of selected insecticides and expression of stress proteins (HSP, GRP) in cultured human cells: Effects of commercial formulations versus pure active molecules. *Biochimica et Biophysica Acta* 2006, 1760:95-103.
99. Paul, E.A.; Simonin, H.A.; Tomajer, T.M. *Arch. Environ. Contam. Toxicol.* 2005, 48:251-259.
100. Thompson, H.M. *Ecotoxicology* 1996, 5, 59-81.
101. Denton, D.; Wheelock, C.; Murray, S.; Deanovic, L.; Hammock, B.; Hinton, D. *Environ. Toxicol. Chem.* 2003, 22:336-341.
102. Belden, J.B.; Lydy, M.J. *Environ. Toxicol. Chem.* 2006, 25(2):623-629.
103. Corbel, V.; Raymond, M.; Chandre, F.; Darriet, F.; Hougard, J.M. *Pest Management Science* 2004, 60 (4): 375-380.
104. California Department of Pesticide Regulation (DPR), Ecotox Database; URL http://dpr.ca.gov, 2007.
105. PAN (Pesticide Action Network) Pesticides Database; URL http://www.pesticideinfo.org/Index.html, 2007.
106. Schroer, A.F.W.; Belgers, J.D.M.; Brock, T.C.M.; Matser, A.M.; Maund, S.J.; Van den Brink, P.J. *Arch. Environ. Contam. Toxicol.* 2004, 46 (3): 324-335.
107. Anderson, B.S.; Phillips, B.M.; Hunt, J.W.; Worcester, K.; Adams, M.; Kapellas, N.; Tjeerdema, R.S. *Environ. Toxicol. Chem.* 2006, 25:1160-1170.
108. Paul, E.A.; Simonin, H.A. Toxicity of three mosquito insecticides to crayfish. *Bull. Environ. Contam. Toxicol.* 2006, 76:614-621.
109. Little, E.E.; Dwyer, F.J.; Fairchild, J.F.; DeLonay, A.J.; Zajicek, J.L. *Environ. Toxicol. Chem.* 1993, 12:871-878.
110. Hansen, D.J.; Goodman, L.R.; Moore, J.C.; Higdon, P.K. *Environ. Toxicol. Chem.* 1983, 2:251-258.

Chapter 15

Aquatic Fate and Effects of *Lambda*-Cyhalothrin in Model Ecosystem Experiments

Stephen J. Maund[1], René P. A. Van Wijngaarden[2], Ivo Roessink[2], Jacqueline S. Warinton[3], Paul J. Van den Brink[2,4], and Theo C. M. Brock[2]

[1]Syngenta Crop Protection AG, 4002 Basel, Switzerland
[2]Alterra Green World Research, Wageningen University and Research Centre, Wageningen, The Netherlands
[3]Syngenta Jealott's Hill International Research Centre, Bracknell, United Kingdom
[4]Department of Aquatic Ecology and Water Quality Management, Wageningen University, Wageningen, The Netherlands

The fate and effects of the synthetic pyrethroid *lambda*-cyhalothrin in aquatic model ecosystem experiments are reviewed. In laboratory studies, *lambda*-cyhalothrin is highly toxic to fish and invertebrates. Its physico-chemical and laboratory fate properties indicate that it will dissipate rapidly from the water phase, reducing exposure for organisms in the water-column. For European aquatic risk assessments, where exposure models predict that spray drift is the main entry route from agricultural uses, this has been a key factor in refining higher-tier risk assessments for water-column organisms. Modified exposure studies in the laboratory confirmed that rapidly reduced exposure mitigates effects on fish and invertebrates. Eight aquatic model ecosystem experiments have been conducted with *lambda*-cyhalothrin in a variety of indoor and outdoor test systems. These were of differing trophic status, and ranged in size from 0.43 to 450 m^3. The timing of application of test substance also varied between studies. The fate of the compound in the various experiments was consistent, typified by rapid dissipation (and degradation) in the water phase with median dissipation times (DT_{50}) of typically a day or less. Only 5-22% (where quantifiable) of the chemical applied to the water column reached the sediment, and it was not possible to calculate sediment DT_{50} values. Effects in the studies were driven by population responses of macrocrustacea and certain insects, along with zooplanktonic microcrustacea. Considering the range of test systems, the variety of locations, different trophic status of the test systems, differences in season of application, and differences in numbers of applications, the effects thresholds observed in the studies were remarkably consistent, with no to slight effects occurring consistently at initial nominal treatment concentrations up to 10 ng/L. The effects threshold values for clear effects with recovery were more variable than the no to slight effects, but still reasonably consistent, with thresholds between initial nominal treatment concentrations of 16 and 50 ng/L. This gives considerable confidence in the potential to extrapolate the effects observed in one study to a different situation, at least in this case where effects tend to be of an acute nature, and the dissipation and degradation of the compound is rapid.

Introduction

Synthetic pyrethroids are highly toxic to aquatic invertebrates and fish in the laboratory (*1, 2*). However, under field conditions, their rapid dissipation from the water column is cited as a mitigation of potential for effects under field conditions (*1*). For this reason, over the last three decades, many aquatic field studies have been performed on a variety of synthetic pyrethroids to measure their effects on populations and communities of aquatic organisms (*3, 4*).

Beginning in the late 1970s and continuing into the early 1980s, the first field studies on pyrethroids were conducted in farm ponds, mainly in the USA (*5*). While these studies had the advantage of being realistic because they were conducted in natural water bodies, they had a number of disadvantages. The experimentation was difficult (e.g. finding appropriate sites of adequate similarity for a treated and untreated system) and expensive, there were no treatment replicates so there was a lack of statistical power, there was high temporal and spatial variability, and there was no dose-response relationship since only one treatment was applied. Consequently, it was difficult to establish cause and effect.

From these early field studies, the mesocosm evolved in the late 70s and mid 1980s. Mesocosms are defined (6) as bounded and partially enclosed outdoor experimental units that closely simulate the natural environment. During the late 1980s and early 1990s, mesocosm studies were required by the United States Environmental Protection Agency (EPA) for synthetic pyrethroid registration submissions. The mesocosms used for these studies were large, using experimental ponds of around 400 m^3, and a range of endpoints were evaluated including plankton, macroinvertebrates, macrophytes, and fish growth and reproduction (7). Though mesocosms arguably moved the science forward from farm pond studies, it was generally recognized that the results were still difficult to interpret for a number of reasons (8). The ponds were stocked with adult bluegill sunfish (Lepomis macrochirus) whose progeny of up to 20 000 young-of-year fish at the end of the study could themselves have a substantial effect on the ponds. The influence of these young fish could be seen in the control and treatment data, as numbers of arthropods consistently declined through the course of the study, presumably due to predation. Also, since the ponds used for such studies were large, there were often environmental gradients across study sites, leading to quite high variability for some endpoints. The studies were also very expensive, typically taking two to three years to complete, and costing several million dollars.

In 1992, the EPA introduced the 'New Paradigm' in which it was recognized that evaluation of ecotoxicological field studies was problematic and time intensive. Since that time, mesocosm data have not been routinely used in pesticide registration in the USA, though they remain a higher-tier option. In Europe however, mesocosms, and their smaller counterparts, microcosms, continued to be an option for higher-tier risk assessment under the European Union (EU) plant protection product directive, 91/414/EEC. During the 1990s, a variety of different test systems were established, and the methodologies for microcosms and mesocosms developed substantially, resulting in a number of reviews and guidance documents (*9, 10, 11*) and ultimately leading to the production of an Organisation for Economic Cooperation and Development (OECD) guidance document in 2006 (*12*). The use and interpretation of aquatic model ecosystem experiments in the EU continue to be the focus of discussions (*13*).

Though the methodology for conducting aquatic model ecosystem studies was well-established by the late 1990s, a number of questions remained regarding their interpretation and implementation in risk assessment (*11*). Four uncertainties that were identified were the extent to which aquatic model ecosystem data generated in one location could be applied to another situation; the potential influence of mixtures of chemicals or stressors; whether the timing (season) of application would influence the outcome of the study; and whether differences in ecosystem properties (e.g. trophic status) might influence the results. Here we review the fate and ecological threshold levels of *lambda*-cyhalothrin in eight indoor and outdoor aquatic model ecosystem experiments under a wide range of experimental conditions in light of these uncertainties.

For *lambda*-cyhalothrin aquatic risk assessment in Europe, concerns for aquatic organisms have focused on the potential exposure of water bodies by spray drift. Based on the results of European exposure modeling approaches, spray drift is identified as the major route of entry of *lambda*-cyhalothrin. from agricultural uses. This review therefore focuses on water-column endpoints. For other regions or uses, different sources or routes of exposure may also be important for pyrethroids (*14, 15*).

Summary of Laboratory Fate and Effects Profile

In common with other pyrethroids, *lambda*-cyhalothrin (1:1 mixture of Z(1R,3R,αS) and Z(1S,3S,αR), esters of α-cyano-3-phenoxybenzyl 3-(2-chloro-3, 3, 3-trifluoroprop-1-enyl)-2, 2-dimethylcyclopropane-carboxylate) has a range of physico-chemical characteristics that can have a substantial influence on its fate in the environment, particularly in aquatic ecosystems. It has a low water solubility of 5 ug/L and a high octanol:water partition coefficient (log K_{ow} = 7.0), resulting in highly lipophilic properties (*16*). This means that the compound readily adsorbs to soils and sediments, and soil- and sediment-water partition coefficients normalized for organic carbon content (K_{oc}) are reported to be typically in the range 200 000 to 350 000 (*16*), and partition coefficients to humic substances in water have been reported in the range 400 000 to 800 000 (*17*). Consequently, under field conditions, *lambda*-cyhalothrin would be expected to partition rapidly and substantially from the water phase to sediment and other organic materials. In laboratory soil and water-sediment degradation studies, *lambda*-cyhalothrin has been shown to be readily degraded. Average soil and aquatic sediment half lives under aerobic laboratory conditions have been reported to be 43 and 22 days respectively (*16*).

Another important physico-chemical property of *lambda*-cyhalothrin is its lability under alkaline aqueous conditions. At higher pH values, the ester bond is readily hydrolysed, and the mean degradation time (DT_{50}) at pH 9 is reported to be 8.7 days (*16*). In small edge of field surface waters that are the protection aim for European risk assessment, pH values of this order are not uncommon (see for example the UK National Pond Survey (*18*)). It therefore might be anticipated that hydrolytic degradation may play a role in the fate of *lambda*-cyhalothrin in such water bodies.

In standard laboratory tests (with maintained exposure concentrations), *lambda*-cyhalothrin is highly acutely and chronically toxic to fish and aquatic invertebrates, but of low toxicity to algae, indicating negligible risks to aquatic plants (Figure 1). When the standard European uncertainty factors of 100 and 10 for acute and chronic assessments respectively are applied to these data, comparison to exposure concentrations triggers further refinement of the risk assessment *(19)*.

Three principal factors have been proposed that could be investigated to refine the assumptions made in the lower tier risk assessments (*19*). Firstly, as mentioned above, it has long been noted that the fate properties of pyrethroids mean that standard, maintained exposure laboratory studies are likely to somewhat overestimate the potential for effects in the field, where the compounds will tend to dissipate rapidly. Secondly, the preliminary risk assessments with *Daphnia magna* and fish in Europe include the use of an uncertainty factor to account for potentially more sensitive species. In the European Union this value is 100 for acute and 10 for chronic assessments. Consequently, if these species are at the sensitive end of the species sensitivity spectrum, the assessment may be conservative. Thirdly, standard laboratory tests do not take into account important processes such as population recovery through reproduction and re-invasion, or avoidance (which may be important for larger organisms that can move quickly).

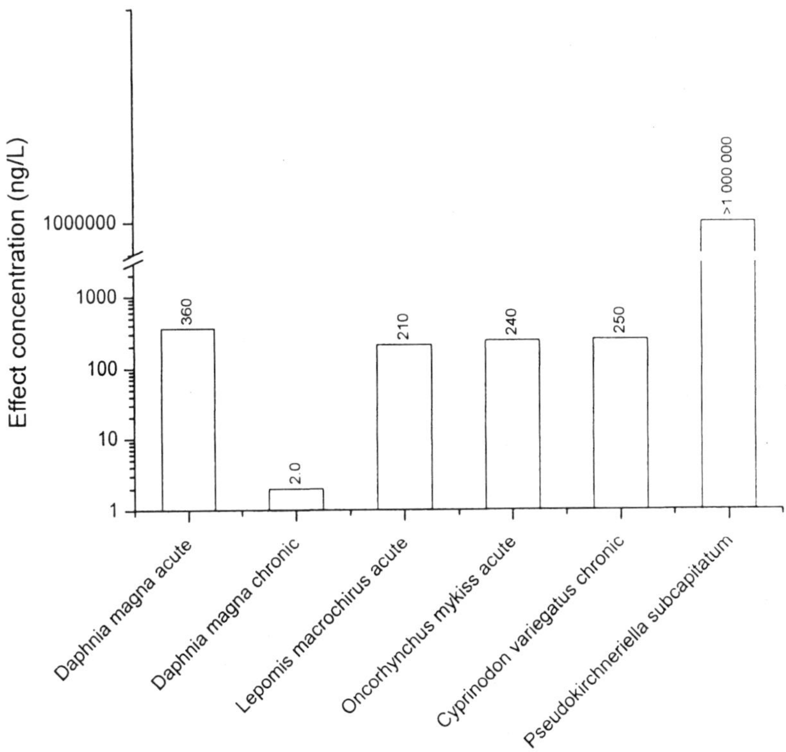

Figure 1. Acute and chronic toxicity of lambda-*cyhalothrin to standard test species. Labels on the column are the effect concentration in ng/L based on measured concentrations (19)*.

The rapid dissipation of *lambda*-cyhalothrin from the water phase in water-sediment systems has been shown to reduce apparent toxicity compared to water-only studies (*19*). In studies where cyhalothrin (the unresolved form of which *lambda*-cyhalothrin is one of the paired isomers) was applied to static water-sediment systems, there was a three- to four-fold reduction in toxicity compared to water-only for fish and *Daphnia*. Schroer *et al. (20)* also found differences in the shape and steepness of laboratory and field species sensitivity distributions.

Similarly, short durations of exposure have been shown to result in substantially less severe effects than maintained, long-term exposures. In studies with a sensitive malacostracan crustacean species *Gammarus pulex* (*19*), there was a significant reduction in toxicity with decreasing exposure times, with one hour exposures to a certain concentration being around eighteen times less toxic than those after ninety-six hours of exposure.

Species sensitivity distributions of fish and aquatic arthropods invertebrates (Figure 2) with *lambda*-cyhalothrin have been reported by several authors (*19, 20, 21*). Broadly speaking, fish tend to be less sensitive to *lambda*-cyhalothrin than arthopods. Within the invertebrates, generally speaking, crustacean and certain insect arthropod taxa tend to be among the most sensitive, with non-arthropod invertebrates being at the less sensitive end of the distribution.

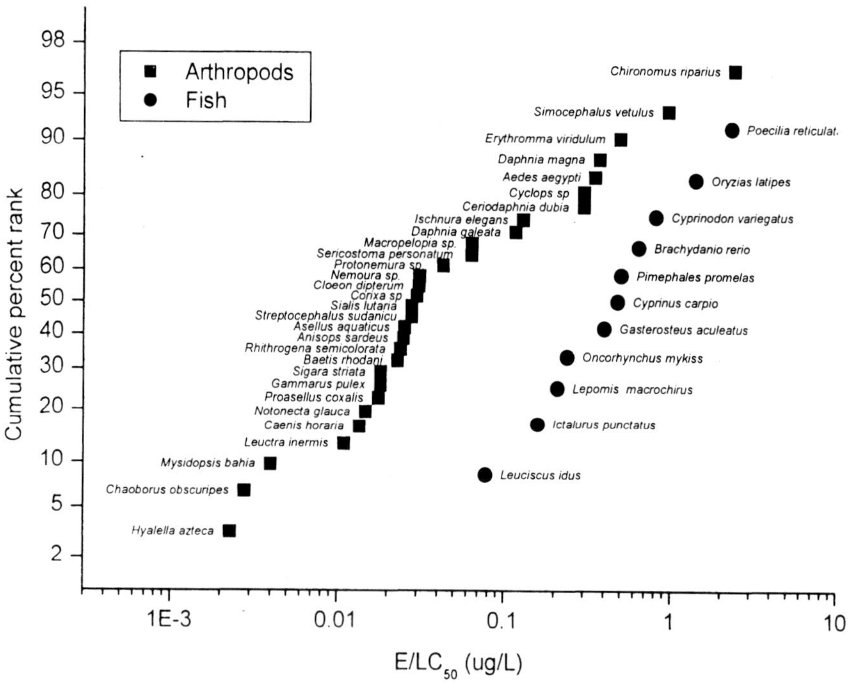

Figure 2. Species sensitivity distributions based on acute laboratory toxicity data for aquatic arthropods (48 h EC50) and fish (96 h LC50)

The results of these studies lead to some important implications for the anticipated effects of *lambda*-cyhalothrin under field conditions. Firstly, short-duration exposure may mitigate effects predicted on the basis of standard laboratory data. Considering that exposure is likely to be of short duration in the aqueous phase, potential for recovery for organisms that can recolonize or have resting stages is likely to be high. The organisms that are most likely to be affected in the field are crustacea and certain insect taxa. Further discussion of these aspects as studied in the field can be found below.

Review of Aquatic Model Ecosystem Studies with *Lambda*-cyhalothrin

An overview of the studies reviewed here is shown in Table I, and each study is numbered in order to facilitate comparisons in later tables. A brief summary of each study design is provided below, along with an overview of the fate and effects of *lambda*-cyhalothrin in subsequent sections. Analytical residue data from the studies are not presented in detail here. In each aquatic model ecosystem study, treatment solutions were analyzed to confirm that the required amount of test substance had been added to the test system. Due to incomplete mixing at the time of application and subsequent rapid dissipation, initial measured concentrations may be a poor indication of what was actually applied (*11*). Treatments were therefore confirmed by measuring the application solution (which was then completely emptied into the test system), and then expressing the results as the initial nominal treatment concentration. All of the studies described below were considered to have been treated as intended by this approach, and so results are expressed as the initial nominal treatment concentration in the water phase. The initial concentration decreases over the course of the experiment due to dissipation and degradation processes. It is therefore important to recognize that this description of the treatment concentration is different from laboratory studies where measured concentrations over the course of the experiment are usually used to express the effect concentration.

Summary of experimental design for studies simulating spray drift and runoff

The first mesocosm study with *lambda*-cyhalothrin was carried out in Greensboro, North Carolina, USA during 1985 and 1986 (*22, 23*). A total of sixteen 15 x 30 m variable depth (0.15 to 2.0 m) pond mesocosms (total water volume of 450 m^3) were used and each of three treatments was replicated four times with four cosms used as controls. Each pond had a 10 cm deep sandy loam sediment and was filled with water from an established pond. Twenty-five adult bluegill sunfish (*Lepomis macrochirus*) were added to each mesocosm. As was the case for most early mesocosms, the experiment followed a 'simulation' (*11*) type of experimental design, with treatments attempting to simulate spray drift and runoff entry into farm ponds.

Table I. Overview of Aquatic Model Ecosystem Studies Performed with *Lambda*-cyhalothrin

Study number	*Loca -tion*	*Test system*	*Initial nominal treatment conc. range (ng/L)*	*No.of applica-tions*	*Refer-ences*
Lambda-cyhalothrin applied alone					
1	USA	Pond mesocosm ($450\ m^3$)	1.6 – 160 [a] 4.7 – 470 [b]	12 [a] 6 [b]	*22, 23*
2	UK	Pond mesocosm ($25\ m^3$)	17 -170	4	*24*
3a	NL	Phyto-plankton dominated ditch enclosure ($0.43\ m^3$)	10 – 250	3	*25, 26*
3b	NL	Macrophyte-dominated ditch enclosure ($0.43\ m^3$)	10 – 250	3	*25, 26*
4	NL	Macrophyte-dominated ditch enclosure ($0.43\ m^3$)	10 – 250	3	*26, 27*
Lambda-cyhalothrin applied with other pesticides					
5	NL	Indoor microcosm ($0.6\ m^3$)	10 – 240	5	*28*
6	NL	Ditch mesocosm ($55\ m^3$)	4 – 85	2	*29*
Lambda-cyhalothrin fate only studied					
7	UK	Indoor microcosm ($0.60\ m^3$)	2300	1	*30*

NOTE: [a] Spray drift [b] Run-off

A total of twelve applications with an emulsifiable concentrate formulation simulating spray drift were made at one week intervals. Six runoff-simulating applications (*lambda*-cyhalothrin mixed into a soil-water slurry) were made to the same mesocosms at two-week intervals, with the first one three days after the first spray drift application. The application rates were:

- 12 x 0.017 g ai/ha (spray-drift) + 6 x 0.05 g ai/ha (run-off slurry).
- 12 x 0.17 g ai/ha (spray-drift) + 6 x 0.5 g ai/ha (run-off slurry).
- 12 x 1.7 g ai/ha (spray-drift) + 6 x 5 g ai/ha (run-off slurry).

Assuming total mixing in the water column, these applications would have resulted in initial nominal water concentrations of 1.7 and 4.7 ng/L; 17 and 47 ng/L; and 170 and 470 ng/L respectively. However, the results of these studies can be difficult to interpret as standard 'toxicological' (*11*) effect concentrations since treatment to the water surface probably would have resulted in a concentration gradient during the early part of the study (higher than nominal water concentrations in the upper layers of the water column). For the purposes of comparison with later 'toxicological' studies, the exposure concentrations from this study were expressed as the median of the nominal concentration from the spray drift and run-off applications.

A second spray drift 'simulation' study was conducted in 1986 in Bracknell, Berkshire, UK (*24*) using outdoor experimental ponds. Each pond was 5.0 m x 5.0 m and 1.3 m deep, and contained 1.0 m depth of water over 0.15 m of sediment (total volume 25 m^3). An emulsifiable concentrate formulation was applied with a spray boom at 0.17 and 1.7 g ai/ha on four occasions at two-week intervals. Resulting nominal treatment concentrations (again, potentially underestimating exposure in the upper water layers soon after treatment) were equivalent to 17 and 170 ng/L.

Both of these early US and UK mesocosm studies were 'simulation' type studies, where the test compound was applied to the water surface of the mesocosms either as a spray or as a slurry application in order to simulate spray drift and/or runoff entry. Interpretation of these data as concentrations (as opposed to application rates in mass per unit area) can be problematic. While the simplest approach to interpreting this is to calculate a nominal concentration based on the total loading divided by the water depth, this ignores the fact that for a short time after application, concentration gradients may exist while the test compound mixes through the water column. Indeed, previous studies have shown that for pyrethroids, application to the water surface may (depending on the method of application) result in concentrations in the upper layers that are higher than those based on a calculation of amount nominally applied divided by water depth (*3, 31*). Some evidence of concentration stratification with water depth was reported by Farmer *et al.* (*24*). Consequently, for organisms inhabiting the upper layers of the water column, the use of nominal concentrations to assign effect concentrations from these studies may be quite conservative (i.e., the effects attributed to a lower nominal concentration actually occurred due to exposure to a higher stratified concentration).

Summary of experimental design for studies evaluating the influence of trophic status and season of application

The studies reported by Roessink *et al.* (*25*) and Schroer *et al.* (*20*) investigated the influence of the trophic status of the test system on the effects of *lambda*-cyhalothrin. Experimental ditches were used that had been established under different regimes of macrophyte growth and nutrient supply over several years to produce distinctive, stable ecosystems: one "macrophyte dominated" and the other "phytoplankton dominated". Van Wijngaarden *et al.* (*27*) conducted studies in the macrophyte-dominated ditches and compared the effects of *lambda*-cyhalothrin on aquatic communities following different application regimes, one in spring and one in late summer.

In all of these experiments, multiple applications of *lambda*-cyhalothrin were made to enclosures (cylinders of 1.1 m diameter and 0.90 m height) placed in the ditches and embedded in the sediment (sandy loam) to a depth of about 0.15 m. The water depth was 0.50 m resulting in a water volume of 0.43 m^3. Two replicate enclosures in each ditch were dosed directly with *lambda*-cyhalothrin ('toxicological' design) as an aqueous solution of a 10% capsule suspension formulation at initial nominal treatment concentrations of 10, 25, 50, 100, and 250 ng/L. Additionally, two enclosures in each ditch were used as controls and dosed with water only. Each enclosure was dosed three times at one-week intervals. The applications were made to the enclosures in macrophyte- and phytoplankton-dominated ditches in spring (May 2000) and to a further macrophyte-dominated ditch in late summer (August 2000).

Summary of experimental design for studies with applications of multiple pesticides

An indoor microcosm study including applications of *lambda*-cyhalothrin was performed at Alterra, Wageningen University and Research Centre, The Netherlands to investigate the ecological impacts of pesticides used in a typical crop protection programme for tulip culture (*28*). Twelve indoor microcosms (simulating aquatic communities typical of macrophyte-dominated Dutch drainage ditches) were used in the experiment. Each microcosm was 1.1 m x 1.1 m x 1.0 m deep, with a sediment layer (sandy loam) of 0.10 m, a water column of 0.50 m, and a water volume of approximately 0.6 m^3. The microcosms were maintained in a climate room with a daily photoperiod of 14 h and a constant temperature of approximately 20°C and acclimatised for two months, during which time the water was circulated through all twelve systems. Multiple applications of *lambda*-cyhalothrin and the three other pesticides (fluazinam, asulam and metamitron) were made to the microcosms at four application rates, equivalent to spray drift entry of 0.2%, 0.5%, 2% and 5% of the label recommended use rates and at the recommended frequencies for each pesticide. In this case, the treatments were mixed into the water column in a 'toxicological' design. *Lambda*-cyhalothrin was applied five times at one week intervals, resulting in mean initial nominal treatment concentrations of 10, 24, 90 and 250 ng/L (based on measured dosage concentrations for each of the applications divided by the water volume of the microcosm). Since *lambda*-cyhalothrin was the most toxic to invertebrates of the four pestsicides applied in the study, it was considered that results could be attributed to the test concentrations of *lambda*-cyhalothrin with reasonable certainty.

In 2002, Arts *et al.* (*29*) also used a 'crop-based' treatment regime to investigate the effects of different spray drift rates from a typical crop protection treatment programme for potatoes in The Netherlands. The experiment was performed in twelve large ditch mesocosms which were 40 m long, 3.3 m wide at the water surface and 1.6 m wide at the sediment surface, and had a water depth of 0.5 m, a sediment (sandy loam) depth of 0.25 m, and a total volume of approximately 55 m^3. In addition to *lambda*-cyhalothrin, the herbicides prosulfocarb and metribuzin, and the fungicides fluazinam and chlorothalonil were applied at rates equivalent to spray drift at 0.2, 1 and 5% of label-recommended rates, with the lowest treatment duplicated, the two higher treatment triplicated, and four untreated control ditches. Applications were made by spray boom and then gently mixed following treatment in a toxicological design. *Lambda*-cyhalothrin was applied twice in the study at five and nine weeks after the start of treatment resulting in initial nominal treatment concentrations of 4, 16 and 85 ng/L. Again since *lambda*-cyhalothrin was the most toxic compound to invertebrates applied in the study, it was considered that results could be attributed to the test concentrations of *lambda*-cyhalothrin with reasonable certainty.

Summary of experimental design for indoor, radiolabelled, aquatic model ecosystem fate study

In 1998, Hand *et al.* (*30*), studied the dissipation and degradation of ^{14}C-radiolabelled *lambda*-cyhalothrin in an indoor aquatic microcosm. Within a glasshouse, a large glass tank (2 m x 1 m x 0.5 m high) was placed in a surrounding tank through which water cooled to 13°C was pumped. Sediment (sandy clay loam) and pond water were obtained from an established ponds at Jealott's Hill Research Station, Bracknell, UK. Sediment was added to the microcosm to a depth of 10 cm, and over this a water column of 30 cm was added. The total volume of the test system was approximately 0.60 m^3. Plant and animal communities were allowed to establish prior to treatment. *Lambda*-

cyhalothrin was applied evenly across the water surface drop-wise with a pipette to provide an initial nominal treatment concentration of 2.3 ug/L. Subsequent to application, radiochemical residue measurements were made in samples of water, plants and sediment taken at intervals from the test system.

Overview of the Fate Profile of *Lambda*-cyhalothrin in Aquatic Model Ecosystem Studies

A summary of the fate profile of *lambda*-cyhalothrin in the various aquatic model ecosystem experiments is presented in Table II. In all studies, the dissipation of *lambda*-cyhalothrin from the water column was rapid. Results between the different test systems were consistent, with water phase DT_{50} values of approximately a day or less. In most cases, residues in the water phase declined to detection limits within a period of four to five days after treatment, and there was no accumulation of residues in the water column resulting from multiple applications. This therefore indicates that under field conditions, water column exposure resulting from spray drift is likely to be of a short-pulsed nature, with residues declining rapidly from their peak values.

Measurements of *lambda*-cyhalothrin in sediments were made less often in the reported studies than those in the water column. Only a small proportion of the compound applied ever reached the sediment (see Table II) in the studies where residues were measured. In two cases, none reached the sediment, and in the remainder where values were reported, only 5 to 22% was detected in the sediment. Because of the low concentrations, and also the short period after application for which residues were measured in most cases, sediment DT_{50} values could not be estimated reliably and were typically not reported.

The results of these studies with *lambda*-cyhalothrin are consistent with those observed for other pyrethroids in aquatic model ecosystem studies. A review of pyrethroid studies (*3*) concluded that of the 38 aquatic model ecosystems reviewed, there was also a rapid decline in water residues, with the DT_{50} in most cases being less than two days. Similar dissipation profiles were noted irrespective of the type or size of test system. Sediment data were also generally not sufficient to make calculations.

In the studies described above, plants seem to have played a significant role in the dissipation and degradation of *lambda*-cyhalothrin, perhaps explaining why a relatively small proportion of the applied residue reached the sediment. Another study confirms these fundings. Wendt-Rasch (*32*) observed 50% dissipation times for *lambda*-cyhalothrin of 0.87 and 2.4 days in *Elodea* (submerged macrophyte) dominated and *Lemna* (floating macrophyte) dominated experimental ponds of similar dimensions (coated concrete walls).

This observation again indicates that the presence of submerged macrophytes is important in the dissipation of lambda-cyhalothrin from water. The method of application of pyrethroid solution to the water column may have had an influence on the dissipation profile of *lambda*-cyhalothrin observed in the studies reviewed here. Due to its highly hydrophobic nature, the compound will adsorb readily onto any available surface such as plants. A different picture may have emerged if the compound was applied as runoff, i.e. already bound to soil particles. In this case, higher sediment residues might be expected unless there was substantial desorption.

Table II. Summary of Dissipation and Distribution of *Lambda*-cyhalothrin in Aquatic Model Ecosystem Studies

Study No. [a]	*Test System*	*Water DT_{50} (days)*	*Max..% of Applied in Sediment*	*Whole System DT_{50} (days)*
1	Pond mesocosm (450 m^3)	c. 1 (20-25% of applied after 2 d)	- [b]	-
2	Pond mesocosm (25 m^3)	< 1 (23% after 24 h)	22	-
3a	Phytoplankton dominated ditch enclosure (0.43 m^3)	< 1 (37% of applied after 24 h)	< limit of quantification	-
3b	Macrophyte dominated ditch enclosure (0.43 m^3)	< 1 (23% of applied after 24 h)	< limit of quantification	-
4	Macrophyte dominated ditch enclosure (0.43 m^3)	< 1 (3-4% after 24 h)	17	-
5	Indoor microcosm (0.60m^3)	0.7 – 1.2	-	-
6	Ditch mesocosm (55 m^3)	0.9 – 1.2	-	-
7	Indoor microcosm (0.60 m^3)	< 0.13	5	< 0.13

NOTES: [a] Study numbers refer to the studies listed in Table II. [b] a dash indicates that data were not reported.

Comparison of Effects of *Lambda*-cyhalothrin in Aquatic Model Ecosystem Experiments

Characteristics of observed effects

The patterns of effects that emerge in the different test systems are reasonably consistent, considering the differences in size, location and experimental design of the studies. In all studies, malacostracan crustaceans and certain insect species (Diptera and Ephemeroptera) were among the most sensitive, reflecting the distribution of sensitivities that were seen in the laboratory. Recovery of affected insect species tended to be reasonably rapid, most probably due to reseeding of aquatic model ecosystems by flying adult stages. For the Malacostraca, at higher concentrations where effects were substantial, recovery tended to be slow or did not occur. However, considering that these organisms would usually recover by immigration from unaffected sites (none present in these enclosed test systems), this result is not too surprising, and has also been observed with a range of insecticides where crustaceans are sensitive (*33*). Effects on zooplankton species tended to occur at higher concentrations than those that affected macroinvertebrates, and effects on zooplankton tended to be followed by rapid recovery, due to the presence of resting stages and the short life-cycle of these organisms. As would be expected, there were generally no direct effects on aquatic plants in these studies, although occasionally, indirect effects (short-term blooms of algae) could be observed due to decreases in grazing pressure from the direct effects of the chemical.

Effects thresholds

In order to compare the ecological effects thresholds observed in these studies, an effect classification system was used (34). The measured endpoints in the studies were assigned to one of eight groups. These groups comprised one functional category (community metabolism) and seven structural categories (microcrustaceans; macrocrustaceans; insects; fish; other zooplankters; other macroinvertebrates; algae and macrophytes). The functional category community metabolism refers to dynamics of dissolved oxygen (DO), pH, inorganic carbon and nutrients in the water column or decomposition as studied by a litter bag technique. The structural categories refer to changes in species composition or population densities and biomass.

To facilitate comparisons, the most sensitive endpoint within each category was selected for each exposure concentration studied, resulting in a more or less worst-case evaluation of the studies. The classification of the categories above was mainly based on univariate analysis of the measurement endpoints. The studies also provided a multivariate analysis of the data which allows an evaluation at the community level. The responses observed for the most sensitive endpoint within each category and at each exposure concentration were assigned to the following effect classes (Table III) based on these criteria:

1. No effects demonstrated: No consistent adverse effects are observed as a result of the treatment. Any observed differences between treated test systems and controls do not show a clear causality.
2. Slight effects: Confined to responses of sensitive endpoints (e.g., partial reduction in abundance of sensitive arthropods). Effects observed on individual samplings only and/or of a short duration directly after treatment.
3. Clear short-term effects, lasting < 8 weeks: Convincing direct and/or indirect effects on measurement endpoints. Recovery, however, takes place within eight weeks after the last treatment. Transient effects reported on both sensitive and less sensitive endpoints. Effects observed on a sequence of samplings.

Table III. Effects Threshold Concentrations (ng/L) from Various Aquatic Model Ecosystem Studies with *Lambda*-cyhalothrin. Results are Expressed as Initial Nominal Treatment Concentrations Applied to the Test System.

		Effect Class				
Study No.[a]	*Test system*	1	2	3	4	5
1	Pond mesocosm (450 m^3)[b]	2.7				27
2	Pond mesocosm (25 m^3)					17
3a	Phytoplankton dominated ditch enclosure (0.43 m^3)		10		25	
3b	Macrophyte dominated ditch enclosure (0.43 m^3)		10	50		
4	Macrophyte dominated ditch enclosure (0.43 m^3)[c]		10	25	50	
5	Indoor microcosm (0.60m^3)[c]	4.0		16		85
6	Ditch mesocosm (55 m^3)		10		25	

NOTES: [a] Study numbers refer to the studies listed in Table I. [b] Experiment was characterized by both spray drift and run-off applications. As exposure concentrations, the median between nominal spray drift and run-off applications was used. [c] Several pesticides applied.

4. Clear effects, recovery not studied: Clear effects are demonstrated (e.g., severe reductions of sensitive taxa over a sequence of samplings), but the duration of the study is too short to demonstrate complete recovery within eight weeks after the last treatment.
5. Clear long-term effects, lasting > 8 weeks: Convincing effects on measurement endpoints that last longer than 8 weeks after the last application.

It is apparent from the aquatic model ecosystem experiments performed with *lambda*-cyhalothrin that regardless of type of test system, initial nominal treatment concentrations specifically in the range of no to slight and transient effects (Effect Class 1 - 2) are consistent (Table III). This consistency in the findings indicates that the threshold level for 'no to slight effects' can be used with confidence as an indicator of safe concentrations in the field under the exposure conditions investigated in the studies (at least, when studies contain representatives of sensitive taxonomic groups and when exposure regimes are more or less similar). The range of concentrations at which there were clear short-term effects with recovery were more variable than the no to slight effects category, but were still reasonably consistent, with effects concentrations ranging from 16 to 50 ng/L, around a factor of 3. Note however, that in studies 1 and 2, class 5 effects were observed at nominal concentrations of 27 and 17 ng/L, respectively. However, as discussed above, the effect thresholds determined for studies 1 and 2 should be treated with some caution, since the nominal treatment concentrations in these studies may have underestimated the actual exposure concentrations.

Considering that the variability in effect class 1-2 responses observed in aquatic model ecosystem experiments is comparable to that one might observe for between study variation in the laboratory, these data are remarkably consistent. Some differences in recovery would to be expected in these types of studies, depending on the type of organism affected and its life history. The margin between the effect classes 2 and 3 (i.e. giving some indication of the steepness of the dose response curve between slight and clear effects) is around a factor 3 to 5. As discussed above, the effect thresholds determined for studies 1 and 2 should be treated with some caution, since the initial nominal treatment concentrations in these studies may have underestimated the actual exposure concentrations.

The results observed in these studies are consistent with those observed for other pyrethroids in aquatic model ecosystem stuides. A 2005 review by Van Wijngaarden et al. (*33*) on the effect thresholds for a range of insecticides including pyrethroids demonstrated a similar pattern of effects to those shown here, with microcrustaceans and insects among the more sensitive taxa.

Discussion

From the eight aquatic model ecosystem studies reviewed here, a consistent pattern of fate and effect of *lambda*-cyhalothrin emerges. As typifies synthetic pyrethroids, dissipation from the water phase was rapid, with DT_{50} values typically of around one day or less. The results were consistent irrespective of the size, location and type of system, trophic status, or season of application. Most of the studies were not designed to distinguish between dissipation and degradation, as only the concentration of the parent molecule was tracked through time in the environmental compartments. However, findings from the microcosm study of Hand *et al.* (*30*), with application of ^{14}C-radiolabelled *lambda*-cyhalothrin demonstrated that degradation via ester hydrolysis was occurring in the test system. It was proposed that one reason for this rapid degradation was the influence of plants, providing a substrate for partitioning of *lambda*-cyhalothrin from the water phase and possibly sites where degradation of the compound is facilitated. The precise mechanism by which this occurs has yet to be determined, but the study does emphasize the importance of considering the influence of aquatic plants on pesticide fate – an environmental component that is not usually considered in laboratory fate experiments or modeling. Aquatic plants have also been demonstrated to be an important factor in the fate of *lambda*-cyhalothrin in agricultural drainage ditches (*15*)

Generally speaking, only a small percentage of the applied lambda-cyhalothrin was detected in sediments, probably due to a combination of adsorption to plants and degradation in the water column and the method of application of the test substance (as solution into the water column). Since sediment residue levels were low, and sampling was only carried out for a short time, it was not possible to calculate sediment half-lives in any of the studies. Degradation rates of pyrethroids in laboratory aquatic systems have been reported to range from 7 to 80 days (16). Future studies to better define the fate of pyrethroids in sediment under field conditions would be a useful addition to the current database. Catchment monitoring studies indicate that pyrethroid residues can be found in sediments (14, 35), but in mixed landuse catchments it can be difficult to establish the relative importance of the various potential sources of these residues.

The effects observed in field studies with *lambda*-cyhalothrin are consistent with those observed in studies with other pyrethroids (*3, 4, 33*), with population responses being driven by effects on macrocrustacea and certain insects, followed by zooplanktonic microcrustacea. Considering the range of test systems, the range of locations, different trophic status of the system, differences in season of application, and differences in numbers of applications, the effects thresholds (no to slight effects) observed in the studies were remarkably consistent. This gives considerable confidence in the potential to extrapolate the no to slight effects observed in one aquatic model ecosystem study to a different situation, at least in this case where effects tend to be of an 'acute' nature, and the dissipation and degradation of the compound is rapid. The effects threshold values for clear effects with recovery were more variable than the no to slight effects, but still reasonably consistent, considering the different ecosystems that were studied.

References

1. Hill, I. R. *Pestic. Sci.* **1989**, *27*, 429-457.
2. Bradbury, S. P.; Coats, J. R. *Environ. Toxicol. Chem.* **1989**, *8*, 373-380.
3. Hill, I. R.; Shaw, J. L.; Maund, S.J. In *Freshwater Field Tests for Hazard Assessment of Chemicals;* Hill I. R.; Heimbach, F.; Leeuwangh, P.; Matthiessen, P., Eds.; Lewis Publishers: Boca Raton, FL, 1994, pp 249-271.
4. Giddings, J.M.; Solomon, K. R.; Maund S. J. *Environ. Toxicol. Chem.* **2001**, *20*, 660-668.
5. Hobson, J. F.; Sherman, J. W.; Palmieri, M. A. In : *Aquatic Toxicology and Hazard Assessment: 12th Volume*; Cowgill, U. M., Ed.; American Society of Testing and Materials, Philadelphia, PA, 1989, pp 113-119.
6. Odum, E. P. *Biosci.* **1984**, *34*, 558-562.
7. Touart, L. W. *Aquatic mesocosm tests to support pesticide registrations.* United States Environmental Protection Agency Technical Guidance Document EPA 540/09-88-035, Washington, DC, 1988. 35 pp.
8. Heimbach, F. In *Freshwater Field Tests for Hazard Assessment of Chemicals;* Hill I. R.; Heimbach, F.; Leeuwangh, P.; Matthiessen, P., Eds.; Lewis Publishers: Boca Raton, FL, 1994, pp 141-149.
9. *European Workshop on Freshwater Field Tests (EWOFFT) Summary and Recommendations;* Crossland N. O.; Heimbach, F.; Hill, I. R.; Boudou, A.; Leeuwangh, P.; Matthiessen, P.; Persoone, G., Eds.; SETAC Europe, Brussels, Belgium, 1993, 37 pp.
10. *Guidance Document on Higher-tier Aquatic Risk Assessment for Pesticides (HARAP);* Campbell, P. J.; Arnold, D. J. S.; Brock, T. C. M.; Grandy, N. J.; Heger, W.; Heimbach, F.; Maund, S. J.; Streloke, M., Eds.; SETAC Europe, Brussels, Belgium, 1999, 179 pp.
11. *Community-level Aquatic Systems Studies - Interpretation Criteria*; Giddings, J. M.; Brock, T. C. M.; Heger, W.; Heimbach, F.; Maund, S. J.; Norman, S. M.; Ratte, H.-T.; Schäfers, C.; Streloke, M., Eds.; SETAC Europe, Brussels, Belgium, 2002, 44 pp.
12. *Test Guideline 53: Guidance Document on Simulating Freshwater Lentic Field Test (Outdoor Microcosms and Mesocosms)*, Organisation for Economic Cooperation and Development, Paris, France, 2006, 37 pp.

13. Van den Brink, P. J. *Integr. Environ. Assess. Manage.* **2006**, *2*, 202-203.
14. Amweg, E. L.; Weston, D. P.; Ureda N. M. *Environ. Toxicol. Chem.* **2005**, *24*, 966-972.
15. Bennett, E. R.; Moore, M. T.; Cooper, C. M.; Smith Jr., S.; Shields Jr., D. S.; Drouillard, K. G.; Schulz, R. *Environ. Toxicol. Chem.* **2005**, *24*, 2121-2127.
16. Laskowski, D. A. Physical and chemical properties of pyrethroids. *Rev. Environ. Contam. Toxicol.* **2002,** *174*, 49-170.
17. Zhou, J. L.; Rowland, S. J. ; Mantoura, R. F. C. *Water Res.* **1995**, *29*, 1023-1031.
18. Biggs, J.; Williams, P.; Whitfield, M.; Nicolet, P.; Weatherby, A. *Conserv. Mar. Freshw. Ecosystems* **2005**, *15*, 693-714.
19. Maund, S. J.; Hamer, M. J.; Warinton, J. S.; Kedwards, T. J. *Pestic. Sci.* **1998**, *54*, 408-417.
20. Schroer, A. F. W.; Belgers, D.; Brock, T. C. M.; Matser, A.; Maund, S. J.; Van den Brink, P. J. *Arch. Environ. Contam. Toxicol.* **2004**, *46*, 324-335.
21. Maltby, L.; Brock, T. C. M.; Blake, N.; Van den Brink, P. J. *Addressing interspecific variation in sensitivity and the potential to reduce this source of uncertainty in ecotoxicological assessments (report PN0932)*. Ministry of Agriculture Fisheries and Food, London, England, 2002; 22 pp.
22. Hill, I.R.; Runnalls, J. K.; Kennedy, J. H.; Ekoniak, P. In *Freshwater Field Tests for Hazard Assessment of Chemicals;* Hill I. R.; Heimbach, F.; Leeuwangh, P.; Matthiessen, P., Eds.; Lewis Publishers: Boca Raton, FL, 1994, pp 345-360.
23. Hadfield, S. T.; Sadler, J. K.; Bolygo, E.; Hill, S.; Hill, I. R. *Pestic. Sci.* **1993**, *38*, 283-294.
24. Farmer, D.; Hill, I. R.; Maund, S. J. *Ecotoxicol.* **1995**, *4*, 219-224.
25. Roessink, I.; Arts, G. H. P.; Bransen, F.; Maund, S. J.; Belgers, D.; Brock, T. C. M. *Environ. Toxicol. Chem.* **2005**, *24*, 1684-1696.
26. Leistra, M.; Zweers, A. J.; Warinton, J. S.; Crum, S. J.; Hand, L. H.; Beltman, W. H. J.; Maund, S. J. *Pest Manage. Sci.* **2003**, *60*, 75-84.
27. Van Wijngaarden, R. P. A.; Brock, T. C. M.; Van den Brink, P. J.; Gylstra, R.; Maund, S. J. *Arch. Environ. Contam. Toxicol.* **2006**, *50*, 220-239.
28. Van Wijngaarden, R. P. A.; Cuppen, J. G. M.; Arts, G. H. P.; Crum, S. J. H.; Van den Hoorn, M. W.; Van den Brink, P. J.; Brock, T. C. M. *Environ. Toxicol. Chem.* **2004**, *23*,1479-1498.
29. Arts, G. H. P.; Buijse-Bogdan, L. L.; Belgers, J. D. M.; Van Rhenen-Kersten, C. H.; Van Wijngaarden, R. P. A.; Roessink, I.; Maund, S. J.; Brock, T. C. M. *Integr. Environ. Assess. Manage.* **2006**, *2*, 105-125.
30. Hand, L. H.; Kuet, S. F.; Lane, M. C. G.; Maund, S. J.; Warinton, J. S.; Hill, I. R. *Environ. Toxicol. Chem.* **2001**, *20*, 1740-1745.
31. Muir, D. C. G.; Rawn, G. P.; Grift, N. P. *J. Agric. Food Chem.* **1985**, *33*, 603-609.

32. Wendt-Rasch, L.; Van den Brink, P. J.; Crum, S. J. H.; Woin, P. *Ecotoxicol. Environ. Saf.* **2004**, *57*, 383-398.
33. Van Wijngaarden, R.P.A.; Brock, T.C.M.; Van den Brink, P. J. *Ecotoxicol.* **2005**, *14*, 353-378.
34. Brock, T. C. M.; Lahr, J.; Van den Brink, P. J. *Ecological risks of pesticides in freshwater ecosystems. Part 1: Herbicides. Final Report 088.* Alterra Green World Research, Wageningen, The Netherlands, 2000; 127 pp.
35. Budd, R.; Bondarenko, S.; Haver, D.; Kabashima, J.; Gan, J. *J. Environ. Qual.* **2007**, *36*, 1006-1012.

Chapter 16

Patterns of Pyrethroid Contamination and Toxicity in Agricultural and Urban Stream Segments

C. M. Ng[1], D. P. Weston[1], J. You[2], and M. J. Lydy[2]

[1]Department of Integrative Biology, University of California, 3060 Valley Life Sciences Building, Berkeley, CA 94720–3140
[2]Department of Zoology, Fisheries and Illinois Aquaculture Center, Southern Illinois University, Life Science II Building, Room 173, Carbondale, IL 62901

Pyrethroid insectides are in widespread use in both agricultural and urban environments. In order to understand if there are systematic differences in the composition of pyrethroid mixtures found in sediment arising from runoff from these two land uses, and to compare their toxicological effects, sediment samples were collected from three creeks in and around Salinas, California. Pyrethroids were present in sediments from both agricultural and urban reaches of all three creeks. Sediment from all sampling locations in both agricultural and urban areas was toxic to *Hyalella azteca*, an amphipod commonly used for sediment testing, and, in all cases there was sufficient mass of pyrethroid present in the sediment to explain the measured toxicity. The organophosphate chlorpyrifos likely contributed to toxicity in one instance. While the compositional differences in sediment pyrethroid mixtures between the land uses were not dramatic, there was a tendency for cyfluthrin and cypermethrin to be typical of urban areas, and lambda-cyhalothrin to be found in agricultural reaches. Bifenthrin and permethrin were somewhat characteristic of urban and agricultural areas, respectively, though either land use could be a potential source.

Introduction

Pesticides are used widely in urban areas, where insects are both nuisances and, in some cases, vectors for disease. Due to the withdrawal of some of the most widely used organophosphates, pyrethroid pesticides are now used extensively in urban environments, whether applied by homeowners or professional pest controllers. Over 327,000 kg of pyrethroids were used by professional applicators for structural pest control and landscape maintenance in California in 2005 (www.cdpr.ca.gov/docs/pur/purmain.htm), and although data are not publicly reported, retail sales to homeowners can be assumed to be considerable. Recent research points to bifenthrin, cyfluthrin and cypermethrin, as the greatest cause for concern in creeks within residential areas, being the most frequent contributors to aquatic toxicity in streams in and around Sacramento, California (*1, 2*).

While a dramatic increase in commercial and home use of pyrethroids has been reported, agricultural use of pyrethroids has been relatively steady in California over the past decade, ranging from a low of 105,000 kg in 1999 up to 142,000 kg in 2005 (the most current data available). However, some industry segments, like almond and stone fruit production, have reported a reduction in organophosphate use with an increased use of pyrethroids *(3)*. As a result of this widespread use, agriculture-affected water bodies may contain pyrethroid residues in the sediments, with permethrin the most commonly found *(4, 5)*. However, since permethrin is among the least toxic of the pyrethroids to aquatic life (*6*), the pyrethroids bifenthrin, lambda-cyhalothrin and esfenvalerate are more frequently found at concentrations associated with toxicity. Pyrethroids are believed to be responsible for toxicity in agricultural-affected sediment samples in about 60% of the instances when toxicity to the standard toxicity testing species, *Hyalella azteca*, is observed (*5*).

Thus, pyrethroids are widely used in both agriculture and urban areas, and both uses have resulted in sediment contamination of creeks within the watersheds. However, when both agricultural and urban areas are in close proximity to one another, it may be difficult to distinguish the sources of the pesticides. Downstream toxicity may not be traceable to a single well-defined source since both urban and agricultural subwatersheds can deliver runoff into the same waterbody, and thus contribute pyrethroids and aquatic toxicity to that water body. In order to make informed management decisions, take regulatory action, or inititate mitigation, it is necessary to be able to discriminate among the potential pyrethroid sources.

While research has been conducted on pyrethroids from agricultural and urban land uses, comparisons between the two different uses have not been made. This study explores the relative toxicological impact and compositional differences in the pyrethroid mixtures of urban and agricultural areas. If such differences exist, it may be possible to develop characteristic "fingerprints" of pyrethroids from both land uses in order to guide management actions.

Materials and Methods

Study area

Salinas, California was chosen as the study site because of the close juxtaposition of urban and agricultural land uses. Salinas is the county seat of Monterey County, and a major urban center with a population of approximately 150,000 people. In addition to residential housing, the city includes associated commercial and industrial development, much of which supports the agricultural industry. The farmland surrounding the city produces salad vegetables (e.g., lettuce, spinach) as well as many other fruits and vegetables (e.g., strawberries, broccoli, cauliflower, celery, artichokes, and wine grapes). Agricultural production is heavily dependent on irrigation, for annual rainfall is approximately 33 cm, and largely limited to November through April. The city itself is surrounded by agricultural lands, but is unique in that is has a flood control basin in the center of the city, Carr Lake Regional Park, which is also used for agriculture (http://www.ci.salinas.ca.us).

Three creeks run through the city: Gabilan Creek, Natividad Creek, and Alisal Creek, the later being renamed Reclamation Ditch at a point southeast of Salinas where the water course turns to the northwest (Figure 1). All three creeks originate in undeveloped land in hills northeast of the city, flow through agricultural lands, through the city, and then back in to agricultural lands. The three water courses join in the Carr Lake area. The combined flow from all three water courses leaves Carr Lake via the Reclamation Ditch, which flows northwest and finally empties into Tembladero Slough and ultimately into the Pacific Ocean in Monterey Bay. Flow into the creeks varies dramatically with season. During the winter, large storm events produce the greatest amount of flow through the creeks (e.g., up to about 200 cfs at US Geological Survey gage in Reclamation Ditch; http://waterdata.usgs.gov/ca/nwis). During the summer, flow rates are very low (about 3 cfs in Reclamation Ditch) and the minimal water present is return flow from irrigated agricultural fields or urban runoff from landscape irrigation.

Sampling Procedures

Background samples, intended to have little or no pesticide residue, were taken from Gabilan and Alisal Creeks upstream of any urban or agricultural development (Table I; Stations SG1 and SA1). There was no comparable background site accessible on Natividad Creek. Two to three additional sampling sites were established along each watercourse as they passed through agricultural lands, and then through the city of Salinas. When possible, particular effort was made to establish sites just upstream of transition points between agricultural and urban land uses, so that those sites would be indicative of the integrated effects of the upstream land use (e.g., agricultural) and just

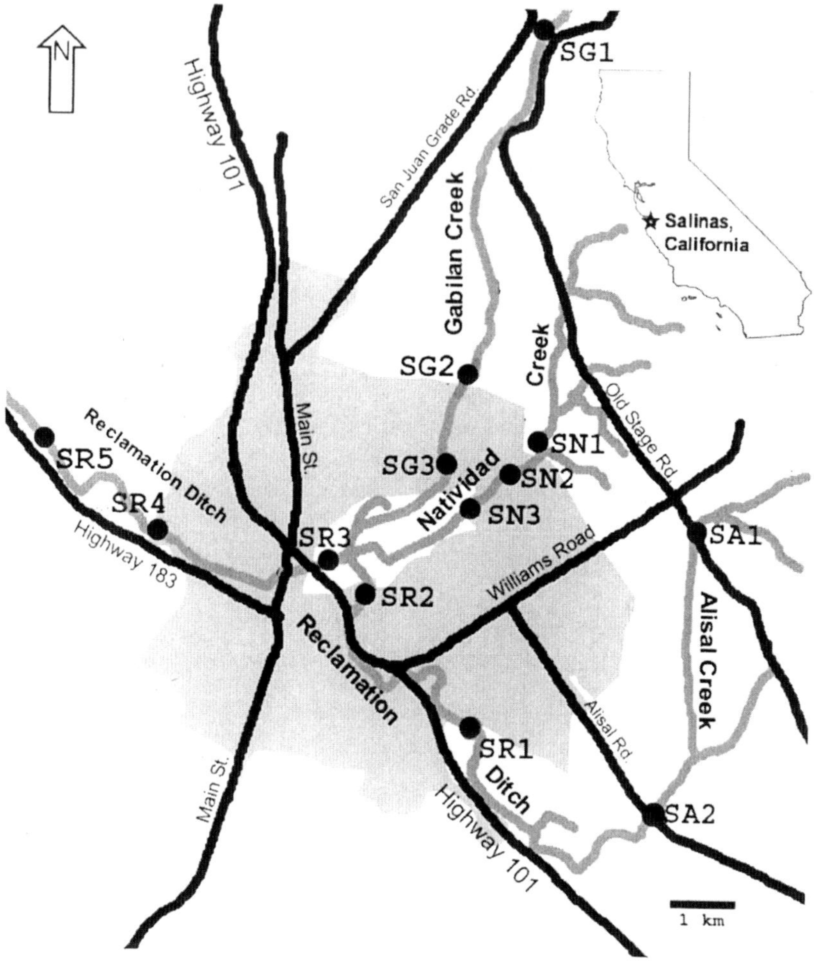

Figure 1. Map of Salinas (a) illustrating the creeks sampled, Alisal Creek (SA1, SA2) and Reclamation Ditch (SR1, SR2, SR3, SR4, SR5), Gabilan Creek (SG1, SG2, SG3), and Natividad Creek (SN1, SN2, SN3). Flow is generally from the east to the west. The urban areas are shaded gray. The white agricultural area in the center of the city represents Carr Lake.

Table I. Locations and descriptions of sampling sites along the three water courses in and around Salinas

Site name	N Latitude W Longitude	Site description	Surrounding land use
Alisal Creek/Reclamation Ditch			
SA1	36.69238 121.56915	Alisal Creek @ Old Stage Rd.	Point where creek transitions from undeveloped land to agricultural land
SA2	36.64567 121.57698	Alisal Creek @ Alisal Rd.	Agriculture
SR1	36.65858 121.61379	Reclamation Ditch @ Moffett St.	Agricultural area, just prior to creek entering commercial district
SR2	36.67978 121.63735	Reclamation Ditch @ Cesar Chavez Park	Mixed commercial and residential
SR3	36.68507 121.64772	Reclamation Ditch @ Sherwood Dr.	Edge of agricultural Carr Lake, just prior to creek entering commercial district
SR4	36.68426 121.66735	Reclamation Ditch @ Victor St. and Victor Way	Commercial district, just prior to creek entering agricultural lands
SR5	36.70475 121.70525	Reclamation Ditch @ San Jon Rd.	Agriculture
Natividad Creek			
SN1	36.70202 121.60262	Natividad Creek @ Boronda Rd.	Agriculture just prior to creek entering residential area
SN2	36.69887 121.61067	Natividad Creek @ Freedom Pkwy.	Residential
SN3	36.69020 121.62151	Natividad Creek @ Gee St.	Residential
Gabilan Creek			
SG1	36.78040 121.58541	Gabilan Creek @ Old Stage Rd.	Undeveloped land
SG2	36.71553 121.61643	Gabilan Creek @ Boronda Rd.	Agriculture just prior to creek entering residential area
SG3	36.70030 121.62196	Gabilan Creek @ Independence Blvd. and Lexington Dr.	Residential

prior to the inputs from the downstream land use (e.g., urban). Urban portions of Natividad and Gabilan creeks consisted largely of single-family residential development, with only minor commercial influence. Urban sites along Reclamation Ditch were a mix of residences, commercial establishments, and industry.

Samples were collected on September 23, 2005, prior to the onset of the winter rains. At each site, the upper one centimeter of the surface sediment in the creek beds was skimmed off with a stainless steel scoop and transferred into solvent-cleaned glass jars. The finest-grained sediments (silts and clays) available at each site were collected since pyrethroids are strongly hydrophobic and associate with the organic fractions of the sediment. In the lab, sediment was homogenized by hand mixing, and then held at 4°C for toxicity samples, and -20°C for chemistry samples.

Analytical Methods

Chemical analysis of the sediment was done using the methods outlined in You et. al. (*7*). Briefly, the sediment sample was sonicated with 50 ml of a 50:50 mixture of acetone and methylene chloride. Three extractions were done, with the extracts combined and solvent exchanged to hexane. Clean-up was performed using Florisil (Thermo Fisher Scientific, Waltham, MA), deactivated with distilled water, and elution from the column with 30% diethyl ether in hexane. Florisil extracts were solvent exchanged to hexane, reduced to 1 ml, 25 mg of primary/secondary amine (PSA) was added, and the samples shaken for 2 min. Following centrifugation, the supernatant was analyzed on an Agilent 6890 series gas chromatograph with an Agilent 7683 autosampler and an electron capture detector (Agilent Techologies, Palo Alto, CA). Two columns from Agilent, a HP-5MS, and a DB-608 were used. The seven pyrethroids quantified were: bifenthrin, lambda-cyhalothrin, esfenvalerate, deltamethrin, permethrin, cyfluthrin, and cypermethrin. Analytes also included one organophosphate, chlorpyrifos, and 21 organochlorines, including: alpha-, beta-, delta-, and gamma-BHC, heptachlor, heptachlor epoxide, alpha- and gamma-chlordane, alpha- and beta-endosulfane, endosulfan sulfate, p,p'-DDE, p,p'-DDD, p,p'-DDT, aldrin, dieldrin, endrin, endrin aldehyde, endrin ketone, and methoxychlor.

Grain size was determined using wet sieving, and total organic carbon was measured using a CE-440 Elemental Analyzer from Exeter Analytical (Chelmsford, MA), following acid vapor treatment to remove inorganic carbon.

Toxicity Testing

Ten-day toxicity tests were performed using 7-10 day old freshwater amphipods, *H. azteca,* according to standard U.S. Environmental Protection

Agency protocols *(8)*. Using 8 replicates for each sediment sample, about 50-75 mL of sediment, and about 250 mL of overlying water were added to 400 ml glass beakers. Tests were conducted at 23°C, with a 16 h light: 8 h dark cycle, with feeding of 1 ml of yeast/cerophyll/trout chow per beaker per day. Fresh water was delivered with an automatic water delivery system that provided two volume additions (500 ml) daily using Milli-Q purified water, made moderately hard by added salts. Water samples for pH, conductivity, alkalinity, hardness, and ammonia were taken at the beginning and end of the test; dissolved oxygen and temperature were monitored regularly. Mortality of amphipods was determined by sieving sediment on a 425 μm screen, and determining the proportion of the initial 10 amphipods per beaker that survived the 10-d exposure.

Toxicity data was analyzed using ToxCalc 5.0 software (Tidepool Scientific Software, McKinleyville, CA). Each batch of test sediments tested included a control sediment from San Pablo Dam Reservoir (Orinda, CA), and survival in test sediments was statistically compared to the control using a t-test with arcsine transformation. Control survival ranged from 86-95%.

The concentrations of each pyrethroid in the sediments were used to calculate toxic units (TU) with respect to *H. azteca* as:

$$TU = \frac{\text{Actual concentration of pyrethroid in sediment}}{\text{Known 10-d LC50 for } H.\ azteca}$$

Since pyrethroids are strongly hydrophobic, both the actual concentration and the LC50 were organic carbon (oc) normalized. The reported 10-d sediment LC50 values were as follows: cypermethrin = 0.38 μg/g oc, lambda-cyhalothrin = 0.45 μg/g oc, bifenthrin = 0.52 μg/g oc, deltamethrin = 0.79 μg/g oc, cyfluthrin = 1.08 μg/g oc, esfenvalerate = 1.54 μg/g oc, permethrin = 10.83 μg/g oc (*9, 10*). Pyrethroid TUs were assumed to be additive due to the common mode of action of compounds within the class.

Results

The sediment samples consisted of fine-grained material ranging from 20-85% fines (silts and clays combined) with a median of 41% fines. The percent total organic carbon of the sediment samples ranged from 0.6 – 4.4% with a median of 2.0%.

All sediments were tested for acute toxicity to *H. azteca*, and only minimal mortality was seen in the designated background sites, prior to the creeks entering agricultural lands (Table I: SA1 and SG1). SA1 had only 4% mortality; SG1 had 14% mortality. While the later value was statistically different (probability < 0.05) from the concurrent control sample with 5% mortality, the mortality rate in a later control test was comparable to the SG1 sample, and the 14% mortality seen at SG1 is not considered to be a meaningful difference.

The remaining 11 other sediment samples collected in the study were significantly toxic, with mortality rates ranging from 31-100% (Figure 2). The highest mortality was seen in two urban sites; the mixed use urban site of SR2 (100% mortality) and the residential area of SN3 (96%). Substantial toxicity was seen in agricultural sites as well, with 90% mortality at SN1, 84% mortality at SA2, and 84% mortality at SG2. All three of these sites were in agricultural reaches of their respective creeks, prior to the creeks entering any urban development. There was little overall difference in the toxicity of agricultural and urban reaches, with a median mortality of 68% among the urban sites and 75% among the agricultural sites.

The two background sites contained no detectable pyrethroids (Table II). However, pyrethroids were present at every other site, whether in areas of agricultural or urban land use. Permethrin was the dominant pyrethroid, and generally typified the agricultural reaches of the creeks. However, it was also found in some urban areas (e.g. SR2, SN2, SN3). It can not be conclusively determined from the existing data whether the permethrin residues in urban areas represent input from the surrounding urban landscape, or transport from more upstream agricultural areas. At only one site (SA2) was the permethrin concentration above the estimated 10-d sediment LC50 for *H. azteca*. At several sites concentrations were about one-third that threshold.

Bifenthrin was present at most sites, and its concentration reached at least half the *H. azteca* LC50 at five sites (SR1, SR5, SN2, SN3, SG3). On Natividad and Gabilan Creeks, the compound was clearly associated with urban land uses, with no measurable bifenthrin in sediments from agricultural regions, but then increasing to over 10 ng/g in urban areas. In Reclamation Ditch the data suggest both urban and agricultural bifenthrin sources.

Among the other pyrethroids, lambda-cyhalothrin tended to be associated with agricultural reaches, and attained concentrations at least half the LC50 at three sites (SA2, SR1, SR2). Cypermethrin and cyfluthrin attained their highest concentrations in urban reaches. Esfenvalerate concentrations were far below the LC50, and the compound was not clearly associated with one particular land use.

Sediment concentration data for the non-pyrethroid analytes are not shown, but concentrations were generally not toxicologically significant at least with respect to explaining *H. azteca* mortality results. Chlorpyrifos was nearly always below 20 ng/g, which would represent about one-third of a TU given the median organic carbon content in the samples (2.0%) and the reported chlorpyrifos LC50 to *H. azteca* (2.97 μg/g oc; (*11*)). The sole exception was SR5 where chlorpyrifos reached 68 ng/g, or 1.6 TU given the organic carbon content at this site (1.4%). The organochlorine pesticides or their degradation products were frequently detected but well below acutely toxic concentrations to *H. azteca*. The most commonly detected were DDE (maximum 254 ng/g), DDD (max. 234 ng/g), DDT (max. 152 ng/g), dieldrin (max. 40 ng/g), endrin (max. 14.9 ng/g),

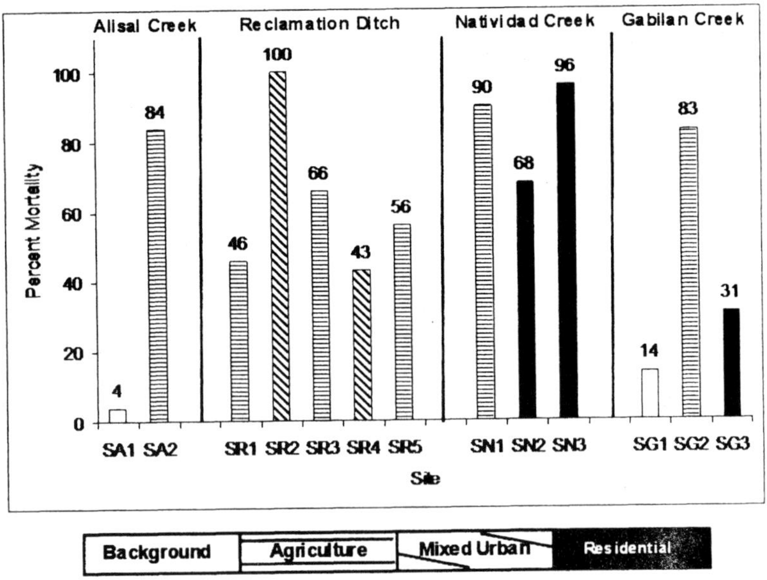

Figure 2. Percent mortality of Hyalella azteca when exposed to sediment from the sampling sites. Shading indicates the classification of each site into background, agricultural, mixed urban (residential/commercial/industrial), or residential.

alpha-chlordane (max. 8.5 ng/g) and gamma-chlordane (max. 7.0 ng/g). These concentrations were all below 0.1 TU given the LC50 estimates of Weston et al., (*4*).

The pyrethroid concentrations alone showed a strong relationship to *H. azteca* mortality as observed in the sediment toxicity tests (Figure 3). Not only did mortality show a significant increase concurrently with increasing pyrethroid TUs, but the overall pattern suggested 50% mortality occurred at about one TU (0.6-1.4 depending on sample), precisely the relationship that would be expected if pyrethroids were the dominant contributor to toxicity. Assuming additive toxicity among the pyrethroids such that the compound-specific TUs could be added to derive a total pyrethroid TU, every site, excluding the two background locations, contained at least 0.5 TU. Six of the eleven sites reached or exceeded one TU. Even without the additivity assumption, a strong pyrethroid contribution to toxicity is still suggested with eight of the eleven sites reaching at least 0.5 TU and three reaching one TU.

There is also limited evidence from toxicity identification evaluation (TIE) procedures for a contributing role of pyrethroids. Sample SR2, which contained potentially toxic concentrations of lambda-cyhalothrin and cypermethrin, was

Table II. Pyrethroid concentrations (ng/g, dry weight basis) in the sediments at the sampling sites, with sites shaded based upon surrounding land use. ND indicates not detected (<1 ng/g). Deltamethrin was among the analytes but was never detected at any site. The number of Hyalella azteca toxic units (TU) each concentration value represents, given the sediment organic carbon content, is shown in parentheses. Bifenthrin = Bif, Cyfluthrin = Cyf, Cypermethrin = Cyp, Esfenvalerate = Esf, Lambda-cyhalothrin= Lam, Permethrin = Per

Background	Agricultural	Mixed urban	Residential

Site and land-use	Total organic carbon (%)	Bif	Cyf	Cyp	Esf	Lam	Per
Alisal Creek/Reclamation Ditch							
SA1	3.57	ND	ND	ND	ND	ND	ND
SA2	0.56	ND	ND	ND	ND	1.6 (0.6)	72.0 (1.2)
SR1	2.51	7.4 (0.6)	ND	2.1 (0.2)	4.3 (0.1)	5.5 (0.5)	14.1 (0.1)
SR2	1.84	4.0 (0.4)	3.5 (0.2)	7.0 (1.0)	3.4 (0.1)	6.8 (0.8)	82.5 (0.4)
SR3	1.99	3.4 (0.3)	ND	ND	1.6 (0.1)	2.0 (0.2)	67.8 (0.3)
SR4	2.00	1.2 (0.1)	1.1 (0.1)	2.7 (0.4)	ND	ND	8.1 (<0.1)
SR5	1.39	4.0 (0.6)	ND	ND	1.0 (0.1)	ND	ND
Natividad Creek							
SN1	0.93	ND	ND	ND	ND	6.8 (1.6)	9.0 (0.1)
SN2	2.99	10.5 (0.7)	3.7 (0.1)	5.1 (0.5)	1.4 (<0.1)	3.0 (0.2)	11.3 (<0.1)
SN3	2.15	8.8 (0.8)	ND	4.6 (0.6)	1.0 (<0.1)	ND	9.3 (<0.1)
Gabilan Creek							
SG1	4.40	ND	ND	ND	ND	ND	ND
SG2	1.87	ND	ND	2.0 (0.3)	ND	1.0 (0.1)	68.8 (0.3)
SG3	4.02	10.7 (0.5)	ND	1.0 (0.1)	ND	ND	5.4 (<0.1)

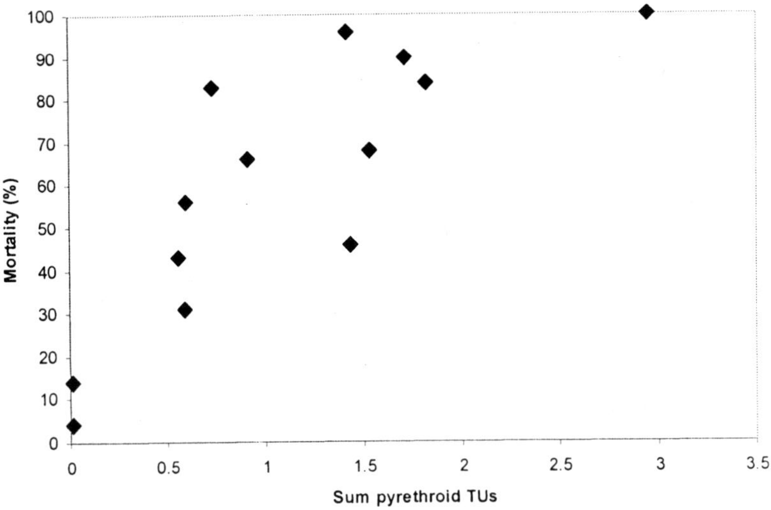

Figure 3. Graph of the percent mortality of Hyalella azteca at each site in Salinas, CA in relation to the sum of pyrethroid toxic units (TU).

tested with addition of an esterase enzyme to the overlying water (*5*). The enzyme is intended to cleave the ester bond present in pyrethroids, substantially reducing the toxicity. Without esterase, the SR2 sediment caused near complete mortality; with esterase 38% of the *H. azteca* survived (*12*), supporting the suspected role of pyrethroids in explaining the toxicity. Results from TIE manipulation of a second sample containing lambda-cyhalothrin at probable toxic concentrations are less conclusive. Sediment SN1 was tested in a dilution series with addition of piperonyl butoxide (PBO) to the overlying water, a procedure which makes pyrethroids more toxic. Without PBO in the overlying water the LC50 of SN1 sediment was 29.8% (expressed as percent original sediment when diluted with control sediment; 95% confidence interval of 23.3-38.2%) (*11*). With PBO, the LC50 was reduced to 19.8% (16.2-24.3%). While the PBO did increase the toxicity (decreasing the LC50) as expected if lambda-cyhalothrin were the toxicant, the decrease was not as dramatic as usually seen with PBO, and the LC50 confidence intervals did slightly overlap. Thus, the results from the SN1 sample were inconclusive and the presence of another unidentified toxicant in the sample remains a possibility.

Discussion

It is clear that pyrethroids from both agricultural and urban uses are reaching the creeks of Salinas, and that they are usually present in the fine sediment in these creeks at concentrations acutely toxic to *H. azteca*, a species widely used for sediment toxicity assessment. This observation is consistent with prior studies in agricultural areas of California (*4, 5, 13, 14*) and work in urban creeks of the Sacramento, California and San Francisco Bay areas (*1, 2*). Toxicity testing with *H. azteca* of both agricultural and urban reaches of Salinas creeks commonly showed acute mortality, and there is evidence from both toxic unit analysis and TIE procedures that pyrethroids were the major contributor to this toxicity. The organophosphate chlorpyrifos was also likely a contributor at one agricultural site. Although this compound no longer has appreciable use in urban environments, it is still widely used in agriculture.

This study indicates that the differences between an agricultural pyrethroid 'fingerprint' and an urban one are not dramatic, but yet some distinctions could be made. Cyfluthrin and cypermethrin were characteristic of urban-affected stream reaches. Bifenthrin was also commonly found and attained highest concentrations in sediments located in urban areas, though it has agricultural sources and uses as well. On the other hand, lambda-cyhalothrin was distinctly found in agriculture-affected samples. Permethrin was characteristic of sediments found in areas with both land uses.

California is unique in that commercial use of pesticides requires reporting of that use to the California Department of Pesticide Regulation, including the compound applied and the amount used. For the most part, the pesticide-related land use distinctions made on the basis of the Salinas creek data are supported by reported use data from California as a whole, and specifically from Monterey County in which Salinas is located (Table III). The usage data as well as the environmental monitoring both support the primarily urban sources of cyfluthrin, the agricultural sources of lambda-cyhalothrin, and the dual sources of permethrin and bifenthrin. The only significant difference between the Salinas findings and pesticide use data is cypermethrin, for which dominant urban use is suggested by the creek sediment data and from statewide use statistics, but in Monterey County use is primarily agricultural.

The use data (Table III) also suggests that some other pyrethroids are distinctly urban or agricultural, though those distinctions could not be made with the Salinas creek data set. The presence of deltamethrin would be a clear marker of urban sources, since its agricultural use is negligible. Similarly esfenvalerate and fenpropathrin are likely to be from agricultural sources because of very limited non-agricultural use. It should, however, be recognized that Table III excludes retail sales, as that data are not tracked by California agencies with the level of detail available for commercial pesticide applications. For example, esfenvalerate can be found in some retail products sold for home and garden use. Finally, it should be recognized that these distinctions apply only to pyrethroid use in California. There are likely to be regional differences in crops produced and pesticides applied which prevent broad national generalizations. A pyrethroid that may have only non-agricultural uses in one area of the country could be a significant agricultural insecticide in another, and thus the distinctions made here would have to be reassessed in other locations.

Table III. Relative agricultural and non-agricultural commercial use of pyrethroids in California as a whole and in Monterey County in which Salinas is located (2005 data; www.cdpr.ca.gov/docs/pur/purmain.htm). Non-agricultural use consists largely of applications by professional pest control firms, and figures do not include retail sales to homeowners for which comparable data are not available. The table also shows whether the compound was characteristic of urban or agricultural stream segments in the current study.

Pyrethroid	Statewide agricultural use (kg)	Statewide non-agricultural use (kg)	Monterey County agricultural use (kg) and primary crop	Monterey County non-agricultural use (kg)	Finding in current Salinas study
Bifenthrin	9,439	18,748	297 Strawberries	175	Largely urban but some agricultural
Cyfluthrin	7,810	14,526	11 Lettuce	63	Urban
Cypermethrin (including S-cypermethrin	14,070	92,068	3138 Lettuce	140	Urban
Deltamethrin	38	6,238	0	19	Not detected
Esfenvalerate	14,780	118	1555 Artichokes, lettuce, broccoli	0	Undeter-mined
Fenpropathrin	17,940	3	2295 Grapes, strawberries	0	Not measured
Lambda-cyhalothrin	10,296	6,298	1390 Lettuce	2	Agricultural
Permethrin	67,796	183,110	9900 Lettuce, spinach, celery	519	Both urban and agric.

The very fact that there were differences in pyrethroid composition among the sampling sites suggests that the sediments on which the pyrethroids are adsorbed may be transported fairly limited distances. Water-soluble pesticides can travel considerable distances (*15*), but being particle-associated, pyrethroid dispersal may be more limited. Our most downstream site, SR5, located approximately 4 km downstream of Salinas contained only esfenvalerate and bifenthrin, with no evidence of the lambda-cyhalothrin, permethrin, cypermethrin and cyfluthrin known to be present in more upstream locations. This conclusion may be a consequence of the timing of sampling. Sediments were collected in September, near the end of the dry season when flow is very low and limited to irrigation runoff. Major winter storm events, typically beginning in December in the Salinas area, may be important in promoting sediment transport over greater distances, and blurring the land use distinctions evident in dry season sampling.

Acknowledgements

This work would not have been possible without funding from the Surface Water Ambient Monitoring Program (SWAMP) of the California State Water Resources Control Board. We particularly wish to thank Karen Worcester and Mary Adams of the Central Coast Regional Water Quality Control Board for their advice and support.

References

1. Weston, D.P.; Holmes, R.W.; You, J.; Lydy, M.J. Aquatic toxicity due to residential use of pyrethroid insecticides. *Environ. Sci. Technol.* **2005**, *39*, 9778-9784.
2. Amweg E. L.; Weston, D. P.; You, J.; Lydy, M. J. Pyrethroid insecticides and sediment toxicity in urban creeks from California and Tennessee. *Environ. Sci. Technol.* **2006**, *40*, 1700-1706.
3. Epstein, L.; Bassein, S.; Zalom, F.G. Almond and stone fruit growers reduce OP, increase pyrethroid use in dormant sprays. *California agriculture.* **2000**, *54* (6), 14-19.
4. Weston, D. P.; You, J.; Lydy, M. J. Distribution and toxicity of sediment-associated pesticides in agriculture-dominated water bodies of California's Central Valley. *Environ. Sci. Technol.* **2004**, *38*, 2752-2759.
5. Weston, D.P.; You, J.; Amweg, E.L.; Lydy, M.J. Sediment toxicity in agricultural areas of California and the role of hydrophobic pesticides. This volume.
6. Solomon, K.R.; Giddings. J.M.; Maund, S.J. Probabilistic risk assessment of cotton pyrethroids: I. Distributional analysis of laboratory aquatic toxicity data. *Environ. Toxicol. Chem.* **2001**, *20*, 652-659.
7. You, J.; Weston, D.P.; Lydy, M.J. A sonication extraction method for the analysis of pyrethroid, organophosphate, and organochlorine pesticides from sediment by gas chromatography with electron-capture detection. *Arch. Environ. Contam. Toxicol. Chem.* **2004**, *24*, 141-147.

8. U.S. EPA. *Methods for Measuring the Toxicity and Bioaccumulation of Sediment-Associated Contaminants with Freshwater Invertebrates;* EPA Publication 600/R-99/064; U.S. Environmental Protection Agency: Washington DC, **2000**. 192 pp.
9. Amweg E.L.; Weston, D.P.; Ureda, N.M. Use and toxicity of pyrethroid pesticides in the Central Valley, California, U.S.A. *Environ. Toxicol. Chem.* **2005**, *24*, 966-972 (with correction in *Environ. Toxicol. Chem.* **2005**, *24*, 1300-1301).
10. Maund, S.J.; Hamer, M.J.; Lane, M.C.G.; Farrelly, E.; Rapley. J.H.; Goggin, U.M.; Gentle, W.E. Partitioning, bioavailability, and toxicity of the pyrethroid insecticide cypermethrin in sediments. *Environ. Toxicol. Chem.* **2002**, *21*, 9-15.
11. Amweg, E.L.; Weston, D.P. Whole sediment toxicity identification evaluation tools for pyrethroid insecticides: I. Piperonyl butoxide addition. *Environ. Toxicol. Chem.* (In press).
12. Weston, D.P.; Amweg, E.L. Whole sediment toxicity identification evaluation tools for pyrethroid insecticides: II. Esterase addition. *Environ. Toxicol. Chem.* (In press).
13. Anderson, B.S.; Phillips, B.M.; Hunt, J.W.; Worcester, K.; Adams, M.; Kapellas, N.; Tjeerdema, R.S. Evidence of pesticide impacts in the Santa Maria River watershed, California, USA. *Environ. Toxicol. Chem.* **2006**, *25*, 1160-1170.
14. Phillips, B.M.; Anderson, B.S.; Hunt, J.W.; Huntley, S.A.; Tjeerdema, R.S.; Kapellas, N.; Worcester, K. Solid-phase sediment toxicity identification evaluation in an agricultural stream. *Environ. Toxicol Chem.* **2006**, *25*, 1671-1676.
15. Kuivila, K.M.; Foe, C.G. Concentrations, transport and biological effects of dormant spray pesticides in the San Francisco Estuary, California. *Environ. Toxicol. Chem.* **1995**, *14*, 1141-1150.

Chapter 17

Recent Advances in Sediment Toxicity Identification Evaluations Emphasizing Pyrethroid Pesticides

B. S. Anderson[1,3], B. M. Phillips[1], J. W. Hunt[1], J. Voorhees[1], S. Clark[1], A. Mekebri[2], D. Crane[2], and R. S. Tjeerdema[1]

[1]Department of Environmental Toxicology, University of California at Davis, Davis, CA 95616
[2]California Department of Fish and Game, Water Pollution Control Laboratory, Rancho Cordova, CA 95814
[3]Current address: c/o Marine Pollution Studies Laboratory, 34500 Highway 1, Monterey, CA 93940 (email: anderson@ucdavis.edu)

Recent monitoring in California has indicated that sediment toxicity associated with pyrethroid pesticides is of concern in urban and agriculture-dominated watersheds. In some cases, waterbodies that are listed as impaired due to sediment toxicity may require development of Total Maximum Daily Load (TMDL) allocations to regulate chemicals of concern. This process will require identification of specific chemicals causing toxicity, and sediment toxicity identification evaluation procedures (TIEs) are one of the primary tools used in this process. This paper provides an overview of sediment TIE methods with an emphasis on those which have been demonstrated to be useful in resolving toxicity caused by pyrethroid pesticides. These include use of media for extracting non-polar organic chemicals from sediment and interstitial water and use of a carboxylesterase enzyme for hydrolyzing pyrethroids. Additional methods that have proven useful for identifying pyrethroid toxicity include addition of the metabolic inhibitor piperonyl butoxide, and manipulations of sample toxicity test temperatures. Methods are described for both solid-phase and interstitial water TIEs. In addition to discussing specific TIE methods, examples of studies that have successfully used the different procedures are cited. Results of two sediment TIE case studies which incorporate the majority of these methods are also provided to illustrate the utility of using a weight-of-evidence in the TIE approach. Results from the case study experiments conducted using samples from Westley Wasteway Creek (WWNCR) and an agriculture tailwater pond in the Salinas Valley (SV03) demonstrate that sediment toxicity at WWNCR was likely due to L-cyhalothrin and bifenthrin, and that toxicity at SV03 was likely due to L-cyhalothrin and cypermethrin. The results suggest that current solid-phase and interstitial water TIE methods are adequate for identifying toxicity caused by pyrethroid pesticides. Refinement of these methods will improve our ability to resolve causes of toxicity in sediments contaminated by complex chemical mixtures.

Introduction

Management of polluted sediments in the U.S. is guided by the Federal Clean Water Act (CWA), which requires the establishment of standards to protect water quality. Because of the complex geochemical processes that govern bioavailability of sediment-associated contaminants, few numeric sediment quality standards have been promulgated. In place of numeric standards, many states use narrative standards to determine whether sediment contaminants are impacting beneficial uses of surface waters. Narrative standards are based on the CWA requirement that waters be free of toxic substances in toxic amounts. Sediment toxicity tests are often used to determine whether sediments have the potential to impair aquatic life in contaminated areas. Once water bodies are listed as impaired due to sediment toxicity, states may be required to develop Total Maximum Daily Load (TMDL) allocations to regulate chemicals of concern. A key step in the TMDL process is the identification of chemicals responsible for toxicity because this ensures that mitigation resources are directed at reducing loads of appropriate chemicals of concern. Sediment toxicity identification evaluation procedures (TIEs) are one of the primary tools used in this process.

A number of recent studies have suggested pyrethroid pesticide contamination is common in freshwater drainages in California, including in the central valley (*1,2*), the central coast (*3,4,5*) and in the San Francisco Bay area (*6*). These studies have sometimes indicated that pyrethroids are a source of sediment toxicity to amphipods. Sediment toxicity associated with pyrethroids has also recently been demonstrated in several marine sites in southern California (*7*). As monitoring proceeds in California it can be presumed that sites will be listed as impaired due to sediment toxicity associated with pyrethroid pesticides, and this could lead to imposition of TMDLs in the associated watersheds. Because this process will be facilitated by identifying specific chemicals causing toxicity, this paper is intended to provide an overview of sediment TIE methods with an emphasis on those which have been demonstrated to be useful in resolving toxicity caused by pyrethroid pesticides. In addition to providing a description of specific methods, examples of studies that have successfully used the different procedures are cited. The final section includes results of two recent sediment TIE case studies. These sediments were collected from two sites in central California: Westley Wasteway Creek (WWNCR) located in the northwestern San Joaquin Valley, and an agriculture tailwater pond (SV03) located in the Salinas valley. The case studies incorporated the majority of the TIE methods specific for pyrethroid pesticides and demonstrate the utility of using a weight-of-evidence in the TIE approach. The results of these and other studies are used to illustrate areas of future research.

Methods

Background

The sediment TIE procedures described below were compiled primarily from U.S. EPA methods. Key reports describing these methods include guidance documents by Ankley et al. (*8*), and Ho et al. (*9*). Additional detailed methods are provided in a recent sediment TIE report produced by the Water Environment Research Foundation (*7*). References for additional methods that provide useful lines of evidence of pyrethroid-associated toxicity are described in the subsections below.

TIEs are designed to proceed in three phases. Phase I manipulations characterize the classes of chemicals causing toxicity and typically differentiate between toxicity caused by organic chemicals, metals, or ammonia. Phase II TIE manipulations identify the individual chemicals causing toxicity, and Phase III TIEs are designed to confirm the Phase II chemical identification. TIE method development for sediments has followed two tracks: 1) identification of toxicants in sediment interstitial water (e.g., *10*), and 2) identification of toxicants in solid-phase sediment (e.g., *11,12*). Interstitial water is thought to a likely medium through which organisms are exposed to many chemicals, including higher Kow compounds when they are present at high concentrations (9,*13*,*14*). There is evidence to suggest interstitial water is a likely source of pyrethroid exposure to benthic corganisms (*15*), and recent TIEs from a number of sites have shown that pyrethroids in interstitial water are responsible for toxicity to amphipods (*7*). Interstitial water TIEs have the advantage of using an aqueous medium that is more amenable to standardized effluent TIE techniques. Solid-phase sediment TIEs have the advantage of maintaining more realistic conditions for oxidation state, pH, and other factors affecting contaminant partitioning, but the medium is less amenable to TIE manipulations. Fewer TIE techniques are available for solid-phase TIEs, although the pyrethroid-specific procedures are amenable for use in both matrices. The TIE methods described below are divided into separate sections for solid-phase and interstitial water.

Toxicity Test Methods

Sediment TIEs have been conducted using a number of freshwater and marine species. Because pyrethroid pesticides are highly toxic to arthropods, particularly amphipods, the following procedures emphasize this taxanomic group. Freshwater pyrethroid TIEs are described for the amphipod *Hyalella azteca* using the U.S. EPA 10d survival and growth protocol (*16*). Marine pyrethroid TIEs are described for the amphipod *Eohaustorius estuarius* using the U.S. EPA 10d survival protocol (*17*). Modifications of these protocols for TIE applications are briefly summarized in the section on solid-phase TIE treatments (below). All methods described in this paper are provided in detail in Anderson et al. (*7*).

TIE Methods

Surficial sediment (upper 2 cm) for toxicity testing and TIE was sampled using either scoops or coring devices that allowed for collection of fine-grained, organic carbon-rich samples. Sufficient sediment was sampled for both solid-phase and interstitial water procedures. Depending on the number of procedures employed, a complete TIE may require as much as 40 l of sediment for all toxicity and chemical analyses. An initial toxicity test was conducted on the sample to determine the magnitude of toxicity, and based on the results of this test, TIE procedures for the case study samples were conducted using either 100% sample (solid-phase TIE), or a dilution series of interstitial water prepared by mixing interstitial water with control water at a number of concentrations.

For each solid-phase treatment described below, a treatment blank was employed to confirm that no artifacts were introduced by the different TIE manipulations. Treatment blanks consisted of laboratory control sediment that had undergone the same manipulations as the sample. In the experiments described in this paper, a formulated sediment was used as a control sediment. The formulated sediment was prepared using equal parts Salinas River, California reference site sediment and clean, kiln-dried sand (#60, RMC Pacific Materials, Monterey, CA, USA). The sediment was amended with 0.75% organic peat moss (Uni-Gro, Chino, CA, USA). One kilogram (dry weight) of formulated sediment was prepared by combining 500g reference sediment, 500g sand, and 7.5g peat with 350 mL dilution water consisting of clean laboratory well (7). In the case of the solid-phase TIEs where 10% (by wet weight) amendment was added to the sediment, an additional dilution blank was used to verify that reduction of toxicity was not simply due to dilution of the sediment. The dilution blank consisted of adding a volume of formulated sediment equal to the amendment mass.

Solid-Phase Treatments

Phase I (characterization) TIE treatments consisted of additions of amendments to the sediment, or treatments of the sediment overlying water. Sediment amendments included addition of carbonaceous resins such as Ambersorb 563, Amberlite XAD4, or addition of powdered coconut charcoal (PCC) to reduce bioavailability of organic chemicals. A cation chelating resin such as SIR-300 was added to reduce bioavailability of cationic metals, and zeolite is an option for removing unionized ammonia. Phase II (identification) TIE procedures consisted of separating the Ambersorb or SIR-300 resins from the sediment, extracting them with either acetone or acid solvent, as appropriate, and spiking control water with the acetone or acid eluate to verify that chemicals sorbed to the resins could be eluted in toxic concentrations. Toxicity of the spiked eluates was then tested using amphipods in water-only exposures using methods described in the section on interstitial water TIEs (below). Chemical concentrations were measured in the eluate-spiked water. Overlying water treatments consisted of addition of carboxylesterase enzyme and bovine serum albumin (BSA) in separate treatments to identify toxicity due to pyrethroid pesticides, and addition of piperonyl butoxide (PBO) to differentiate between pyrethroid and organophosphorus pesticides. In addition to these treatments, toxicity tests can also be performed at a colder temperature, because colder temperatures enhance pyrethroid toxicity. Phase III (confirmation) procedures consisted of comparing concentrations of chemicals in sediments or in the solvent eluates to known toxicity thresholds. These methods are briefly described below, and citations are provided where these techniques have proven useful for identifying toxicity due to pyrethroid pesticides. The procedures are briefly summarized in Table 1. Most of the solid-phase TIE procedures were used in the case study TIEs, but a subset of the interstitial water treatments that specifically apply to non-polar organic chemicals were used in the interstitial water TIEs. (Note: aeration, filtration/centrifugation, EDTA addition, zeolite addition, and pH adjustment were not used). TIE methods for characterizing and identifying toxicity caused by ammonia were not used in the case studies because concentrations of unionized ammonia were low in these samples.

In baseline tests, amphipods were exposed to un-manipulated sample to determine the magnitude of toxicity. In the case studies described below, five replicate 250-mL beakers were used for the baseline tests. Each beaker contained approximately 50g sediment and 200 mL clean dilution water. Each beaker was inoculated with 10 amphipods, and survival was assessed after a 10d exposure. This same experimental design was used with each of the solid-phase treatments described below. These exposure methods are essentially a miniaturized adaptation of the 10-d freshwater sediment protocol for *Hyalella azteca* (*16*). In addition to using smaller sediment volumes and fewer replicates, the TIE exposures are conducted under static conditions without overlying water renewals (*7*).

Ambersorb 563® (Rohm and Haas, Spring House, PA, USA), is a carbonaceous, non-polar resin added to reduce bioavailability of organic chemicals such as pyrethroid pesticides. In the case-studies discussed below, this resin was prepared by rinsing it thoroughly with Nanopure® water prior to adding it to the sediment. Ten percent Ambersorb by wet weight was added to sediment (*18,19*). Treated sediment was homogenized for 24 hours on a roller apparatus and loaded into exposure chambers. A dilution blank was created by combining test sediment with 10% formulated sediment, and an Ambersorb blank was created by adding 10% Ambersorb to formulated sediment. At test termination the sediment was sieved through a series of screens ranging from 250-400 μm to retain the Ambersorb for Phase II TIE solvent elution procedures. In this step, the Ambersorb was eluted by loading a column with approximately 7.5g of the resin and pumping 10 mL of acetone through the column at a rate of 1 mL per minute. Post-column acetone was collected in a 50 mL beaker and evaporated to a final volume of one mL. The final volume was combined with 100 mL clean dilution water to create the eluate sample for toxicity testing with *H. azteca*. The 100 mL water volume was chosen because *H. azteca* are tolerant of 1% acetone, and this step is designed to maximize contaminant concentrations and toxicity. The magnitude of toxicity and the concentrations of contaminants in the Ambersorb eluate sample are used in the weight-of-evidence for determining the cause of toxicity. The process of separating resin amendments from the sediments and eluting them with solvents as part of a Phase II solid-phase TIE has not been described in earlier U.S. EPA guidance documents and has only been recently developed for sediment TIEs (*7*). This is a crucial step in solid-phase TIEs because it allows analysts to add the solvent eluates to clean dilution water to demonstrate that toxic concentrations of specific chemicals were removed by the resins (e.g., cationic metals or non-polar organics). When combined with chemical analysis of the resin eluate, this Phase II TIE step provides evidence of specific chemicals causing toxicity, because concentrations of chemicals eluted from the resins can

Table 1. Brief Description of Solid-phase and Interstitial Water TIE Methods (see text for descriptions of the methods).

Solid-Phase Method	Description	Citations
Zeolite (SIR-600)	Addition of 10-20% zeolite to sediment reduces interstitial and overlying water ammonia	*42,43*
Chelating Resin (SIR-300)	Addition of 10% SIR-300 chelates heavy metal ions and reduces metal bioavailability	*11*
Coconut Charcoal (PCC)	Addition of 10-15% PCC to sediment reduces toxicity caused by organic contaminants	*7,12,21,22*
Ambersorb 563	Addition of 10% Ambersorb to sediment reduces toxicity caused by organic contaminants	*7,12,18,19,20*
Carboxylesterase (enzyme)	Addition of enzyme to overlying water reduces toxicity caused by pyrethroid pesticides by breaking down compounds into non-toxic forms	*7,20,25*
Piperonyl Butoxide (PBO)	Addition of PBO to overlying water can reduce toxicity caused by organophosphate pesticides, and increase toxicity caused by pyrethroid pesticides and DDT	*3,7,20,29*
Temperature Reduction	Testing warm water organisms at colder temperatures can increase the toxicity of pyrethroids and DDT	*7,31*

Interstitial Water Method	Description	Citations
Aeration	The sample is aerated for one hour to determine if toxicity is caused by volatile compounds or surfactants	*32,40,43*
Filtration or Centrifugation	Reduces toxicity that is particle-related. Also used as a pre-treatment step for the column treatments	*32,40*

EDTA	Organic chelating agent that preferentially binds with divalent metals, such as copper, nickel, lead, zinc, cadmium, mercury, and other transition metals to form non-toxic complexes	*32, 40*
pH Adjustment and Volatilization	Reduces ammonia in sample by converting it to the unionized fraction through pH adjustment and volatilizing it by stirring	*32, 40, 41*
Graduated pH	Determines if pH-dependent toxicants are responsible for the observed toxicity. The toxicity of ammonia, sulfide and some metals change with pH	*32, 40, 44*
Zeolite Column Solid-Phase Extraction	Reduces toxicity caused by ammonia	*32, 42*
Cation Column Solid-Phase Extraction and Elution	Removes metals from the sample. Column can be eluted with 1N hydrochloric acid (HCl) and resulting eluate tested to determine if substances removed by the column were toxic	*36, 40, 43*
Non-Polar Organic Column Solid-Phase Extraction and Elution (HLB)	Removes non-polar organic compounds. Column can be eluted with solvent to determine if substances removed by the column were toxic	*7,32, 40, 44*
Carboxylesterase (enzyme)	Addition of enzyme to sample reduces toxicity caused by pyrethroid pesticides by breaking down compounds into non-toxic forms	*3,7,23,24*
Piperonyl Butoxide (PBO)	Addition of PBO can reduce toxicity caused by organophosphate pesticides, and increase toxicity caused by pyrethroid pesticides and DDT	*3,7,27,32,33,34*
Temperature Reduction	Testing warm water organisms at colder temperatures can increase the toxicity of pyrethroids and DDT	*3,7,27,33,34*

then be compared to toxicity thresholds derived from dose-response experiments as part of a Phase III TIE. In the case studies, this allows the analyst to reduce bioavailability of chemicals below toxic thresholds in the solid-phase exposures, and to confirm that concentrations of chemicals sorbed to the resins exceeded toxicity thresholds after they were spiked into control water. While it is recognized that the mass balance relationships between spiked, sorbed and eluted chemicals have not yet been determined (see discussion below), the results demonstrate that the Phase II TIE elution procedures provide a useful additional line of evidence in the TIE process. It should also be noted that Ambersorb 563 is no longer available from the manufacturer, but similar resins are available and these have proven to be equally effective (*7*).

An Ambersorb elution blank was prepared by performing the above treatments on Ambersorb that had been combined with formulated sediment. A 1% acetone blank was also tested. In addition to the examples provided in the case studies below, these procedures have been used to identify toxicity due to pyrethroid pesticides in studies reported by Anderson et al. (*7*), and Phillips et al. (*20*).

Addition of powdered coconut charcoal (PCC) has also been demonstrated to be an effective Phase I solid-phase TIE treatment (*13,21,22*). Because of its larger surface area to volume ratio, PCC is often more effective at binding non-polar organic chemicals than carbonaceous resins. However, there is currently no method for extracting PCC from treated sediment, and therefore it is not possible to conduct Phase II TIE elution steps using PCC. Previous research has shown that PCC addition provides a useful line-of-evidence of toxicity due to non-polar organics and is especially effective when used in conjunction with Ambersorb addition because the latter treatment allows for Phase II elution procedure (*7*). Addition of 10% (by wet weight) PCC was included in the SV03 case study. The PCC was hydrated prior to mixing with SV03 sediment using methods described in Anderson et al. (*7*). Treatment and dilution blanks were included as described above for the Ambersorb.

SIR-300 (ResinTech, West Berlin, NJ) is a cation exchange resin which has chelating properties for heavy metal ions. After preparation, SIR-300 is mixed into sediment to reduce cationic metal bioavailability (*11*). In the WWNCR case study, ten percent SIR-300 (wet weight) was added to the sediment in a 500 mL mixing jar. Treated sediment was homogenized for 24 hours on a roller apparatus, and loaded into exposure chambers. A dilution blank was created by combining test sediment with 10% formulated sediment, and an SIR-300 blank was created by adding 10% SIR-300 to formulated sediment. At test termination the sediment was sieved through a series of screens ranging from 250-400 μm to retain the SIR-300. The SIR-300 was then eluted by loading a column with approximately 7.5g resin and pumping 10 mL of 1N hydrochloric acid through the column at a rate of 1 mL per minute. Post-column acid was combined with 100 mL clean dilution water and neutralized to create the eluate sample for toxicity testing with *H. azteca*. An SIR-300 elution blank was prepared by performing the above treatments using SIR-300 combined with formulated sediment. An acid elution blank was also tested. These procedures duplicate those described above for Ambersorb and are useful for identifying toxicity due to cationic metals. Since sediments may contain mixtures of metal and organic contaminants, previous research has shown it is useful to use the Ambersorb and SIR-300 resins in combination to provide lines-of-evidence separating metal and non-polar organic toxicity in the TIE process (*7*).

A porcine carboxylesterase enzyme (Sigma-Aldrich, St. Louis, MO) can be used to hydrolyze ester-containing compounds such as pyrethroids to their corresponding acid and alcohol, which are generally not toxic (*23,24*). In the case studies, carboxylcarboxylesterase (500x) was added to the overlying water on the day of test initiation, six hours prior to the addition of amphipods, and this allowed interaction between the enzyme and pyrethroids. The enzyme was added based on units of activity. One 'x' of enzyme activity equals 0.0025 units of enzyme per mL of sample, therefore at 500x, 1.25 units per mL were added. Enzyme strength is unique for each lot purchased (*23*). To control for reduced toxicity due to binding of contaminants to the protein base of the enzyme, a separate set of replicates was treated with bovine serum albumin (BSA; after Wheelock et al. *24*). Reduction of toxicity by the enzyme, and not the BSA, helps characterize toxicity due to pyrethroids. Since carboxylesterase activity is known to diminish with time (*24*), daily additions of enzyme were made to the beakers to restore nominal activity, as well as addition of an equal mass of BSA to the BSA treatments. This method was originally developed for use with water and interstitial water samples (*3,7,23*), but has proven to be useful for identifying pyrethroid pesticide toxicity as a treatment of sediment overlying water (Phillips *7,20,25*).

Piperonyl butoxide (PBO; Sigma-Aldrich, St. Louis, MO) inhibits the cytochrome P450 mixed-function oxidase system, a key system for toxicant detoxication in invertebrates. In TIE applications, PBO addition is used to block the metabolic activation of acetylcholinesterase-inhibiting organophosphate pesticides (*26*). It is also a potent synergist of pyrethroid pesticide toxicity, because it inhibits their metabolism (*27,28*). In the case studies, the PBO treatment contained 500 μg/L of PBO in the water overlying the sediment. Decreased toxicity with the addition of PBO suggests the presence of organophosphate pesticides. Increased toxicity with the addition of PBO suggests the presence of pyrethroids. PBO additions to sediment overlying water have been reported in several solid-phase TIE studies, including Amweg and Weston (*29*), Phillips et al. (*20*), and Anderson et al. (*7*).

A number of researchers have noted that pyrethroid pesticides are toxic at lower temperatures (e.g., *27,30*). This negative temperature coefficient can be used as a TIE treatment to provide an additional line-of-evidence of pyrethroid-associated toxicity. Temperature manipulation is particularly useful in tests with *H. azteca,* because this protocol is conducted at a relatively high temperature (23 °C) and this species tolerates a wide range of temperatures. In the WWNCR case study, amphipod survival was compared in duplicate tests conducted at 23 and 15°C. Evidence of pyrethroid pesticides associated with increased amphipod mortality in tests conducted at lower test temperatures has been reported in solid-phase TIE studies by Anderson et al. (*7*), and Weston (*31*).

Interstitial Water Treatments

Because interstitial water is thought to be a likely route of exposure of pyrethroid pesticides to benthic infauna (*14*), and is amenable to TIE manipulations, interstitial water TIEs have been demonstrated to provide important lines-of-evidence of pyrethroid toxicity (*7*). In the case studies described below, interstitial water was extracted from sediment using a refrigerated centrifuge (2500 rpm for 30 minutes at 4 °C). The Phase I and II TIE treatments were performed on a dilution series of each sample following US EPA methods (*32*), except where noted. Sample concentrations in the initial test and TIE treatments were 0, 10, 25, 50, and 100% interstitial water. The dilution water and treatment blank was laboratory control water that underwent the same manipulation as the sample.

An interstitial water baseline toxicity test was conducted on a dilution series of unmanipulated sample to determine the magnitude of toxicity, and TIE treatments were conducted on interstitial water concentrations of 0, 10, 25, 50 and 100%. In the case studies, three replicate 20 mL scintillation vials were used for the baseline tests. Each vial contained 15 mls of interstitial water. Five amphipods were placed in each vial, and survival was assessed after a 96h exposure. Test solutions were not renewed and the amphipods were not fed. Blanks for all TIE treatments consisted of control water subjected to the same manipulations.

A column treatment designed to remove non-polar organic compounds from the sample was conducted by passing interstitial water through an OASIS HLB (= hydrophilic-lipophilic balance; Waters Corp., Milford, MA, USA) solid-phase extraction column. This treatment uses reverse phase liquid chromatography to extract nonionic organic chemicals from the interstitial water. Toxicity of the column rinsate was assessed in the Phase I TIE, then the column was eluted with methanol to provide test solution for a Phase II procedure. In the Phase II TIE, amphipods were exposed to the eluate to determine if toxic concentrations of non-polar organic chemicals were eluted from the HLB column. As a Phase III TIE procedure, a second sample of the solvent eluate was analyzed using GC-MS to measure specific organic chemicals present. The chemical concentrations were then compared to their toxicity thresholds for *H. azteca* to confirm chemicals which were likely responsible for toxicity. These procedures have been previously reported by Anderson et al. (*7*) in interstitial water TIEs involving pyrethroid pesticides.

An extraction column treatment designed to remove cationic metals from the sample was conducted by passing interstitial water through a Supelco LC-WCX cation exchange column (Bellefonte, PA, USA). This test is analogous to the SIR-300 solid-phase TIE described above. The column rinsate was tested for toxicity, and then the column was eluted with 1N hydrochloric acid (HCl) to provide test solution for a Phase II TIE. In this procedure, the acid eluate was spiked into control water, this was neutralized with 1N sodium hydroxide, and amphipods were exposed to the spiked eluate to determine if toxic concentrations of acid-soluble cations were eluted from the column.

In samples where mixtures of metal and organic chemicals may be causing toxicity, the cation and HLB column treatments may be used in sequence to separate the relative contributions of these constituents to interstitial water toxicity. The rinsate is tested for toxicity after passing through both columns, then each column is eluted (cation column with acid, HLB column with methanol) to determine if chemicals removed by the columns were toxic when spiked into control water (*7*).

In the case studies, carboxylesterase enzyme was added to interstitial water samples to hydrolyze pyrethroid pesticides using methods described in Wheelock et al. (*24*). In these tests, BSA was added in a separate treatment to control for the binding of contaminants to the protein base of the enzyme. Addition of carboxylesterase enzyme has provided useful lines-of-evidence in recent studies investigating the role of pyrethroid pesticides in sediment interstitial water toxicity (e.g., *3,7,24*).

Addition of the metabolic inhibitor PBO to interstitial water has proven to be useful for separating the relative roles of pyrethroid and organophosphate pesticides in sediment interstitial waters. Methods in the case studies followed those described in U. S. EPA (*31*). PBO addition has been reported in a number of recent interstitial water TIE studies concerning pyrethroid and organophosphate pesticides in sediment (e.g., *3,7,33,34*).

As in the solid-phase TIEs described above, an additional line-of-evidence that has proven useful in interstitial water TIEs emphasizing pyrethroids is comparing toxicity of the sample at lower(15 °C) and higher (23 °C) test temperatures (*3,7,33,34*).

Physical and Chemical Measurements

Water quality parameters of dissolved oxygen, pH and conductivity were measured in the case study TIEs using a Hach SensION© selective ion meter with appropriate electrodes; and ammonia was measured using a Hach 2010 spectrophotometer. Temperature was measured using a continuously recording thermograph and thermometer. Concentrations of the organophosphate pesticides chlorpyrifos and diazinon were measured in interstitial water using enzyme-linked immunosorbent assays (ELISA, Strategic Diagnostics Inc,

Newark, DE). The ELISA reporting limits for chlorpyrifos and diazinon were, 100 and 60 ng/L, respectively. The associated ELISA quality assurance and quality control procedures followed those described in Anderson et al. 2007 (7).

Sediment samples were analyzed for organochlorine compounds, pyrethroids, and organophosphates, following U.S. EPA methods 8081, 1660, and 8141, respectively. Reporting limits for organochlorine, pyrethroids, and organochlorine pesticides ranged from 1-5 ng/g, 1-8 ng/g, and 10 ng/g, respectively. All analyte identifications were confirmed by gas chromatography-mass spectroscopy or liquid chromatography- mass spectroscopy, and all sediment analyses met prescribed U.S. EPA quality assurance and quality control guidelines. Acetone eluates of the Ambersorb and methanol eluates of the HLB column treatments (for the WWNCR case study sample) were also analyzed for the same three classes of pesticides. Ambersorb eluates were further analyzed using direct injection of the acetone into the gas chromatograph.

Data Interpretation

Effects of the solid-phase TIE treatments were assessed by comparing the magnitude of toxicity to that of the baseline sample. Treatment blanks were evaluated to determine if sample manipulations added toxic artifacts. In the interstitial water TIEs, treatment data were compared to baseline toxicity using the toxic unit (TU) approach. Toxic units (TU) were calculated by dividing 100 by the LC50 calculated from each TIE treatment dilution series. A lower toxic unit value was used to indicate a treatment had been effective in reducing toxicity.

Results

Case Studies

Sample Handling

Sediment for the first case study was collected from Westley Wasteway Creek (WWNCR), on October 10, 2005. The initial toxicity test was conducted on November 18, 2005. Once the magnitude of the initial toxicity was determined, interstitial water TIEs were initiated on December 16, 2005 and solid-phase TIEs were initiated on December 17, 2005. Sediment for the second case-study was collected from an agricultural tailwater pond (SV03) which is used as a vegetated treatment system. This sediment was collected on November 25, 2006, and the initial sediment and porewater toxicity tests were conducted on November 17 and December 1, respectively. Solid-phase and interstitial water TIEs were conducted on December 8 and December 15, 2006, respectively.

Initial Tests

Survival in Westley Wasteway Creek sediment and interstitial water was 0% (Table 2). WWNCR interstitial water contained 2.2 TUs, based on toxicity measured in the five interstitial water concentrations. Both diazinon and chlorpyrifos were measured in the interstitial water using enzyme-linked immunosorbent assays (ELISA), and the concentrations of both were below reporting limits. Survival in SV03 sediment and interstitial water was also 0% (Table 2). No chlorpyrifos was detected in this interstitial water, while 234 ng/L of diazinon was detected. Water quality parameters for all TIEs from both sediments were within acceptable limits for the test organism (e.g., ammonia, pH, dissolved oxygen, conductivity, hardness and alkalinity; data not shown). Because toxicity was observed in both interstitial water and solid-phase samples, TIEs were conducted on both sediment matrices.

Westley Wasteway Creek TIEs

Survival of amphipods was 0% in untreated WWNCR sediment (baseline), and did not increase with addition of Ambersorb. Addition of the carboxylesterase enzyme to sediment overlying water increased survival in the solid-phase TIE from 0% to 48% (Table 2). Addition of BSA had a minimal effect on survival (survival= 8%). These results suggest toxicity was partially caused by a pyrethroid pesticide. Decreasing test temperature did not increase toxicity because complete mortality was observed in the baseline sediment (data not shown).

While addition of Ambersorb to WWNCR sediment did not reduce toxicity, the resin was recovered from the sediment and eluted with acetone, and the eluate was highly toxic when it was spiked into control water (0% survival, Table 2). This provides an important line of evidence supporting the conclusion that a non-polar organic chemical was responsible for WWNCR toxicity. Analysis of this eluate showed a 1279 ng/L of L-cyhalothrin, and this was the only chemical detected in the eluate at a concentration sufficient to account for toxicity (Table 3).

WWNCR sediment contained low concentrations of DDE (p,p') and elevated concentrations of two pyrethroids (Table 3). The sediment contained just over one TU of bifenthrin and 31.1 TUs of L-cyhalothrin, based on organic carbon-corrected concentrations of these pyrethroids (*35*).

It is not clear how the concentration of L-cyhalothrin measured in the Ambersorb eluate sample relates to its concentration in the sediment or in the interstitial water. Interstitial water concentrations were not measured in this study because there was insufficient sample volume. Rather than use the eluate concentration as a predicter of interstitial water concentration, detection of toxic concentrations of L-cyhalothrin in the Ambersorb eluate is used to provide an

Table 2. Mean percent survival and standard deviation (SD) of Hyalella azteca in 10d solid-phase Westley Wasteway Creek (WWNCR) TIE.

Treatment	Solid-Phase Treatments		Acetone Eluate Treatments	
	Mean	SD	Mean	SD
Baseline WWNCR	0	0		
WWNCR (10% Ambersorb)	0	0	0	0
Blank (10% Ambersorb + Formulated)	98	4	88	11
WWNCR (10% SIR-300)	0	0		
Blank (10% SIR-300 + Formulated)	94	5		
WWNCR (10% Control)	0	0		
WWNCR (Enzyme)	48	37		
Blank (Enzyme + Formulated)	100	0		
WWNCR (BSA)	8	11		
Blank (BSA + Formulated)	82	8		
WWNCR (PBO)	0	0		
Blank (PBO + Formulated)	96	9		
Elution Blank			81	2
Formulated Control	94	9	93	12

Table 3. Concentrations of detected pesticides in Westley Wasteway Creek (WWNCR) sediment. Sediment LC50 values. ND indicates not detected. NA indicates not analyzed.

Chemical	WWNCR Sediment ng/g dry wt.	Sediment LC50[a] ng/g dry wt.	WWNCR Ambersorb Eluate ng/L	Water LC50 ng/L
DDE (p,p')	67.1		1.12	1660 [b]
Bifenthrin	5.08	12.9	ND	9.3 [c]
Bifenthrin µg/g oc	0.63	0.52		
(Es)Fenvalerate	ND	41.8	ND	
(Es)Fenvalerate µg/g oc	ND	1.54		
L-cyhalothrin	113.7	5.6	1279	
L-cyhalothrin µg/g oc	14.0	0.45		
Permethrin	ND	201	ND	
Permethrin µg/g oc	ND	10.8		
Total Organic Carbon	0.81%			

[b] *(45)*; [c] *(46)*

additional line-of-evidence that this pyrethroid played a roll in toxicity of WWNCR sediment.

Toxicity of the WWNCR interstitial water was reduced by treatment with both the cation exchange and HLB solid-phase extraction columns (Figure 1; Table 4). The cation column reduced toxicity from 5.6 TU to 1.5 TU, but there was no toxicity in the cation acid eluate after it was spiked into control water. This suggests that metals were likely not the source of toxicity. It is possible the cation column reduced toxicity through binding of organic chemicals (*7,36*). We note that control survival in this interstitial water TIE was 76%, which is less than the 80% survival criterion listed for the 10-d amphipod solid-phase protocol. There are no acceptability criteria for interstitial water TIEs, and since the results of the WWNCR interstitial water TIE demonstrated considerable differences between many of the treatments, these results were included as part of the weight-of-evidence.

Treatment of interstitial water with the HLB column reduced toxicity to 1.9 TUs, and when the column eluate was spiked into control water 6.5 TUs were recovered. This indicates that the cause of toxicity was an organic contaminant. Organochlorine, organophosphate and pyrethroid pesticides were analyzed in the HLB methanol eluate, but none were detected. Lack of detection of chemicals in the eluate was likely due to incomplete elution with methanol, or dilution during the liquid-liquid extraction of the sample which was conducted prior to gas chromatography (see discussion below).

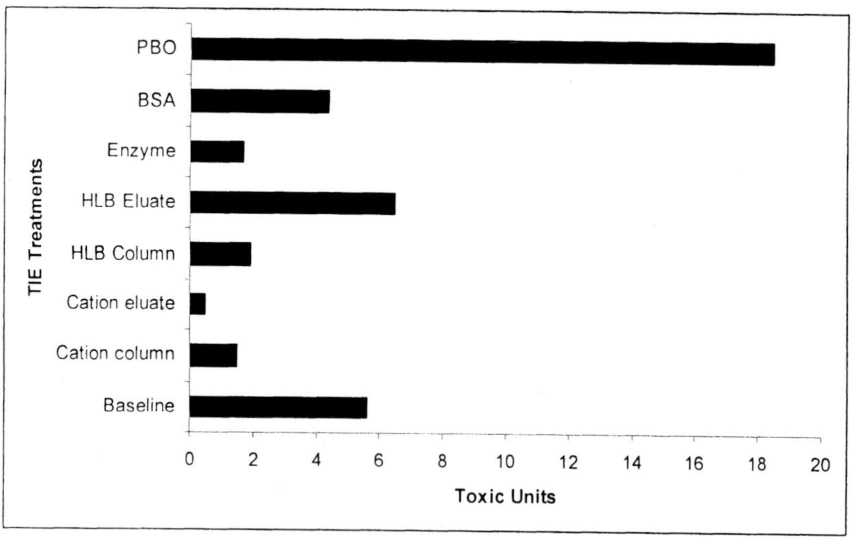

Figure 1. Results of 96h Hyalella azteca TIE using Westley Wasteway Creek interstitial water (see text for explanation of Toxic Unit calculation)

Table 4. Mean percent survival and standard deviation (SD) of Hyalella azteca in 96h Westley Wasteway Creek (WWNCR) interstitial water TIE.

Treatment	Percent Sample										Toxic Units
	0%		10%		25%		50%		100%		
	Mean	SD	Mean	SD	Mean	SD	Mean	SD	Mean	SD	
Baseline	76	8	100	0	0	0	13	12	0	0	5.6
Cation Column	100	0	93	12	100	0	93	12	0	0	1.5
Cation Eluate	93	12	100	0	100	0	100	0	93	12	<1
HLB Column	93	12	100	0	67	23	87	12	0	0	1.9
HLB Eluate	93	12	88	11	0	0	0	0	0	0	6.5
Sequential Cation HLB	100	0	100	0	73	31	73	12	0	0	2.1
Sequential Cation Eluate	93	12	47	42	100	0	100	0	93	12	<1
Sequential HLB Eluate	93	12	100	0	80	20	100	0	27	12	1.3
Enzyme	100	0	100	0	87	12	87	23	0	0	1.7
BSA	93	12	100	0	33	23	13	23	0	0	4.4
PBO	93	12	7	12	0	0	0	0	0	0	18.5

Addition of carboxylesterase enzyme reduced interstitial water toxicity to 1.7 TUs, while addition of BSA had a negligible effect (4.4 TUs). These results provided an additional line-of-evidence that toxicity of WWNCR was caused by a pyrethroid. This conclusion was supported with the observation of an increase in interstitial water toxicity from 5.6 to 18.5 TUs with addition of the metabolic inhibitor PBO (Table 4). PBO is a potent synergist for pyrethroid pesticides.

Agriculture Tailwater Treatment Pond TIEs

Survival of amphipods in sediment from SV03 was 0% in the baseline sample and did not increase with the addition of Ambersorb (Table 5). However, amphipod survival increased to 50% with the addition of 10% PCC, indicating toxicity was caused by an organic chemical (Table 5). While Ambersorb did not reduce sample toxicity, the resin was recovered from the sediment and eluted with acetone, and the eluate was highly toxic when it was spiked into control water (0% survival, Table 5). This step provides an additional line-of-evidence that toxic concentrations of organic chemicals were present in this sediment. It should be noted, however, that some toxicity was also observed in the eluate blank (43% survival), which measured toxicity of acetone after elution of the control sediment. As discussed above, there are no

Table 5. Mean percent survival and standard deviation (SD) of Hyalella azteca in 10d solid-phase Salinas Valley tailwater pond (SV03) TIE

Treatment	Solid-Phase TIE		Amendment Elution	
	Mean	SD	Mean	SD
SV03 A	0	0		
SV03 A (10% Ambersorb)	0	0	0	0
Blank (10% Ambersorb + Form.)	96	5	43	6
SV03 A (10% PCC)	50	19		
Blank (10% PCC + Form.)	100	0		
SV03 A (10% SIR-300)	0	0		
Blank (10% SIR-300 + Form.)	98	4		
SV03 A (Enzyme)	78	15		
Blank (Enzyme + Form.)	92	8		
SV03 A (BSA)	0	0		
Blank (BSA + Form.)	98	4		
SV03 A (10% Control)	0	0		
Sediment Control	96	9		
Ambersorb Elution Control			100	0

Form. = Formulated sediment

strict criteria for blank performance in sediment TIEs. The control acceptability criterion for survival applies only to control survival in the initial solid-phase test. While 57% mortality in the elution blank is of concern in this TIE, the results still provide evidence of organic chemical toxicity when compared to the 100% mortality observed in the acetone eluate. These data were therefore included in the weight-of-evidence. Chemical analysis of the Ambersorb eluate is pending.

In separate solid-phase treatments, addition of carboxylesterase enzyme to the overlying water increased amphipod survival to 78%, while the addition of BSA did not reduce toxicity. This suggests that toxicity was caused by a pyrethroid. Three pyrethroids were detected in this sediment, and concentrations of cypermethrin and L-cyhalothrin exceeded their respective LC50s (*35*). SV03 interstitial water contained 234 ng/L diazinon (Table 6), which is considerably less than the diazinon LC50 for *H. azteca* (LC50 = 6210 ng/L; *37*). In addition to these pesticides, a number of organochlorine pesticides w7ere detected in SV03 sediment, including dieldrin (Probable Effects Concentration - PEC = 61.8 μg/kg; 38), endrin (PEC = 207 μg/kg; 38), chlordane (PEC =17.6 μg/kg; 38), and DDT (LC50 = 371 ug/g oc; 39). Of these, only chlordane was present at a concentration that exceeding the PEC sediment quality guideline value.

Baseline toxicity of interstitial water extracted from SV03 sediment showed 6.3 TUs (Figure 2; Table 7). Treatment of SV03 interstitial water with the HLB column reduced the toxicity to 2.4 TU, and the HLB methanol eluate was toxic when it was spiked into control water. Addition of carboxylesterase enzyme to the interstitial water completely removed toxicity, while the addition of BSA reduced toxicity by 1.5 TUs. Addition of PBO increased toxicity to 20 TUs, and the addition of carboxylesterase and PBO in combination reduced the toxicity to 3.5 TUs (Figure 1). Toxicity reduction by carboxylesterase and potentiation by PBO provide two strong lines of evidence implicating a pyrethroid pesticide as the cause of toxicity in this sediment sample.

Discussion

Sediment TIE methods include a suite of procedures that are useful for identifying toxicity caused by pyrethroid pesticides. In the case studies described here, several lines-of-evidence provided by solid-phase and interstitial water TIEs were compiled into a weight-of-evidence to support the conclusion that toxicity of these sediments was caused primarily by pyrethroid pesticides.

In the WWNCR TIE, evidence from the interstitial water proved to be valuable in confirming the cause of toxicity, because the magnitude of toxicity in this sample overwhelmed many of the solid-phase TIE treatments. The only treatment that reduced toxicity of the solid-phase sample was addition of carboxylesterase enzyme. The volume of Ambersorb added to this sample was

Table 6. Selected organic pesticides detected in Salinas Valley tailwater pond (SV03) sediments. ND = not detected.

		SV03 A
Organochlorines		
Dieldrin	ng/g	20.1
Endrin	ng/g	6.64
Total Chlordane	ng/g	20.87
Total DDT	ng/g	546.2
Total DDT/g oc	μg/g oc	23.05
DBOB	ng/g	68.3
DBCE	ng/g	100
Organophosphates		
Chlorpyrifos	ng/g	ND
Diazinon	ng/g	ND
Triphenyl phosphate	ng/g	109
Pyrethroids		
Cypermethrin	ng/g	66.4
(Es)Fenvalerate	ng/g	14.7
(Es)Fenvalerate / OC	ug/g oc	0.62
L-cyhalothrin	ng/g	18.8
L-cyhalothrin / OC	ug/g oc	0.79
Dibromooctafluorobiphenyl	ng/g	98.2
Total Organic Carbon	%	2.37
Grain Size Distribution		
Fines	%	92.31

apparently insufficient to completely adsorb available pyrethroid, and therefore no reduction in toxicity was observed with this treatment. Because of its superior binding capacity addition of powdered coconut charcoal provides additional supporting evidence for toxicity due to organic chemicals in highly toxic sediments (*21*). PCC addition was not included in the WWNCR TIE because of the limited amount of sediment available. However, despite the lack of toxicity reduction with Ambersorb, the observation that the acetone eluate from the Ambersorb was toxic due to high concentrations of L-cyhalothrin provided evidence of toxicity due to this pyrethroid. This was confirmed by several lines-of-evidence from the interstitial water TIEs: reduction of toxicity using HLB column filtration, observed toxicity in the HLB column eluate, reduction of toxicity with carboxylesterase addition, and an increase in toxicity with the addition of the metabolic inhibitor PBO. When these lines-of-evidence were combined with the measurement of 31 TUs of L-cyhalothrin in this sediment, the

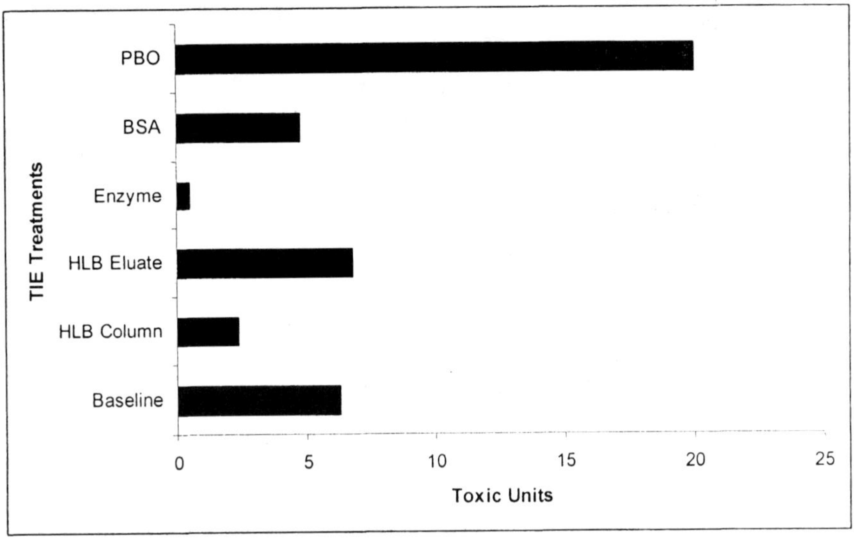

Figure 2. Results of 96h Hyalella azteca TIE using SV03 interstitial water (see text for explanation of Toxic Unit calculation)

weight-of-evidence indicated that this pesticide was the primary cause of toxicity, in combination with one TU of bifenthrin.

As discussed above, the fact that no pyrethroids were detected in the HLB column methanol eluate from the WWNCR interstitial water TIE was likely due two factors: use of a moderately polar solvent (methanol), and dilution during sample preparation for chemical analysis. Recent research by the United States Geologic Survey has shown that methanol is inefficient at recovering pyrethroids sorbed to HLB columns (percent recoveries = 10% to 60% for 13 pyrethroids), and that a more polar solvent, such as acetone, is required (percent recoveries = 75% to 100% for 13 pyrethroids; personal communication, K. Smalling, USGS, Sacramento, CA). In the case study experiments, use of methanol likely resulted in only partial elution of L-cyhalothrin from the HLB column used to treat the interstitial water. While there was sufficient pyrethroid present in the methanol eluate to cause amphipod mortality, the total L-cyhalothrin mass was probably not eluted from the column, and the eluted pesticide may have been lost during subsequent sample cleanup steps. Prior to gas chromatography, the methanol eluate was spiked into 1L of water, then this water was extracted using dichloromethane. This step resulted in a five-fold dilution of the methanol eluate, resulting in non-detection of pesticides. Subsequent research has indicated better recoveries of pyrethroids when TIE extraction media such as

Table 7. Mean percent survival and standard deviation (SD) of Hyalella azteca in 96h Salinas Valley tailwater pond (SV03) interstitial water TIE.

SV03	Percent Sample											
	0%		10%		25%		50%		100%		Toxic	Diazinon
Treatment	Mean	SD	Mean	SD	Mean	SD	Mean	SD	Mean	SD	Units	ng/L
Baseline	93	12	93	12	0	0	0	0	0	0	6.3	234
HLB Column	100	0	100	0	100	0	27	31	0	0	2.4	<RL
HLB Eluate	93	12	67	31	20	20	0	0	0	0	6.8	177
Carboxylcarboxylesterase	93	12	87	23	94	10	100	0	87	12	<1	133
BSA	100	0	100	0	33	42	0	0	0	0	4.8	163
PBO	100	0	0	0	0	0	0	0	0	0	20.0	198
Carboxylcarboxylcstcrasc/PBO	100	0	100	0	47	23	30	26	0	0	3.5	189

Ambersorb and HLB columns are eluted with acetone and injected directly into the chromatograph after a florisil clean-up step (personal communication, A. Mekebri, California Department of Fish and Game, Sacramento, CA; see also reference *7*).

Results of solid-phase and interstitial water TIEs with samples from the agriculture treatment pond (SV03) were used in a weight-of-evidence approach to implicate pyrethroid pesticides. Lines of evidence from the solid-phase TIE included reduction of toxicity with PCC addition to the solid phase, and with carboxylesterase enzyme addition to the sediment overlying water. As in the other case study, Ambersorb did not reduce SV03 sediment toxicity, but toxicity was observed when the Ambersorb eluate was spiked into control water. Lines of evidence from the interstitial water TIE included reduction of toxicity with HLB treatment and recovery of toxicity in the HLB eluate. SV03 interstitial water toxicity was eliminated with addition carboxylesterase and dramatically increased with addition of the metabolic inhibitor PBO, two lines of evidence that indicate pyrethroid toxicity. Pyrethroid toxicity was confirmed by measuring 18 TUs of cypermethrin and 2 TUs of L-cyhalothrin in this sediment.

These case studies illustrate the utility of compiling evidence from solid-phase and interstitial water TIEs to determine causes of sediment toxicity. In these studies, the magnitude of toxicity apparently overwhelmed the capacity of Ambersorb to reduce toxicity of either sample of solid-phase sediment. In the case of SV03 sediment, addition of PCC provided additional evidence of organic chemical toxicity. TIEs of highly toxic sediments can also be conducted using a dilution series of solid-phase samples to improve efficacy of individual treatments (e.g., *7,29*). This can also be achieved using treatments applied to a dilution series of interstitial water. This latter approach is particularly appropriate with pyrethroids because interstitial water is a likely route of exposure for many sediment-dwelling species, especially under conditions of high contaminant loading, and dilution allows better resolution of treatment responses. While it takes a considerable volume of sediment to provide sufficient interstitial water for TIEs, interstitial water, once extracted, is easy to work with and provides a practical sediment matrix for TIEs.

The sediments investigated in these case-studies were collected from agriculture areas and were contaminated primarily by pesticides. Because the sediments were dominated by pyrethroids, addition carboxylesterase and PBO were particularly effective TIE treatments. This supports previous research conducted using solid-phase and interstitial water samples (Table 1). Manipulation of test temperature has also been demonstrated to provide a useful line of evidence in sediments dominated by pyrethroids (Table 1). The abbreviated procedures used in the case studies did not include all of the TIE treatments available. For example, low ammonia concentrations were measured in these samples, so treatments designed to identify ammonia toxicity were excluded (e.g., pH manipulation, zeolite addition). Methods were also not included for oxidants, volatiles, and surfactants.

Because many sediments contain more complex mixtures, it is important to emphasize that TIEs are most effective when the complete suite of treatments described in standard EPA procedures are employed. These include Phase I (characterization), Phase II (identification) and Phase III (confirmation) steps. This is particularly important in highly urbanized watersheds and marine habitats at the bases of these systems, where sediments may be contaminated by toxic concentrations of ammonia, metals, organochlorines, and PAHs in addition to pyrethroids (e.g., *7,9*). As discussed above, the Phase I TIE treatments that facilitate characterization of toxicity due to chemical mixtures in solid-phase samples include the addition of resins (e.g., Ambersorb for organics and SIR-300 for cationic metals), and the analogous procedures for interstitial water (e.g., HLB filtration and cation exchange columns). These procedures can provide evidence of toxicity due to mixtures of non-polar organics and metals, and the resins and solid-phase extraction columns are also amenable to Phase II TIE identification procedures, because they can be eluted with solvents to allow separation of specific chemicals. Once eluted, the chemicals can then be spiked into clean water and their toxicity tested. When this step is combined with chemical analyses of the toxic eluates as a Phase III TIE step, it provides additional evidence to confirm chemicals causing toxicity. An important topic for future research is quantifying the relationship between concentrations of toxicants in the original sediments, and concentrations in TIE resin eluates. In these experiments, the acetone eluates from the resin amendments were added to 100 mL of control water for toxicity testing. The volume of water into which the acetone eluate fractions were added was arbitrary and could not be related to the concentrations of contaminants in the sediment or interstitial water of the spiked sediments. This differs from the procedure followed in the interstitial water TIE using the cation, C18 and HLB solid phase extraction columns, where the acid or solvent eluate fraction is reconstituted at the same volume as the amount of sample passed through the column. In the case of the resin amendments used in the whole-sediment TIEs, it also cannot be assumed that the resin amendment had 100% adsorption efficiency, that all of the resin was fully recovered from the sediment, or that there was complete elution of contaminants from the resin. When adjusting the final volume of water that the solvent eluate fraction is added to, the analyst can potentially increase or decrease the toxicity of the eluate. The current experiments used 100 ml as a standard volume for eluate toxicity testing because that volume was the minimum amount necessary to conduct water only exposures with amphipods and this volume contained 1% acetone or acid solvent, the maximum tolerated by the amphipods. By adding the solvent to a minimal amount of water in the eluate treatments, the toxicity and chemistry signals were maximized, which is especially helpful in TIEs involving highly hydrophobic chemicals. Further research should quantify the mass balance relationships between chemicals in the four relevant sediment compartments in whole sediment TIEs: solid-phase sediment, sediment porewater, the resin amendment, and the solvent eluate of the resin.

Quantification of these relationships will be important for interpreting the toxicity and chemical concentration data in sediment TIEs (7).

Because pyrethroids are highly hydrophobic and are toxic at very low concentrations, their elution and analysis is difficult using conventional techniques. Therefore, the methanol elution step recommended in EPA TIE guidance documents (*32,40*) is less effective than when using a more polar solvent such as acetone. Similarly, analysis of pyrethroids in solvent eluates and sediment interstitial water may require some modification of standard methods. For example better recoveries of pyrethroids have been observed when acetone eluates are injected directly for gas chromatography, rather than reconstituting the eluate in water and using a liquid-liquid extraction procedure (*7*). In addition, relatively large volumes of interstitial water are required to obtain toxicologically relevant detection limits.

One challenge for solid-phase TIEs with Ambersorb is incomplete reduction of toxicity in highly toxic samples. In the, case studies, the largest Ambersorb volume used was 10%, and this was mixed with sediment for 24 hours, following methods described by Anderson et al. (*7*). While this procedure has often reduced toxicity in lower dilutions of ambient sediment samples, it is less effective in less diluted, highly toxic sediments (e.g., WWNCR). Although additional evidence may be produced by conducting TIEs using dilutions of sediment, it is preferable to include undiluted sediment in solid-phase TIEs, both to avoid overlooking toxicologically important contaminants that might be reduced to non-toxic concentrations at lower dilutions, and to minimize other confounding factors introduced by dilution. Additional research needs to be devoted to increasing the effectiveness of carbonaceous resins in solid-phase TIEs. This should include determination of the optimal volume of Ambersorb for use in solid-phase TIEs, and to determine appropriate equilibration times necessary to maximize sorption of chemicals from sediments highly contaminated by non-polar organics. This is particularly important given that nonpolar organic compounds may be bound by different carbon sources in sediments, and that desorption rates vary depending on the nature of the carbon present in ambient samples (e.g., soot or black carbon-like materials vs. natural organic matter). A number of papers have suggested that desorption rates vary from hours and days to months, depending on the chemical contaminants and carbon constituents present in bedded sediments (*41*).

As sediment TIEs evolve, several procedures that apply specifically to identification of pyrethroids require additional research. In addition to those described above, these include methods to optimize recovery of pyrethroids from sediment interstitial water, refinement of carboxylesterase enzyme procedures for TIE applications, and methods to separate relative toxicities of pyrethroids mixtures, and of pyrethroids in combination with other contaminants. Because the majority of TIE studies emphasizing pyrethroid toxicity have involved tests with amphipods, additional studies should be conducted using other taxa (e.g., chironomids, mysids, daphnids).

As water quality management agencies proceed with TMDL development for waterbodies listed as impaired due to sediment toxicity, it will be necessary to identify specific chemicals causing toxicity so that management resources are properly applied to address sources of key chemicals of concern. The solid-phase and sediment interstitial water TIE procedures described here comprise an adequate suite of tools to allow application of a weight-of-evidence approach to identify causes of sediment toxicity. Recent development of methods for specific identification of pyrethroid pesticide toxicity provide useful additions to the TIE framework. Refinement of these methods will improve our ability to resolve causes of toxicity in sediments contaminated by complex chemical mixtures

References

1. Weston, D.P., You, J., Lydy, M.J. 2004. Distribution and toxicity of sediment-associated pesticides in agriculturally-dominated water bodies of California's Central Valley. Environ Sci Technol 39: 9778-9784.
2. Weston, D.P., Holmes, R.W., You, J., Lydy, M.J. 2005. Aquatic toxicity due to residential use of pyrethroid pesticides. Environ Sci Technol 39: 9778-9784.
3. Anderson, B.S., Phillips, B.M., Hunt, J.W., Richard, N., Connor, V., Tjeerdema, RS. 2006 Evidence of pesticide impacts in the Santa Maria River watershed (California, USA). Environ. Toxicol. Chem 25:1160 – 1170.
4. Hunt, J.W., Anderson, B.S., Phillips, B.M., Tjeerdema, R.S., Richard, N., Connor, N., Worcester, K., Angelo, M., Bern, A., Fulfrost, B., Mulvaney, D. 2006. Spatial relationships between water quality and pesticide applications in agricultural watersheds. Environ Monit Assess. 121:245-262.
5. Kelley, K., Starner, K., 2004. Monitoring Surface Waters and Sediments in the Salinas and San Joaquin River Basins for Organophosphate and Pyrethroid Pesticides (Technical memorandum, November 16, 2004). California Department of Pesticide Regulation, Sacramento, California, 41 pp.
6. Lowe, S., Anderson, B.S., Phillips, B.M. 2006. Investigations of sources and effects of pyrethroid pesticides in watersheds of the San Francisco Estuary. Final report - PRISM GRANT: 041355520. State Water Resources Control Board, SWRCB, Sacramento, CA.
7. Anderson, B.S., Hunt, J.W., Phillips, B.M., Tjeerdema, R.S. 2007. Navigating the TMDL Process: Sediment Toxicity. Final Report- 02-WSM-2. Water Environment Research Federation. 181 pp.
8. Ankley, G.T., Schubauer-Berigan, M.K., Dierkes, J.R. 1991. Sediment toxicity identification evaluation: Phase I (characterization), Phase II (identifica-tion), and Phase III (confirmation) modifications of effluent procedures. National Effluent Toxicity Assessment Center. Technical Report 08-91.

9. Ho, K.T., Burgess, R., Mount, D., Norberg-King, T., Hockett, R.S. 2007. Sediment toxicity identification evaluation: interstitial and whole methods for freshwater and marine sediments. U.S. EPA, Atlantic Ecology Division/Mid-Continental Ecology Division, Office of Research and Development, Narragansett, RI.
10. Ho, K., McKinney, R.A., Kuhn, A., Pelletier, M.C., Burgess, R. 1997. Identification of acute toxicants in New Bedford Harbor sediments. *Environ Toxicol Chem* 16: 551-558
11. Burgess, R.M., Cantwell, M.G., Pelletier, M.C., Ho, K.T., Serbst, J.R., Cook, H., Kuhn, A. 2000. Development of a toxicity identification evaluation procedure for characterizing metal toxicity in marine sediments. *Environ Toxicol Chem* 19: 982-991.
12. Lebo, J.A., Huckins, J.N., Petty, J.D. 2000. Selective removal of organic contaminants from sediments: a methodology for toxicity identification evaluations (TIEs). *Chemosphere* 40: 811-819.
13. Ho, K.T., Burgess, R.M., Pelletier, M.C., Serbst, J.R., Ryba, S.A., Cantwell, M.G., Kuhn, A., Raczelowski, P. 2002. An overview of toxicant identification in sediments and dredged materials. *Mar Poll Bull* 44: 286-293.
14. Williamson, B., Burgess, R. 2003. Sediment and porewater chemistry. In: Carr, RS, Nipper, M (eds): *Pore water toxicity testing: biological chemical and ecological considerations – methods, applications, and recommendations for future areas of research.* SETAC Press, Pensacola, FL
15. Maund, S.J, Hamer, M.J., Lane, M.C.G., Farrelly, E., Rapley, J.H., Goggin, U.M., Gentle, W.E. 2002. Partitioning, bioavailability, and toxicity of the pyrethroid insecticide cypermethrin in sediments. *Environ Toxicol Chem* 21:9-15.
16. U.S. Environmental Protection Agency. 2000. Methods for measuring the toxicity and bioaccumulation of sediment-associated contaminants with freshwater invertebrates. EPA/600/R-99/064. Office of Research and Development, Washington DC, U.S.
17. U.S. Environmental Protection Agency. 1994. Methods for assessing the toxicity of sediment-associated contaminants with estuarine and marine amphipods. EPA/600/R-94/025. Office of Research and Development, Washington DC, U.S.
18. Kosian, P.A., West, C.W., Pasha, M.S., Cox, J.S., Mount, D.R., Huggett, R.J., Ankley GT. 1999. Use of nonpolar resin for reduction of fluoranthene bioavail-ability in sediment. *Environ Toxicol Chem* 18: 201-206.
19. West, C.W., Kosian, P.A., Mount, D.R., Makynen, E.A., Pasha, M.S., Sibley, P.K., Ankley, G.T. 2001. Amendment of sediments with a carbonaceous resin reduces bioavailability of polycyclic aromatic hydrocarbons. *Environ Toxicol Chem* 20: 1104-1111.
20. Phillips, B.M., Anderson, B.S., Hunt, J.W., Huntley, S., Tjeerdema, R.S., Kapellas, N., Worcester, K. 2006. Solid-phase sediment toxicity identification evaluation in an agricultural stream. Environ Toxicol Chem.25: 1671-1676.

21. Ho, K.T., Burgess, R.M., Pelletier, M.C., Serbst, J.R., Cook, H., Cantwell, M.G., Ryba, S.A., Perron, M.M., Lebo, J., Huckins, J., Petty, J. 2004. Use of powdered coconut charcoal as a toxicity identification and evaluation manipulation for organic toxicants in marine sediments. *Environ Toxicol Chem* 23: 2124-2131.
22. Lebo, J.A., Huckins, J.N., Petty, J.D., Ho, K.T. 1999. Removal of organic contaminant toxicity from sediments--early work toward development of a toxicity identification evaluation (TIE) method. *Chemosphere* 39: 389-406.
23. Wheelock, C.E., Miller, J.L., Miller, M.J., Gee, S.J., Shan, G., Hammock, B.D. 2004. Development of toxicity identification evaluation procedures for pyrethroid detection using carboxylesterase activity. *Environ Toxicol Chem.* 23: 2699-2708.
24. Wheelock, C., Miller, J.L., Miller, M.J., Phillips, B.M., Huntley, S.A., Gee, S.J., Tjeerdema, R.S., Hammock, B.D. 2006. Use of carboxylcarboxylesterase activity to remove pyrethroid-associated toxicity to *Ceriodaphnia dubia* and *Hyalella azteca* in toxicity identification evaluations (TIEs). *Environ Toxicol Chem* 25: 973-984.
25. Weston, D.P., Amweg, E.L. 2007. Whole sediment Toxicity Identification Evaluation tools for pyrethroid insecticides: II. Esterase addition. Environ Toxicol Chem. 26;2397-2404.
26. Ankley, G.T., Collyard, S.A. 1995. Influence of Piperonyl Butoxide on the Toxicity of Organophosphate Insecticides to Three Species of Freshwater Benthic Invertebrates. *Comp Biochem Physiol C* 110: 149-155.
27. Ware, G.W. 1989. The Pesticide Book. Thompson Publications, Fresno, CA. 339 pp.
28. Kakko, I., Toimela, T., Tähti, H. 2000. Piperonyl butoxide potentiates the synaptosome ATPase inhibiting effect of pyrethrin. *Chemosphere* 40: 301-305.
29. Amweg, E.L., Weston, D.P. 2007. Whole sediment Toxicity Identification Evaluation tools for pyrethroid insecticides: II. Piperonyl Butoxide addition. Environ Toxicol Chem. 26; 2389-2396.
30. Coats, J.R, Symonik, D.M., Bradbury, S.P., Dyer, S.D., Timson, L.K., Atchison, G.J. 1989. Toxicology of synthetic pyrethroids in aquatic organisms: an overview. *Environ Toxicol Chem* 8: 671-679.
31. Weston, D.P. 2006. Temperature dependence of pyrethroid toxicity: TIE applications and environmental consequences. Conference proceedings, Society of Environmental Toxicology and Chemistry (SETAC) 27th annual meeting, Montreal, Quebec, Canada.
32. U.S. Environmental Protection Agency. 1991. Methods for aquatic toxicity identification evaluations. Phase I Toxicity Characterization Procedures. EPA 600/6-91/003. Office of Research and Development. Washington, DC.
33. Phillips, B.M., Anderson, B.S., Hunt, J.W., Nicely, P.A., Kosaka, R., de Vlaming, V., Connor, V., Richard, N., Tjeerdema, R.S. 2004. *In situ* water and sediment toxicity in an agricultural watershed. Environ. Toxicol. Chem. 23: 435-442.

34. Anderson, B.S., Hunt, J.W., Phillips, B.M., Nicely, P.A., Gilbert, K.D., de Vlaming, V., Connor, V., Richard, N., Tjeerdema, R.S. 2003. Eco-toxicologic impacts of agriculture drainwater in the Salinas River (California, USA). Environ. Toxicol. Chem. 22: 2375-2384
35. Amweg, E.L., Weston, D.P., Ureda, N.M. 2005. Use and toxicity of pyrethroid pesticides in the Central Valley, CA, USA. *Environ Toxicol Chem* 24: 966-972. Erratum 24:1300-1301.
36. Burgess, R.M., Charles, J.B., Kuhn, A., Ho, K.T., Patton, L., McGovern, D. 1997. Development of cation-exchange methodology for marine toxicity identification evaluation applications. *Environ Toxicol Chem* 16: 1203-1211.
37. Phipps, G.L., Mattson, V.R., Ankley, G.T. 1995. The relative sensitivity of three benthic test species to ten chemicals. *Arch Environ Toxicol Chem* 28: 281-286.
38. Macdonald, D.D., Ingersoll, C.G., Berger, T.A. 2000. Development and evaluation of consensus-based sediment quality guidelines for freshwater ecosystems. *Arch Environ Contam Toxicol* 39 (1): 20-31.
39. Nebeker, A.V., Schuytema, G.S., Griffis, W.L., Barbitta, J.A., Carey, L.A. 1989. Effect of sediment organic carbon on survival of *Hyalella azteca* exposed to DDT and endrin. *Environ Toxicol Chem* 8: 705-718.
40. U.S. Environmental Protection Agency. 1993. Methods for aquatic toxicity identification evaluations. Phase II Toxicity Identification Procedures for Samples Exhibiting Acute and Chronic Toxicity. EPA 600/R-92/080. Office of Research and Development. Washington, DC.
41. Cornelissen, G., Rigternink, H., van Noort, P.C.M., Govers, H.A.J. 2000. Slowly and very slowly desorbing organic compounds in sediments exhibit Langmuir-type sorption. Environ. Toxicol. Chem. 19: 1532-1540.
42. Besser, J.M., Ingersoll, C.G., Leonard, E.N., Mount DR. 1998. Effect of zeolite on toxicity of ammonia in freshwater sediments: implications for toxicity identification evaluation procedures. *Environ Toxicol Chem* 17: 2310-2317.
43. Burgess, R.M., Pelletier, M.C., Ho, K.T., Serbst, J.R., Ryba, S.A., Kuhn, A., Perron, M.M., Raczelowski, P., Cantwell, M.G. 2003. Removal of ammonia toxicity in marine sediment TIEs: a comparison of *Ulva lactuca*, zeolite and aeration methods. *Mar Poll Bull* 46: 607-618.
44. U.S. Environmental Protection Agency. 1996. Marine Toxicity Identification Evaluation (TIE), Phase I Guidance Document. Office of Research and Development. EPA/600/R-95/054. Washington, DC.
45. Hoke, R.A., Ankley, G.T., Cotter, A.M., Goldenstein, T., Kosian, P.A., Phipps, G.L., VanderMeiden, F.M. 1994. Evaluation of equilibrium partitioning theory for predicting acute toxicity of field-collected sediments contaminated with DDT, DDE, and DDD to the amphipod *Hyalella azteca*. *Environ Toxicol Chem* 13: 157-166.
46. Anderson, B.S., Phillips, B.M., Hunt, J.W., Richard, N., Connor, V., Tjeerdema, R.S. 2006. Identifying primary stressors impacting macroinvertebrates in the Salinas River (California, USA): relative effects of pesticides and suspended particles. Environmental Poll. 141: 402-408.

Chapter 18

Chemical Analysis and Enantioselective Toxicity of Pyrethroids

Mae Grace Nillos[1,2], Jay Gan[1], and Daniel Schlenk[1]

[1]Departments of [1]Environmental Sciences and [2]Environmental Toxicology Program, University of California, Riverside, CA 92521

Chirality is an important consideration in the risk assessment of pyrethroids. In addition, the current use patterns and the wide-spectrum aquatic toxicity of pyrethroids make them an emerging ecotoxicological concern. A number of studies have reported the occurrence of enantioselective degradation in some pyrethroids. However, enantioselectivity in degradation has not been adequately linked with toxicity. In general, the acute or chronic effects of pyrethroid enantiomers on the survival or endocrine disruption endpoints of aquatic organisms are still largely unexplored. Only a handful of studies have looked into enantiomer-specific toxicity of popularly used pyrethroids. These studies reported large differences in acute toxicities between pyrethroid enantiomers. The availability of suitable analytical techniques are crucial in the determination of enantioselectivity in pyrethroid toxicity. Chiral separation and analysis has rapidly improved in recent years. Chiral HPLC and GC are among the most commonly adopted analytical techniques for the separation and quantification of the stereoisomers.

Introduction

A relatively large proportion of current-use pesticides have active ingredients with chiral structures (*1,2*). Most of these compounds are applied and released to the environment as mixtures of enantiomers and risk assessment is generally not carried out for individual components of the mixture. Physical and chemical processes in the environment (e.g. leaching, air-water exchange, sorption, and other non-selective chemical reaction) do not differentiate between the enantiomers of the same compound (*3,4*). However, chirality is a property of matter found throughout the biological system. Thus, the interactions of chiral compounds with protein receptors and enzymes in biologically-mediated environmental processes are expected to be stereoselective (*5*).

Pyrethroids are among the most environmentally important class of modern pesticides. Their widespread use and high potential for contamination in the aquatic environment make them an emerging ecotoxicological concern (*6-9*). Chirality exists in all pyrethroids, and most have multiple chiral centers resulting in up to 4 to 8 enantiomers in any given compound (*10*).

A number of recent studies have reported the occurrence of enantioselective degradation of permethrin, bifenthrin and cypermethrin (*11-13*). However, the significance of these findings would be of limited consequence if the enantiomers have comparable toxicity to non-target organisms, particularly in the aquatic environment. Thus, in order to properly assess the risks associated with the use of pyrethroids, the potential for enantioselectivity in their bio-transformation and toxicity should be concurrently evaluated.

In most chiral pesticides only one of the two (or more) enantiomers is usually responsible for most, if not all, of its pesticidal activity (*14,15*). The other enantiomer(s), although inert to the target pest, end up as undesirable chemical load to the environment and could be toxic to non-target organisms (*2,11*). A recent study reported enantioselectivity in acute toxicity of the phenyl-pyrazole broad-spectrum insecticide, fipronil, to the aquatic invertebrate *Ceriodaphnia dubia* (*2,16*). Studies with rats have also demonstrated that (-)-*o,p'*-DDT enantiomer is more active as an estrogen-mimic than the (+)-enantiomer (*17*). However, except for a yeast-based assay (*18*), similar investigations have not been conducted in other biological systems or for other chiral pesticides.

Moreover, enantioselective bioaccumulation of chiral contaminants has been shown in a range of organisms (*19-21*). For example, in a recent study on the current-use insecticide fipronil, the (-) enantiomer was consistently more abundant in the tissues of rainbow trout (*Oncorhynchus mykiss*) exposed to racemic fipronil (*21*). This observation suggests greater accumulation rate of the (-) enantiomer, or faster biotransformation of the (+) enantiomer of fipronil, or both. Isomer conversion or enantiomerization may also occur during enantioselective biotransformation, potentially influencing the activity and contributing to side effects of chiral pesticides in the environment (*3,22*). All these studies demonstrate occurrence of enantioselectivity in pesticides in various biological processes.

Enantioselectivity in Pyrethroid Toxicity

The primary function of pyrethroid insecticides is the alteration of the gating kinetics of the voltage-gated sodium channels which mediate the transient increase in the sodium permeability of the nerve membrane in target insects (*23*). The mode of action of pyrethroids is reportedly similar in insects and mammals (*24*). Studies on the stereospecific binding of various pyrethroids to the sodium channel revealed significant differences in the ability of stereoisomers to cause depolarizing afterpotentials, repetitive firing, and membrane depolarization in crayfish giant axon (*25*). These results suggest that the nerve membrane response is largely dependent on the configuration of the pyrethroid.

In addition to the acute lethality afforded by pyrethroids to aquatic invertebrates and fish (*26*), the U.S. Environmental Protection Agency (EPA) has listed this insecticide class as potential endocrine disrupters in humans and wildlife (*27*). Estrogenic and antiestrogenic effects have been observed in terrestrial invertebrates and mammalian cell lines (*28-30*). Permethrin, fenvalerate, cypermethrin and deltamethrin were estrogenic in the MCF-7 cell-line assay (*29,31*). Some pyrethroid insecticides were also shown to inhibit the binding of estradiol to the estrogen receptor, while others induced mRNA expression that is transcriptionally induced by estrogen (*31*). Utilizing the human estrogen receptor in the Yeast Estrogen Screen (YES), permethrin and its hydrolytic metabolites, elicited both estrogen and anti-estrogenic responses (*30*). Inspite of these, the enantiomer-specific acute or chronic effects of pyrethroids on aquatic organisms are still largely unexplored.

Enantioselective Insecticidal Activity

The significance of the overall molecular shape in the mode of action of pyrethroids is evident in the stereospecificity of its insecticidal action (*23*). Insecticidal activity has been shown to be exclusive to those stereoisomers with the *1R* configuration on the ring of the cyclopropane carboxylic esters (*32*). For instance, the activity (median lethal dose; LD50) of the stereoisomers of resmethrin in topical application assays against the housefly, *Musca domestica* were compared. The *1R-trans* (LD50 = 13 ng/insect) and *1R-cis* (LD50 = 40 ng/insect) were approximately 100 times more potent than the *1S-trans* (LD50 = 1,770 ng/insect) and *1S-cis* (LD50 = 4,000 ng/insect) enantiomers (*33*). Elliott et al. (*34*) synthesized individual isomers of permethrin and showed that the *1R-trans* and *1R-cis* isomers were more active than their corresponding *1S* enantiomers against housefly (*M. domestica*) and mustard beetle (*Phaedon cochliariae*). In pyrethroids with 4 or 8 enantiomers, only one or two enantiomers are insecticidal. For instance, of the eight enantiomers in cypermethrin or cyfluthrin, only *1R-cis-αS* and *1R-trans-αS* are known to have insecticidal activity, while the rest are essentially inactive toward the target pest (*14,15*). In *cis*-bifenthrin, only the *1R-3R* enantiomer is active, while the *1S-3S* enantiomer is inactive (*14*).

Findings of enantioselectivity in insecticidal activity led to the development of enantiopure products, including esfenvalerate (*S,S-)* and deltamethrin (*1R-cis-αS*), or isomer-enriched products, such as α- (*1R-cis-αS* + *1S-cis-αR*), β- (*1R-cis-αS* + *1S-cis-αR* and *1R-trans-αS* + *1S-trans-αR*), and θ-(*1R-trans-αS and 1S-trans-αR*) cypermethrin (*12, 35*).

Enantioselective Mammalian Toxicity

Pyrethroids have relatively low toxicity to birds and mammals due to their rapid biotransformation and elimination. In general, the initial metabolism of the parent pyrethroid compound involves the attack of either esterases at the central ester bond (or to a lesser extent, of cytochrome-P450, at one or more sites in the acid and alcohol moieties). Cleavage is followed by further oxidation of the primary alcohol moieties (e.g., 3-phenoxybenzyl alcohol) to the corresponding carboxylic acids, while the α-cyano-substituted alcohols lose the cyanide constituent forming the corresponding aldehyde. This is followed by hydroxylation of a portion of the cleavage product and hydrolysis of the hydroxylated ester metabolites. Further metabolism involves conjugation with amino acids, sugars, sugar acids or sulfate prior to excretion (*23, 36*).

The chemical structure and stereochemistry of pyrethroids determine their metabolic pathway and rate of phase I biotransformation (*37*). Previous investigations indicated that the biotransformation of *cis*- and *trans*- isomers of pyrethroids follow different pathways (oxidative and hydrolytic) (*38*). Several studies have also shown greater persistence and toxicity of *cis*-permethrin over *trans*-permethrin in rats as a result of differences in the rate of hydrolysis between the two diastereomers (*39*). Moreover, a recent study reported species differences in oxidative and hydrolytic metabolism of pyrethroids (*37*).

Stereoselective toxicokinetics of *θ*-cypermethrin in rats was recently reported. Wang et al. (*35*) found that the *S*-(+)-enantiomer was less prevalent than the *R*-(-)-enantiomer in the plasma, heart, liver, kidney and fat tissues of rats intravenously treated with racemic dose of *θ*-cypermethrin. In addition, *in vivo* chiral conversion in the plasma was also reported. For the *θ*-cypermethrin enantiomers, the (+)-enantiomer was converted to the (-)-enantiomer in a unidirectional manner. These findings suggest enantioselective redistribution and/or biochemical processing of θ-cypermethrin in rats *in vivo*.

Similarly, *in vitro* biotransformation of *1R-cis-* and *1R-trans-*cyphenothrin by various rat tissues indicated that enantioselectivity could be tissue-dependent (*40*). Liver esterase hydrolyzed the *trans-*isomer more rapidly, than did the lung and kidney esterases. However, hydrolysis of both isomers proceeded at almost equal rates in the plasma. In addition, when the results with cyphenothrin was compared to results from similar experiments using phenothrin (with no cyano group on the alcohol moiety), the *trans* isomers from both pyrethroids exhibited similar rates of hydrolysis but *cis-*cyphenothrin underwent ester cleavage three times the rate of *cis-*phenothrin (*40*). Also, no significant difference in metabolite profiles between enantiomer pairs of phenothrin and tetramethrin were observed in rats (*40, 41*).

Since the mid 1970s, the stereoselective metabolism of pesticides has been investigated on insecticides that could potentially cause adverse health effects in humans (*42*). Biotransformation of most classes of pesticides generally leads to detoxification. However, bioactivation can result in the production of highly potent metabolites. For instance, the biotransformation of fenvalerate produced a toxic metabolite in experimental animals. Investigation into the enantioselectivity in this effect indicated that this metabolite was only formed from the non-insecticidally active enantiomer. These findings resulted in the development and marketing of the active enantiopure esfenvalerate (*10*).

Stereoselective Aquatic Toxicity

Pyrethroids are highly toxic to fish and aquatic invertebrates (*26*). This could be due the high affinity of pyrethroids to the molecular target in these species and/or significant differences in the pathways and rates of metabolism, from that in mammals (*43*). Because of their high acute aquatic toxicity, pyrethroids are ideal for understanding the importance of enantioselectivity in aquatic toxicity, because exposure does not require large amounts of enantiopure compounds to elicit biological effects. However, studies on enantiomer-specific ecotoxicological effects of pyrethroids are even more rare than investigations into their enantioselective environmental degradation. Only a few studies have looked into enantiomer-specific toxicity of popularly used pyrethroids such as permethrin, bifenthrin, cypermethrin and cyfluthrin (*11,12,44-45*). Results from these investigations are summarized in Tables I and II.

Acute aquatic toxicities of racemates and individual enantiomers of *cis-*bifenthrin, permethrin, cypermethrin and cyfluthrin were evaluated in standard 96-h tests with the aquatic invertebrates, *Ceriodaphnia dubia* and *Daphnia*

Table I. Enantioselective acute toxicity in aquatic invertebrates of some pyrethroids

		D. pulex[2]	*D. magna*[3]		*C. dubia*[3]	
cis-Bifenthrin	*Selectivity ratio*[1]	14.0	22.0		18.0	
	Active enantiomer	(-)	*1R-cis*		*1R-cis*	
Permethrin	*Selectivity ratio*[1]	-	>15.5	>19.5	>38.5	>30.5
	Active enantiomer	N.D.[4]	*1R-cis*	*1R-trans*	*1R-cis*	*1R-trans*
Cypermethrin	*Selectivity ratio*[1]	-	-		>10	>8
	Active enantiomer	N.D.[4]	N.D.[4]		*1R-cis-αS*	*1R-trans-αS*
Cyfluthrin	*Selectivity ratio*[1]	-	-		>96	>47
	Active enantiomer	N.D.[4]	N.D.[4]		*1R-cis-αS*	*1R-trans-αS*

[1]Selectivity ratio is the ratio of the LC50 (or TLM) of less active enantiomer(s) to that of the active enantiomer(s); [2][44]; [3][11] ; [4]N.D. (no data)

magna. A significant difference in the median lethal effect concentration (LC_{50}) between the two enantiomers of *cis*-bifenthrin was observed. The *1R-cis* enantiomer was 18–22 times more toxic than the corresponding *1S-cis* enantiomer (Table I). A similar pattern was observed for *cis*- and *trans*-permethrin. Therefore, out of the four enantiomers of permethrin, only the two with the *1R* configuration (*1R-cis* and *1R-trans*) were acutely toxic, while the two other enantiomers with the *1S* configuration were essentially inactive to the aquatic invertebrates. Both *cis*-bifenthrin and permethrin are pyrethroid compounds with chirality originating only from the cyclopropane ring in the acid moiety.

In cypermethrin, only two of the eight enantiomers, the *1R-cis-αS* and *1R-trans-αS*, were significantly toxic against *C. dubia* (*12*). The same trend was observed for cyfluthrin, in which the *1Rcis-αS* and *1R-trans-αS* enantiomers were 47–96 times more toxic to *C. dubia* than the six other stereoisomers (Table I). Thus, for pyrethroids with an *α*-carbon chiral center in their structures, it appears that a combination of the *1R* and *αS* configurations is essential for the observed toxicity. The consistency in enantioselectivity between insecticidal and aquatic toxicity further suggests a common mode of action shared between the target pests and aquatic invertebrates.

The acute toxicity of *cis*-bifenthrin enantiomers has also been evaluated on carp *(Cyprinus carpio)* and tilapia (*Tilapia* spp.) (*44*). The 96-h LC50 for carp indicated enantioselective acute toxicity with the (-)-bifenthrin as the more toxic of the two enantiomers. A similar trend was observed for tilapia, with (-)-bifentrhin about 4 times more toxic than the (+) enantiomer. Additionally, stereoselective toxicity to Japanese medaka (*Oryzias latipes*) was also reported for several pyrethroids (*46*). The (+)-*cis* and (+)-*trans* enantiomers of permethrin were significantly more toxic than the corresponding (-)-*cis* and *(-)-trans* enantiomers. A similar trend was also observed for the resmethrin and fenothrin enantiomers (Table II).

Enantioselectivity in the chronic toxicity and biotransformation of pyrethroids is essentially unknown in aquatic organisms. The U.S. EPA has listed this insecticide class as potential endocrine disrupters in humans and wildlife. At present however, there is no clear consensus on the estrogenic and/or antiestrogenic activities of pyrethroids in fish. Estrogenic and antiestrogenic effects of pyrethroids have been observed in mammalian cell lines by several researchers (*28-29*). Recent studies also demonstrated the ability of several pyrethroid metabolites to mimic 17β-estradiol interaction with estrogen receptors (*9*). Enantioselectivity in these effects has not yet been investigated, particularly in fish. Previous studies however, have shown that endocrine disruption by some chiral pesticides (e.g. *o,p'*-DDT) could be enantioselective (*18*). Significant difference in vitellogenin induction in adult male Japanese medaka was recently reported for *cis*-bifenthrin enantiomers (*47*). The 1*S*-*cis* enantiomer was found to elicit 123 times greater response than 1*R*-*cis*-bifenthrin.

In general however, the acute or chronic effects of pyrethroid enantiomers on the survival or endocrine disruption endpoints of aquatic organisms are still largely unexplored. In addition, having focused almost entirely on differences in the intrinsic biological activity of enantiomers, the importance of differential biotransformation of enantiomers of chiral pesticides, like pyrethroids, remains poorly characterized. However, if enantioselectivity does occur in the biotransformation and toxicity of chiral pesticides, their overall environmental risk will depend on the behavior of the active enantiomer(s), rather than the total chemical concentrations.

There are very few studies investigating differences in metabolic fates of pyrethroid enantiomers in aquatic organisms (*39, 47*). Significant selectivity in the uptake of *cis*-bifenthrin enantiomers in the liver of Japanese medaka was reported recently (*47*). However, the mechanism of enantioselective biotransformation of pyrethroids is still essentially unknown, particularly in aquatic organisms. Glickman et al. (*39*) reported on the preferential hydrolysis

Table II. Enantioselective acute toxicity in fish of some pyrethroids

		Carp[2]	*Tilapia*[2]	*Japanese medaka*[3]	
cis-Bifenthrin	*Selectivity ratio*[1]	2.0	4.0	-	
	Active enantiomer	(-)	(-)	N.D.[4]	
Permethrin	*Selectivity ratio*[1]	-	-	>588	>769
	Active enantiomer	N.D.[4]	N.D.[4]	(+)-*trans*	(+)-*cis*
Resmethrin	*Selectivity ratio*[1]	-	-	>625	>437
	Active enantiomer	N.D.[4]	N.D.[4]	(+)-*trans*	(+)-*cis*
Fenothrin	*Selectivity ratio*[1]	-	-	>83	>59
	Active enantiomer	N.D.[4]	N.D.[4]	(+)-*trans*	(+)-*cis*

[1]Selectivity ratio is the ratio of the LC50 (or TLM) of less active enantiomer(s) to that of the more active enantiomer(s); [2][44]; [3][46]; [4]N.D. (no data)

by esterases of *trans*-permethrin over *cis*-permethrin in carp (*Cyprinus carpio*) and trout (*Salmo gairdeneri*) liver microsomes. Previous studies also suggest pyrethroid metabolism and elimination in fish as substantially lower and qualitatively different from that reported in mammals (*26,39*).

Data Gaps

As pyrethroids tend to partition into the sediment phase, toxicity to sediment-dwelling organisms could be important. However, no study has attempted to characterize enantioselectivity in sediment toxicity for pyrethroids. In addition, little is known about the enantioselectivity in biotransformation, particularly in fish. In general, very few published data describe the metabolism of chiral pestcides. It is therefore difficult to evaluate the significance of enantioselective biotransformation in the toxicology of chiral pesticides like pyrethroids.

Enantiomer Separation and Analysis

The stereoselective behavior of chiral compounds has been recognized for almost a century, but investigations into the mechanisms of their differential effects were not possible until recent decades (*2*). That is because several factors need to be considered when addressing the need for enantiomer-specific toxicological data for chiral pesticides, including availability of suitable analytical techniques and appropriate assay methods.

One of the biggest challenges in studying enantioselectivity in chiral pesticides has been the separation and analysis of enantiomers. Separation of enantiomers can only be achieved in chiral environments, mostly on chiral HPLC or GC columns containing a chiral agent (e.g. cyclodextrin derivatives). In addition, enantiomeric separation of pyrethroids can be challenging since they contain 2 or 3 stereogenic centers, and therefore would consist of 4 or 8 enantiomers. The difficulty in chiral analysis is further complicated by the lack of enantiomer standards that are hard to synthesize and purify. Under certain circumstances, enantiomers may also undergo isomer inversion or racemization, which further complicates quantitative analysis of enantiomers. The lack of adequate analytical techniques has contributed to the limited research activity on the chirality of pyrethroids.

Fortunately, chiral separation technology has undergone significant developments over the past 20 years, facilitating the routine separation of enantiomers of a number of chiral pesticides (*4*). Typical instrumentation that is currently being used includes high performance liquid chromatography (HPLC) and gas chromatography (GC) in conjunction with a suitable chiral separation column. Both of these instruments can use mass spectrometry detection to enable more sensitive quantitation of analytes.

Chiral HPLC methods have been reported for analysis of several pyrethroids (*48-49*). However, while chiral HPLC is ideal for preparative work and for laboratory experiments, GC is more desirable for environmental analysis. Overall, separation and identification of enantiomers remain as the bottleneck for understanding the environmental behaviors of chiral pesticides. Wong (*4*) provides an extensive summary of the chiral analytical techniques that have been developed and used to quantify enantiomer composition in environmental matrices.

Chiral HPLC Separation Techniques

High performance liquid chromatography methods have been developed for the separation and analysis of enantiomers of a number of pyrethroids. The advantage of using chiral HPLC techniques is that individual enantiomers could be recovered following the analysis, and used in toxicological experiments that require separated enantiomers (Table III).

The isolation of individual enantiomers of pyrethroids are mostly done with normal-phase chiral HPLC (*11,12*). This technique has also been used in enantiomer quantitation in biological tissues (*35*). Reversed-phase chiral HPLC and capillary electrophoresis techniques have also been used with several pyrethroids (*50*), as well as two-dimensional HPLC through coupling of achiral and chiral columns (*51*).

Normal-phase chiral HPLC methods were recently developed for the separation and analysis of enantiomers of a number of pyrethroids (*12,13*). In the method development process, various commercially available chiral stationary phase columns, including Sumichiral OA-2500 (Sumika Chemical Service, Tokyo, Japan), Chirex 00G-3019-OD (Phenomenex, CA), Chiralcel OD-R, ChiralPak IA and Chiralcel OJ (Daicel Chemical Industries, Tokyo, Japan) columns were tested. Optimal resolution and isolation of enantiomers of permethrin and bifenthrin was achieved using a Sumichiral OA-2500-I column. Separation of enantiomers of cypermethrin and cyfluthrin has been relatively successful in two Chirex 00G-3019-OD columns in series (Phenomenex, CA).

Lisseter and Hambling (*49*) have also used a Pirkle Type 1A-C (Technicol, Stockport, U.K.) chiral phase to separate enantiomers of cyhalothrin, cypermethrin and cyfluthrin. However, most peaks were not well separated at the baseline under optimal conditions. Oi et al. (48) used Sumichiral OA-2500I and OA-4700 columns to resolve a number of pyrethroid insecticides. Fenpropathrin was separated at the baseline. Cypermethrin was almost separated to the baseline when using the two columns in series.

Chiral GC Analysis

Chiral analytical techniques are determined by the properties of the chiral molecules (*52*). Gas chromatography is primarily used in the analysis of volatile and thermally stable samples. Chiral GC has the advantage of high efficiency, sensitivity and reproducibility. In addition, auxilliary systems, such as mass spectrometer (MS) and electron capture detector (ECD) can be coupled with chiral GC for the analysis of enantiomers in complicated matrices including environmental, biological and agricultural samples (*11-13, 51, 53-55*). Some chiral analysis of pyrethroids in various environmental matrices are summarized in Table IV.

Pyrethroids are semivolatile nonpolar compounds that are typically analyzed by chiral GC with either ECD or MS detection. Gas chromoatography methods have been developed to separate the diastereomers and enantiomers of several pyrethroids. An achiral HP-5MS column (Agilent, Wilmington, DE) has been

Table III. Chiral HPLC techniques for the preparation of enantiopure pyrethroids

	# of isomers	Separation and detection	CSP[1]	Mobile Phase[2]	Ref
cis-Bifenthrin	2	HPLC/UV	OA-2500I	I	(*11, 12*)
Permethrin	4	HPLC/UV	OA-2500I	I	(*11, 12*)
Cyfluthrin	8	HPLC/UV	00G-3019-OD	II	(*12*)
Cypermethrin	8	HPLC/UV	00G-3019-OD	II	(*12*)
		HPLC/UV	OA-4600 + OA-4700	II	(*48*)
Fenvalerate	4	HPLC/UV	OA-2500I	III	(*48*)
Fenpropathrin	2	HPLC/UV	OA-4600	III	(*48*)
Resmethrin	4	HPLC/UV	OA-2500I	I	(*48*)
Allethrin	8	HPLC/UV	OA-4600	II	(*48*)
Phenothrin	4	HPLC/UV	OA-2500I	I	(*48*)
Terallethrin	2	HPLC/UV	OA-4000	II	(*48*)
Tetramethrin	4	HPLC/UV	OA-2500I	II	(*48*)
Cyhalothrin	4	HPLC/UV	Pirkle 1A - C	IV	(*49*)

[1] CSP, Chiral Stationary Phase;

[2] I. Hexane/1,2-Dichloroethane (500:1); II. Hexane/1,2-Dichloroethane-ethanol (500:10:0.05); III. Hexane/1,2-Dichloroethane-ethanol (500:30:0.15); IV. Hexane/2-propanol (500:0.75)

Table IV. Chiral analytical techniques used to quantify enantiomer composition of pyrethroids in environmental matrices

	Sample Matrix	CSP[1]	Analysis	Detection	Ref.
cis-Bifenthrin	sediment	BGB-172	GC	ECD	(*11,13*)
	water	BGB-172	GC	ECD	(*12,54*)
Permethrin	sediment	BGB-172	GC	ECD	(*11,13*)
	water	BGB-172	GC	ECD	(*12,53*)
Cyfluthrin	sediment	BGB-172	GC	ECD	(*54*)
	water	BGB-172	GC	ECD	(*54*)
Cypermethrin	sediment	BGB-172	GC	ECD	(*54*)
	water	BGB-172	GC	ECD	(*54*)
θ-Cypermethrin	tissue	CDMPC	HPLC	UV	(*35*)
Fenvalerate	Soil	OD-H	HPLC	UV	(*51*)
	milk	Bakerbond	HPLC	UV	(*56*)

[1] CSP, Chiral Stationary Phase

used to separate the diastereomers, and a derivatized β-cyclodextrin chiral column, BGB-172 (BGB Analytik, Adliswil, Switzerland) was used for the separation of the enantiomers of bifenthrin, permethrin, cypermethrin and cyfluthrin (*53-54*). All diastereomers of both cypermethrin and cyfluthrin were separated on the achiral column. The enantiomers of all *cis*-diastereomers were resolved, while those of the *trans*-diastereomers were not separated at the enantiomer level.

A potential limitation of chiral GC analysis is the possibility of isomerization of pyrethroids such as cypermethrin and cyfluthrin that have an asymmetric α-carbon cyano substituent. Enantiomerization may occur in the presence of heat, polar solvents, or light (*16*). Liu et al. (*55*) reported about 9% chiral conversion of cypermethrin and cyfluthrin enantiomers, likely caused by the heated GC inlet when operated at 260°C. However, isomerization was relatively insignificant when on-column injection was used, or when the inlet temperature was lowered to 180°C. There was no evidence of isomerization of bifenthrin and permethrin during GC analysis. Both of these pyrethroids lack a cyano substituents at the chiral α-carbon center. Consequently, caution should be exercised in the analysis of certain pyrethroids in order to avoid abiotic isomerization, and inaccurate interpretation of enantiomer data.

Conclusion

The current use patterns and the wide-spectrum aquatic toxicity of pyrethroids make them an emerging ecotoxicological concern. Studies so far have not adequately linked enantioselectivity in degradation with that in toxicity. This could be attributed to the limited research characterizing enantioselectivity in pyrethroid toxicity to non-mammalian and non-target aquatic organisms. The significant enantioselectivity observed in the acute aquatic toxicity studies discussed in this chapter suggests that the environmental behavior of the active enantiomers, instead of the racemate bears more relevance to the ecotoxicological importance of pyrethroids.

Undoubtedly, chirality is an important consideration in the risk assessment of pyrethroids. The availability of suitable analytical techniques and toxicity assays are crucial in the determination of enantioselectivity in ecological toxicity of pyrethroids. Separation and analysis of enantiomers has rapidly improved in recent years. This technology has developed to a point where we are allowed several options for the development of enantiomer resolution and preparation techniques. Chiral HPLC and GC are at present the most reliable and commonly adopted analytical tools for the separation and quantitation of pyrethroid stereoisomers.

References

1. Williams, A. *Pestic. Sci.* **1996**, *46*, 3-9
2. Garrison, A. W. *Environ. Sci. Technol.* **2006**, *40*, 16-23.
3. Jarman, J. L.; Jones, W. J.; Howell, L. A.; Garrison, A. W. *J. Agric. Food Chem.* **2005**, *37*, 2668-2674.
4. Wong, C. *Anal. Bioanal. Chem.* **2006**, *386*, 544-558.
5. Hegeman, W.; Laane, R. *Rev. Environ. Contam. Toxicol.* **2002**, *173*, 85-116.
6. Weston, D. P.; You, J.; Lydy, M. J. *Environ. Sci. Technol.* **2004**, *38*, 2752-2759.
7. Weston, D.P.; Holmes, R.; You, J.; Lydy M.J. *Environ. Sci. Technol.* **2005**, *39*, 9778-9784.
8. Amweg, E.; Weston D.P.; Lydy, M.J. *Environ. Sci. Technol.* **2006**, *40*, 1700-1706.
9. McCarthy, A.; Thomson, B.; Shaw, I.; Abell A. *J. Environ. Monitor.* **2006**, *8*, 197-202.
10. Kurihara, N.; Miyamoto, J.; Paulson, G.; Zeeh, B.; Skidmore, M.; Hollingworth, R.; Kuiper, H. *Pure & Appl. Chem.* **1997**, *69*, 2007-2025.
11. Liu, W. P.; Gan, J. Y.; Schlenk, D.; Jury, W. A. *PNAS* **2005**, *102*, 701-706.
12. Liu, W.P.; Gan, J.; Qin, S. *Chirality* **2005**, *17*, S127-S133.
13. Qin, S.; Budd, R.; Bondarenko, S.; Liu, W.; Gan, J. *J. Agric. Food Chem.* **2006**, *54*, 5040 -5045.
14. *Synthetic pyrethroid insecticides: Structures and properties;* Naumann, K. Eds; Springer-Verlag Berlin Heidelberg: New York, **1990**.
15. Leicht, W.; Fuchs, R.; Londershausen, M. *Pestic. Sci.* **1996**, *48*, 325-332.
16. Konwick, B.J.; Fisk, A.T.; Garrison, A.W.; Avants, J.K.; Black, M.C. *Environ. Toxicol. Chem.* **2005**, *24*, 2350–2355.
17. McBlain, W. *Life Science* **1987**, *40*, 215-221.
18. Hoekstra, P.; Burnison, B.; Neheli, T.; Muir, D. *Toxicology Letters* **2001**, *125*, 75-81.
19. Buser, H.; Muller, M. *Environ. Sci. Technol.* **1998**, *32*, 626-633
20. Bidleman, T.; Falconer, R. *Environ. Sci. Technol* **1999**, *33A*, 206-209.
21. Konwick, B.J. ; Garrison, A.W. ; Black, M.C. ; Avants, J.K. ; Fisk, A.T. *Environ. Sci. Technol.* **2006**, *40*, 2930-2936.
22. Muller, M.; Buser, H. *Environ. Sci. Technol.* **1997**, *31*, 1953-1959.
23. Soderlund, D.; Clark, J.; Sheets, L.; Mullin, L.; Piccirillo, V.; Sargent, D.; Stevens, J.; Weiner, M. *Toxicology* **2002**, *171*:3-59.
24. Casida, J.; Gammon, D.; Glickman, A.; Lawrence, L. *Ann. Rev. Pharmacol. Toxicol.*, **1983**, *23*, 413-438.
25. Lund, A.E.; Narahashi, T. *Pestic. Biochem. Physiol.* **1983**, *20*, 203-216.
26. Bradbury, S.; Coats, J. *Environ. Toxicol. Chem.* **1989**, *8*, 373-380.

27. U.S. Environmental Protection Agency, 1997, *Report EPA/630/R-96/012* U.S. EPA: Washington, D.C.
28. Taylor, D.; Chinze, Y.; Ito, K.; Higuchi, N.; Ando, K. *J. Med. Entomol.*, **1991**, *28*, 322-329.
29. Go, V.; Garey, J.; Wolff, M.; Pogo, B. *Environ Health Perspec* **1999**, *107*, 173-177.
30. Tyler, C. R.; Beresford, N.; Woning, M. V.; Sumpter, J. P.; Thorpe, K. *Environ. Toxic. Chem.* **2000**, *4*, 801-809.
31. Chen, H.; Xiao, J.; Hu, G.; Zhou, J.; Xiao, H.; Wang, X. *J Toxicol Environ Health* **2002**, *A65*, 1419-1435.
32. Chamberlain, K.; Matsuo, N.; Kaneko, H.; Khambay, B. In: Kurihara N., Miyamoto, J., editors. *Chirality in agrochemicals*. Chichester, U.K.:Wiley; **1998**, 9-84.
33. Wickham, J. *Pestic. Sci.* **1976**, 273-237.
34. Elliot, M. E.; Farnham, A. W.; Janes, N. F.; Soderland, D. M. *Pestic. Sci.* **1978**, *9*, 112-116.
35. Wang, Q.; Qui, J.; Zhu, W.; Jia, G.; Li, J.; Bi, C.; Zhou, Z. *Environ Sci Technol* **2006**, *40*, 721-726.
36. Wheelock, C.; Shan, G.; Ottea, J. *J. Pestic. Sci.* **2005**, *30*, 75-83.
37. Godin, S.; Scollon, E.; Hughes, M.; Pooter, P.; DeVito, M.; Ross, M. *DMD*, **2006**, *34*, 1764-1771
38. Shono, T.; Unia, T.; Casida, J.E. *Pestic. Biochem. Physiol.* **1978**, *9*, 96-106.
39. Glickman, A.; Shono, T.; Casida, J., Lech, J. *J. Agric. Food. Chem.* **1979**, *5*,1038-1041.
40. Kaneko, H.; Izumi, T.; Ueda, Y.; Matsuo, M.; Miyamoto, J. *J. Pestic. Sci.*, **1984**, *9*, 249-258.
41. Izumi, T.; Kaneko, H.; Matsuo, M.; Miyamoto, J. *J. Pestic. Sci.*, **1984**, *9*, 259-267.
42. Soderlund, D.; Casida, J. *Pestic. Biochem. Physiol.*, **1977**, *7*, 391-401.
43. Huckle, K.; Milburn, P. IN: *Environmental Fate of Pestcides*, Hutson, D.; Roberts, T., editors, John Wiley and Sons, NY, **1990**; 175-243.
44. Liu, T.; Wang, Y.; Yen, J. *J Food Drug Anal*, **2005**, *13*, 357-360.
45. Chamberlain, K.; Matsuo, N.; Kaneko, H.; Khambay, B. IN: Chirality in Agrochemicals, Kurihara, N.; Miyamoto, J., editors, John Wiley and Sons, Chichester, **1998**; 9-83.
46. Miyamoto, J. *Environ. Health Perspec.*, **1976**, *14*, 15-28.
47. Wang, W.; Liu, W.; Yang, C.; Pan, Z.; Gan, J.; Xu, C.; Zhao, M.; Schlenk, D.; *Environ. Sci. Technol.* **2007**, *41*, 6124-6128.
48. Oi, N.; Kitahara, H.; Kira, R. *J. Chromatogr*, **1990**, *515*, 441-450.
49. Lisseter, S.; Hambling, S. *J. Chromatogr*, **1991**, *539*, 207-210.
50. Sevcik, J.; Lemr, K.; Stransky, Z.; Vecera, T.; Hlavac J. *Chirality*, **1997**, *9*, 162-166.

51. Faraoni, M.; Messina, A.; Polcaro, C. M.; Aturki, Z.; Sinibaldi, M.; *J. Liquid Chromatogr Rel Technol.* **2004**, *27*, 995-1012.
52. He, L.; Beesley, T. J. *J. Liquid Chromatogr Rel Technol.* **2005**, *28*, 1075-1114.
53. Liu, W.; Gan, J.; Lee, S.; Werner, I. *Environ. Toxicol. Chem.* **2005**, *24*, 1861-1866.
54. Liu, W.; Gan, J. *J. Agric. Food Chem.* **2004**, *52*, 736-741.
55. Liu, W.; Qin, S.; Gan, J. *J. Agric. Food Chem.* **2005**, *53*, 3814-3820.
56. Papadopoulou-Mourkidou, E. *Chromatographia*, **1985**, *20*, 376-378.

Risk Mitigation and Stewardship

Chapter 19

Mitigation of Permethrin in Irrigation Runoff by Vegetated Agricultural Drainage Ditches in California

D. L. Denton[1], M. T. Moore[2], C. M. Cooper[2], J. Wrysinski[3], W. M. Williams[4], J. L. Miller[5], K. Reece[6], D. Crane[7], and P. Robins[3]

[1]U.S. Environmental Protection Agency, Region IX, Sacramento, CA 95814
[2]National Sedimentation Laboratory, U.S. Department of Agriculture, Agricultural Research Service, Oxford, MS 38655
[3]Yolo County Resource Conservation District, 221 W Court Street, Suite 1, Woodland, CA 95695
[4]Waterborne Environmental Inc., 897-B Harrison Street, SE, Leesburg, VA 20175
[5]AQUA-Science, 17 Arboretum Terrace, Davis, CA 95616
[6]Aquatic Toxicology Laboratory, University of California at Davis, Davis, CA 95616
[7]California Department of Fish and Game, Rancho Cordova, CA 95670

As organophosphate use has decreased in California, a concomitant increase in their replacement insecticides (pyrethroids) has occurred. Although the probability of off-site movement of pyrethroids is less than their predecessors (organophosphates), transport of pyrethroids to aquatic receiving systems is still a potential threat. To mitigate possible harm, several in-field and edge-of-field management practices have been proposed, including conservation tillage, stiff grass hedges, riparian buffers, and constructed wetlands. By incorporating several individual components of these management practices, vegetated agricultural drainage ditches (VADD) have been proposed as a potential economical and environmentally efficient management practice to mitigate effects of pesticides in irrigation and storm runoff. A field trial was held in Yolo County, California, where three ditches (U-shaped vegetated; V-shaped vegetated; and V-shaped unvegetated) were constructed and amended for 8 h each with a mixture of permethrin and suspended sediment simulating an irrigation runoff event. Spatial and temporal collections of water, sediment, and plant samples were analyzed for cis and trans permethrin concentrations. Because the cis- isomer of permethrin is considered more toxic than the trans- isomer, only cis-permethrin results are reported herein. Cis-permethrin half-lives in water were similar between ditches ranging from 2.4-4.1 h. The differences between half-distances (distance required to reduce initial pesticide concentration by 50%) among the V-shaped vegetated and unvegetated ditches were two times more efficient with vegetation, indicating importance of vegetation in mitigation. Cis-permethrin half-distances ranged from 22 m (V-vegetated) to 50 m (V-unvegetated). These studies are being used to validate a computer simulation model that is being developed to design VADD for site-specific implementation. Utilizing features already present in the agricultural landscape, such as drainage ditches, will provide farmers with an economical alternative that still is protective of the receiving aquatic environment.

Introduction

Pyrethroid use in California has increased since 1992 (*1,2*). Approximately 241,570 kg (active ingredient) of the pyrethroid permethrin was applied to 64 crops in 2005 compared to 171,790 kg on 50 crops in 1992 (*2*). Although the probability of off-site movement of pyrethroids is less than their predecessors (organophosphates), transport of pyrethroids to aquatic receiving systems is still a potential threat. Recently, sediment toxicity to pyrethroids has been documented in urban waterways and agriculturally dominated waterways (*3,4*). While pesticide efficacy has greatly improved over the last several decades, there is still a void in research on management practices to decrease the likelihood of non-point source pollution. Lee and Jones-Lee (*5*) urged the need for quantitative information on best management practice (BMP) efficiency for agricultural runoff, particularly within California's Central Valley.

Since the early to mid 1990s, there has been increased emphasis on the 303 (d) provision of the Clean Water Act, focusing on the total maximum daily load (TMDL) process. In California, pesticides are the leading cause of impairments to waterbodies (*6*). In 2002, USEPA published their "Twenty Needs Report" on how research can enhance the TMDL process (*7*). Current research described here addresses many of those needs, including "Improve information on BMPs, restorations or other management practice effectiveness, and the related processes of system recovery." Several BMPs currently promoted by the USDA's Natural Resource Conservation Service (USDA-NRCS) include, but are not limited to, buffer and filter strips, riparian buffers, grassed waterways, and constructed wetlands. Each of these practices requires farmers to remove acreage from production landscape to meet physical BMP requirements. An economical alternative is needed that will allow production acreage to remain intact, but still accomplish the necessary environmental tasks for water quality improvements. Moore et al. (*8*) demonstrated the usefulness of vegetated agricultural drainage ditches as one such alternative to traditional BMPs.

Drainage ditches are a common part of the agricultural landscape, but are often considered of little value other than for movement of excess water from the field. These unique ditch ecosystems provide a host of potential services other than water conveyance, including sediment trapping and mitigation of nutrients and pesticides (*8*).

The current study involved a field trial to determine efficiency of recently-constructed vegetated drainage ditches for mitigation of permethrin-associated runoff from tomato fields. The concept was based on earlier studies that resulted in substantial sorption of pyrethroid insecticides (lambda-cyhalothrin, bifenthrin, and esfenvalerate) by ditch vegetation from agricultural fields in Mississippi (*8,9,10,11*). For example, three hours following initiation of simulated storm events, 97% of lambda-cyhalothrin was associated with plant material. Of the measured bifenthrin and esfenvalerate, 52% and 66%, respectively, were associated with vegetation.

Three main objectives involved in the current study were to 1) evaluate mitigation efficiency of two types of ditch design—U (typical in Mississippi Delta) versus V (typical in California)—with the pyrethroid permethrin; 2) evaluate the benefit of vegetation in a typical California V ditch by comparing its permethrin mitigation efficiency to an unvegetated V-ditch as a control; and 3) determine permethrin mass distribution within the water, sediment, and (if applicable) plants located in the U-vegetated, V-vegetated, and V-unvegetated ditches to estimate permethrin half-lives, half-distances, while providing data for modeling efforts.

Materials and Methods

Ditch Design

Three ditches, each 116 m in length, were constructed on a farm in Yolo County, California. Two different ditch designs ("U" and "V") were employed in this research. Although the "V" ditch design is most common throughout Yolo County, researchers also wanted to compare the broader "U" shape design for potential improved permethrin mitigation efficiency. One U-shaped ditch was constructed with a 3 m top bank width, and a maximum holding capacity water depth of 0.37 m. Two, V-shaped ditches were identically constructed with top bank widths of 1.8 m and a maximum holding capacity water height width and water depth of 0.6 m and 0.24 m, respectively. Vegetation in U- and V-ditches was similar in density and distribution through a cross-section of both vegetated ditches. Because of sandy field soil conditions, no significant outflow occurred from either U- or V-ditches. One V-ditch remained unvegetated to serve as a control ditch. The other V-ditch and single U-ditch were planted with *Hordeum vulgare* (barley) and *Lolium multiflorum* (annual ryegrass). Lamb's quarter (*Chenopodium album*) was an invasive prevalent weed within the vegetated ditches, and it served as an unexpected source of organic material within the ditch systems. Identical sampling sites were established within all three ditches at the simulated runoff inlet (0 m) (site 1), 42 m (site 2), 51 m (site 3), 88 m (site 4), and 108 m (site 5), sampling sites were delineated to reflect different types of vegetation within the ditch. Prior to the initiation of the simulated irrigation runoff event (24 h), ditch vegetative cover and dominant plant species were determined by sampling three 0.23 m^2 quadrants at each sampling site. The study was designed such that any runoff leaving the ditch was routed into a vegetated sump pond to prevent direct release into the aquatic receiving system; however, the small volume of water entering the sump filtered through the soil column within 16 hours of entry.

Simulated Irrigation Runoff Event

In July 2005, a simulated irrigation runoff event was delivered into each of the three constructed drainage ditches. A mixture of permethrin (Pounce® 3.2 EC), and 45 kg of dry soil was added to a 3800 L steel water tank filled with ground water and kept in suspension using a small submersible pump. The concentration of permethrin in simulated runoff (0.02 mg/L) was based on the recommended Pounce® 3.2 EC application rate (0.37 L/ha) for a 32-ha contributing area of tomatoes and an assumed 0.09% permethrin runoff with a targeted discharge of 7 L/s into experimental ditches (*12*). Using an Atwood™ 450 submersible pump (1703 L/h maximum flow) and 1.9 cm tubing, simulated runoff was pumped from the tank into calibrated values entering ditch inflows. Irrigation pipe (30-cm diameter) carried dilution water (approximately 198,000 L per ditch) from a nearby pump to the ditches.

Collection of Water, Sediment, and Plant Samples

Velocity (m/s), temperature (°C), pH, dissolved oxygen (mg/L) and electrical conductivity (μS/cm) were measured with calibrated hand-held field meters at inflow (site 1) and near the outflow (site 5) of each of the three constructed ditches at times 0, 0.5 h, 1 h, 4 h, 8 h, and 16 h. Grab samples of water were collected in pre-cleaned, certified 1 L amber Boston round, narrow mouth glass bottles with Teflon® lined closures at 0 h, 0.5 h, 1 h, 4 h, 8 h, and 16 h, post-application from each site. Sediment samples were collected in 120-mL wide mouth glass bottles with Teflon® lined closures at times identical to water collection (including 24 h, 48 h and 120 h samples). Plant samples were also collected along the same time schedule as sediments. Sediment samples were obtained from the top 1 cm using solvent-rinsed stainless steel spatulas, while plant materials were collected with solvent-rinsed scissors. Only plant material exposed in the water column (between sediment-water surface) was collected for analysis. Plant samples were wrapped in aluminum foil and placed in pre-labeled 3 L freezer bags. All samples were preserved on wet ice from collection through transport to the Aquatic Toxicology Laboratory (ATL) at the University of California, Davis. Water samples were kept in the dark at 4°C prior to transport to the California Department of Fish and Game Water Pollution Control Laboratory (DFG-WPCL). Sediment samples (also transported to DFG-WPCL) were frozen and kept in the dark until transport. Plant material was frozen immediately upon receipt at the ATL and shipped overnight to the USDA Agricultural Research Service National Sedimentation Laboratory (NSL) for sample preparation. Upon arrival at the NSL, plant samples were dried and ground using a Thomas-Wiley Model 4 laboratory mill. After preparation, samples were placed in glass vials and shipped to DFG for permethrin analyses.

Permethrin Extraction – Water

Water samples were extracted within 7 days, according to USEPA Method 3510C – Separatory Funnel Liquid-Liquid Extraction. One-liter water samples were fortified with triphenyl phosphate and dibromooctafluorobiphenyl to monitor extraction proficiency and extracted twice with dichloromethane (DCM) using a mechanical rotating extractor. Extracts were dried using sodium sulfate, concentrated, and solvent exchanged with petroleum ether (PE) using Kuderna-Danish (K-D) evaporative glassware equipped with a 3-ball Snyder column followed with a micro-Snyder apparatus and adjusted to a final volume of 2 mL in iso-octane. Concentrations of cis and trans isomers of permethrin were generated separately. Since cis-permethrin is generally considered more toxic than trans-permethrin, results discussed herein will only focus on the cis- isomer.

Permethrin Extraction and Cleanup – Sediment and Vegetation

Sediment and vegetation sample extraction followed USEPA Method 3545A – Pressurized Fluid Extraction. Homogenized sediment (10 g) and dried vegetation (2.5 g) samples were mixed with pre-extracted Hydromatrix® (7 g,

Varian Corporation) and fortified with triphenyl phosphate, dibromo-octafluorobiphenyl and dibutylchlorendate. Samples were extracted twice with acetone/DCM (50/50, v/v) using a Dionex® Accelerated Solvent Extractor (ASA 200, 100°C, 1500 psi). Extracts were dried using sodium sulfate, concentrated and solvent exchanged with PE using K-D evaporative glassware equipped with a 3-ball Snyder column followed with a micro-Snyder apparatus and adjusted to final volume of 2 mL in iso-octane. Clean up of sulfur, chlorophyll and other matrix interferences followed USEPA Method 3600C guidelines, as needed.

Instrument Analysis

Water, sediment, and vegetation sample final extracts were analyzed for permethrin using USEPA 8081B guidelines for permethrin analysis. Permethrin was analyzed using dual column high resolution gas chromatography equipped with electron capture detectors. The aqueous reporting limit for cis-permethrin was 0.005 μg/L, while The sediment reporting limit (dry weight) was 4.00 ng/g. Vegetation reporting limit (fresh weight) was 5.00 ng/g for cis-permethrin.

Data Analysis

Ordinary least-squares linear regression analyses (*13*) were used to fit curves to log-transformed permethrin water concentrations (y) versus the log of the distance down ditch from the inlet (x). Mass balances were performed using data on water, plant and sediments collected along transects of the ditch length for each sample time point (0.5 h, 1 h, 4 h, 8 h, 16 h, 24 h, 48 h, and 120 h).

Ditch chemical depuration rate constants (k_2) were determined for water in each of the three ditches. This was accomplished by plotting the ln (total concentration) as a function of time and, through linear regression analysis, determining the slope. Pesticide half lives ($t_{½}$) in water were estimated using the equation $\ln(2) / k_2$. Using the same premise, ditch half-distances were determined by plotting the ln (total concentration) as a function of ditch sample distance, determining the slope, and using the $\ln(2) / k_2$ equation.

Results

Although all ditch delivery systems were calibrated and re-checked prior to the simulated irrigation event, variability of inflow concentrations of cis-permethrin still occurred. Samples collected from the inflow pipe at time 0 (test initiation), indicated cis-permethrin concentrations of 27.0 μg/L, 225 μg/L, and 117 μg/L for the U-ditch, V-vegetated, and V-unvegetated ditches, respectively. At V-vegetated ditch site 5 (108 m down-ditch), final cis-permethrin concentration decreased 80% from the 1 h sampling to the 8 h sampling (Table I). Due to changes in pump pressure, constant flows were difficult to maintain. Average inflow measurements for the U-ditch, V-vegetated, and V-unvegetated ditches were 3.59±0.56 m/s, 3.95±0.53 m/s, and 3.11±0.27 m/s, respectively. Flow measurements were also taken at site 5 (outflow), approximately 5 m from the actual slotted board riser drain pipe. Average values for the U-ditch, V-vegetated, and V-unvegetated ditches were 0.45±0.27 m/s, 0.56±0.38 m/s, and 0.49±0.37 m/s, respectively.

Table I. Selected aqueous pesticide concentrations (μg/L) in inflow and outflow (site 5) of three experimental drainages ditches following a simulated irrigation event in Yolo County, CA.

	U-vegetated	*V-vegetated*	*V-unvegetated*
Inflow (0 h)	27.0	225	117
Site 5 (1 h)	18.2	9.63	13.5
Site 5 (4 h)	6.80	1.37	1.35
Site 5 (8 h)	1.08	1.89	1.17
Site 5 (16 h)	0.981	*	*

* indicates no water available for sampling

Even though initial inflow water concentrations of cis-permethrin differed between ditches, by converting concentration to mass, ditches can be compared to one another. A mass balance shift for cis-permethrin in water occurred from the 1 h sample compared to the 8 h sample. In the U-ditch, 26±11% of measured cis-permethrin mass was located in the water at 1 h; however, only 4±1% of the mass was in the water column at the 8 h sample. Similar trends were evident for the same time periods in the V-vegetated (31±8% and 9±3%) and V-unvegetated (32±7% and 17±3%) ditches. Examination of each ditch indicated 14±6%, 16±8%, and 20±6% of measured cis-permethrin mass during the 8 h dose was located in water of the U-ditch, V-vegetated, and V-unvegetated ditches, respectively. Using all time and distance sediment measurements, percent mean measured mass (± SE) of cis-permethrin in sediment was 64(5), 52(2), and 80(6), respectively, for the U and, V-vegetated, and V-unvegetated ditches. Cis-permethrin mean percent masses (± SE) in plants were 23(7) and 33(5) for the U and V-vegetated ditches, respectively. Total cis-permethrin masses measured during the 8 h exposure and additional samples collected at 16 h, 24 h, 48 h, and 5 d post-exposure (including water, sediment, and plants) ranged from 225-1901 mg for the U ditch, 192-843 mg for the V-vegetated ditch, and 206-2149 mg for the V-unvegetated ditch. Mass estimates are that 65%, 56%, and 47% of cis-permethrin applied were accounted for in the U-, V-vegetated and V-unvegetated ditches.

Several sites in the drainage ditches were dry before the 16 h sampling. As a result, sediment-permethrin masses shifted. When examining the overall experiment, cis-permethrin mass percentage in sediment for the U-ditch, V-vegetated ditch, and V-unvegetated ditch was 70±3%, 58±6%, and 86±6% respectively. By analyzing data where no water was present, the cis-permethrin sediment mass percentages change to 75±3%, 72±3%, and 100±0% respectively, for the U-ditch, V-vegetated, and V-unvegetated ditches. Cis-permethrin half-lives in ditch water were 4.1 h (U ditch), 2.4 h (V-vegetated) and 3.5 h (V-unvegetated). Half-distances for cis-permethrin were 169 m (U ditch), 22 m (V-vegetated) and 50 m (V-unvegetated).

Discussion

Pesticide entry into receiving waters following storm or irrigation events depends on several factors, such as pesticide chemistry, rainfall intensity, time of application, and surrounding soil properties. In efforts to reduce the possibility of this occurring, management practices have been suggested to mitigate pesticide runoff. Vegetation plays a significant role in many suggested BMPs. Stiff grass hedges, grassed waterways, and riparian filter strips are just three examples of incorporating vegetation into runoff mitigation strategies. Vegetation has been documented to assist in mitigation of permethrin. Filter strips containing trees, shrubs, and grasses at widths of 7.5 m and 15 m reduced permethrin-associated contaminants 27-83% (*14*). Vegetated drainage ditches are becoming increasingly popular among farmers and landowners with little available production acreage to set aside for potential mitigation purposes.

A mathematical model is being developed as part of this study as both a design tool to determine the ideal properties (e.g., length and width) for a particular farm system, and as an analysis tool to evaluate the efficacy that might be obtained with an under sized ditch. The tool will be able to estimate ditch performance for different chemicals, soils, plant species, and climatologic conditions.

The Vegetated Filter Ditch Model (VFDM) simulates pesticide fate and transport from agricultural fields through a vegetative filter ditch based on water, sediment, and pesticide mass balance. Water mass balance accounts for inflow, precipitation, evaporation, seepage, and outflow. Sediment mass balance accounts for settlement and resuspension. Pesticide mass balance can accommodate dilution; volatilization; partitioning between water, sediment, and foliage; decay in water, sediment, and foliage; uptake by plants; resuspension from sediment and foliage; and outflow from overflow or drainage.

Model input includes boundary condition loadings in terms of time series influx of water, sediment, and pesticide; ditch geometry including length, width, depth, and riser height; chemical properties including solubility, degradation rates (water, wet & dry sediment, foliage), adsorption coefficients (sediment, foliage), and uptake by plants; plant properties related to plant biomass and growth; and sediment properties including field capacity, wilting point, porosity, bulk density, and initial soil moisture. Model output includes time series outflow of water, sediment, and pesticide; and chemical concentrations in water, sediment, foliage.

An existing pesticide fate model, RICEWQ version 1.7.2 (*15*) was the starting point for the generation of the VFDM. This model was selected because of its pesticide chemistry (degradation, partitioning, and ability to simulate metabolites) and water balance (variable inflow rates, water levels, and drainage) algorithms. Pesticide application routines were replaced by boundary condition inflow files for water, sediment, and pesticide. Geometry changes were made to allow channels of varying configuration. Mass balance algorithms were changed to track pesticide residues in sediment and foliar in multiple vertical compartments.

Preliminary model predictions are encouraging but not conclusive because of the number of assumptions required to configure the model. The assumptions relate to uncertainty regarding the variability of pesticide and sediment dose over time and plant uptake and adsorption. Model validation will be assessed using information from additional field studies being conducted as part of this research study. Additional research is being conducted within the context of this study on the role of pesticide uptake and adsorption by plants. Additional field studies are being conducted that involve the implementation of VADD on working farms in Yolo County receiving permethrin application to tomato and alfalfa fields.

Conclusions

The use of vegetative ditches is effective for the mitigation of pesticides, and particularly pyrethroids, as demonstrated in this project and previous studies (*8,9,10,11*). Since pyrethroids have shorter environmental half-lives than organochlorines and many of the OP insecticides, there is less concern for pesticide accumulation in ditch water, sediment, and plants. Distances needed to reduce initial cis-permethrin concentrations by 50% were two times less in the V-vegetated ditch (i.e., more efficient) than the V-unvegetated ditch. When comparing the V-vegetated to U-vegetated ditch, cis-permethrin half-distances were eight times more in the U-vegetated ditch, thus making the V-vegetated the most efficient of the three ditches. Research into the significant differences reported between U- and V-vegetated ditches is one possible avenue for further study; however, it is beyond the time and financial resources available for the current study. Although an effective BMP, vegetated ditches should be considered one tool of many available options for mitigation of pesticides, including constructed wetlands, sediment retention ponds, grassed buffers, etc. Site specific needs routinely call for multiple BMPs in sequence to sufficiently address the non-point source problem.

Acknowledgements

This project was made possible by Proposition 13, Pesticide Research and Identification of Source Mitigation (PRISM) funds through the California State Water Resources Control Board. Special thanks to Blake Harlan for access and field site development. Without the field and laboratory assistance of Sam Testa, Kyle Wooldrige, Jennifer Drewitz, William Spong, Charissa Codina, Dan Riordan, Elizabeth Kelber, Kevin Goding, and Abdu Mekebri, the project could not have been completed.

References

1. Epstein, L.; S. Bassein, S.; Zalom, F.G. Almond and stone fruit growers reduce OP, increase pyrethroid use in dormant sprays. *California Agric.* **2000**, *54*, 14-19.
2. California Pesticide Information Portal Database, URL http://calpip.cdpr.ca.gov/ (accessed September 2007).
3. Amweg, E.L.; Weston, D.P.; You, J.; Lydy, M.J. Pyrethroid insecticides and sediment toxicity in urban creeks from California and Tennessee. *Environ Sci Tech.* **2006**, *40*, 1700-1706.
4. Weston, D.P.; You, J.; Lydy, M.J. Distribution and toxicity of sediment-associated pesticides in agriculture-dominated waterbodies of California's Central Valley. *Environ Sci Technol.* **2004**, *38*, 2752-2759.
5. Lee, G.F.; Jones-Lee, A. Review of management practices for controlling the water quality impacts of potential pollutants in irrigated agriculture stormwater runoff and tailwater discharges. Report TP 02-05. California Water Institute, California State University – Fresno: Fresno, CA, 2002.
6. U.S. Environmental Protection Agency. 2002 Section 303(d) List Fact Sheet for California, URL http://iaspub.epa.gov/waters/ state_rept.control?p_state=CA#IMP (accessed May 2007).
7. U.S. Environmental Protection Agency. The 20 Needs Report: How research can improve the TMDL program. EPA 841-B-02-2002. Washington, DC, 2002.
8. Moore, M.T.; Bennett, E.R.; Cooper, C.M.; Smith, Jr.; S.; Shields, Jr., F.D.; Farris, J.L.; Milam, C.D. Transport and fate of atrazine and lambda-cyhalothrin in an agricultural drainage ditch in the Mississippi Delta, USA. *Agric. Ecosys. Environ.* **2001**, *87*, 309-314.
9. Bennett, E.R.; Moore, M.T.; Cooper, C.M.; Smith, Jr., S.; Shields, Jr., F.D.; Drouillard, K.G.; Schulz, R. Vegetated agricultural drainage ditches for the mitigation of pyrethroid-associated runoff. *Environ. Toxicol. Chem.* **2005**, *24*(9), 2121-2127.

10. Cooper, C.M.; Moore, M.T.; Bennett, E.R.; Smith, Jr., S.; Farris, J.L. Innovative uses of vegetated drainage ditches for reducing agricultural runoff. Proceedings of the 6th International Conference on Diffuse Pollution, Amsterdam, The Netherlands, 2002, pp. 119-126.
11. Cooper, C.M.; Moore, M.T.; Bennett, E.R.; Smith, Jr., S.; Farris, J.L. Alternative environmental benefits of agricultural drainage ditches. *Verh. Internat. Verein. Limnol.* **2002**, *28*, 1-5.
12. Spencer, W.F.; Cliath, M.M. Pesticide losses in surface runoff from irrigated fields. In: *Chemistry for the Protection of the Environment;* Pawlowski, L., et al., Eds: Plenum Press: New York, NY, 1991, pp 277-289.
13. Sokal, R.R.; Rohlf, F.J. *Biometry*, 2nd ed. W.H. Freeman: New York, NY, USA, 1981.
14. Schmitt, T.J.; Dosskey, M.G.; Hoagland, K.D. Filter strip performance and processes for different vegetation, widths, and contaminants. *J. Environ. Qual.* **1999**, *28*(5), 1479-1489.
15. Williams, W.M.; Ritter, A.M.; Zdinak, C.E.; Cheplick, J.M. RICEWQ: Pesticide runoff model for rice crops – user's manual and program documentation, version 1.7.2. Waterborne Environment, Inc.: Leesburg, VA, 2004.

Chapter 20

Reduction of Pyrethroid Runoff from a Commercial Nursery

W. J. Lao[1], G. Arye[1], F. Ernst[1], Y. P. Xu[1], S. Bondarenko[1], D. Haver[2], J. Kabashima[2], and J. Gan[1]

[1]Department of Environmental Sciences, University of California, 2258 Geology Building, Riverside, CA 92521
[2]University of California Cooperative Extension, Orange County, Irvine, CA 92618

Nurseries are heavy users of pesticides. Irrigation or storm runoff from nurseries generally contains high levels of organic matter that serves as a carrier for hydrophobic pesticides such as pyrethroids. We continuously monitored runoff flow rates and pesticide levels at a 125-acre commercial nursery and evaluated the effectiveness of several BMPs for reducing the offsite movement of pyrethroids such as bifenthrin. This study experienced two relatively dry seasons. However, pesticide exports in storm runoff far exceeded pesticide runoff during dry months. Under dry weather conditions, improved irrigation practices, water retention and reuse, and the isolation of production areas were found to essentially eliminate pesticide runoff export. Sediment cleaning before the arrival of the rainy season was recommended as a highly inexpensive and feasible BMP for reducing pyrethroid loads in storm runoff.

Introduction

Newport Bay is the second largest estuarine embayment in Southern California and is the location of a state ecological reserve containing many endangered species. Located next to Newport Bay is San Joaquin Marsh, the largest coastal freshwater marsh in Southern California. San Diego Creek is the primary freshwater input into Newport Bay and also provides a corridor for wildlife movement between the Bay, Marsh and upland areas. This grouping of diverse habitats makes this area ecologically important in the urban landscape of Southern California.

Pesticide runoff to water bodies in the Newport Bay watershed has been identified as a cause for observed acute and chronic toxicity to aquatic life. In particular, elevated concentrations of diazinon and chlorpyrifos have been associated with numerous toxic events in the watershed. Recently, monitoring studies showed an almost widespread presence of trace levels of various pyrethroid compounds in sediments from both urban and agricultural streams in California (*1-6*). A study on sediment toxicity in the Newport Bay suggests a potential role of pyrethroids in the observed acute toxicity (*7*).

Nursery and urban uses of pesticides have been identified as the two main contributing sources for the elevated pesticide concentrations in the Newport Bay/San Diego Creek Watershed. In the Orange County region, the use of pyrethroid insecticides such as bifenthrin is a mandatory quarantine requirement for controlling the red imported fire ant (RIFA) for commercial nurseries. For instance, Talstar®, a granular formulation of bifenthrin, is incorporated into potting mix for all nursery containers in the Orange County area. Previous monitoring studies by California Department of Pesticide Regulation (CDPR) and University of California Riverside have shown high levels of pyrethroids in runoff from selected nurseries in the San Diego Creek/Newport Bay Watershed (*8,9*).

The main objective of this study was to characterize pesticide runoff from commercial nurseries, and to further evaluate the effectiveness of various best management practices (BMPs) for reducing pesticide runoff from nurseries under dry and wet weather conditions. To achieve this objective, a large nursery located in Orange County was selected as the study site, where the runoff flow rates and pesticide transport were monitored on an almost continual basis for over 20 months. This chapter is a review of monitoring activities, BMP implementation, and evaluation of the effectiveness of various BMPs from this large-scale collaborative effort.

Materials and Methods

Site Description

A large commercial nursery was selected for this study based on its location, pesticide use history, and the occurrence of surface runoff at the time of this study. The nursery has 125 acres of production area and is involved mainly in the cultivation of container plants in outdoor areas. The nursery plants are irrigated by drip irrigation, sprinklers, micro-sprinklers and/or by hand. Most of the runoff water flows into an unlined flood control channel that runs along the border on the west side of the nursery. Some runoff flows through a small natural creek along the border on the east side of the property. The west side channel has a number of retention basins formed by large rocks, contours, and a check dam.

Measurement of Runoff Flow Rates

The monitoring of runoff flow rates and pesticide concentrations was carried mostly at the beginning (inlet) and exit (outlet) points of the west side drainage channel. At the inlet, there was a continuous runoff flow from upstream areas that were mostly commercial and residential areas. The incoming flow entered the west side drainage channel via an underground concrete drainage conduit. During the first phase of this project, the flow rate at the inlet was recorded with a pressure transducer installed in a portable metal flume connected to a data logger (Model WL15, Global Water Instrumentation, Gold River, CA). Due to the limited capacity of this flume, the flume at the inlet was removed prior to rain events in the 2005/2006 rainy season. Therefore, no storm runoff rates were measured at this site in the rainy season of 2005/2006. In October 2006, a large permanent V-shaped flume was installed, which allowed flow rate measurement during storm events in the 2006/2007 rainy season.

The outlet is downstream from the inlet and is the runoff discharge point for the west side drainage channel. The flow rate was measured by flow meters until November 8, 2005. A large "V"-shaped flume was installed in November 2005. A pressure transducer and a data logger (Model WL400, Global Water Instrumentation, Gold River, CA) were used to record the flow rate starting December 7, 2005. This setup allowed for the measurement of flow rates during storm events in both the 2005/2006 and 2006/2007 rainy seasons.

The principle of flumes is to force water flow through a cross section area with a known geometric configuration, and from the water depth measured by a pressure transducer, to calculate the flow rate for the water passing through the flume. The water depth in the flume as measured by the pressure transducer was recorded with a data logger and then downloaded onto a computer using an interfacing program that was supplied by the vendor. The runoff volume was measured in cubic meter (m^3), and the runoff rates were calculated as m^3/day, m^3/week, or m^3/month.

Collection of Runoff and Sediment Samples

Runoff water samples were taken on a weekly basis at the two monitoring sites when there was measurable flow. The weekly sampling was done by a "grab" method, and water was collected by hand into a 1-liter amber glass narrow-mouth bottle. In the 2005-2006 rainy season, autosamplers were installed and programmed for collecting water samples in an attempt to sample water throughout a storm. Storm runoff was also manually collected using amber narrow-mouth bottles whenever possible during or immediately after a rain event. Runoff samples were transported to the analytical laboratory in Riverside within 4 hours of the time of sample collection, or kept in a refrigerator (4 °C) at the nursery and then transported to the analytical laboratory in Riverside for preparation and analysis.

Sediment samples were taken from the drainage channels in October 2006 when the accumulated sediment was excavated as a management practice to reduce pesticide runoff during storm events. The sediment samples were collected in mason jars, transported to the analytical laboratory in Riverside, and kept at 4 °C until analysis.

Pesticide Analysis

Water samples were prepared and analyzed using procedures consistent with EPA Method 8081A. Briefly, a 1.0 L "unfiltered" water sample was measured out by weighing, and transferred into a 2-L glass separatory funnel. Methylene chloride (60 mL) was added into the separatory funnel, and then mixed vigorously by hand for 1 min. The water-solvent mixture was allowed to separate on a stand, and the solvent phase was drained into a glass beaker. The same step was repeated two additional times, and the solvent extracts were combined. The extracts were then passed through 50 g anhydrous sodium sulfate to remove the residual water, and then concentrated to near dryness on a vacuum rotary evaporator at 35 °C. The residue was recovered with 1.0 mL and transferred to a brown GC vial. The recovery of this procedure was determined by the addition of a surrogate (PCB 209) before extraction and was found to be >90% in all cases.

For analysis, an aliquot of the final extract was injected into an Agilent 6890N-gas chromatography (GC) (Agilent Technologies, Palo Alto, CA) equipped with an electron capture detector (ECD) and a HP-5MS capillary column (30 m × 0.25 mm × 0.5 μm). The GC conditions were such that both organophosphate compounds (diazinon and chlorpyrifos) and pyrethroids were screened and quantified. Analysis for diazinon and chlorpyrifos was discontinued after the initial analyses suggested no presence of these compounds in the runoff. Quantitation was completed using an external standard method with a six level multi-point calibration.

The sediment samples were extracted by mixing a known amount of dried sediment with dichloromethane by sonication and filtration. The filtrate was concentrated to a small volume, and an aliquot was analyzed on GC under the same conditions as described above to determine the pesticide concentrations on a dry weight basis.

Results and Discussion

Precipitation Rates and Patterns

The meteorological data from the California Irrigation Management Information System (CIMIS) Meteorological Station No. 75 were used to calculate the daily and cumulative precipitation rates from 2002 to early 2007 (Figure 1). The CIMIS station was close to the study site, and therefore the precipitation data closely represented the conditions of the study site. This region displays distinctive dry and wet seasons, with precipitation events occurring mostly from October to early March each year. The last five wet seasons showed greatly differing rain distribution patterns and rates. This study only covers the time period from July 2005 through March 2007, and thus included only 2 rainy seasons. These two seasons were relatively dry compared to an average year in this region. The 2005/2006 rainy months produced a total of 8.8 inches of precipitation, while the 2006/2007 rainy season produced only 2.1 inches. The great variations in rainfall patterns and rates make the analysis of trends extremely difficult for a short-term study such as this project.

Runoff Rates and Volumes

Figures 2 and 3 show the daily runoff rates (m^3/day) recorded from the inlet and outlet sites during this study. Included in each figure (upper x-axis) is the distribution of precipitation on the same time scale. It is apparent that for both locations, the hydrograph of runoff corresponds closely with the occurrence of the rain events. It must be noted again that at the inlet, the large flume was not installed until October 2006 and the measured flow rate data were incomplete due to a lack of accurate flow rate measurements during the 2005/2006 rainy season. Therefore, daily runoff rates at the inlet did not show large spikes as they did for the outlet (Figure 3).

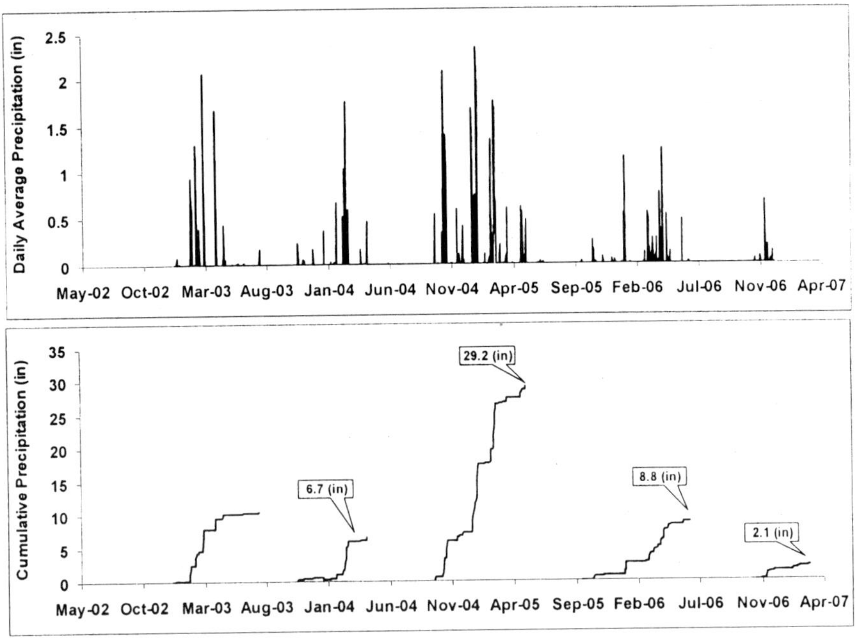

Figure 1. Daily precipitation distribution (upper graph) and cumulative precipitation (lower graph) for 2002-2007 at the study site

To facilitate the comparison of runoff patterns between the dry and wet seasons, in the following discussion, the wet and dry seasons considered in this study are referred to as "dry1" (7/18/2005 – 12/28/2005), "wet1" (12/29/2005 – 6/15/2006), "dry2" (6/16/2006 – 11/12/2006) and "wet2" (11/13/2006 -3/16/ 2007). Note that "dry1" includes only a partial dry season of 2005 because monitoring was started on 6/1/2005 at the outlet and on 7/8/2005 at the inlet. Also note that for "wet2", the duration includes only the time period up to March 16, 2007, when the monitoring was stopped. The beginning of a wet season is operationally defined by the occurrence of the first significant storm for that year, and the end of a wet season is the last storm occurrence. For the rainy season of 2005/2006, the wet season was extended into June due to a lingering storm in May 2006.

Dry Weather Conditions

The total runoff volumes (m^3) measured during "dry1" are 9814 and 7268 m^3 at the inlet and outlet, respectively. During "dry2", the respective values are 20377, and 12186 m^3. A significant observation can be made from the recorded flow rates and volumes in dry months. The cumulative discharge volume at the

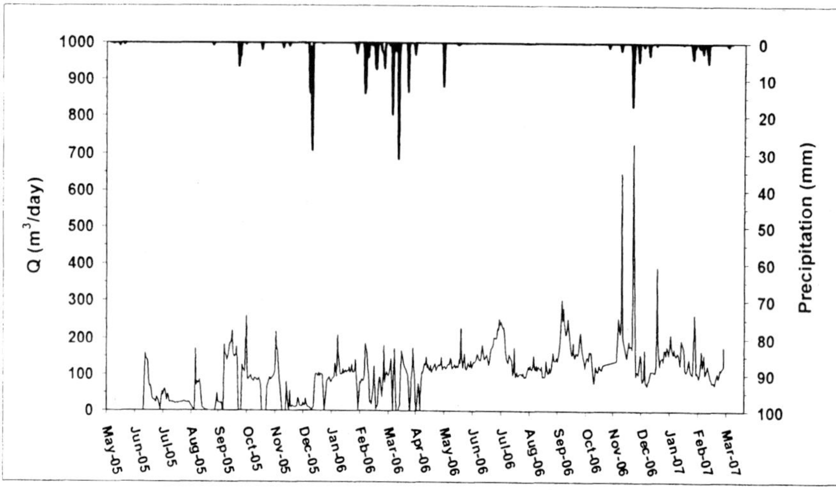

Figure 2. Runoff flow rate (m³/day) and daily precipitation (mm) monitored at the inlet during the course of the project

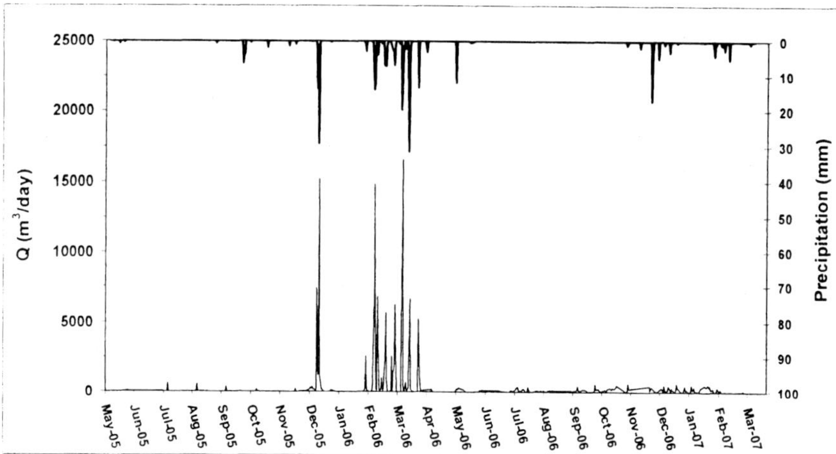

Figure 3. Runoff flow rate (m³/day) and daily precipitation (mm) monitored at the outtlet during the course of the project

outlet during dry seasons was consistently smaller than that at the inlet for the same time period. Cumulatively, during "dry1", the total discharge at the outlet was 7268 m^3, which was 74% of the amount of water that was released onto the property via the inlet. During "dry2", the total discharge volume at the outlet was 12186 m^3, which was 60% of the amount of water that was released onto the property via the inlet. Therefore, under dry weather conditions, the nursery acted as a sink for the runoff water entering the property. As described later in this chapter, this was possible because of the implementation of a number of BMPs.

Dry vs. Wet Weather Conditions

The total runoff volume for the outlet was 7268 m^3 for "dry1" and 114893 m^3 for "wet1". Therefore, for the 2005-2006 monitoring year, the dry weather runoff was only 6% of the total runoff at the outlet. This difference highlights the overwhelming contribution from rain storms to the total runoff in this region. Compared to storm-induced runoff, irrigation-induced runoff becomes negligible in an average year. In the 2006/2007 monitoring year, the cumulative dry weather runoff at the outlet was 1701 m^3, and the total wet weather runoff was 5604 m^3, which translates into a relative contribution of about 23% by the dry weather runoff to the total runoff. While this suggests significant variations between years, it is apparent that even for a very dry year such as the 2006/2007 season, the contribution from precipitation to the total runoff can still be overwhelming. This observation suggests that in order to achieve appreciable reductions in runoff volumes and pollutant loads, it is essential to effectively manage storm runoff.

Pesticides Concentrations and Loads

No detectable levels of diazinon and chlorpyrifos were ever found in the runoff samples. Pesticide concentrations in the dry weather runoff showed that bifenthrin was consistently present in the runoff. Fenpropathrin, cyhalothrin, cyfluthrin, and deltamethrin were occasionally present in the runoff. However, fenpropathrin was more frequently found in the runoff than the other pyrethroids, not including bifenthrin. Bifenthrin concentrations measured in the runoff from the inlet and outlet are shown in Figure 4 on normal (upper graph) and logarithmic (lower graph) scales. The results show that in most instances, the bifenthrin concentrations were significantly higher at the outlet than at the inlet. The differences in bifenthrin levels between the inlet and the outlet are generally in the range of one order of magnitude. These observations clearly suggest that the use of bifenthrin products at the nursery had directly contributed to bifenthrin contamination of the runoff water.

The measured bifenthrin concentrations in the inlet water samples were lower or around the LC50 value reported for *Ceriodaphnia dubia* (0.078 μg/L) (*10*). In contrast, many of the concentrations seen at the outlet exceeded the LC50 for *C. dubia*, suggesting that the runoff leaving the property may likely cause aquatic toxicity downstream to water column invertebrates. However, it must be noted that the nursery runoff generally contained high levels of dissolved organic matter (DOM) and suspended solids (SS), and that it is likely

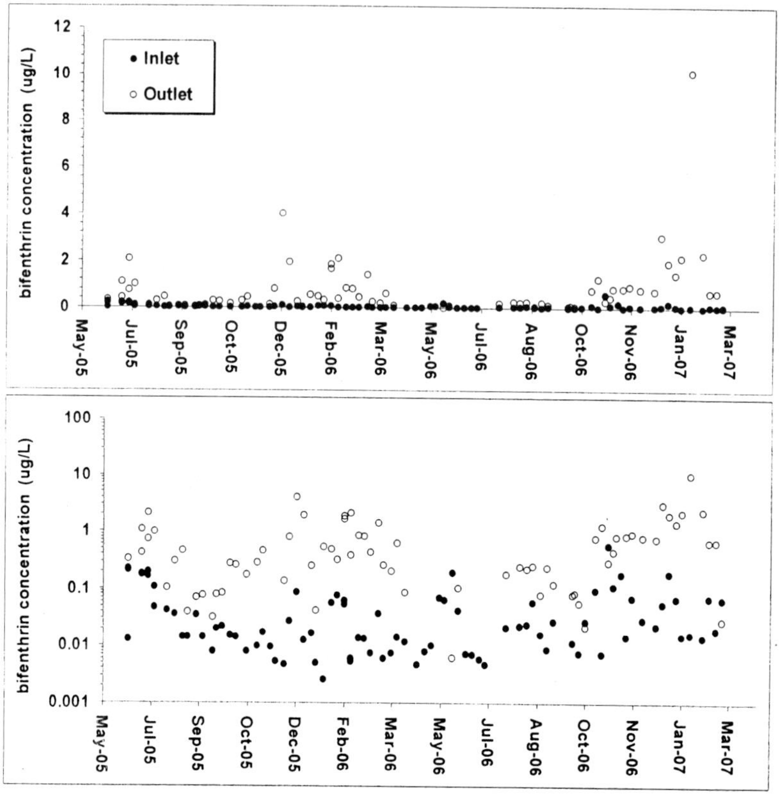

Figure 4. Bifenthrin concentrations (μg/L) as function of time at the inlet and the outlet. (A) normal concentration scale; and (B) logarithmic concentration scale

that the majority of bifenthrin in the nursery runoff was associated with the DOM and SS phases. This distinction is important, as recent studies show that when pyrethroids are associated with DOM or SS, they become less toxic to water column invertebrates (*11-14*). In addition, pyrethroids associated with SS are expected to be relatively immobile in the environment, because the suspended particles may easily settle to the bottom along a runoff path, thus becoming isolated from the moving flow.

To evaluate the cumulative bifenthrin loads during this project, we first calculated the runoff volumes on a weekly basis and further estimated the cumulative amounts of bifenthrin moving through each monitoring site from the weekly representative bifenthrin concentrations (Figure 5). For the inlet, cumulative amounts of bifenthrin are relatively constant and are relatively small compared to the amounts for the outlet for the same time periods. However, as the flow rate measurements were incomplete for the inlet during "wet1", the total

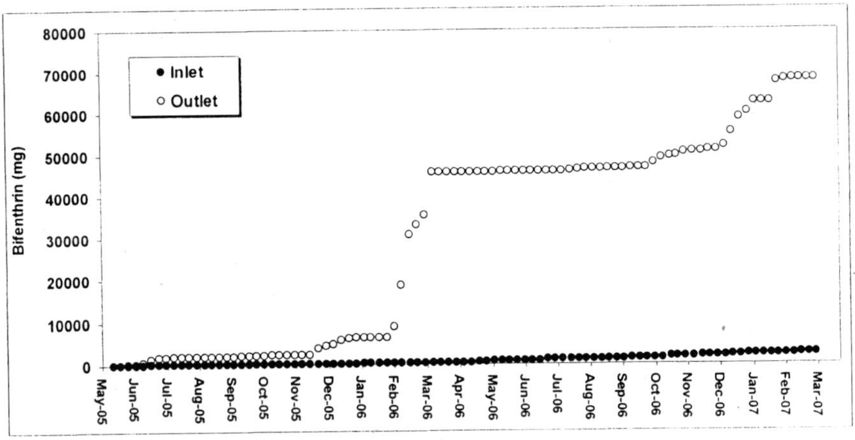

Figure 5. Cumulative bifenthrin loads (mg) at the inlet and outlet

bifenthrin loads for this site were underestimated. Based on the weekly bifenthrin loads at each monitoring site, bifenthrin loads at the outlet were generally much greater than those at the inlet, and the difference between the inlet and the outlet was more than one order of magnitude.

Dry Weather Conditions

The sum of the bifenthrin load (mg) obtained at the inlet was 238 mg for "dry1" which increased to 1022 mg in "dry2", suggesting changes in pesticide use patterns from the upstream urban sources and changes in runoff input. At the outlet, the total bifenthrin load was 2347 mg in "dry1" and 3718 mg in "dry2", reflecting a relatively constant export of bifenthrin under dry weather conditions. Overall, bifenthrin loads from the outlet were much greater than the amount of bifenthrin that had entered the property via urban runoff at the inlet. For instance, for "dry1", the total bifenthrin load for the outlet was 2347 mg, which was about 10 times of that at the inlet. For "dry2", the total bifenthrin load at the outlet was 3718 mg, which was about 3.5 times of that at the inlet. These observations clearly suggest that because of its use of bifenthrin products, the nursery served as a potential for bifenthrin contamination for downstream environments.

Dry vs. Wet Weather Conditions

A comparison between pesticide concentrations and loads again highlights the importance of storm runoff. At the outlet, the total bifenthrin load estimated for "wet1" was 43527 mg, which was 18 times more than that from "dry1" for this site. This clearly suggests the dominant contribution of storm events to bifenthrin loads from the study site. The overwhelming contribution of storm runoff to the total bifenthrin loads may be attributed to two causes. First, the volumes of storm runoff were many times more than that of dry weather runoff, as described previously. Secondly, pesticide levels were generally higher in storm runoff than those seen in dry weather flow (Figure 4). The relatively low levels of bifenthrin in the dry season runoff were partly a result of water retention practices and other BMPs. As the runoff water was retained in the drainage channel, the suspended particles were allowed to settle to the bottom of the channel due to gravity. As bifenthrin was associated mostly with suspended particles, this process led to decreases in bifenthrin levels in the runoff water. In comparison, the swift movement of water during a storm would not allow sedimentation to occur, and consequently, the total bifenthrin level in storm flow would be higher due to the high content of suspended solids and the associated DOM.

Mitigation Practices and Efficacy

Since the inception of this study, in collaboration with the nursery managers and workers, we implemented various BMPs. Some of these BMPs are structural, while others are changes in practices and behaviors, and their effectiveness would be difficult to record or quantify. The following is a brief description of these mitigation practices, along with a discussion of their principles and efficiency at reducing the runoff volume, pesticide loads, or both. As the role of the individual BMPs would be difficult to quantify separately, the collective effect of these BMPs are discussed at the end of this section.

Improved Irrigation Practices

Since the beginning of this study, the nursery has improved their irrigation practices with the aim to improve irrigation efficiency and reduce surface runoff. These changes include the maintenance of irrigation systems, reduced irrigation rates, and improved irrigation uniformity, among others. The net outcome was minimal or no visible surface runoff on most production days under dry weather conditions. Improvements in irrigation practices clearly contributed to the negligible runoff from the nursery. For instance, the cumulative discharge volume at the outlet during dry seasons was consistently smaller than that at the inlet for the same time period. The cumulative runoff volume at the outlet in "dry1" was 7268 m^3, which was only 74% of the amount of water that was released onto the property via the inlet during the same time period. During "dry2", the total discharge volume at the outlet was 12186 m^3, which was 60%

of the amount of water that was released onto the property via the inlet. Therefore, under dry weather conditions, the nursery not only did not contribute to offsite runoff, but it also consumed or absorbed a large portion of the input water flowing onto the property from upstream areas. This is strong evidence that this study encouraged the nursery management to improve their irrigation practices, which directly contributed to the negligible runoff leaving the property under dry weather conditions. The reduced runoff rates were essential for reducing the overall pesticide loads during the dry seasons.

Edge Control of Runoff

Since the beginning of this project, the nursery has adopted various practices to prevent surface runoff from entering the drainage channels. An example of these practices was the construction of physical barriers between the production areas and the drainage channels. The barriers include sand bags and concrete berms. These barriers effectively prevented irrigation runoff and should also reduce/retain runoff from small storms. The edge barriers should be especially effective at preventing loose soil particles and potting mix from entering the drainage channels, as the barriers serve as physical traps to trap and retain most of the loose particles. The reduction in surface erosion is expected to contribute to lower pesticide levels in the runoff, as most pesticides are attached to suspended particles.

Isolation of Potting Mixing Area

Another BMP considered in this study is the installation of barriers around potting mix preparation areas that were prone to runoff from irrigation or storms. In a previous study, it was found that loose potting mix on the soil surface was a significant source for pesticides such as bifenthrin (*8*). For instance, bifenthrin in the formulation of Talstar® is always mixed into the growth media before seeding or transplanting. Therefore, all potting mix would contain high levels of bifenthrin (in the ppm range). The isolation of potting mix handling areas would effectively prevent the contamination of the runoff water by potting mix that could result in high levels of bifenthrin in the runoff water.

Check Dams and Retention Basins

The most important BMP implemented through this study is the construction and use of removable concrete-steel check dams before the exit of the two drainage ditches. The check dams were made of concrete and steel plates, and were removable during large storms to allow the passage of storm water. The check dams also had an overflow mechanism that would allow water to overflow after reaching a certain depth. The check dams had effectively converted the large drainage ditches into multiple retention basins, allowing the "raw" or "primary" runoff to undergo several processes while the runoff water is temporarily retained. First, the retention of runoff water was shown to be highly effective for removing suspended particles from the runoff. Samples of "raw" runoff and the dammed drainage ditch were compared for their levels of bifenthrin. The retention of runoff water caused the suspended solids to settle under gravity, resulting in a decrease of the bifenthrin concentration by >80%. In addition, pesticides in the retained water may adsorb to the sediment or soil on the bottom of the drainage channel. The accumulated pesticides may also undergo microbial degradation during retention, and a previous study showed a widespread distribution of microorganisms capable of degrading pyrethroid compounds in the sediment (*15*).

The check dams should also be ideal for capturing the first flush of small storms, which was shown to be useful for reducing storm water runoff through the east side ditch. In addition, the runoff retention during dry months would lead to an in-channel accumulation of sediment that is rich in pesticides. There is a good chance that the accumulated sediment may be washed away during a storm. However, the use of check dams makes it feasible to clean and remove the accumulated sediment, along with the adsorbed pesticides, before the rainy season.

Vegetated Ditch Banks

The nursery has planted and maintained active vegetation along the sides of the drainage ditch. The plants include mostly papyrus, as well as cacti. The benefits of the vegetated banks are two-fold: bank stabilization and water consumption. Bank stabilization prevents soil erosion and also intercepts lateral surface runoff during storms. Water consumption by the plants results in reduction of runoff volume.

Polyacrylamide (PAM)

While the check dam and retention mechanisms are extremely effective at removing suspended particles from the runoff stream, it was observed that the runoff was brownish and contained high levels of DOM. The DOM was likely from the potting materials, such as bark, making further reduction of pyrethroid levels in the runoff water difficult. We tested the use of polyacrylamide (PAM) to further clean the runoff water in the summer of 2006. However, our observations suggest that PAM was largely ineffective at reducing the DOM level of the retained runoff water. It must be noted that due to the use of the check dam, the runoff was kept under static conditions in the drainage channel, allowing most of the suspended solids to settle to the bottom of the channel. If check dams were not used, or if the runoff contained high levels of suspended solids, the use of PAM may prove to be effective, as found in a previous study (*8*). Therefore, when retention basins are used, the use of PAM should not be recommended, as the effects would be duplicative.

Sediment Removal

As mentioned above, the installation of the check dam was found to be highly effective at retaining the suspended particles within the nursery property. However, the sedimentation and consequently accumulation of the sediment on the ditch floor may act as a potential source for pesticides during a rain event, because the storm flow can easily wash some of the accumulated sediment downstream, contributing to large amounts of pesticides being discharged during a storm. The removal of the accumulated sediment is therefore a management practice that may greatly reduce the pesticide export through runoff, especially from the first few storms. This practice was recommended to the nursery managers and was implemented in early October of 2006 (before the first expected rain event).

The sediment accumulated before the check dam was manually removed by the nursery workers. The volume of the removed sediment was estimated from the number of 53-gallon drums that were used to store the excavated sediment. Samples of the removed sediments were taken to the laboratory and analyzed for their pesticide concentrations. Table 1 shows the concentrations of the pesticides found in the sediments on a dry weight basis (μg/kg). Assuming a gravimetric water content of 44% and a bulk density of 1.44 g/cm^3, the amount of sediment removed was estimated to be 1586 kg for the east side drainage ditch and 433 kg for the west side ditch. The amount of sediment accumulated in the east side ditch was about three fold higher than the amount from the west side ditch. These observations are in agreement with the period of time that the water was dammed in these ditches. Given the amounts of sediment removed and the pesticide concentrations in the sediments, the total amounts of pesticides that were removed by this practice were calculated and shown in Table 2.

From Table 1 and Table 2, it can be seen that many pyrethroid insecticides were present in the sediments excavated from the drainage ditches. This is due to the strong affinity of these compounds for the sediment phase. Among the detected pyrethroids, bifenthrin showed the highest concentrations, which was followed by fenpropathrin. This pattern is in agreement with the pesticide levels found in the runoff water throughout the entire monitoring period. Given the higher pesticide concentrations and the greater sediment mass removed from the east side drainage ditch, the estimated total amounts of pesticides removed were much higher for the east side ditch (Table 2).

Table 1. Pesticide concentrations (ng/g) in sediments removed from the drainage ditches before check dams

Pesticide	*East side ditch*	*West side ditch*
	ng/ g dry sediment	
Bifenthrin	486.6	118.2
Fenpropathrin	61.3	16.1
Lambda-Cyhalothrin	4.2	2.5
cis-Permethrin	6.4	3.9
trans-Permethrin	5.3	6.6
Cyfluthrin	9.5	1.4
Cypermethrin	ND	ND
Esfenvalerate	ND	ND
Deltamethrin	8.7	15.8

Table 2. Amounts of pesticides (mg) removed from the drainage ditches by sediment cleanup

Pesticide	*East side ditch*	*West side ditch*
	mg	
Bifenthrin	772	51
Fenpropathrin	97	7
Lambda-Cyhalothrin	7	1
cis-Permethrin	10	2
trans-Permethrin	8	3
Cyfluthrin	15	1
Cypermenthrin	ND	ND
Esfenvalerate	ND	ND
Deltamethrin	14	7

This practice clearly demonstrates the usefulness of sediment removal for reducing pesticide loads in storm runoff. Various retention practices are often used to curtail runoff volume and pollutant loads during dry months. These practices usually lead to the accumulation of sediment and pollutants on the site. The accumulated sediment and pollutants may be easily washed off by storm water. Therefore, retention practices, if not used properly, may only delay the pollutant export from the dry season until the wet season. This study shows that when retention practices are used, it is important, as well as effective, to clean the retention basins or ditches before the first rain event. This practice is technically simple, as well as inexpensive. The removed sediment can be recycled, as shown in this project, by incorporation into the potting mix, or can otherwise be properly isolated from the storm runoff.

Collective Effect of Mitigation Practices

The use of the above BMPs has collectively contributed to near complete reductions in pesticide runoff under dry weather conditions, and significant reductions under wet weather conditions. In particular, the implementation of the various practices as listed above had led to an essential elimination of pesticide runoff during dry weather months. For instance, compared to the July 2005 bifenthrin export from the outlet, which represents the immediately "before" scenario, bifenthrin export loads from the outlet decreased significantly until the first rain event of the 2005/2006 season (Figure 6). Using the week of July 14 as a reference point, the weekly export of bifenthrin in the following weeks was almost negligible. It is estimated that on average, after the initiation of this project, the reduction in daily bifenthrin export through the outlet was 98.7%.

Conclusions

Results from this study show that it is feasible to eliminate dry weather pesticide runoff from nursery operations through irrigation management, recycling, and the use of retention basins. For dry weather months, the useful management practices, in the order of their effectiveness, include: 1) efficient and uniform irrigation practices; 2) the collection and retention of runoff in large drainage ditches or isolated ponds; 3) the reuse of the retained water; 4) the isolation of potting mix handling areas; 5) the isolation of production areas from runoff channels using curbs and berms; 6) growing vegetation in drainage channels; and 7) in the event that retention practices are not used, PAM treatment to settle the suspended solids and remove pesticides from the runoff water.

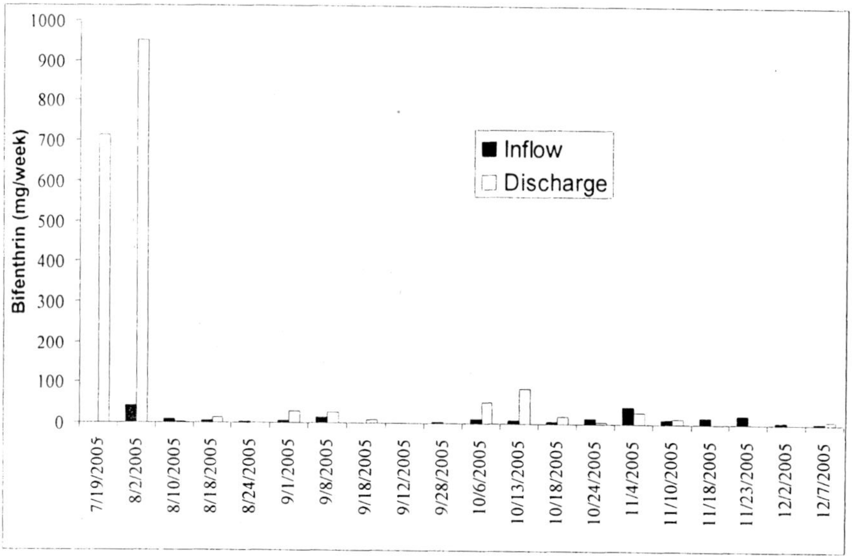

Figure 6. Changes in weekly bifenthrin export (mg/week) through the outlet before and after the implementation of various mitigation practices

This project experienced two relatively dry years with annual precipitation rates considerably below the average value. However, the measurement of runoff rates and volumes suggests that rain-induced runoff still dominated the overall runoff. In the first year, about 94% of the total runoff was from rain events, while the contribution from storm runoff was 77% in the second year. The export of bifenthrin from the nursery through storm runoff was >18 times more than through dry season runoff. These observations highlight the overwhelming contribution of rain events to the overall runoff and pesticide loads in this region. The importance of storm runoff is expected to be even greater for a wetter year. Therefore, in order to achieve appreciable reductions in pollutant loads, attention should be given to the management of storm runoff. As relatively higher levels of pesticides were found in the runoff from the first few storms, the management of the first few storms in the season would be especially important and produce the most benefit. Common stormwater BMPs, such as isolating potting mix preparation areas using sand bags and berms and the use of check dams, are effective in reducing sediment runoff from small storms. Annual clean-out and removal of sediment from detention/retention basins and ditches prior to the rainy season is highly effective in minimizing storm-induced pesticide runoff.

This study also revealed that urban sources contribute to pyrethroid runoff. The levels were below or around the LC50s of the respective pyrethroid compounds, and the detection was constant over time, suggesting sustained sources. Pyrethroids in urban runoff are likely from the widespread use of pyrethroid-containing products at residential homes, retail nurseries, and other commercial entities. The significance of pyrethroids in urban runoff should be further addressed, and mitigation strategies are urgently needed. Although the focus of the present study was on the mitigation of pyrethroid runoff from a nursery, some of the management practices may be equally effective for reducing pyrethroid movement in other environments. In particular, as pyrethroid compounds are highly hydrophobic and are associated with soil/sediment particles and DOM, practices aiming to cause on-site retention of loose particles and organic matter will be effective at reducing pyrethroid loads. It is important, however, to perform sediment cleanup (e.g., in retention ponds, basins, ditches) before the rainy season, as the sediment may be greatly enriched in pyrethroids and its isolation from storm runoff constitutes a highly inexpensive and feasible option for reducing storm-induced runoff of pyrethroids.

References

1. Weston, D.; You, J.; Lydy, M.J. *Environ. Sci. Technol.* **2004**, *38*, 2752-2759.
2. Weston, D.P.; Holmes, R.W.; You, J.; Lydy, M.J. *Environ. Sci. Technol.* **2005**, *39*, 9779-9784.
3. Amweg, E.; Weston, D.; Ureda, N. *Environ. Toxicol. Chem.* **2005**, *24*, 966-972.
4. Amweg, E.; Weston, D.; You, J.; Lydy, M. *Environ. Sci. Technol.* **2006**, *40*, 1700-1706.
5. Bacey, J.; Spurlock, F.; Starner, K.; Feng, H.; Hsu, J.; White, J.; Tran, D. *Bull. Environ. Contam. Toxicol.* **2005**, *74*, 864-871.
6. Budd, R.; Bondarenko, S.; Haver, D.; Kabashima, J.; Gan, J. *J. Environ. Qual.* **2007** (In press).
7. Bay, S.; Greenstein, D.; Brown, J. *Newport Bay sediment toxicity studies: Final report.* Technical Report #433; Southern California Coastal Water Research Project (SCCWRP), 2004, Westminster, CA.
8. Kabashima, J.N.; Lee, S.J.; Haver, D.L.; Wu, L.; Goh, K.; Gan, J. In *Deactivation and Detoxification of Biocides and Pesticides*; Gan, J.; Zhu, P.; Aust, S.A.; Lemley, A.T., Eds.; ACS Symposium Series, ACS: Washington, DC, 2004, 213-230.
9. Gan, J.; Lee, S.J.; Liu, W.P.; Haver, D.L.; Kabashima, J.N. *J. Environ. Qual.* **2005**, *34*, 836-841.
10. Mokry, L.E.; Hoagland, K.D. *Environ. Toxicol. Chem.* **1990**, *9*, 1045-1051.
11. Liu, W.P.; Gan, J.; Lee, S.J.; Kabashima, J.N. *Environ. Toxicol. Chem.* **2004**, *23*, 7-11.
12. Yang, W.C.; Spurlock, F.; Liu, W.P.; Gan, J. *J. Agric. Food Chem.* **2006**, *54*, 3967-3972.

13. Yang, W.C.; Hunter, W.; Gan, J.; Spurlock, F.C. *Environ. Toxicol. Chem.* **2006**, *25*, 1585-1591.
14. Yang, W.C.; Spurlock, F.C.; Liu, W.P.; Gan, J. *Environ. Toxicol. Chem.* **2006**, *25*, 1913-1919.
15. Lee, S.J.; Gan, J.; Kim, J.S.; Kabashima, J.N.; Crowley, D. *Environ. Toxicol. Chem.* **2004**, *23*, 1-6.

Indexes

Author Index

Subject Index

A

B

D

E

F

G

H

I

J

L

N

O

P

Q

R

S

T

U

V

W

Y

Z